Organic Nanostructures

Edited by
Jerry L. Atwood and Jonathan W. Steed

Related Titles

Vollath, D.

Nanomaterials

An Introduction to Synthesis, Characterization and Processing

approx. 200 pages with approx. 220 figures and approx. 20 tables
Softcover
ISBN: 978-3-527-31531-4

Zehetbauer, M. J., Zhu, Y. T. (eds.)

Bulk Nanostructured Materials

approx. 750 pages with approx. 250 figures and approx. 20 tables
2008
Hardcover
ISBN: 978-3-527-31524-6

Astruc, D. (ed.)

Nanoparticles and Catalysis

approx. 672 pages in 2 volumes with approx. 850 figures
2007
Hardcover
ISBN: 978-3-527-31572-7

Rao, C. N. R., Müller, A., Cheetham, A. K. (eds.)

Nanomaterials Chemistry

Recent Developments and New Directions

420 pages with 190 figures and 9 tables
2007
Hardcover
ISBN: 978-3-527-31664-9

Organic Nanostructures

Edited by
Jerry L. Atwood and Jonathan W. Steed

WILEY-VCH Verlag GmbH & Co. KGaA

The Editors

Prof. Dr. Jerry L. Atwood
University of Missouri–Columbia
Department of Chemistry
125 Chemistry Building
Columbia, MO 65211
USA

Prof. Jonathan W. Steed
University of Durham
Department of Chemistry
South Road
Durham, DH1 3LE
United Kingdom

Cover illustration
The front cover shows a space-filling image illustrating the packing of the ligands in the optically pure cage complex $[Zn_4(L^{o\text{-}Ph*})_6(ClO_4)](ClO_4)_7$ and is adapted from Figure 9.5 with the permission of Michael Ward. The structure is superimposed on an SEM image of the helical fibrous structure of a chiral supramolecular xerogel.

All books published by **Wiley-VCH** are carefully produced. Nevertheless, authors, editors, and publisher do not warrant the information contained in these books, including this book, to be free of errors. Readers are advised to keep in mind that statements, data, illustrations, procedural details or other items may inadvertently be inaccurate.

Library of Congress Card No.: applied for

British Library Cataloguing-in-Publication Data
A catalogue record for this book is available from the British Library.

Bibliographic information published by the Deutsche Nationalbibliothek
Die Deutsche Nationalbibliothek lists this publication in the Deutsche Nationalbibliografie; detailed bibliographic data are available in the Internet at <http://dnb.d-nb.de>.

© 2008 WILEY-VCH Verlag GmbH & Co. KGaA, Weinheim

All rights reserved (including those of translation into other languages). No part of this book may be reproduced in any form – by photoprinting, microfilm, or any other means – nor transmitted or translated into a machine language without written permission from the publishers. Registered names, trademarks, etc. used in this book, even when not specifically marked as such, are not to be considered unprotected by law.

Typesetting Thomson Digital, Noida, India
Printing Strauss GmbH, Mörlenbach
Binding Litges & Dopf GmbH, Heppenheim

Printed in the Federal Republic of Germany
Printed on acid-free paper

ISBN: 978-3-527-31836-0

*In memory of Professor Dimitry M. Rudkevich
(1963–2007)*

Contents

Preface *XIII*
List of Contributors *XV*

1 **Artificial Photochemical Devices and Machines** *1*
Vincenzo Balzani, Alberto Credi, and Margherita Venturi
1.1 Introduction *1*
1.2 Molecular and Supramolecular Photochemistry *2*
1.2.1 Molecular Photochemistry *2*
1.2.2 Supramolecular Photochemistry *4*
1.3 Wire-Type Systems *5*
1.3.1 Molecular Wires for Photoinduced Electron Transfer *5*
1.3.2 Molecular Wires for Photoinduced Energy Transfer *9*
1.4 Switching Electron-Transfer Processes in Wire-Type Systems *11*
1.5 A Plug–Socket Device Based on a Pseudorotaxane *13*
1.6 Mimicking Electrical Extension Cables at the Molecular Level *14*
1.7 Light-Harvesting Antennas *17*
1.8 Artificial Molecular Machines *19*
1.8.1 Introduction *19*
1.8.2 Energy Supply *20*
1.8.3 Light Energy *21*
1.8.4 Threading Dethreading of an Azobenzene-Based Pseudorotaxane *21*
1.8.5 Photoinduced Shuttling in Multicomponent Rotaxanes: a Light-Powered Nanomachine *23*
1.9 Conclusion *27*
References *28*

2 **Rotaxanes as Ligands for Molecular Machines and Metal–Organic Frameworks** *33*
Stephen J. Loeb
2.1 Interpenetrated and Interlocked Molecules *33*
2.1.1 Introduction *33*

2.1.2	Templating of [2]Pseudorotaxanes	33
2.1.3	[2]Rotaxanes	36
2.1.4	Higher Order [n]Rotaxanes	37
2.1.5	[3]Catenanes	40
2.2	Molecular Machines	41
2.2.1	Introduction	41
2.2.2	Controlling Threading and Unthreading	41
2.2.3	Molecular Shuttles	42
2.2.4	Flip Switches	44
2.3	Interlocked Molecules and Ligands	46
2.3.1	[2]Pseudorotaxanes as Ligands	46
2.3.2	[2]Rotaxanes as Ligands	46
2.4	Materials from Interlocked Molecules	48
2.4.1	Metal–Organic Rotaxane Frameworks (MORFs)	48
2.4.2	One-dimensional MORFs	49
2.4.3	Two-dimensional MORFs	51
2.4.4	Three-dimensional MORFs	51
2.4.5	Controlling the Dimensionality of a MORF	54
2.4.6	Frameworks Using Hydrogen Bonding	57
2.5	Properties of MORFs: Potential as Functional Materials	57
2.5.1	Robust Frameworks	57
2.5.2	Porosity and Internal Properties	59
2.5.3	Dynamics and Controllable Motion in the Solid State	59
	References	59
3	**Strategic Anion Templation for the Assembly of Interlocked Structures** 63	
	Michał J. Chmielewski and Paul D. Beer	
3.1	Introduction	63
3.2	Precedents of Anion-directed Formation of Interwoven Architectures	64
3.3	Design of a General Anion Templation Motif	70
3.4	Anion-templated Interpenetration	72
3.5	Probing the Scope of the New Methodology	74
3.6	Anion-templated Synthesis of Rotaxanes	79
3.7	Anion-templated Synthesis of Catenanes	82
3.8	Functional Properties of Anion-templated Interlocked Systems	88
3.9	Summary and Outlook	93
	References	94
4	**Synthetic Nanotubes from Calixarenes** 97	
	Dmitry M. Rudkevich and Voltaire G. Organo	
4.1	Introduction	97
4.2	Early Calixarene Nanotubes	98
4.3	Metal Ion Complexes with Calixarene Nanotubes	99

4.4	Nanotubes for NO$_x$ Gases	*101*
4.5	Self-assembling Structures	*107*
4.6	Conclusions and Outlook	*108*
	References	*109*

5   Molecular Gels – Nanostructured Soft Materials   *111*
David K. Smith

5.1	Introduction to Molecular Gels	*111*
5.2	Preparation of Molecular Gels	*114*
5.3	Analysis of Molecular Gels	*115*
5.3.1	Macroscopic Behavior – "Table-Top" Rheology	*115*
5.3.1.1	Tube Inversion Methodology	*116*
5.3.1.2	Dropping Ball Method	*116*
5.3.2	Macroscopic Behavior – Rheology	*117*
5.3.3	Macroscopic Behavior – Differential Scanning Calorimetry	*117*
5.3.4	Nanostructure – Electron Microscopy	*118*
5.3.5	Nanostructure – X-Ray Methods	*120*
5.3.6	Molecular Scale Assembly – NMR Methods	*120*
5.3.7	Molecular Scale Assembly – Other Spectroscopic Methods	*122*
5.3.8	Chirality in Gels – Circular Dichroism Spectroscopy	*123*
5.4	Building Blocks for Molecular Gels	*124*
5.4.1	Amides, Ureas, Carbamates (–XCONH– Groups, Hydrogen Bonding)	*125*
5.4.2	Carbohydrates (Multiple –OH Groups, Hydrogen Bonding)	*127*
5.4.3	Steroids/Bile Salts (Hydrophobic Surfaces)	*129*
5.4.4	Nucleobases (Hydrogen Bonding and π–π Stacking)	*130*
5.4.5	Long-chain Alkanes (van der Waals Interactions)	*132*
5.4.6	Dendritic Gels	*133*
5.4.7	Two-component Gels	*137*
5.5	Applications of Molecular Gels	*141*
5.5.1	Greases and Lubricants	*142*
5.5.2	Napalm	*142*
5.5.3	Tissue Engineering – Nerve Regrowth Scaffolds	*142*
5.5.4	Drug Delivery – Responsive Gels	*144*
5.5.5	Capturing (Transcribing) Self-assembled Architectures	*145*
5.5.6	Sensory Gels	*147*
5.5.7	Conductive Gels	*147*
5.6	Conclusions	*148*
	References	*148*

6   Nanoporous Crystals, Co-crystals, Isomers and Polymorphs from Crystals   *155*
Dario Braga, Marco Curzi, Stefano L. Giaffreda, Fabrizia Grepioni, Lucia Maini, Anna Pettersen, and Marco Polito

6.1	Introduction	*155*

6.2	Nanoporous Coordination Network Crystals for Uptake/Release of Small Molecules *156*
6.3	Hybrid Organic–organometallic and Inorganic-organometallic Co-crystals *161*
6.4	Crystal Isomers and Crystal Polymorphs *167*
6.5	Dynamic Crystals – Motions in the Nano-world *170*
6.6	Conclusions *172*
	References *173*

7	**Supramolecular Architectures Based On Organometallic Half-sandwich Complexes** *179*
	Thomas B. Rauchfuss and Kay Severin
7.1	Introduction *179*
7.2	Macrocycles *180*
7.3	Coordination Cages *187*
7.3.1	Cyanometallate Cages *187*
7.3.1.1	Electroactive Boxes *189*
7.3.1.2	Defect Boxes $\{[(C_5R_5)M(CN)_3]_4[Cp^*M]_3\}^z$ *190*
7.3.2	Expanded Organometallic Cyano Cages *191*
7.3.3	Cages Based on *N*-Heterocyclic Ligands *193*
7.4	Expanded Helicates *198*
7.5	Clusters *200*
7.6	Conclusions *200*
	References *201*

8	**Endochemistry of Self-assembled Hollow Spherical Cages** *205*
	Takashi Murase and Makoto Fujita
8.1	Introduction *205*
8.2	Biomacromolecular Cages *206*
8.3	Polymer Micelles *207*
8.4	$M_{12}L_{24}$ Spheres *207*
8.4.1	Self-assembly of $M_{12}L_{24}$ Spheres *207*
8.4.2	Endohedral Functionalization of $M_{12}L_{24}$ Spheres *209*
8.4.3	Fluorous Nanodroplets *210*
8.4.4	Uptake of Metal Ions into a Cage *212*
8.4.5	Polymerization in a Nutshell *213*
8.4.6	Photoresponsive Molecular Nanoballs *216*
8.4.7	Peptide-confined Chiral Cages *217*
8.5	Conclusions and Outlook *219*
	References *220*

9	**Polynuclear Coordination Cages** *223*
	Michael D. Ward
9.1	Introduction *223*
9.2	Complexes Based on Poly(pyrazolyl)borate Ligands *225*

9.3	Complexes Based on Neutral Ligands with Aromatic Spacers 227
9.3.1	Complexes Based on $L^{o\text{-}Ph}$ and $L^{12\text{-}naph}$ 227
9.3.2	Larger Tetrahedral Cages Based on L^{biph} 234
9.3.3	Higher Nuclearity Cages Based on Other Ligands 235
9.4	Mixed-ligand Complexes: Opportunities for New Structural Types 243
	References 248

10	**Periodic Nanostructures Based on Metal–Organic Frameworks (MOFs): En Route to Zeolite-like Metal–Organic Frameworks (ZMOFs)** 251
	Mohamed Eddaoudi and Jarrod F. Eubank
10.1	Introduction 251
10.2	Historical Perspective 252
10.2.1	Metal–Cyanide Compounds 252
10.2.2	Werner Complexes 254
10.2.3	Expanded Nitrogen-donor Ligands 255
10.2.4	Carboxylate-based Ligands 258
10.3	Single-metal Ion-based Molecular Building Blocks 261
10.3.1	Discrete, 2D and 3D Metal–Organic Assemblies 262
10.3.2	Zeolite-like Metal–Organic Frameworks (ZMOFs) 264
10.3.2.1	*sod*-ZMOF 265
10.3.2.2	*rho*-ZMOF 266
10.4	Conclusion 270
	References 271

11	**Polyoxometalate Nanocapsules: from Structure to Function** 275
	Charalampos Moiras and Leroy Cronin
11.1	Introduction 275
11.2	Background and Classes of Polyoxometalates 277
11.3	Wells–Dawson $\{M_{18}O_{54}\}$ Capsules 278
11.4	Isopolyoxometalate Nanoclusters 280
11.5	Keplerate Clusters 282
11.6	Surface-Encapsulated Clusters (SECs): Organic Nanostructures with Inorganic Cores 285
11.7	Perspectives 287
	References 287

12	**Nano-capsules Assembled by the Hydrophobic Effect** 291
	Bruce C. Gibb
12.1	Introduction 291
12.2	Synthesis of a Water-soluble, Deep-cavity Cavitand 292
12.2.1	Structure of the Cavitand (What It Is and What It Is Not) 292
12.2.2	Assembly Properties of the Cavitand 294
12.2.3	Photophysics and Photochemistry Within Nano-capsules 299
12.2.4	Hydrocarbon Gas Separation Using Nano-capsules 301

12.3	Conclusions *302*	
	References *303*	

13	**Opportunities in Nanotechnology via Organic Solid-state Reactivity: Nanostructured Co-crystals and Molecular Capsules** *305*	
	Dejan-Krešimir Bučar, Tamara D. Hamilton, and Leonard R. MacGillivray	
13.1	Introduction *305*	
13.2	Template-controlled [2 + 2] Photodimerization in the Solid State *305*	
13.3	Nanostructured Co-crystals *307*	
13.3.1	Organic Nanocrystals and Single Crystal-to-single Crystal Reactivity *308*	
13.4	Self-assembled Capsules Based on Ligands from the Solid State *309*	
13.5	Summary and Outlook *312*	
	References *313*	

14	**Organic Nanocapsules** *317*	
	Scott J. Dalgarno, Nicholas P. Power, and Jerry L. Atwood	
14.1	Introduction *317*	
14.2	First Generation Nanocapsules *317*	
14.3	Second Generation Nanocapsules *320*	
14.4	Third Generation Nanocapsules *323*	
14.5	Fourth Generation Nanocapsules *329*	
14.6	Fifth Generation Nanocapsules *331*	
14.7	Sixth Generation Nanocapsules *339*	
14.8	From Spheres to Tubes *342*	
14.9	Conclusions *344*	
	References *345*	

Index *347*

Preface

Current research in chemistry and materials science is now vigorously pushing the boundaries of the components studied firmly into the multi-nanometer length scale. In terms of traditional "molecules" a nanometer (10^{-9} m) is relatively large. As a result, it is only relatively recent advances in analytical instrumentation capable of delivering a molecular-level understanding of structure and properties in this kind of size regime that have allowed access to and the study of such large molecules and assemblies. The key interest in multi-nanometer-scale structures (nanostructures) is the fact that their size allows them to exhibit a significant degree of functionality and complexity – complexity that is mirrored in biological systems such as enzymes and polynucleotides, Nature's own nanostructures. However, this functionality is compressed into a space that is very small on the human scale, sparking interest in fields such as molecular computing and molecular devices. Thus one of the great opening frontiers in molecular sciences is the upward synthesis, understanding of structure and application of molecules and molecular concepts into the nanoscale.

In compiling this book we have sought to bring together chapters from leading experts working on the cutting edge of this revolution on the nanoscale. Each chapter is a self-contained illustration of the way in which the nanoscale view is influencing current thinking and research across the molecular sciences. The focus is on the "organic" (loosely applied) since it is generally carbon-based building blocks that are the most versatile molecular components that can be induced to link into nanoscale structures. As chapters by Mohammed Eddaoudi and Lee Cronin show, however, hybrid organic–inorganic materials and well-defined inorganic building blocks as just as capable of assembling into well-defined and well-characterized discrete and polymeric nanostructures.

Crucial to the whole field of nanochemistry is the cross-fertilization between researchers from different disciplines that are approaching related structures from very different perspectives. It is with this aspect in mind that we have deliberately mixed together contributions from the solid-state materials community as in Dario Braga's perspective on the crystal engineering or organic nanostructures and from experts in discrete molecular assemblies such as Dimitry Rudkevich, Kay Severin, Thomas Rauchfuss and Bruce Gibb. Of course, nanostructures are not

Organic Nanostructures. Edited by Jerry L. Atwood and Jonathan W. Steed
Copyright © 2008 WILEY-VCH Verlag GmbH & Co. KGaA, Weinheim
ISBN: 978-3-527-31836-0

always so well defined and so these aspects are balanced nicely by David Smith's chapter on gel-phase materials – in some respects a "halfway house" between solution-phase and solid-state assemblies. We also felt it of key importance to illustrate ways to use small-scale molecular concepts in order to "synthesize-up" nanostructures. Chapters by Paul Beer, Steve Loeb and Len MacGillivray provide very different perspectives on templation and assembly in the field, while Makoto Fujita and Mike Ward deal with larger-scale self-assembly. Finally, all-important functional nanostructured devices are illustrated by Vincenzo Balzani's chapter.

Although a book of this size can only be illustrative of such a burgeoning field, it is our sincere hope that the juxtaposition of these different perspectives and systems in one place will stimulate and contribute to the ongoing process of cross-fertilization that is driving this fascinating and emerging area of molecular science. It has certainly been a fascinating and pleasurable experience to work on this project and we thank all of the authors wholeheartedly for their enthusiastic contributions to this project. We are grateful also to Manfred Köhl and Steffen Pauly at Wiley-VCH for their belief in the book and for their help in making it a reality. As this book went to press we learned of the sad and untimely death of Dimitry Rudkevich. We would like to dedicate this book to his memory and legacy to science.

December 2007

Jonathan W. Steed, Durham, UK
Jerry L. Atwood, Columbia, MO, USA

List of Contributors

Jerry L. Atwood
University of Missouri–Columbia
Department of Chemistry
125 Chemistry Building
MO 65211 Columbia
USA

Vincenzo Balzani
Università di Bologna
Dipartimento di Chimica "G. Ciamician"
Via Selmi 2
40126 Bologna
Italy

Paul D. Beer
University of Oxford
Department of Chemistry
Inorganic Chemistry Laboratory
South Parks Road
Oxford OX1 3QR
UK

Dario Braga
Università di Bologna
Dipartimento di Chimica "G. Ciamician"
Via Selmi 2
40126 Bologna
Italy

Dejan Krešimir Bučar
University of Iowa
Department of Chemistry
Iowa City
IA 52245
USA

Michał J. Chmielewski
University of Oxford
Department of Chemistry
Inorganic Chemistry Laboratory
South Parks Road
Oxford OX1 3QR
UK

Alberto Credi
Università di Bologna
Dipartimento di Chimica "G. Ciamician"
Via Selmi 2
40126 Bologna
Italy

Leroy Cronin
University of Glasgow
Department of Chemistry
Glasgow G12 8QQ
UK

Marco Curzi
Università di Bologna
Dipartimento di Chimica "G. Ciamician"
Via Selmi 2
40126 Bologna
Italy

Scott J. Dalgarno
Heriot–Watt University
School of Engineering and Physical Sciences – Chemistry
Edinburgh EH14 4AS
UK

Mohamed Eddaoudi
University of South Florida
Department of Chemistry
4202 East Fowler Avenue (CHE 205)
Tampa
FL 33620
USA

Jarrod F. Eubank
University of South Florida
Department of Chemistry
4202 East Fowler Avenue (CHE 205)
Tampa
FL 33620
USA

Makoto Fujita
The University of Tokyo
School of Engineering
Department of Applied Chemistry
7-3-1 Hongo
Bunkyo-ku
Tokyo 113-8656
Japan

Stefano Luca Giaffreda
Università di Bologna
Dipartimento di Chimica "G. Ciamician"
Via Selmi 2
40126 Bologna
Italy

Bruce C. Gibb
University of New Orleans
Department of Chemistry
New Orleans
LA 70148
USA

Fabrizia Grepioni
Università di Bologna
Dipartimento di Chimica "G. Ciamician"
Via Selmi 2
40126 Bologna
Italy

Tamara D. Hamilton
University of Iowa
Department of Chemistry
Iowa City
IA 52245
USA

Stephen J. Loeb
University of Windsor
Department of Chemistry and Biochemistry
Windsor
Ontario N9B 3P4
Canada

Leonard R. MacGillivray
University of Iowa
Department of Chemistry
Iowa City
IA 52245
USA

Lucia Maini
Università di Bologna
Dipartimento di Chimica "G. Ciamician"
Via Selmi 2
40126 Bologna
Italy

Charalampos Moiras
University of Glasgow
Department of Chemistry
Glasgow G12 8QQ
UK

Takashi Murase
The University of Tokyo
School of Engineering
Department of Applied Chemistry
7-3-1 Hongo
Bunkyo-ku
Tokyo 113-8656
Japan

Voltaire G. Organo
University of Texas at Arlington
Department of Chemistry and
Biochemistry
Arlington
TX 76019-0065
USA

Anna Pettersen
Università di Bologna
Dipartimento di Chimica "G. Ciamician"
Via Selmi 2
40126 Bologna
Italy

Marco Polito
Università di Bologna
Dipartimento di Chimica "G. Ciamician"
Via Selmi 2
40126 Bologna
Italy

Nicholas P. Power
University of Missouri–Columbia
Department of Chemistry
125 Chemistry Building
MO 65211 Columbia
USA

Thomas B. Rauchfuss
University of Illinois
Department of Chemistry
Urbana
IL 61801
USA

Dmitry M. Rudkevich
University of Texas at Arlington
Department of Chemistry and
Biochemistry
Arlington
TX 76019-0065
USA

Kay Severin
École Polytechnique Fédérale de
Lausanne
Institut des Sciences et Ingénierie
Chimiques
CH-1015 Lausanne
Switzerland

David K. Smith
University of York
Department of Chemistry
Heslington
York YO10 5DD
UK

Margherita Venturi
Università di Bologna
Dipartimento di Chimica "G. Ciamician"
Via Selmi 2
40126 Bologna
Italy

Michael D. Ward
University of Sheffield
Department of Chemistry
Dainton Building
Sheffield S3 7HF
UK

1
Artificial Photochemical Devices and Machines
Vincenzo Balzani, Alberto Credi, and Margherita Venturi

1.1
Introduction

The interaction between light and matter lies at the heart of the most important processes of life [1]. Photons are exploited by natural systems as both quanta of energy and elements of information. Light constitutes an energy source and is consumed (or, more precisely, converted) in large amount in the natural photosynthetic process, whereas it plays the role of a signal in vision-related processes, where the energy used to run the operation is biological in nature.

A variety of functions can also be obtained from the interaction between light and matter in artificial systems [2]. The type and utility of such functions depend on the degree of complexity and organization of the chemical systems that receive and process the photons.

About 20 years ago, in the frame of research on supramolecular chemistry, the idea began to arise [3–5] that the concept of macroscopic device and machine can be transferred to the molecular level. In short, a molecular device can be defined [6] as an assembly of a discrete number of molecular components designed to perform a function under appropriate external stimulation. A molecular machine [6–8] is a particular type of device where the function is achieved through the mechanical movements of its molecular components.

In analogy with their macroscopic counterparts, molecular devices and machines need energy to operate and signal to communicate with the operator. Light provides an answer to this dual requirement. Indeed, a great number of molecular devices and machines are powered by light-induced processes and light can also be useful to "read" the state of the system and thus to control and monitor its operation. Before illustrating examples of artificial photochemical molecular devices and machines, it is worthwhile recalling a few basic aspects of the interaction between molecular and supramolecular systems and light. For a more detailed discussion, books [9–15] can be consulted.

Organic Nanostructures. Edited by Jerry L. Atwood and Jonathan W. Steed
Copyright © 2008 WILEY-VCH Verlag GmbH & Co. KGaA, Weinheim
ISBN: 978-3-527-31836-0

1.2
Molecular and Supramolecular Photochemistry

1.2.1
Molecular Photochemistry

Figure 1.1 shows a schematic energy level diagram for a generic molecule that could also be a component of a supramolecular species. In most cases the ground state of a molecule is a singlet state (S_0) and the excited states are either singlets (S_1, S_2, etc.) or triplets (T_1, T_2, etc.). In principle, transitions between states having the same spin value are allowed, whereas those between states of different spin are forbidden. Therefore, the electronic absorption bands observed in the UV–visible spectrum of molecules usually correspond to $S_0 \rightarrow S_n$ transitions. The excited states so obtained are unstable species that decay by rapid first-order kinetic processes, namely chemical reactions (e.g. dissociation, isomerization) and/or radiative and nonradiative deactivations. In the discussion that follows, excited-state reactions do not need to be explicitly considered and can formally be incorporated within the radiationless decay processes. When a molecule is excited to upper singlet excited states (Figure 1.1), it usually undergoes a rapid and 100% efficient radiationless deactivation [internal conversion (ic)] to the lowest excited singlet, S_1. Such an excited state undergoes deactivation via three competing processes: nonradiative decay to the ground state (internal conversion, rate constant k_{ic}); radiative decay to the ground state (fluorescence, k_{fl}); conversion to the lowest triplet state T_1 (intersystem crossing, k_{isc}). In its turn, T_1 can undergo deactivation via nonradiative (intersystem crossing, k'_{isc}) or radiative (phosphorescence, k_{ph}) decay to the ground state S_0. When the molecule contains heavy atoms, the formally forbidden intersystem crossing and

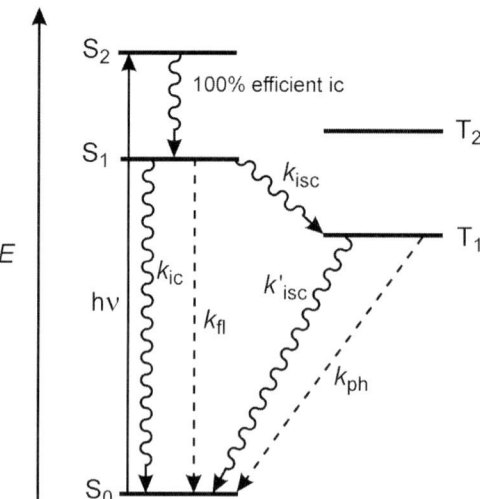

Figure 1.1 Schematic energy level diagram for a generic molecule. For more details, see text.

phosphorescence processes become faster. The lifetime (τ) of an excited state, that is, the time needed to reduce the excited-state concentration by 2.718 (i.e. the basis for natural logarithms, e), is given by the reciprocal of the summation of the deactivation rate constants:

$$\tau(S_1) = \frac{1}{(k_{ic}+k_{fl}+k_{isc})} \quad (1)$$

$$\tau(T_1) = \frac{1}{(k'_{isc}+k_{ph})} \quad (2)$$

The orders of magnitude of $\tau(S_1)$ and $\tau(T_1)$ are approximately $10^{-9} - 10^{-7}$ and $10^{-3} - 10^0$ s, respectively. The quantum yield of fluorescence (ratio between the number of photons emitted by S_1 and the number of absorbed photons) and phosphorescence (ratio between the number of photons emitted by T_1 and the number of absorbed photons) can range between 0 and 1 and are given by

$$\Phi_{fl} = \frac{k_{fl}}{(k_{ic}+k_{fl}+k_{isc})} \quad (3)$$

$$\Phi_{ph} = \frac{k_{ph} \times k_{isc}}{(k'_{isc}+k_{ph}) \times (k_{ic}+k_{fl}+k_{isc})} \quad (4)$$

Excited-state lifetimes and fluorescence and phosphorescence quantum yields of a great number of molecules are known [16].

When the intramolecular deactivation processes are not too fast, that is, when the lifetime of the excited state is sufficiently long, an excited molecule *A may have a chance to encounter a molecule of another solute, B (Figure 1.2). In such a case, some specific interaction can occur leading to the deactivation of the excited state by second-order kinetic processes. The two most important types of interactions in an encounter are those leading to electron or energy transfer. The occurrence of these processes causes the quenching of the intrinsic properties of *A; energy transfer also

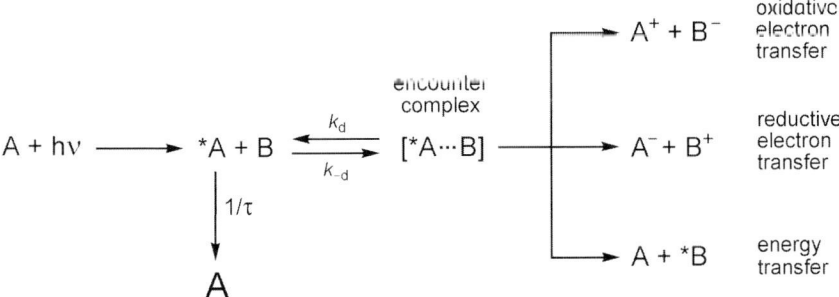

Figure 1.2 Schematic representation of bimolecular electron- and energy-transfer processes that may occur following an encounter between an excited state, *A, and another chemical species, B.

leads to sensitization of the excited-state properties of the B species. Simple kinetic arguments show that only the excited states that live longer than ca. 10^{-9} s may have a chance to be involved in encounters with other solute molecules.

An electronically excited state is a species with completely different properties to those of the ground-state molecule. In particular, because of its higher energy content, an excited state is both a stronger reductant and a stronger oxidant than the corresponding ground state [17]. To a first approximation, the redox potential of an excited-state couple may be calculated from the potential of the related ground-state couple and the one-electron potential corresponding to the zero–zero excited-state energy, E^{0-0}:

$$E(A^+/{}^*A) \approx E(A^+/A) - E^{0-0} \tag{5}$$

$$E({}^*A/A^-) \approx E(A/A^-) + E^{0-0} \tag{6}$$

Detailed discussions of the kinetics aspects of electron- and energy-transfer processes can be found in the literature [11,18–20].

1.2.2
Supramolecular Photochemistry

A supramolecular system can be preorganized so as to favor the occurrence of electron- and energy-transfer processes [10]. The molecule that has to be excited, A, can indeed be placed in the supramolecular structure nearby a suitable molecule, B.

For simplicity, we consider the case of an A–L–B supramolecular system, where A is the light-absorbing molecular unit [Eq. (7)], B is the other molecular unit involved with A in the light-induced processes and L is a connecting unit (often called bridge). In such a system, after light excitation of A there is no need to wait for a diffusion-controlled encounter between *A and B as in molecular photochemistry, since the two reaction partners can already be at an interaction distance suitable for electron and energy transfer:

$$A-L-B + h\nu \rightarrow {}^*A-L-B \quad \text{photoexcitation} \tag{7}$$

$$^*A-L-B \rightarrow A^+-L-B^- \quad \text{oxidative electron transfer} \tag{8}$$

$$^*A-L-B \rightarrow A^--L-B^+ \quad \text{reductive electron transfer} \tag{9}$$

$$^*A-L-B \rightarrow A-L-{}^*B \quad \text{electronic energy transfer} \tag{10}$$

In the absence of chemical complications (e.g. fast decomposition of the oxidized and/or reduced species), photoinduced electron-transfer processes [Eqs. (8) and (9)] are followed by spontaneous back-electron-transfer reactions that regenerate the starting ground-state system [Eqs. 8′ and 9′] and photoinduced energy transfer [Eq. (10)] is followed by radiative and/or nonradiative deactivation of the excited acceptor [Eq. 10′]:

$$A^+-L-B^- \rightarrow A-L-B \quad \text{back oxidative electron transfer} \tag{8′}$$

$A^- - L - B^+ \rightarrow A - L - B$ back reductive electron transfer   (9′)

$A - L - {}^*B \rightarrow A - L - B$ excited state decay   (10′)

In supramolecular systems, electron- and energy-transfer processes are no longer limited by diffusion and occur by first-order kinetics. As a consequence, in suitably designed supramolecular systems these processes can involve even very short-lived excited states.

1.3
Wire-Type Systems

An important function at the molecular level is photoinduced energy and electron transfer over long distances and/or along predetermined directions. This function can be performed by rod-like supramolecular systems obtained by linking donor and acceptor components with a bridging ligand or a spacer.

1.3.1
Molecular Wires for Photoinduced Electron Transfer

Photoinduced electron transfer in wire-type supramolecular species has been extensively investigated [6,10]. The minimum model is a *dyad*, consisting of an electron donor (or acceptor) chromophore, an additional electron acceptor (or donor) moiety and an organizational principle that controls their distance and electronic interactions (and therefore the rates and yields of electron transfer). A great number of such dyads have been constructed and investigated [6,10].

The energy-level diagram for a dyad is schematized in Figure 1.3. All the dyad-type systems suffer to a greater or lesser extent from rapid charge recombination

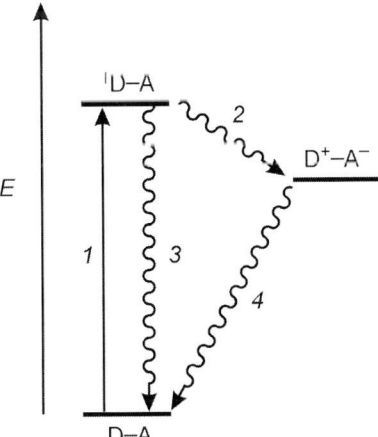

Figure 1.3 Schematic energy-level diagram for a dyad.

[process (4)]. An example of a systematic study on dyads is that performed on compounds 1^{5+}–5^{5+} (Figure 1.4) [21,22]. When excitation is selectively performed in the Ru(II) chromophoric unit, prompt intersystem crossing from the originally populated singlet metal-to-ligand charge-transfer (^{1}MLCT) excited state leads to the long-lived ^{3}MLCT excited state which lies ~2.1 eV above the ground state, can be oxidized approximately at −0.9 V (vs. SCE) and has a lifetime of ~1 μs in deaerated solutions [23]. Before undergoing deactivation, such an excited state transfers an electron to the Rh(III) unit, a process that is then followed by a back electron-transfer reaction.

Comparison of compounds 1^{5+} and 2^{5+} shows that, despite the longer metal–metal distance, the forward electron transfer is faster across the phenylene spacer ($k = 3.0 \times 10^9$ s^{-1}) than across the two methylene groups ($k = 1.7 \times 10^8$ s^{-1}). This result can be related to the lower energy of the LUMO of the phenylene group, which facilitates electronic coupling. In the homogeneous family of compounds 2^{5+}–4^{5+}, the rate constant decreases exponentially with increasing metal–metal distance.

For compound 5^{5+}, which is identical with 4^{5+} except for the presence of two solubilizing hexyl groups on the central phenylene ring, the photoinduced electron-transfer process is 10 times slower, presumably because the substituents increase the twist angle between the phenylene units, thereby reducing electronic coupling.

Photoinduced electron transfer in three–component systems (*triads*) is illustrated in Figure 1.5 [24]. The functioning principles are shown in the orbital-type energy diagrams of the lower part the figure. In both cases, excitation of a chromophoric component (step *1*) is followed by a primary photoinduced electron transfer to a primary acceptor (step *2*). This process is followed by a secondary thermal electron-transfer process (step *3*): electron transfer from a donor component to the oxidized chromophoric component (case a) or electron transfer from the primary acceptor to a secondary acceptor component (case b). The primary process competes with excited-state deactivation (step *4*), whereas the secondary process competes with primary charge recombination (step *5*). Finally, charge recombination between remote molecular components (step *6*) leads the triad back to its initial state.

For case a, the sequence of processes indicated above (*1–2–3*) is not unique. Actually, the alternative sequence *1–3–2* would also lead to the same charge-separated state. In general, these two pathways will have different driving forces for the primary and secondary steps and thus one may be kinetically favored over the other. Occasionally one of the two pathways is thermodynamically allowed and the other is not, although in a simple one-electron energy diagram like that shown in Figure 1.5a this aspect is not apparent.

The performance of a triad for wire-type applications is related to the rate and quantum yield of formation of the charge separated state (depending on the competition between forward and back processes, $\Phi = [k_2/(k_2 + k_4)][k_3/(k_3 + k_5)]$). For energy conversion purposes, important parameters are also the lifetime of charge separation (depending on the rate of the final charge-recombination process, $\tau = 1/k_6$) and the efficiency of energy conversion ($\eta_{\text{en.conv.}} = \Phi \times F$, where F is the fraction of the excited-state energy conserved in the final charge-separated state). To put things in a real perspective, it should be recalled that the "triad portion" of the

1.3 Wire-Type Systems

1[5+] $k = 1.7 \times 10^8$ s^{-1}

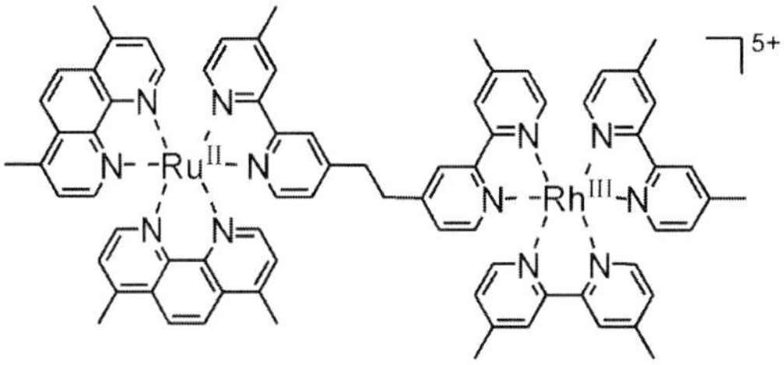

2[5+] $k = 3.0 \times 10^9$ s^{-1}

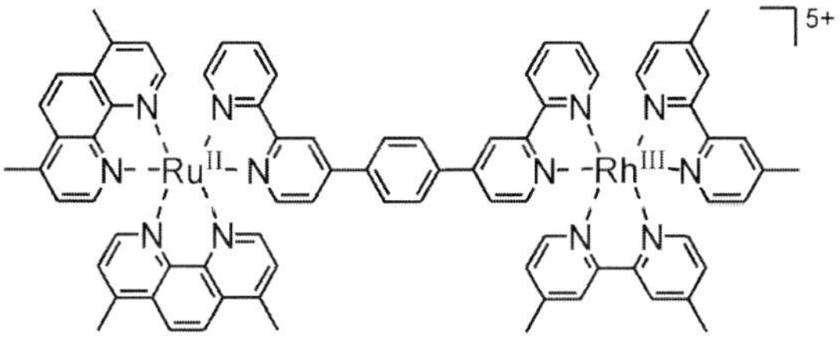

3[5+] $k = 4.3 \times 10^8$ s^{-1}

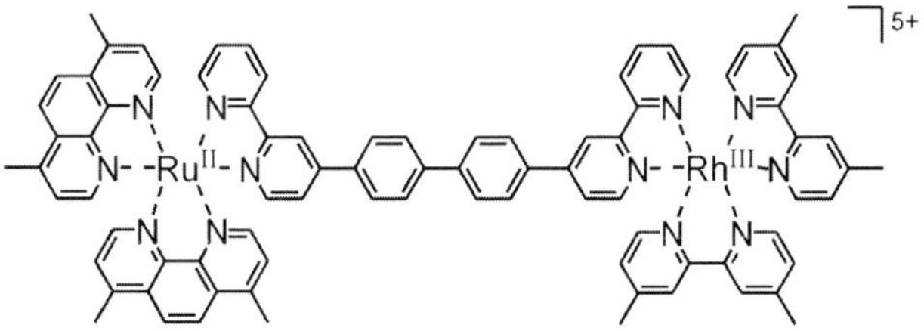

4[5+] $k = 1 \times 10^7$ s^{-1}

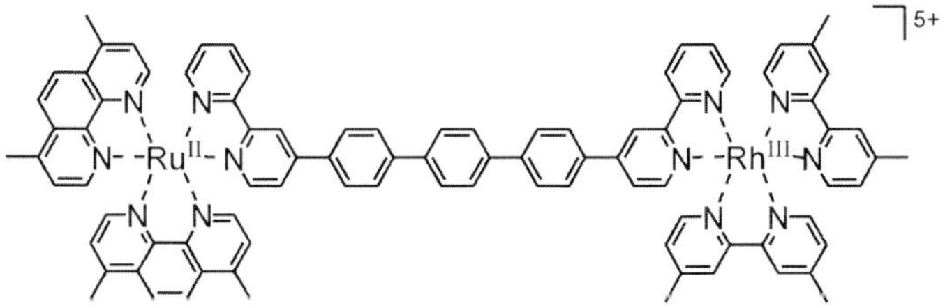

5[5+] $k = 1.1 \times 10^6$ s^{-1}

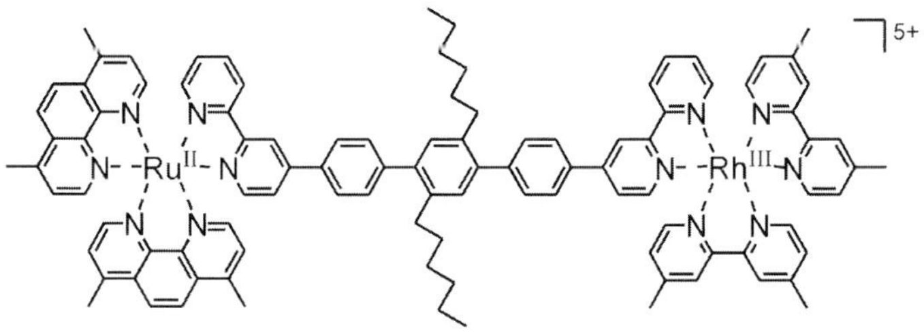

Figure 1.4 Binuclear metal complexes **1**[5+]–**5**[5+] used for photoinduced electron-transfer experiments [21,22].

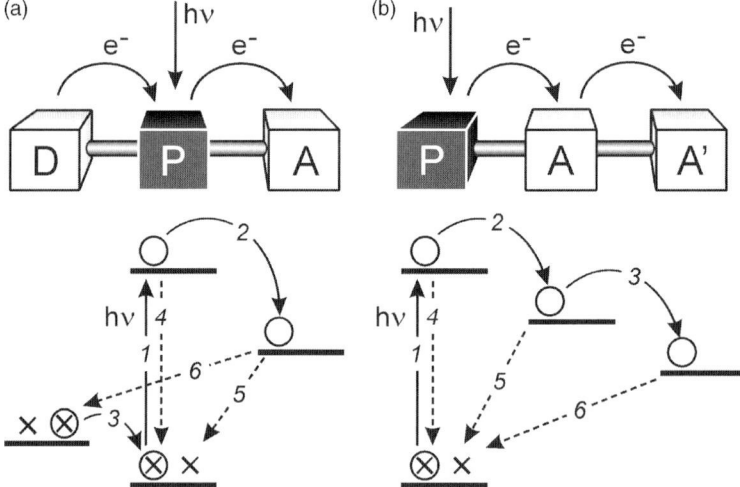

Figure 1.5 Schematic representation of the two possible arrangements for charge-separating triads.

reaction center of bacterial photosynthesis converts light energy with $\tau \approx 10$ ms, $\Phi = 1$ and $\eta_{\text{en.conv.}} \approx 0.6$.

The introduction of further molecular components (*tetrads* and *pentads*) leads to the occurrence of further electron-transfer steps, which, in suitably designed systems, produce charge separation over larger and larger distances [6,10]. As the number of molecular components increases, also the mechanistic complexity increases and charge separation may involve energy-transfer steps.

Several triads have been designed and investigated. A very interesting system is the 4-nm long triad **6**$^{3+}$ shown in Figure 1.6, which consists of an Ir(III) bis-terpyridine

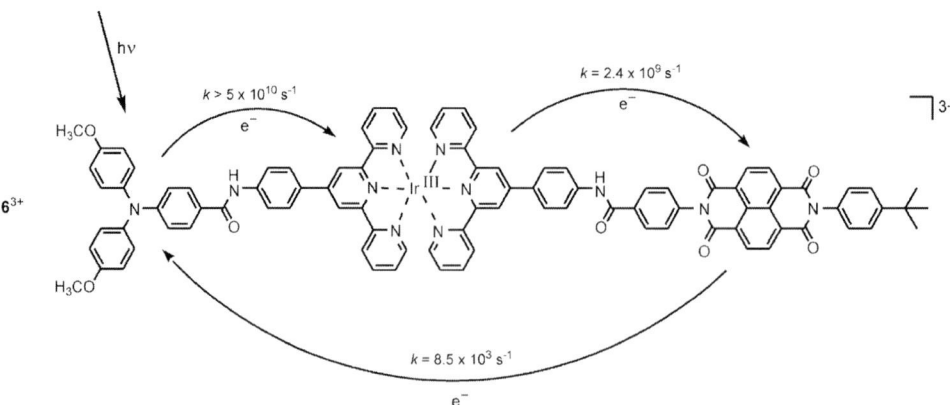

Figure 1.6 Electron-transfer processes in triad **6**$^{3+}$ [25].

complex connected to a triphenylamine electron donor (D) and a naphthalene bisimide electron acceptor (A) [25]. Upon excitation of the electron donor D (or even the Ir-based moiety), a charge separated state D^+-Ir^--A is formed with 100% yield in less than 20 ps that successively leads to D^+-Ir-A^- with 10% efficiency in 400 ps. Remarkably, the fully charge-separated state D^+-Ir-A^- has a lifetime of 120 µs at room temperature in deaerated acetonitrile solution.

1.3.2
Molecular Wires for Photoinduced Energy Transfer

Many investigations on electronic energy transfer in supramolecular species have been performed in the past few years [6,10], a relevant fraction of which have been obtained for systems containing polypyridine metal complexes as donor and acceptor units. Usually, the photoexcited chromophoric group is $[Ru(bpy)_3]^{2+}$ (bpy = 2,2′-bipyridine) and the energy acceptor is an $[Os(bpy)_3]^{2+}$ unit. The excited state of $[Ru(bpy)_3]^{2+}$ playing the role of energy donor is the lowest, formally triplet, metal-to-ligand charge-transfer excited state, ^3MLCT, which, as we have seen above, can be obtained by visible light excitation ($\lambda_{max} \approx 450$ nm), lies ~2.1 eV above the ground state and has a lifetime of ~1 µs in deaerated solutions [23]. This relatively long lifetime is very useful because it permits the study of energy transfer over long distances. The occurrence of the energy-transfer process promotes the ground-state $[Os(bpy)_3]^{2+}$ acceptor unit to its lowest energy excited state ^3MLCT, which lies approximately 0.35 eV below the donor excited state. Both the donor and the acceptor excited states are luminescent, so that the occurrence of energy transfer can be monitored by quenching and/or sensitization experiments with both continuous and pulsed excitation techniques.

Ru(II) and Os(II) polypyridine units have been connected by a variety of bridging ligands and spacers. When the metal-to-metal distance is very short, fast energy transfer occurs by a Förster-type resonance mechanism [26]. In other systems the two photoactive units are separated by a more or less long spacer. When the spacer is flexible [e.g. $-(CH_2)_n-$ chains], the geometry of the system is not well defined and it is difficult to rationalize the results obtained.

These problems are overcome by using rigid and modular spacers to connect the two chromophoric units; the systems so obtained have a well-characterized geometry and the energy transfer can occur over long distances. Interesting examples of this type of systems are the $[Ru(bpy)_3]^{2+}-(ph)_n-[Os(bpy)_3]^{2+}$ (ph = 1,4-phenylene; n = 2, 3, 4, 5) species [27] shown in Figure 1.7. In such compounds, excitation of the $[Ru(bpy)_3]^{2+}$ moiety is followed by energy transfer to the $[Os(bpy)_3]^{2+}$ unit, as shown by the sensitized emission of the latter (CH_3CN, 293 K). The energy-level diagram is shown schematically in Figure 1.7. The lowest energy level of the bridge decreases slightly as the number of phenylene units is increased, but always lies above the donor and acceptor levels involved in energy transfer. A further decrease in the energy of the triplet excited state of the spacer would be expected to switch the energy-transfer mechanism from superexchange-mediated to hopping, similar to what happens for photoinduced electron transfer. In the series of compounds shown in

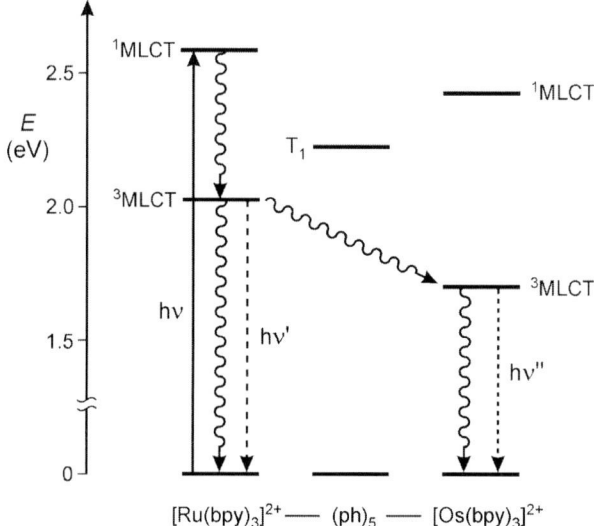

Figure 1.7 Structure of compounds $[Ru(bpy)_3]^{2+}$–$(ph)_n$–$[Os(bpy)_3]^{2+}$ and energy-level diagram for the energy-transfer process [27].

Figure 1.7, the energy-transfer rate decreases with increasing length of the oligophenylene spacer. Such rate constants are much higher than those expected for a Förster-type mechanism, whereas they can be accounted for by a superexchange Dexter mechanism [28]. The values obtained for energy transfer in the analogous series of compounds $[Ru(bpy)_3]^{2+}$–$(ph)_n R_2$–$[Os(bpy)_3]^{2+}$ [29], in which the central phenylene unit carries two hexyl chains, are much lower than those found for the unsubstituted compounds, most likely because the bulky substituents R increase the tilt angle between the phenyl units. A strong decrease in the rate constant is observed when the Ru-donor and Os-acceptor units are linked via an oligophenylene bridge connected in the meta position [30].

In an another family of similar compounds, $[Ir(ppyF_2)_2(bpy)]^+$–$(ph)_n$–$[Ru(bpy)_3]^{2+}$ (ph = 1,4-phenylene; n = 2, 3, 4, 5) [31], the energy-transfer rate constant is much higher and substantially independent of the length of the spacer. The energy-level

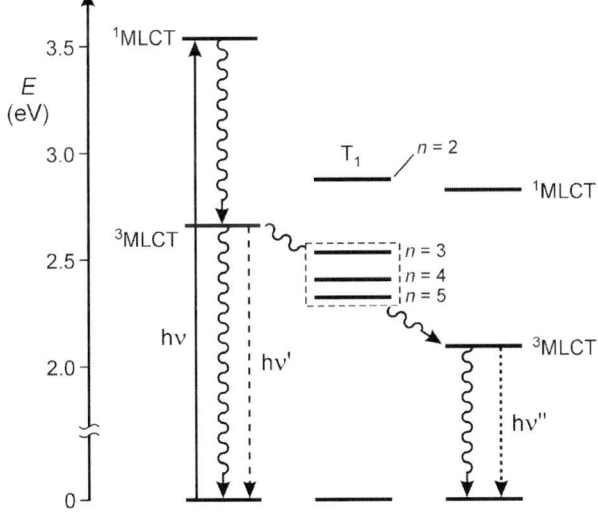

Figure 1.8 Structure of compounds [Ir(ppyF$_2$)$_2$(bpy)]$^+$–(ph)$_n$–[Ru(bpy)$_3$]$^{2+}$ and energy-level diagram for the energy-transfer process [31].

diagram for this family, displayed in Figure 1.8, shows that the energy level of the donor is almost isoenergetic with the triplet state of the spacers. The energy of the Ir-based donor can, therefore, be transferred to the Ru-based acceptor via hopping on the bridging ligand, at least for $n > 2$.

1.4
Switching Electron-Transfer Processes in Wire-Type Systems

A clever choice of molecular components and their assembly in suitable sequences allow the design of very interesting molecular-level photonic switches for photoinduced electron-transfer processes.

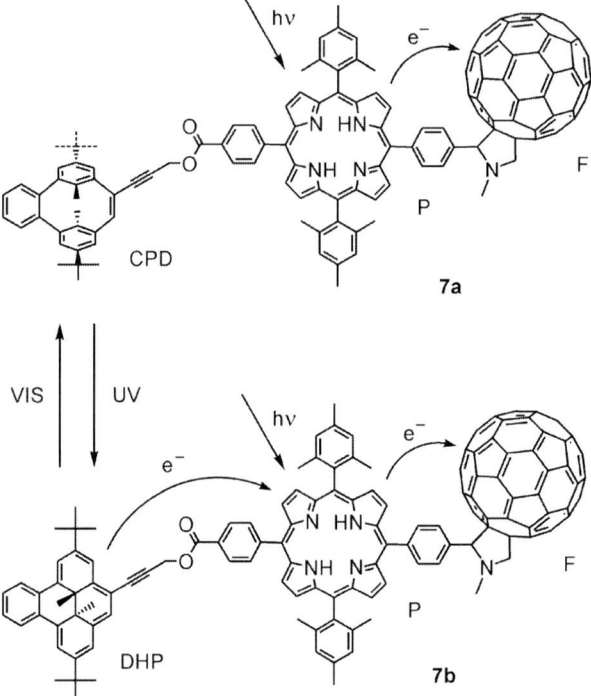

Figure 1.9 A single pole electron-transfer switch. Light-induced isomerization and electron transfer processes in triad **7** [32].

Triad **7** (Figure 1.9) is an example of electron transfer switch generated by the light-induced interconversion between the two forms of a chromophore. This triad, which performs as a single pole molecular switch, consists of a porphyrin unit (P) linked covalently to both a fullerene (F) electron acceptor and a dihydropyrene photochrome [32]. In structure **7a**, the photochrome is in the cyclophanediene (CPD) form, which absorbs light only in the UV region. Excitation of the porphyrin unit leads to CPD–^{1}P–F excited state which undergoes electron transfer yielding the CPD–P$^+$–F$^-$ charge-separated state with unitary efficiency. Such a state then decays to the ground state with time constant 3.3 ns. Irradiation of **7a** with UV light at 254 nm converts the cyclophanediene form of the photochrome into the dihydropyrene form (DHP). The photochemistry of the resulting DHP–P–F species (**7b**) is different from that of **7a**. The DHP–^{1}P–F excited state leads again to charge separation, DHP–P$^+$–F$^-$, but before the charge separated state can recombine to the ground state, an electron migrates from the DHP moiety to the porphyrin, producing DHP$^+$–P–F$^-$ with quantum yield 0.94. This state lives much longer (2.0 μs) than the CPD–P$^+$–F$^-$ species because the charges are much farther apart and, therefore, the electronic coupling is smaller. Reconfiguration of the system to **7a** can be obtained by visible light irradiation.

1.5
A Plug–Socket Device Based on a Pseudorotaxane

Supramolecular species whose components are connected by means of noncovalent forces can be disassembled and re-assembled [33] by modulating the interactions that keep the components together, with the consequent possibility of switching energy-transfer processes. Two-component systems of this type are reminiscent of plug–socket electrical devices because, like their macroscopic counterparts, they are characterized by (i) the possibility of connecting–disconnecting the two components in a reversible way and (ii) the occurrence of an electronic energy flow from the socket to the plug when the two components are connected (Figure 1.10a). Hydrogen-bonding interactions between ammonium ions and crown ethers are particularly

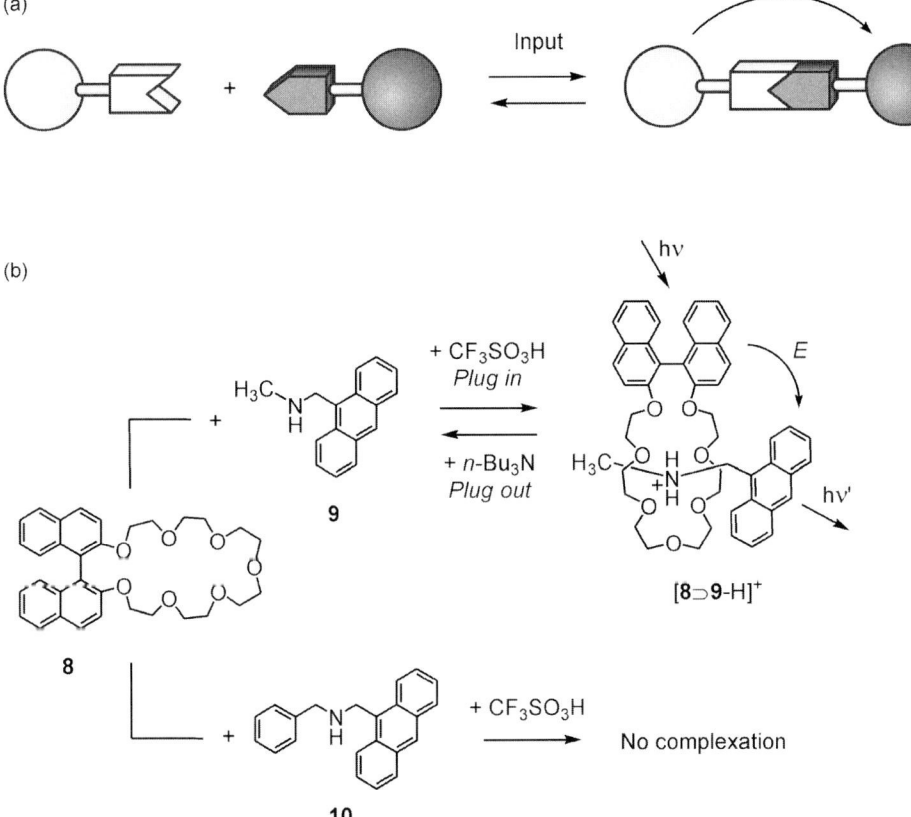

Figure 1.10 (a) Schematic representation of the working mechanism of a plug–socket system. (b) Switching of photoinduced energy transfer by the acid–base-controlled plug in–plug out of binaphthocrown ether **8** and anthracenylammonium ion **9**-H$^+$ [34].

suitable for constructing molecular-level plug–socket devices, since they can be switched on and off quickly and reversibly by means of acid–base inputs.

A plug–socket system which deals with the transfer of electronic energy is illustrated in Figure 1.10b [34]. The absorption and fluorescence spectra of a CH_2Cl_2 solution containing equal amounts of ($\pm$)-binaphthocrown ether **8** and amine **9** indicate the absence of any interaction between the two compounds. Addition of a stoichiometric amount of acid, capable of protonating **9**, causes profound changes in the fluorescence behavior of the solution, namely (i) the fluorescence of **8** is completely quenched and (ii) the fluorescence of **9**-H^+ is sensitized upon excitation with light absorbed exclusively by the crown ether. These observations are consistent with the formation of an adduct between **8** and **9**-H^+, wherein very efficient electronic energy transfer occurs from the photoexcited binaphthyl unit of the crown ether to the anthracenyl group incorporated within the **9**-H^+ component. Such an adduct belongs to the class of pseudorotaxanes, that is, supermolecules made (at the minimum) of a thread-like guest molecule surrounded by a macrocyclic host, because dialkylammonium ions are known [35] to penetrate the cavity of crown ethers such as **8**. The very fast rate constant ($k > 4 \times 10^9 \, s^{-1}$) for the energy-transfer process [34] can be accounted for by a coulombic mechanism, as molecular models show that the maximum distance between binaphthyl and anthracene units in the **8** $\supset$ **9**-H^+ complex ($\sim$15 Å) is much shorter than their Förster radius (26 Å).

The pseudorotaxane **8** $\supset$ **9**-H^+ can be disassembled by the subsequent addition of a stoichiometric amount of base, capable of deprotonating **9**-H^+, thereby interrupting the photoinduced energy flow, as indicated by the restoring of the initial absorption and fluorescence spectra. Moreover, the stability of this pseudorotaxane can be influenced by changing the nature of the counteranion of **9**-H^+ [36]. Interestingly, the plug-in process does not occur when a plug component incompatible with the size of the socket, such as the benzyl-substituted amine **10**, is employed (Figure 1.10b).

1.6
Mimicking Electrical Extension Cables at the Molecular Level

The plug–socket concept described above can be used to design molecular systems which mimic the function played by a macroscopic electrical extension cable. The operation of an extension cable is more complex than that of a plug–socket system, because it involves *three* components that must be held together by *two* connections that have to be controllable *reversibly* and *independently*; in the fully connected system, an electron or energy flow must take place between the remote donor and acceptor units (Figure 1.11).

In the attempt to construct a molecular-level extension cable for electron transfer, the pseudorotaxane shown in Figure 1.12a, made of the three components **11**$^{2+}$, **12**-H^{3+} and **13**, has been obtained and studied [37]. Component **11**$^{2+}$ consists of two moieties: an [Ru(bpy)$_3$]$^{2+}$ unit, which behaves as an electron donor under light excitation, and a dibenzo[24]crown-8 macrocycle, capable of playing the role of a hydrogen-bonding first socket. The dialkylammonium-based moiety of **12**-H^{3+},

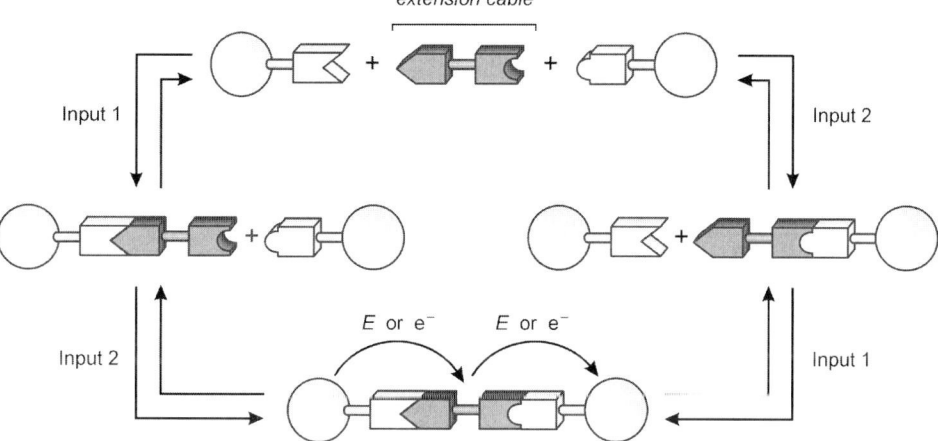

Figure 1.11 (a) Schematic representation of the working mechanism of an electrical extension cable.

driven by hydrogen-bonding interactions, threads as a plug into the first socket, whereas the π-electron accepting 4,4′-bipyridinium unit threads as a plug into the third component **13**, the π-electron rich 1,5-dinaphtho[38]crown-10 macrocycle, which plays the role of a second socket. In CH_2Cl_2–CH_3CN (98 : 2 v/v) solution, reversible connection–disconnection of the two plug–socket junctions can be controlled independently by acid–base and redox stimulation, respectively, and monitored by changes in the absorption and emission spectra, owing to the different nature of the interactions (hydrogen bonding and π-electron donor–acceptor) that connect the components. In the fully assembled triad, $11^{2+} \supset 12\text{-}H^{3+} \subset 13$, light excitation of the $[Ru(bpy)_3]^{2+}$ unit of the component 11^{2+} is followed by electron transfer to the bipyridinium unit of the component $12\text{-}H^{3+}$, which is plugged into component **13**.

It should be noted that in the system described above, the transferred electron does not reach the final component of the assembly. Moreover, a true extension cable should contain a plug and a socket at the two ends, instead of two plugs as component $12\text{-}H^{3+}$. An improved system of that type has been investigated recently (Figure 1.12b) [38]. The electron-source component is again 11^{2+}, whereas the new extension cable $14\text{-}H^+$ is made up [39] of a dialkylammonium ion, that can insert itself as a plug into a dibenzo [24]crown-8 socket, a biphenyl spacer and a benzonaphtho [36] crown-10 unit, which fulfills the role of a π-electron-rich socket. Finally, the 1,1′-dioctyl-4,4′-bipyridinium dication 15^{2+} can play the role of an electron drain plug. As for the previously studied system, the two plug–socket connections $11^{2+} \supset 14\text{-}H^+$ and $14\text{-}H^+ \supset 15^{2+}$ can be controlled by acid–base and redox stimuli, respectively.

In the complete ensemble, $11^{2+} \supset 14\text{-}H^+ \supset 15^{2+}$, light excitation of the Ru-based unit of 11^{2+} is followed by electron transfer to 15^{2+}, with $14\text{-}H^+$ playing the role of an extension cable (Figure 1.12b). The occurrence of this process is confirmed by nanosecond laser flash photolysis experiments, showing a transient absorption

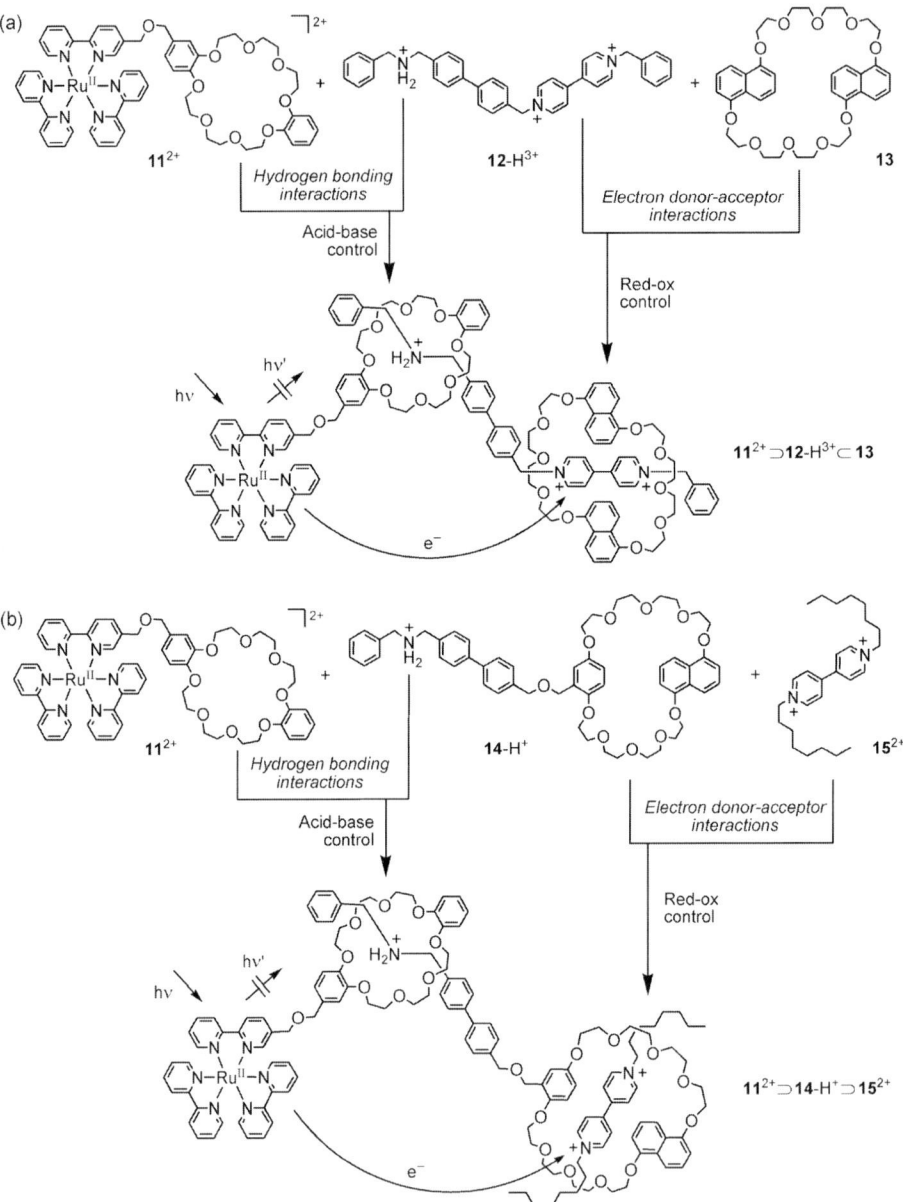

Figure 1.12 First- and second-generation systems for mimicking an electrical extension cable. (a) Structural formulas of the three molecular components **11^{2+}**, **12-H$^+$** and **13**, which self-assemble in solution to give the **11^{2+} ⊃ 12-H^{3+} ⊃ 13** triad. The photoinduced electron-transfer process from the Ru-based unit of **11^{2+}** to the bipyridinium unit of **12-H^{3+}** taking place in the fully connected system is also represented [37]. (b) Structural formulas of the three molecular components **11^{2+}**, **14-H$^+$** and **15^{2+}**, which self-assemble in solution to give the **11^{2+} ⊃ 14-H$^+$ ⊃ 15^{2+}** triad. In the fully connected system, excitation with visible light of the Ru-based unit of **11^{2+}** is followed by electron transfer to **15^{2+}**, with **14-H$^+$** playing the role of an extension cable [38].

signal assigned to the 4,4′-bipyridinium radical cation formed by photoinduced electron transfer within the self-assembled triad. Such a second-generation system exhibits two conceptual and significant advancements: (i) **14**-H^+ consists of a plug and a socket components and thus it really mimics an extension cable; (ii) the photoinduced electron transfer does occur from the first component – the Ru-based unit of **11**$^{2+}$ – to the remote **15**$^{2+}$ moiety, whereas in the previous system the electron receiving bipyridinium unit was a component of the cable.

1.7
Light-Harvesting Antennas

An antenna for light harvesting (Figure 1.13) is an organized multicomponent system in which several chromophoric molecular species absorb the incident light and channel the excitation energy to a common acceptor component [40]. For artificial systems, the term "antenna effect" was first used [41] to discuss the case of strongly emitting but weakly absorbing lanthanide ions surrounded by strongly absorbing ligands, where the luminescence of the lanthanide ion was sensitized by excitation in the ligand-centered excited states. Research in this area is still very active [42]. Antenna systems are widely used by Nature to solve the problem of light-harvesting efficiency in the photosynthetic process where light is converted into chemical energy [43]. Collecting light by an antenna system, however, may also be useful for other purposes, such as signal amplification in luminescence sensors [44], photodynamic cancer therapy [45] and up-conversion processes [46]. A large system, where an array of chromophoric units absorb light and transfer energy to a luminescent center, can also be considered a spatial and spectral energy concentrator ("molecular lens") [47].

The antenna effect can only be obtained in supramolecular arrays suitably organized in the dimensions of time, energy and space. Each molecular component has to absorb the incident light and the excited state so obtained (donor) has to transfer electronic energy to a nearby component (acceptor), before undergoing radiative or nonradiative deactivation (organization in the time dimension). In order

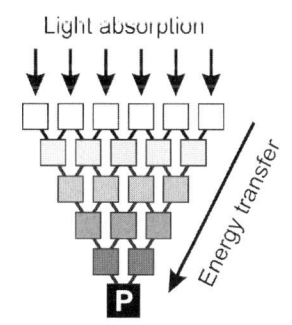

Figure 1.13 Schematic representation of a light-harvesting antenna system. Squares represent light-absorbing molecules. P is the molecule to which excitation energy is channeled. Excited state energy decreases with increasing shade.

for energy transfer to occur, the energy of the acceptor excited state has to be lower or, at most, equal to the energy of the excited state of the donor (organization in the energy dimension). Finally, the successive donor-to-acceptor energy-transfer steps must result in an overall energy-transfer process leading the excitation energy towards a selected component of the array (organization in the space dimension).

In the course of evolution, Nature has succeeded to build up antenna systems that fully satisfy the above requirements. In green plants, such natural antennae collect an enormous amount of solar energy and redirect it as electronic excitation energy to reaction centers where subsequent conversion into redox chemical energy takes place. In recent years, the development of supramolecular chemistry (particularly of dendrimer chemistry) and the high level of experimental and theoretical efficacy reached by photochemistry have enabled scientists to design and construct a number of interesting artificial antenna systems.

Dendrimer **16**$^{2+}$ (Figure 1.14) is a classical example of antenna system [48]. The 2,2′-bipyridine ligands of the [Ru(bpy)$_3$]$^{2+}$-type [49] core carry branches containing

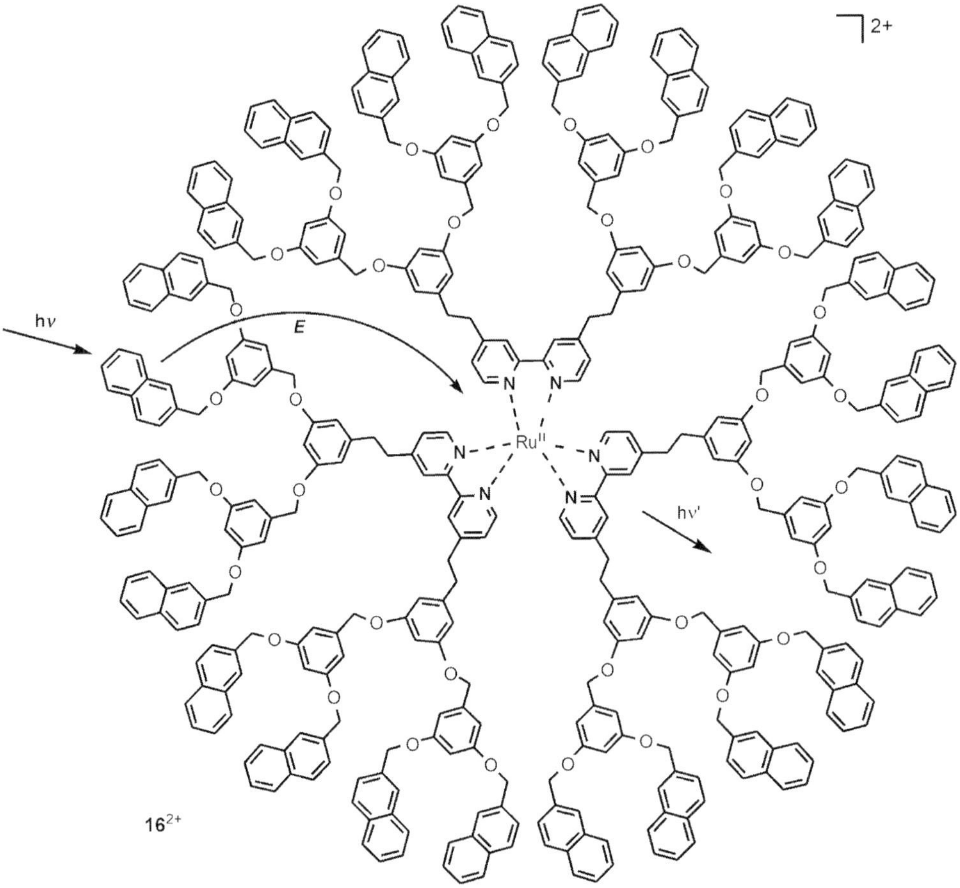

Figure 1.14 Antenna effect in dendrimer **16**$^{2+}$ with [Ru(bpy)$_3$]$^{2+}$ core [48].

1,2-dimethoxybenzene- and 2-naphthyl-type chromophoric units. Because such units (as well as the core) are separated by aliphatic connections, the interchromophoric interactions are weak and the absorption spectrum of the dendrimer is substantially equal to the summation of the spectra of the chromophoric groups that are present in its structures. The three types of chromophoric groups, namely, $[Ru(bpy)_3]^{2+}$, dimethoxybenzene and naphthalene, are potentially luminescent species. In the dendrimer, however, the fluorescence of the dimethoxybenzene- and naphthyl-type units is almost completely quenched in acetonitrile solution, with concomitant sensitization of the luminescence of the $[Ru(bpy)_3]^{2+}$ core ($\lambda_{max} = 610$ nm). These results show that a very efficient energy-transfer process takes place converting the very short-lived (nanosecond time-scale) UV fluorescence of the aromatic units of the wedges to the long-lived (microsecond time-scale) orange emission of the metal-based dendritic core. It should also be noted that in aerated solution the luminescence intensity of the dendrimer core is more than twice as intense as that of the $[Ru(bpy)_3]^{2+}$ parent compound because the dendrimer branches protect the Ru-bpy based core from dioxygen quenching [50]. In conclusion, because of the very high absorbance of the naphthyl groups in the UV spectral region, the high energy-transfer efficiency and the strong emission of the $[Ru(bpy)_3]^{2+}$-type core, dendrimer 16^{2+} (Figure 1.14) exhibits a strong visible emission upon UV excitation even in very dilute (10^{-7} mol L^{-1}) solutions [48].

1.8
Artificial Molecular Machines

1.8.1
Introduction

Natural molecular-level machines and motors are extremely complex systems. Any attempt to construct systems of such complexity by using an artificial bottom-up molecular approach would be hopeless. In the field of artificial systems, we can only construct simple prototypes consisting of a few molecular components, but we can use a chemical toolbox much larger than that used by Nature, exploit innovative ideas and operate in a much wider range of conditions (particularly as far as energy supply is concerned).

It designing artificial molecular devices, it should be recalled that they cannot be "shrunk" versions of macroscopic counterparts, because the operational mechanisms of motion at the molecular level have to deal with phenomena different from those that govern the macroscopic world [51,52]. Gravity and inertia motions that we are familiar with in our everyday experience are negligible at the molecular scale, where the viscous forces resulting from intermolecular interactions (including those with solvent molecules) largely prevail. This means that although we can describe the bottom-up construction of a nanoscale device as an assembly of suitable (molecular) components by analogy with what happens in the macroscopic world, we should not forget that the design principles and the operating mechanisms at the molecular level are different.

Mechanical movements at the molecular level result from nuclear motions caused by chemical reactions. Any kind of chemical reaction involves, of course, some nuclear displacement, but only large-amplitude, nontrivial motions leading to real translocation of some component parts of the system are considered. Particularly interesting nuclear motions from the viewpoint of artificial molecular systems are those related to (i) isomerization reactions involving $-N=N-$, $-C=N-$ and $-C=C-$ double bonds in covalent supramolecular structures, (ii) acid–base or redox reactions causing making or breaking of intermolecular bonds (including hydrogen bonds) and (iii) metal–ligand reactions causing the formation or disruption of coordination bonds.

Like macroscopic systems, mechanical molecular-level systems are characterized by: (a) the kind of energy supplied to make them work; (b) the kind of movement performed by their components; (c) the way in which their operation can be controlled and monitored; (d) the possibility of repeating the operation at will; (e) the time-scale needed to complete a cycle of operation; and (f) the function performed. Particularly interesting is the way in which energy can be supplied.

1.8.2
Energy Supply

To make a molecular machine move, energy must be supplied. The most obvious way of supplying energy to a chemical system is by adding a reactant (fuel) capable of causing a desired reaction. In his famous address "There is Plenty of Room at the Bottom" to the American Physical Society, R.P. Feynman discussed the possibility of constructing molecular-level machines and observed [53]: *"An internal combustion engine of molecular size is impossible. Other chemical reactions, liberating energy when cold, can be used instead"*. This is exactly what happens in our body, in which the chemical energy, ultimately derived from food and oxygen, is used in a long series of slightly exoergonic reactions to power the biological machines that sustain life.

If an artificial molecular-level machine must work by inputs of chemical energy, it will need addition of fresh reactants ("fuel") at any step of its working cycle [54]. It should be noticed that even cycling between two forms of a molecular-level system under the action of chemical inputs implies formation of waste products. For example, if the forward reaction is caused by an acid input, successive addition of a base will return the system to its original form, but the acid–base reaction generates waste products. Accumulation of waste products will inevitably compromise the operation of the machine, unless they are removed from the system, as happens both in natural machines and in macroscopic internal combustion engines.

The need to remove waste products introduces noticeable limitations in the design and construction of artificial molecular machines and motors based on "chemical fuel" inputs [55]. All the proposed systems operating by use of chemical energy become increasingly less efficient on increasing the number of cycles and finally stop working.

There are, however, alternative, more convenient, ways of powering artificial molecular machines.

1.8.3
Light Energy

In green plants the energy needed to sustain the machinery of life is provided by sunlight [43]; in general, light energy is not used as such to produce mechanical movements, but it is used to produce a chemical fuel, namely ATP, suitable for feeding natural molecular machines. Light energy, however, can directly cause photochemical reactions involving large nuclear movements. A simple example is a photoinduced isomerization from the lower energy *trans* to the higher energy *cis* form of a molecule containing $-C=C-$ or $-N=N-$ double bonds; this is followed by a spontaneous or light-induced back reaction [9,10]. Such photoisomerization reactions have indeed been used to make molecular machines driven by light energy inputs [56]. In supramolecular species, photoinduced electron-transfer reactions can often cause large displacement of molecular components [6,7,10,57]. Indeed, working with suitable systems, an endless sequence of cyclic molecular-level movements can in principle be performed making use of light-energy inputs without generating waste products [55,58].

Compared with chemical energy inputs, photonic energy has other advantages, besides the fundamental one of not generating waste products: (i) light can be switched on/off easily and rapidly; (ii) lasers provide the opportunity of working in very small space and very short time domains; (iii) photons, besides supplying the energy needed to make a machine work, can also be useful to "read" the state of the system and thus to control and monitor the operation of the machine. For all these reasons, photonic energy is extensively used to power artificial molecular machines.

Here we will briefly describe two examples: the first is based on a photoisomerization reaction, whereas the second relies on photoinduced electron-transfer processes.

1.8.4
Threading–Dethreading of an Azobenzene-Based Pseudorotaxane

Pseudorotaxanes are interesting in the context of molecular machinery, because the assembly–disassembly of the thread-like and macrocyclic components reminds one of the threading–dethreading of a needle. They can hardly be used to make unimolecular machines because of the chemical equilibrium between the components, but they represent good models for the development of rotaxane- and catenane-based systems.

An example of a pseudorotaxane exhibiting threading–dethreading motions based on a photoisomerization process is shown in Figure 1.15 [59]. The thread-like species *trans*-**17**, which contains a π-electron rich azobiphenoxy unit, and the π-electron-deficient macrocycle **18**$^{4+}$ self-assemble very efficiently to give a

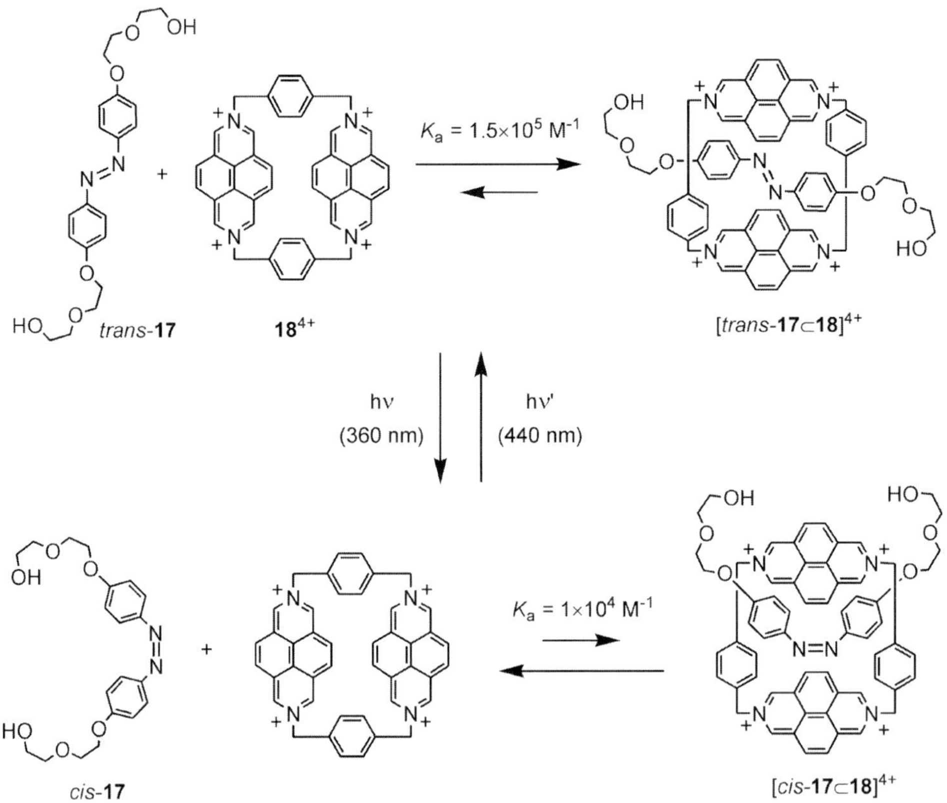

Figure 1.15 Threading–dethreading of **17** and **18**$^{4+}$ as a consequence of the *cis–trans* photoisomerization of the azobenzene-type unit contained in the thread-like component **17** [48].

pseudorotaxane, stabilized by electron donor–acceptor interactions. The association constant, obtained by fluorescence titration in acetonitrile solution at room temperature, is $K_a = (1.5 \pm 0.2) \times 10^5$ L mol^{-1}. In the pseudorotaxane structure, the intense fluorescence characteristic of free **18**$^{4+}$ ($\lambda_{max} = 434$ nm, Figure 1.16) is completely quenched by the donor–acceptor interaction.

Irradiation of an acetonitrile solution containing 1.0×10^{-4} mol L^{-1} *trans*-**17** and **18**$^{4+}$ (ca. 80% complexed species) with 365-nm light – almost exclusively absorbed by the *trans*-azobiphenoxy unit – causes strong absorption spectral changes, as expected for the well-known *trans* → *cis* photoisomerization of the azobenzene-type moiety. Such spectral changes are accompanied by a parallel increase in the intensity of the fluorescence band with $\lambda_{max} = 434$ nm (Figure 1.16), characteristic of free **18**$^{4+}$ (see above). This behavior shows that photoisomerization is accompanied by dethreading (Figure 1.15), a result which is confirmed by the finding that the association constant of **18**$^{4+}$ with *cis*-**17**, $K_a = (1.0 \pm 0.1) \times 10^4$ L mol^{-1}, is much smaller than that with

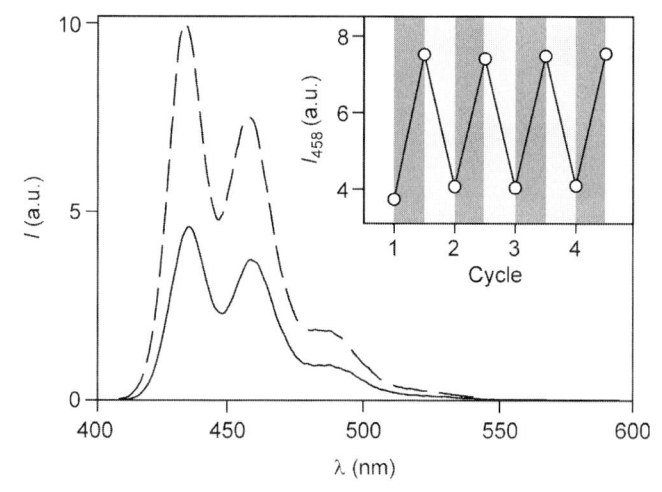

Figure 1.16 Fluorescence spectrum of an equimolar mixture (1.0×10^{-4} mol L^{-1}) of trans-**17** and **18**$^{4+}$ in acetonitrile at room temperature (full line) and fluorescence spectrum of the same mixture after irradiation at 365 nm until a photostationary state is reached (dashed line). The inset shows the changes in intensity of the fluorescence associated with the free macrocyclic ring **18**$^{4+}$ upon consecutive trans $\rightarrow$ cis (irradiation at 365 nm, dark areas) and cis trans $\rightarrow$ cis trans (irradiation at 436 nm, light areas) photoisomerization cycles. Excitation is performed in an isosbestic point at 411 nm [59].

trans-**17**. On irradiation at 436 nm or by warming the solution in the dark, the trans isomer of **17** can be reformed. This process is accompanied by a parallel decrease in the fluorescence intensity at $\lambda_{max} = 434$ nm, indicating that the trans-**17** species rethreads through the macrocycle **18**$^{4+}$.

Although this system is a rudimentary attempt towards the making of light-driven molecular machines, it should be noted that it exhibits a number of valuable features. First, threading-dethreading is controlled exclusively by light energy, without generation of waste products. Furthermore, owing to the reversibility of the photoisomerization process, the light-driven dethreading–rethreading cycle can be repeated at will (Figure 1.16, inset). Another relevant feature of this system is that it exhibits profound changes of a strong fluorescence signal.

1.8.5
Photoinduced Shuttling in Multicomponent Rotaxanes: a Light-Powered Nanomachine

Rotaxanes are appealing systems for the construction of molecular machines because the mechanical binding of the macrocyclic host with its dumbbell-shaped substrate leaves the former free to displace itself along and/or around the latter without losing the system's integrity. Two interesting molecular motions can be envisaged in rotaxanes, namely (i) rotation of the macrocyclic ring around the thread-like portion of the dumbbell-shaped component and (ii) translation of the ring along the same portion. The molecular components of a rotaxane usually exhibit some kind of interaction originating from complementary chemical properties, which is also exploited in the template-directed synthesis of such systems. In rotaxanes containing two different recognition sites in their thread-like portion, it is possible to switch the position of the ring between these two "stations" by an external stimulus. Systems of this type, termed molecular shuttles [60], probably constitute the most common examples of artificial molecular machines. Interestingly, the dumbbell component of a molecular shuttle exerts a restriction on the ring motion in the three dimensions of space, similar to that imposed by the protein track for linear biomolecular motors kinesin and dynein [61].

On the basis of the experience gained with pseudorotaxane model systems [62], the rotaxane 19^{6+} (Figure 1.17) was specifically designed [63] to achieve photoinduced ring shuttling in solution. This compound has a modular structure; its ring component R is a π-electron-donating bis-p-phenylene [34]crown-10, whereas its dumbbell component is made of several covalently linked units. They are a Ru(II)-polypyridine complex (P^{2+}), a p-terphenyl-type rigid spacer (S), a 4,4′-bipyridinium (A_1^{2+}) and a 3,3′-dimethyl-4,4′-bipyridinium (A_2^{2+}) π-electron-accepting stations and a

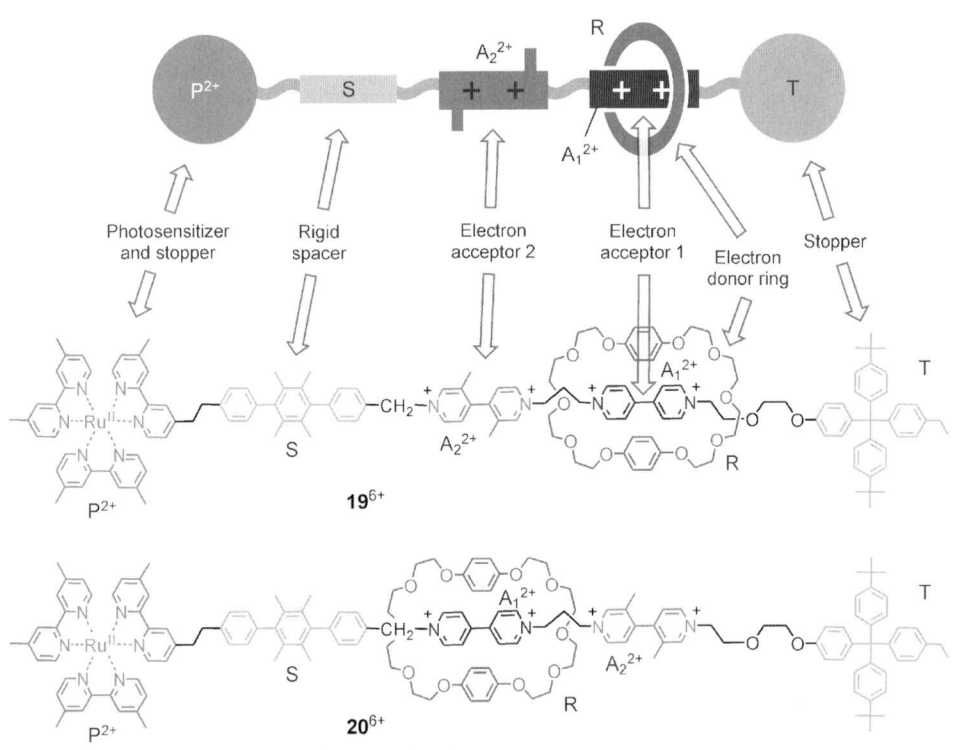

Figure 1.17 Structural formulas of multicomponent rotaxanes 19^{6+} [63] and 20^{6+} [66], designed to work as photochemically driven molecular shuttles. A cartoon representation of 19^{6+} is also shown.

tetraarylmethane group as the terminal stopper (T). The Ru-based unit plays the dual role of a light-fueled power station and a stopper, whereas the mechanical switch consists of the two electron-accepting stations and the electron-donating macrocycle. The stable translational isomer of rotaxane 19^{6+} is the one in which the R component encircles the A_1^{2+} unit, in keeping with the fact that this station is a better electron acceptor than the other.

The strategy devised in order to obtain the photoinduced shuttling movement of the macrocycle between the two stations A_1^{2+} and A_2^{2+} is based on the following "four-stroke" synchronized sequence of electronic and nuclear processes (Figure 1.18):

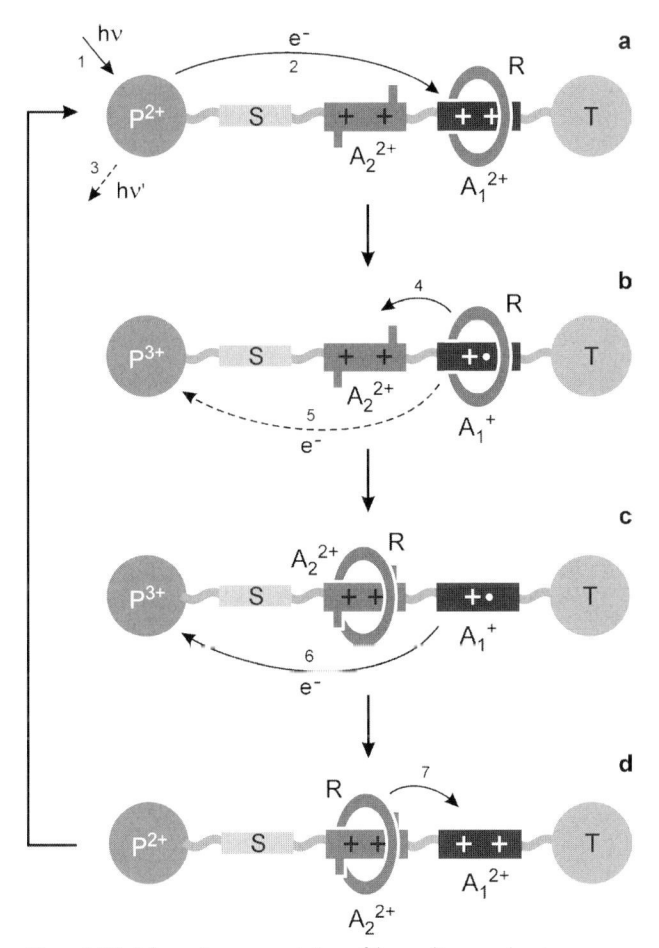

Figure 1.18 Schematic representation of the working mechanism of rotaxane 19^{6+} as an autonomous "four-stroke" molecular shuttle powered by visible light [64]. See Figure 1.17 for the legend for the cartoons.

(a) *Destabilization of the stable translational isomer:* light excitation of the photoactive unit P^{2+} (process *1*) is followed by the transfer of an electron from the excited state to the A_1^{2+} station, which is encircled by the ring R (process *2*), with the consequent "deactivation" of this station; such a photoinduced electron-transfer process competes with the intrinsic decay of the P^{2+} excited state (process *3*).

(b) *Ring displacement:* the ring moves (process *4*) for 1.3 nm from the reduced station A_1^+ to A_2^{2+}, a step that is in competition with the back electron-transfer process from A_1^+ (still encircled by R) to the oxidized unit P^{3+} (process *5*).

(c) *Electronic reset:* a back electron-transfer process from the "free" reduced station A_1^+ to the oxidized unit P^{3+} (process *6*) restores the electron acceptor power to such a station. At this point the machine is reset and the ring has been "pumped" into an energetically higher state.

(d) *Nuclear reset:* as a consequence of the electronic reset, thermally activated back movement of the ring from A_2^{2+} to A_1^{2+} takes place (process *7*).

Steady-state and time-resolved spectroscopic experiments together with electrochemical measurements in acetonitrile solution showed [64] that the absorption of a visible photon by **19**$^{6+}$ can cause the occurrence of a forward and back ring movement, that is, a full mechanical cycle according to the mechanism illustrated in Figure 1.18 [65]. It was estimated that the fraction of the excited-state energy used for the motion of the ring amounts to $\sim$10% and the system can generate a mechanical power of about 3×10^{-17} W per molecule. The somewhat disappointing quantum efficiency for ring shuttling (2% at 30 °C) is compensated for by the fact that the investigated system gathers together the following features: (i) it is powered by visible light (in other words, sunlight); (ii) it exhibits autonomous behavior, like motor proteins; (iii) it does not generate waste products; (iv) its operation can rely only on intramolecular processes, allowing in principle operation at the single-molecule level; (v) it can be driven at a frequency of about 1 kHz; (vi) it works in mild environmental conditions (i.e. fluid solution at ambient temperature); and (vii) it is stable for at least 10^3 cycles.

The molecular shuttle **19**$^{6+}$ can also be operated, with a higher quantum yield, by a sacrificial mechanism [63] based on the participation of external reducing (triethanolamine) and oxidizing (dioxygen) species and by an intermolecular mechanism [64] involving the kinetic assistance of an external electron relay (phenothiazine), which is not consumed. However, operation by the sacrificial mechanism does not afford an autonomous behavior and leads to consumption of chemical fuels and formation of waste products. On the other hand, the assistance by an electron relay affords autonomous operation in which only photons are consumed, but the mechanism is no longer based solely on intra-rotaxane processes.

Owing to its modular design, the rotaxane **19**$^{6+}$ is amenable to structural modification in an attempt to improve its performance as a light-driven molecular shuttle. For instance, the rotaxane **20**$^{6+}$ (Figure 1.17), which differs from **19**$^{6+}$ only in the exchange of the position of the two electron-accepting stations along the dumbbell-shaped component, has been recently synthesized and its photochemical properties investigated [66]. It has been found that the shorter distance of the electron-transfer photosensitizer P^{2+} to the better (A_1^{2+}) of the two electron acceptors in 20^{6+} results in an increase in the rate – and hence the efficiency – of the photoinduced electron-transfer step compared with 19^{6+}. The rate of the back electron transfer, however, also increases. As a consequence, such a second-generation molecular shuttle performs better than 19^{6+} in a sacrificial mechanism, but much worse when it is powered by visible light (e.g. sunlight) alone. Another interesting difference between these two parent rotaxanes lies in the fact that the macrocyclic ring R, which initially surrounds the A_1^{2+} station, moves in opposite directions upon light excitation, i.e. towards the photosensitizer P^{2+} in 19^{6+} and towards the stopper T in 20^{6+}.

This study shows that the structural and functional integration of different molecular subunits in a multicomponent structure is a powerful strategy for constructing nanoscale machines [67]. Nevertheless, the molecular shuttle 19^{6+} in its present form could not perform a net mechanical work in a full cycle of operation [68] (as for any reversible molecular shuttle, the work done in the "forward" stroke would be cancelled by the "backward" stroke) [69]. To reach this goal, a more advanced design of the molecular machine and/or a better engineering of its operating environment (e.g. a surface or a membrane) are required [6].

1.9
Conclusion

One of the most interesting aspects of supramolecular (multicomponent) systems is their interaction with light. The systems described here show that, in the frame of research on supramolecular photochemistry, the design and construction of nanoscale devices capable of performing useful light-induced functions can indeed be attempted.

The potential applications of photochemical molecular devices and machines are various – from energy conversion to sensing and catalysis – and, to a large extent, still unpredictable. As research in the area is progressing, two interesting kinds of unconventional applications of these systems begin to emerge: (i) their behavior can be exploited for processing information at the molecular level [70] and, in the long run, for the construction of chemical computers [71]; and (ii) their mechanical features can be utilized for transportation of nano-objects, mechanical gating of molecular-level channels and nanorobotics [72].

However, it should be noted that the species described here, as most multicomponent systems developed so far, operate in solution, that is, in an incoherent fashion and without control of spatial positioning. Although the solution studies are of fundamental importance to understand their operation mechanisms and for some use (e.g. drug delivery), it seems reasonable that before such systems can find applications in many fields of technology, they will have to be interfaced with the macroscopic world by ordering them in some way. The next generation of multicomponent molecular species will need to be organized so that they can behave coherently and can be addressed in space. Viable possibilities include deposition on surfaces, incorporation into polymers, organization at interfaces or immobilization into membranes or porous materials. Recent achievements in this direction [73–76] suggest that useful devices based on functional (supra)molecular systems could be obtained in the not too distant future.

Apart from foreseeable applications related to the development of nanotechnology, investigations on photochemical molecular devices and machines are important to increase the basic understanding of photoinduced reactions and other important processes such as self-assembly, and also to develop reliable theoretical models. This research also has the important merit of stimulating the ingenuity of chemists, thereby instilling new life into chemistry as a scientific discipline.

Acknowledgments

Financial support from the European Union (STREP "Biomach" NMP2-CT-2003–505487), MIUR (PRIN 2006034123-001), Regione Emilia-Romagna (NANOFABER) and the University of Bologna is gratefully acknowledged.

References

1 Hader, D.-P. and Tevini, M. (1987) *General Photobiology*, Pergamon Press, Oxford.
2 Nalwa, H.S. (ed.) (2003) *Handbook of Photochemistry and Photobiology*, American Scientific Publishers, Stevenson Ranch, CA, vols. 1–4.
3 Joachim, C. and Launay, J.P. (1984) *Nouv. J. Chem.*, **8**, 723.
4 Balzani, V., Moggi, L. and Scandola, F. (1987) *Supramolecular Photochemistry* (ed. V. Balzani), Reidel, Dordrecht, p. 1.
5 Lehn, J.-M. (1988) *Angew. Chem. Int. Ed.*, **27**, 89.
6 Balzani, V., Credi, A. and Venturi, M. (2003) *Molecular Devices and Machines – a Journey into the Nano World*, Wiley-VCH, Weinheim.
7 Balzani, V., Credi, A., Raymo, F.M. and Stoddart, J.F. (2000) *Angew. Chem. Int. Ed.*, **39**, 3348.
8 Kay, E.R., Leigh, D.A. and Zerbetto, F. (2007) *Angew. Chem. Int. Ed.*, **46**, 72.
9 Gilbert, A. and Baggott, J. (1991) *Essentials of Molecular Photochemistry*, Blackwell Science, London.
10 Balzani, V. and Scandola, F. (1991) *Supramolecular Photochemistry*, Ellis Horwood, Chichester.
11 Balzani, V. (ed.) (2001) *Electron Transfer in Chemistry*, Wiley-VCH, Weinheim, vols. 1–5.
12 Nalwa, H.S. (ed.) (2003) *Handbook of Photochemistry and Photobiology*, American Scientific Publishers, Stevenson Ranch, CA, vols. 1–4.
13 Michl, J. (2006) in *Handbook of Photochemistry*, 3rd edn (eds M. Montalti, A. Credi, L. Prodi and M.T. Gandolfi) CCR, Taylor and Francis, New York. p. 1.
14 Balzani, V. (2006) in *Handbook of Photochemistry* 3rd edn (eds M. Montalti, A. Credi, L. Prodi and M.T. Gandolfi) CCR, Taylor and Francis, New York, p. 49.
15 Lakowicz, J.R. (2006) *Principles of Fluorescence Spectroscopy*, 3rd edn Springer, Singapore.
16 Montalti M., Credi A., Prodi L. and Gandolfi M.T. (eds) (2006) *Handbook of Photochemistry*, 3rd edn CCR, Taylor and Francis, New York.

17 Balzani, V., Bolletta, F., Gandolfi, M.T. and Maestri, M. (1978) *Top. Curr. Chem.*, **75**, 1.

18 *Adv. Chem. Phys.* (1999). pp. 106–107 special volumes (eds M. Bixon, J. Jortner) on Electron Transfer: from Isolated Molecules to Biomolecules.

19 May, V. and Kühn, O. (2000) *Charge and Energy Transfer Dynamics in Molecular Systems*, Wiley-VCH, Weinheim.

20 Gray, H.B. and Winkler, J.R. (2003) *Q. Rev. Biophys.*, **36**, 341.

21 Indelli, M.T., Bignozzi, C.A., Harriman, A., Schoonover, J.R. and Scandola, F. (1994) *J. Am. Chem. Soc.*, **116**, 3768.

22 Indelli, M.T., Chiorboli, C., Flamigni, L., De Cola, L. and Scandola, F. (2007) *Inorg. Chem*, **46**, 5630.

23 Juris, A., Balzani, V., Barigelletti, F., Campagna, S., Belser, P. and von Zelewsky, A. (1988) *Coord. Chem. Rev.*, **84**, 85.

24 Balzani, V. and Scandola, F. (1996) *Comprehensive Supramolecular Chemistry* (eds J.L. Atwood, J.E.D. Davies, D.D. MacNicol and F. Vögtle), Pergamon Press, Oxford, vol. 10, p. 687.

25 Flamigni, L., Baranoff, E., Collin, J.-P. and Sauvage, J.-P. (2006) *Chem. Eur. J.*, **12**, 6592.

26 Th. Förster, *Discuss. Faraday Soc.*, **1959**, *27*, 7. Note that in this paper there is a misprint since in Eq. (29) π^6 is used instead of π^5. See (a) Förster, Th. (1965) *Modern Quantum Chemistry* (ed. O. Sinanoğlu) Academic Press New York Part III: Action of Light and Organic Crystals,93. (b) Barigelletti, F. and Flamigni, L. (2000) *Chem. Soc. Rev.*, **29**, 1 (c) Scholes, G.D. (2003) *Annu. Rev. Phys. Chem.*, **54**, 57.

27 Welter, S., Salluce, N., Belser, P., Groeneveld, M. and De Cola, L. (2005) *Coord. Chem. Rev.*, **249**, 1360.

28 Dexter, D.L. (1953) *J. Chem. Phys.*, **21**, 836.

29 Schlicke, B., Belser, P., De Cola, L., Sabbioni, E. and Balzani, V. (1999) *J. Am. Chem. Soc.*, **121**, 4207.

30 D'Aleo, A., Welter, S., Cecchetto, E. and De Cola, L. (2005) *Pure Appl. Chem.*, **77**.1035.

31 Welter, S., Lafolet, F., Cecchetto, E., Vergeer, F. and De Cola, L. (2005) *ChemPhysChem*, **6**, 2417.

32 Liddell, P.A., Kodis, G., Andréasson, J., de la Garza, L., Bandyopadhyay, S., Mitchell, R.H., Moore, T.A., Moore, A.L. and Gust, D. (2004) *J. Am. Chem. Soc.*, **126**, 4803.

33 Balzani, V., Credi, A. and Venturi, M. (2002) *Proc. Natl. Acad. Sci. USA*, **99**, 4814.

34 Ishow, E., Credi, A., Balzani, V., Spadola, F. and Mandolini, L. (1999) *Chem. Eur. J.*, **5**, 984.

35 (a) Kolchinski, A.G., Busch, D.H. and Alcock, N.W. (1995) *J. Chem. Soc., Chem. Commun.*, 1289. (b) Ashton, P.R., Campbell, P.J., Chrystal, E.J.T., Glink, P.T., Menzer, S., Philp, D., Spencer, N., Stoddart, J.F., Tasker, P.A. and Williams, D.J. 1865 *Angew. Chem. Int. Ed.*, **1995**, *34*,

36 Clemente-León, M., Pasquini, C., Hebbe-Viton, V., Lacour, J., Dalla Cort, A. and Credi, A. (2006) *Eur. J. Org. Chem.*, 105.

37 Ballardini, R., Balzani, V., Clemente-Leon, M., Credi, A., Gandolfi, M.T., Ishow, E., Perkins, J., Stoddart, J.F., Tseng, H.-R. and Wenger, S. (2002) *J. Am. Chem. Soc.*, **124**, 12786.

38 Ferrer, B., Rogez, G., Credi, A., Ballardini, R., Gandolfi, M.T., Balzani, V., Liu, Y., Tseng, H.-R. and Stoddart, J.F. (2006) *Proc. Natl. Acad. Sci. USA*, **103**, 18411.

39 Rogez, G., Ferrer Ribera, B., Credi, A., Ballardini, R., Gandolfi, M.T., Balzani, V., Liu, Y., Northrop, B.H. and Stoddart, J.F. (2007) *J. Am. Chem. Soc.*, **129**, 4633.

40 Balzani, V., Credi, A. and Venturi, M. (1997) *Curr. Opin. Chem. Biol.*, **1**, 506.

41 Alpha, B., Balzani, V., Lehn, J.-M., Perathoner, S. and Sabbatini, N. (1987) *Angew. Chem. Int. Ed. Engl.*, **26**, 1266.

42 Bünzli, J.-C.G. and Piguet, C. (2005) *Chem. Soc. Rev.*, 34.1048.

43 Blankenship, R.E. (2002) *Molecular Mechanism of Photosynthesis*, Blackwell Science, Oxford.

44 Balzani, V., Ceroni, P., Gestermann, S., Kauffmann, C., Gorka, M. and Vögtle, F. (2000) *Chem. Commun.*, 853.

45 Oar, M.A., Dichtel, W.R., Serin, J.M., Fréchet, J.M.J., Rogers, J.E., Slagle, J.E., Fleitz, P.A., Tan, L.-S., Ohulchanskyy, T.Y. and Prasad, P.N. (2006) *J. Am. Chem. Soc.*, **18**, 3682.
46 Yan, X., Goodson, T., III Imaoka, T. and Yamamoto, K. (2005) *J. Phys. Chem. B*, **109**, 9321.
47 Hecht, S. and Fréchet, J.M.J. (2001) *Angew. Chem. Int. Ed.*, **40**, 75.
48 Pleovets, M., Vögtle, F., De Cola, L. and Balzani, V. (1999) *New J. Chem.*, **23**, 63.
49 Balzani, V., Bergamini, G., Marchioni, F. and Ceroni, P. (2006) *Coord. Chem. Rev.*, **250**, 1254.
50 The dendrimer branches protect the excited state of the core also from quenching by other species: Vögtle, F., Plevoets, M., Nieger, M., Azzellini, G.C., Credi, A., De Cola, L., De Marchis, V., Venturi, M. and Balzani, V. (1999) *J. Am. Chem. Soc.*, **121**, 6290.
51 Goodsell, D.S. (2004) *Bionanotechnology – Lessons from Nature*, Wiley, Hoboken, NJ.
52 Jones, R.A.L. (2004) *Soft Machines, Nanotechnology and Life*, Oxford University Press, Oxford.
53 (a) Feynman, R.P. (1960) *Eng. Sci.*, **23**, 22. (b) Feynman, R.P. (1960) *Saturday Rev.*, **43**, 45. See also http://www.feynmanonline.com.
54 See, e.g.: (a) Balzani, V., Credi, A., Langford, S.J., Raymo, F.M., Stoddart, J.F. and Venturi, M. (2000) *J. Am. Chem. Soc.*, **122**, 3542. (b) Jimenez-Molero, M.C., Dietrich-Bucheker, C. and Sauvage, J.-P. (2002) *Chem. Eur. J.*, **8**, 1456. (c) Badji,ć J.D., Balzani, V., Credi, A., Silvi, S. and Stoddart, J.F. (2004) *Science*, **303**, 1845. (d) Leigh, D.A., Lusby, P.J., Slawin, A.M.Z. and Walker, D.B. (2005) *Chem. Commun.*, 4919
55 Ballardini, R., Balzani, V., Credi, A., Gandolfi, M.T. and Venturi, M. (2001) *Acc. Chem. Res.*, **34**, 445.
56 See, e.g.: (a) Shinkai, S., Nakaji, T., Ogawa, T., Shigematsu, K. and Manabe, O. (1981) *J. Am. Chem. Soc.*, **103**, 111. (b) Irie, M. and Kato, M. (1985) *J. Am. Chem. Soc.*, **107**, 1024. (c) van Delden, R.A., Koumura, N., Schoevaars, A., Meetsma, A. and Feringa, B.L. (2003) *Org. Biomol. Chem.*, **1**, 33. (d) Leigh, D.A., Wong, J.K.Y., Dehez, F. and Zerbetto, F. (2003) *Nature*, **424**, 174. (e) Qu, D.-H., Wang, Q.-C. and Tian, H. (2005) *Angew. Chem. Int. Ed.*, **44**, 5296. (f) Muraoka, T., Kinbara, K. and Aida, T. 2006 *Nature*, **440**, 512
57 Credi, A. (2006) *Aust. J. Chem.*, **59**, 157.
58 Ballardini, R., Balzani, V., Credi, A., Gandolfi, M.T. and Venturi, M. (2001) *Int. J. Photoenergy*, **3**, 63.
59 Balzani, V., Credi, A., Marchioni, F. and Stoddart, J.F. (2001) *Chem. Commun.*, 1861.
60 Bissell, A., Córdova, E., Kaifer, A.E. and Stoddart, J.F. (1994) *Nature*, **369**, 133.
61 Schliwa M. (ed.) (2003) *Molecular Motors*, Wiley-VCH, Weinheim.
62 (a) Ballardini, R., Balzani, V., Gandolfi, M.T., Prodi, L., Venturi, M., Philp, D., Ricketts, H.G. and Stoddart, J.F. (1993) *Angew. Chem. Int. Ed.*, **32**, 1301. (b) Ashton, P.R., Ballardini, R., Balzani, V., Constable, E.C., Credi, A., Kocian, O., Langford, S.J., Preece, J.A., Prodi, L., Schofield, E.R., Spencer, N., Stoddart, J.F. and Wenger, S. (1998) *Chem. Eur. J.*, **4**, 2413. (c) Ashton, P.R., Ballardini, R., Balzani, V., Kocian, O., Prodi, L., Spencer, N. and Stoddart, J.F. 1998 *J. Am. Chem. Soc.*, **120**, 11190
63 Ashton, P.R., Ballardini, R., Balzani, V., Credi, A., Dress, R., Ishow, E., Kleverlaan, C.J., Kocian, O., Preece, J.A., Spencer, N., Stoddart, J.F., Venturi, M. and Wenger, S. 2000 *Chem. Eur. J.*, **6**, 3558.
64 Balzani, V., Clemente-León, M., Credi, A., Ferrer, B., Venturi, M., Flood, A.H. and Stoddart, J.F. (2006) *Proc. Natl. Acad. Sci. USA*, **103**, 1178.
65 For a related example of a photochemically driven molecular shuttle, see: Brouwer, A.M., Frochot, C., Gatti, F.G., Leigh, D.A., Mottier, L., Paolucci, F., Roffia, S. and Wurpel, G.W.H. (2001) *Science*, **291**, 2124.
66 Balzani, V., Clemente-León, M., Credi, A., Semeraro, M., Venturi, M., Tseng, H.-R., Wenger, S., Saha, S. and Stoddart, J.F. (2006) *Aust. J. Chem.*, **59**, 193.

67 Balzani, V., Credi, A., Silvi, S. and Venturi, M. (2006) *Chem. Soc. Rev.*, **35**, 1135.
68 Kay, E.R. and Leigh, D.A. (2006) *Nature*, **440**, 286.
69 Chatterjee, M.N., Kay, E.R. and Leigh, D.A. (2006) *J. Am. Chem. Soc.*, **128**, 4058.
70 For a representative recent example, see: Margulies, D., Felder, C.E., Melman, G. and Shanzer, A. (2007) *J. Am. Chem. Soc.*, **129**, 347.
71 (a) Rouvray, D. (2000) *Chem. Br.*, **36** (12), 46. (b) Ball, P. (2000) *Nature*, **406**, 118.
72 Requicha, A.A.G. (2003) *Proc. IEEE*, **91**, 1922.
73 Berná J., Leigh, D.A., Lubomska, M., Mendoza, S.M., Pérez, E.M., Rudolf, P., Teobaldi, G. and Zerbetto, F. (2005) *Nature Mater.*, **4**, 704.
74 Kocer, A., Walko, M., Meijberg, W. and Feringa, B.L. (2005) *Science*, **309**, 755.
75 de Silva, A.P., James, M.R., Mckinney, B.O.F., Pears, D.A. and Weir, S.M. (2006) *Nature Mater.*, **5**, 787.
76 Bhosale, S., Sisson, A.L., Talukdar, P., Furstenberg, A., Banerji, N., Vauthey, E., Bollot, G., Mareda, J., Roger, C., Würthner, F., Sakai, N. and Matile, S. (2006) *Science*, **313**, 84.

2
Rotaxanes as Ligands for Molecular Machines and Metal–Organic Frameworks
Stephen J. Loeb

2.1
Interpenetrated and Interlocked Molecules

2.1.1
Introduction

Interlocked molecules such as rotaxanes and catenanes (Figure 2.1) contain a fundamentally unique feature in chemical bonding: the *mechanical link*. As such, these threaded molecular species are not constrained by the normal conformational limits of standard organic molecules. The separate components of these mechanically linked organic nanostructures have an inherent freedom of motion that allows them to participate in large amplitude changes in overall structure. This concept has lead to the design of a vast array of molecular machines and nanoscale devices [1].

A variety of strategies exist for the synthesis of mechanically linked molecules. Probably the most common is the utilization of supramolecular assistance, which involves noncovalent interactions between a linear "axle" and a cyclic "wheel" to form an interpenetrated adduct known as a [2]pseudorotaxane (Figure 2.2) [2]. This initial assembly can be converted to a permanently interlocked [2]rotaxane by capping with bulky end groups or to a [2]catenane by linking the two ends of the linear axle. A wide variety of complementary components are capable of [2]pseudorotaxane formation in this manner, but there are only a handful of systems that are efficient at threading and can withstand significant structural modification. One of these, the [1,2-bis(pyridinium)ethane] ⊂ (24-crown-8) template is the subject of this chapter [3].

2.1.2
Templating of [2]Pseudorotaxanes

Examination of CPK and computer models suggested that the interaction of a 1,2-bis (pyridinium)ethane dication as the axle and a 24-membered crown ether as the wheel would give rise to an interpenetrated [2]pseudorotaxane structure. Indeed, detailed

Organic Nanostructures. Edited by Jerry L. Atwood and Jonathan W. Steed
Copyright © 2008 WILEY-VCH Verlag GmbH & Co. KGaA, Weinheim
ISBN: 978-3-527-31836-0

[2] rotaxane   **[2] catenane**

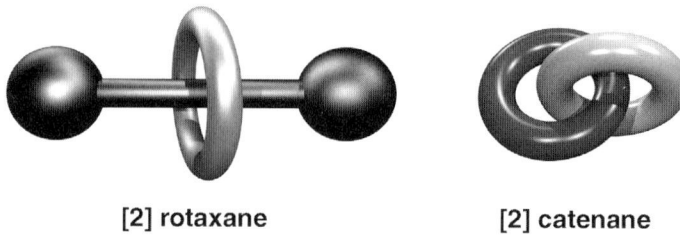

Figure 2.1 Cartoon depictions of simple examples of interlocked molecules: a [2]rotaxane (left) and a [2]catenane (right).

experiments showed that the combination of this new type of linear axle and a crown ether wheel such as 24-crown-8 (24C8), dibenzo-24-crown-8 (DB24C8) or dinaphtho-24-crown-8 (DN24C8) was an efficient and versatile templating method for the formation of [2]pseudorotaxanes. Nuclear magnetic resonance (^{1}H NMR) solution studies and solid-state X-ray structures showed clearly how the interpenetrated adducts are held together by $N^+\cdots O$ ion–dipole interactions, a series of eight $CH\cdots O$ hydrogen bonds and, when available, significant π-stacking interactions between electron-poor pyridinium rings and electron-rich catechol rings (Figure 2.3) [4].

The strength of the noncovalent interactions could be controlled by varying the substituents on the axle pyridinium rings. It was demonstrated that an

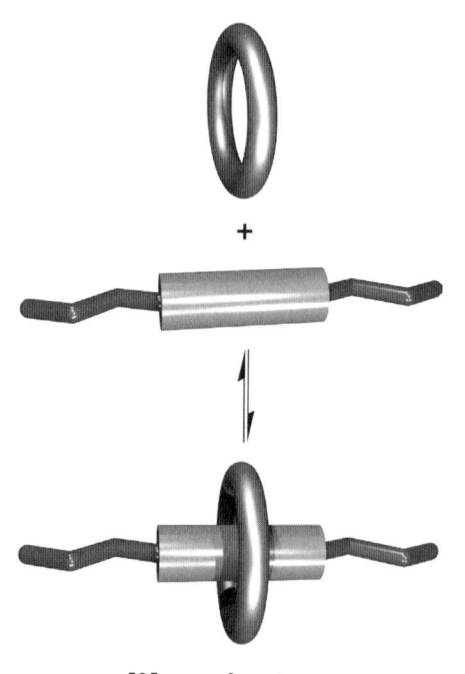

[2]pseudorotaxane

Figure 2.2 Cartoon outlining the equilibrium between an axle and wheel pair and an interpenetrated [2]pseudorotaxane.

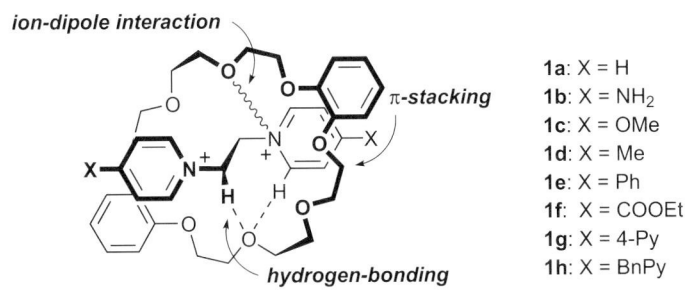

1a: X = H
1b: X = NH$_2$
1c: X = OMe
1d: X = Me
1e: X = Ph
1f: X = COOEt
1g: X = 4-Py
1h: X = BnPy

Figure 2.3 A schematic example of a [2]pseudorotaxane showing the noncovalent interactions that hold the axle and crown ether wheel together in an interpenetrated arrangement. The strength of these interactions can be fine tuned by the addition of EWG or EDG (X) and the presence of aromatic groups (benzo, naphtho) on the crown ether.

electron-withdrawing group provides increased hydrogen bonding and electrostatic interactions and therefore an increase in the stability of the adduct. Unlike the alkylammonium axles studied by Stoddart's group, which show their highest association constants with 24C8, the 1,2-bis(pyridinium)ethane axles exhibit significant contributions from π-stacking and thus higher association constants with DB24C8. Association constants for a series of [2]pseudorotaxanes ranged from 10 to 4700 M^{-1} in MeCN-d_3 at 298 K [4]. Figure 2.4 shows the variation in association

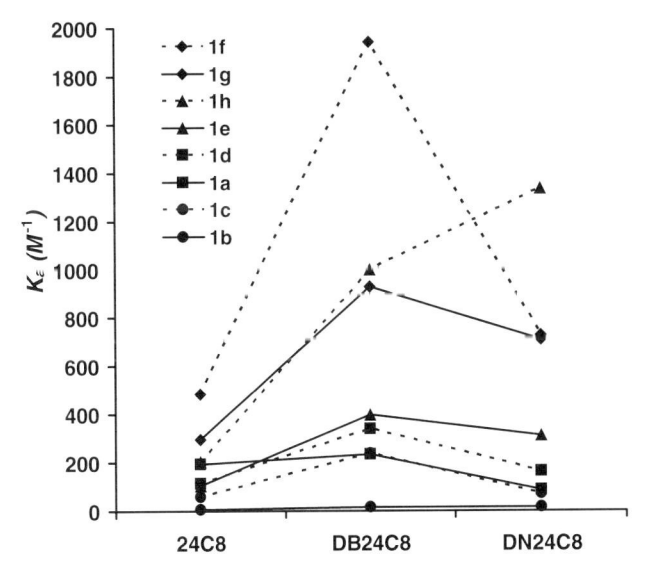

Figure 2.4 A plot showing the association constant for [2] pseudorotaxane formation (2 × 10^{-3} M, MeCN solution, 298 K) with variation in substituent at the 4-position on the pyridinium ring and crown ether.

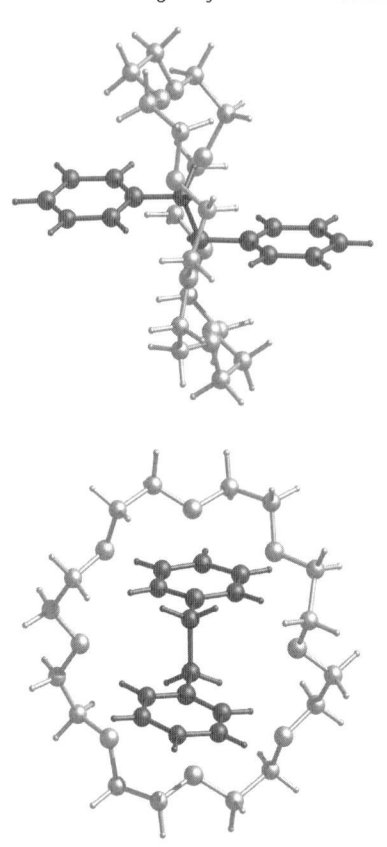

Figure 2.5 Two views of the interpenetration of axle and wheel in the X-ray structure of the simplest [2]pseudorotaxane, [(**1a**) ⊂ (24C8)]$^{2+}$.

constant as a function of crown ether and substituent at the 4-position of the pyridinium ring. Figure 2.5 shows two views of the X-ray structure of the simplest [2]pseudorotaxane [(**1a**) ⊂ (24C8)]$^{2+}$ containing the parent axle **1a**$^{2+}$(R = H) and 24C8.

2.1.3
[2]Rotaxanes

The formation of [2]rotaxanes from [2]pseudorotaxanes involves the incorporation of bulky capping groups to prevent unthreading of the axle component from the wheel unit (Figure 2.6). This "threading-followed-by-capping" method has given rise to a number of synthetic capping strategies such as alkylation of amines and phosphines, ester, carbonate and acetal formation, oxidative coupling, cycloaddition and Wittig reactions [5].

2.1 Interpenetrated and Interlocked Molecules

[2]rotaxane

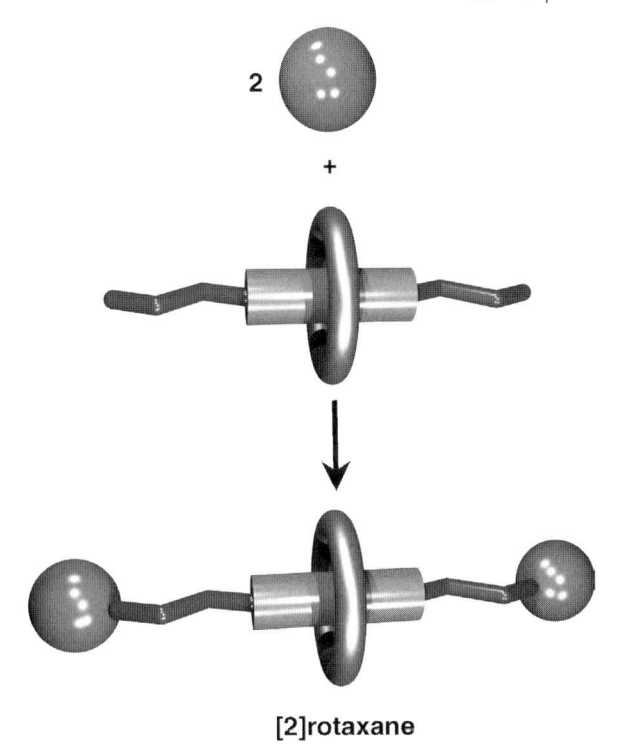

Figure 2.6 Cartoon outlining the conversion of an interpenetrated [2]pseudorotaxane into a permanently interlocked [2]rotaxane by capping with bulky groups.

We prepared the axle 1,2-bis(4,4′-bipyridinium)ethane, $1g^{2+}$, which contains terminal pyridine groups. This allowed for the straightforward incorporation of a capping group by direct alkylation of the terminal pyridines to give axle 2^{4+} and form the permanently interlocked [2]rotaxane $[(2) \subset (DB24C8)]^{4+}$. The *tert*-butylbenzyl group was employed as the cap as it was shown previously [4] that a pyridinium ring with a *t*-Bu substituent could not pass through DB24C8 [6]. The X-ray structures of the precursor [2]pseudorotaxane, $[(1g) \subset (DB24C8)]^{2+}$, and the resulting [2]rotaxane, $[(2) \subset (DB24C8)]^{4+}$, are shown in Figure 2.7.

The same methodology of threading-followed-by-capping can also be utilized to stopper the single terminal end of an axle that already contains a built-in bulky group. For example, [2]rotaxanes could also be prepared with a large triethylphosphonium group at one end; axle 3^{3+}. Figure 2.8 shows the X-ray structure of such a species [7].

2.1.4
Higher Order [*n*]Rotaxanes

Polyrotaxanes and polycatenanes are supramolecular polymers which contain macromolecular architectures built with mechanical linkages. Interest in these systems can be attributed to the fundamental role that interlocked components might play in

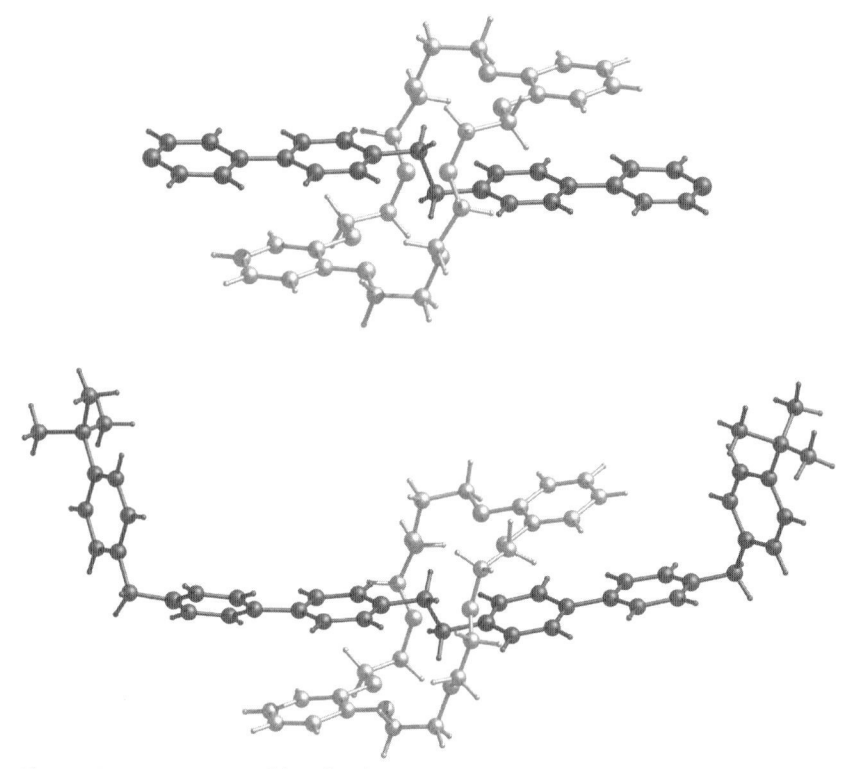

Figure 2.7 X-ray structures of the related [2]pseudorotaxane [(**1g**) ⊂ (24C8)]$^{2+}$ and [2]rotaxane [(**2**) ⊂ (DB24C8)]$^{4+}$.

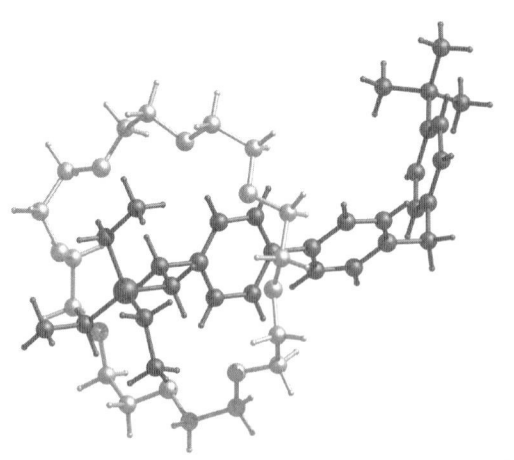

Figure 2.8 X-ray structure of the [2]rotaxane [(**3**) ⊂ (24C8)]$^{3+}$ containing a large –(PEt$_3$)$^+$ phosphonium group.

the mechanical properties of polymers. Dendrimers and hyperbranched macromolecules based on interlocked components have also attracted recent attention as models for supramolecular polymers [8].

We were particularly interested in dendrimers in which all the branching points of the macromolecule are mechanical linkages. These can be constructed by either threading multiple ring components on to branches attached to a single core (Type A) or threading multiple axles on to rings appended to a central core (Type B) [9].

Initially we prepared extended axles containing two 1,2-bis(pyridinium)ethane-type binding sites. Threading multiple units of DB24C8 on to the axle followed by capping with *tert*-butylbenzyl groups gave Type A [3]rotaxanes [10].

Type B dendrimeric polyrotaxanes were prepared using multi-site crown ethers and multiple versions of axle 2^{4+}. Figure 2.9 shows an example of a [4]rotaxane prepared in this manner [10]. The product distribution of branched [*n*]rotaxanes

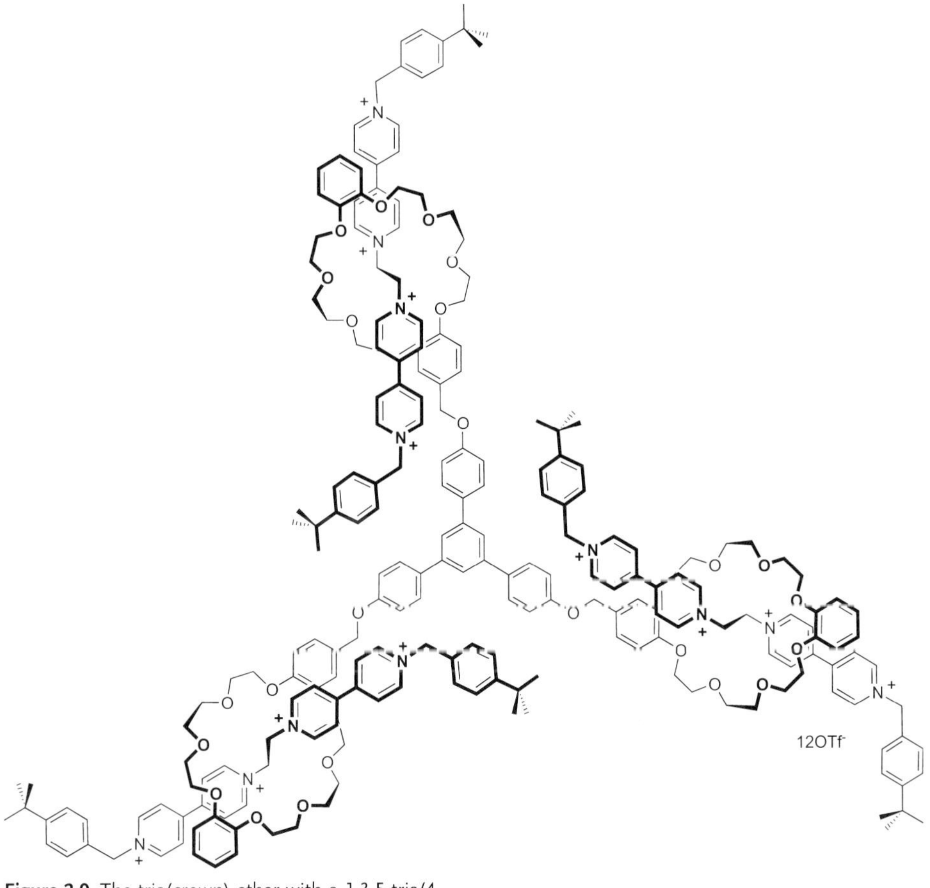

Figure 2.9 The tris(crown) ether with a 1,3,5-tris(4-hydroxyphenyl)benzene spacer allows the formation of a [4]rotaxane using axle 2^{4+} containing *tert*-butylbenzyl capping groups.

observed gave a measure of the efficiency of the threading process. The size of the core could be a single benzene ring but this was only practical for the formation of two branches as the two crowns needed to be positioned away from each other in the 1,4-orientation to minimize inhibitory interactions [11]. The efficiency of the threading steps decreased sequentially and was attributed to three factors: (1) steric interactions between an already complexed crown ether site (pseudorotaxane or rotaxane) and the new incoming axle, (2) electrostatic repulsions between an existing rotaxane axle and the new incoming axle and (3) partial occupation of the unoccupied crown ether recognition elements by a neighboring rotaxane unit [11].

2.1.5
[3]Catenanes

The synthesis of complex interlocked assemblies such as [$n > 2$]catenanes, [$n > 2$] molecular necklaces and [$n > 2$]rotacatenanes remains a considerable challenge for supramolecular chemists [1]. One of the major problems is that, regardless of design, there is always the requirement of forming at least one large ring during the self-assembly process. A potential strategy to aid ring closure involves the use of an external template, for example, a guest for a host catenane. During our studies on the (1,2-bis(pyridinium)ethane) $\subset$ (24-crown-8) templating motif, we discovered a unique one-step, self-assembly procedure for the preparation of [3]catenanes utilizing a terphenyl spacer unit that used DB24C8 as a very efficient template for the assembly of the [3]catenane containing this crown [12]. The X-ray structure in Figure 2.10

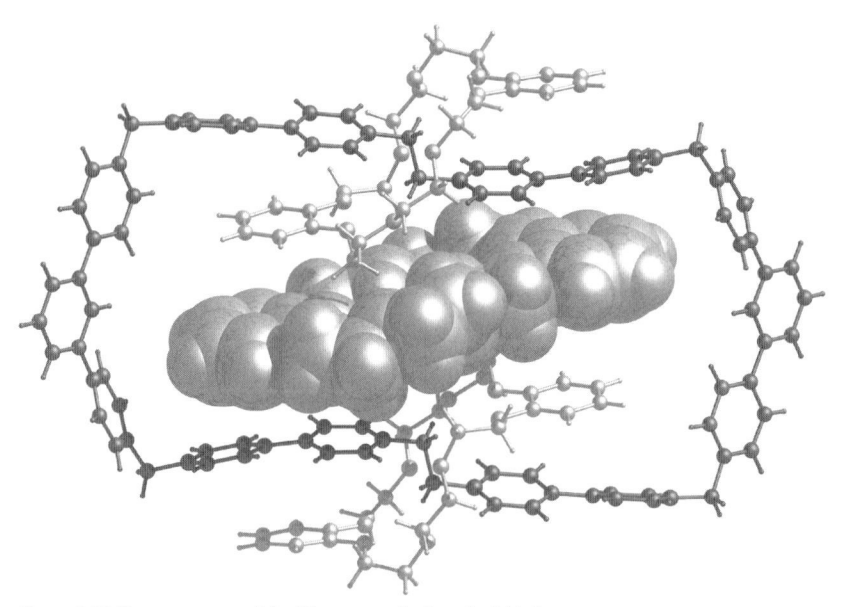

Figure 2.10 X-ray structure of the [3]catenane (ball-and-stick) that acts as host to an equivalent of DB24C8 (space-filling) as the guest.

shows how DB24C8 can act as a guest for the [3]catenane host, an interaction that can act to template the cyclization reaction and produce [3]catenane in good yield. The yield of the [3]catenane with DB24C8 (66%) was approximately twice that with any other crown ether studied, 24C8 (33%), B24C8 (35%), N24C8 (22%), BN24C8 (17%) and DN24C8 (23%), as only DB24C8 was a suitable guest [12].

2.2 Molecular Machines

2.2.1 Introduction

Control over the relative position and motion of components in interpenetrated or interlocked molecules can impart machine-like properties at the molecular level. Examples include threading and unthreading of a [2]pseudorotaxane, translation of the macrocycle in a [2]rotaxane molecular shuttle, rotation of the rings in a [2]catenane and reorientation (flipping, pirouetting) of the cyclic wheel in [2]rotaxanes [13].

2.2.2 Controlling Threading and Unthreading

During our study of [2]pseudorotaxane formation between 1,2-bis(pyridinium) ethane-type axles and 24-membered crown ether wheels, we noted that the presence of a strong EDG, such as NH_2, dramatically reduced the association constant [4]. This was attributed to a reduction in the acidity of the hydrogen bonding groups and a reduction in the charge at the pyridinium nitrogen due to contributions from an unfavorable resonance form. We then prepared an axle which could be represented by two resonance forms having dramatically different structures and charge distributions. This new axle had the structure of an organic D–π-A–π-D chromophore with two terminal donor groups (N,N-dimethylamino) and an inner acceptor group (bispyridinium), giving rise to an intramolecular charge transfer (ICT). It was possible to turn *OFF* the ICT by addition of a Lewis acid. X-ray crystal structures of the neutral and protonated forms as well as DFT (B3LYP) calculations verified the predicted resonance structures and electronic differences [14].

A substantial increase in association constant was observed for the protonated form compared to the unprotonated form. An ICT band with λ_{max} at 447 nm was observed in the absorption spectrum, but this intense orange color was completely eliminated upon addition of acid. The colorless, protonated axle then formed a [2]pseudorotaxane with DB24C8, which gave rise to a pale yellow coloration due to the weak charge-transfer interaction typical of $[(1) \subset (DB24C8)]^{2+}$ pseudorotaxanes (Figure 2.11).

The linking of this mechanical action of [2]pseudorotaxane formation to a significant color change can be described as a *NOT* logic gate since the threading

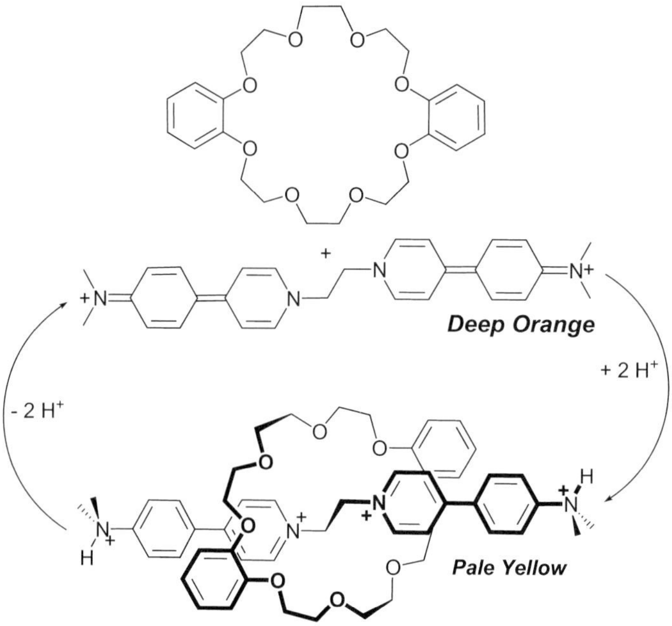

Figure 2.11 Schematic representation of the threading and unthreading process driven by alternating acid and base using an ICT axle.

of the two components to form the interpenetrated molecule is signaled by the loss of the orange color [14].

2.2.3
Molecular Shuttles

The transformation of mechanically linked molecules such as [2]rotaxanes and [2]catenanes into molecular machines requires the synthesis of systems with two or more distinct molecular arrangements (co-conformations) [13]. One of the most widely studied of these is the molecular shuttle pioneered by Stoddart's group [13]. In a molecular shuttle, two different recognition sites are present on the axle for the binding of a single macrocyclic wheel. The two states are translational isomers related by the relative positioning of the two interlocked components (Figure 2.12).

We utilized the axles from our [3]rotaxane study [15] to create molecular shuttles containing two binding sites of the 1,2-bis(pyridinium)ethane type for a single molecule of DB24C8 [16]. Since only one set of ^{1}H NMR resonances was observed for the axles at room temperature, it was concluded that the DB24C8 molecules were undergoing fast exchange between the two binding sites. From VT-NMR spectral data, it was determined that the rate of exchange between the two sites was on the order of 200–300 Hz. For the unsymmetrical species, the ratio of site occupancy was determined as 2:1.

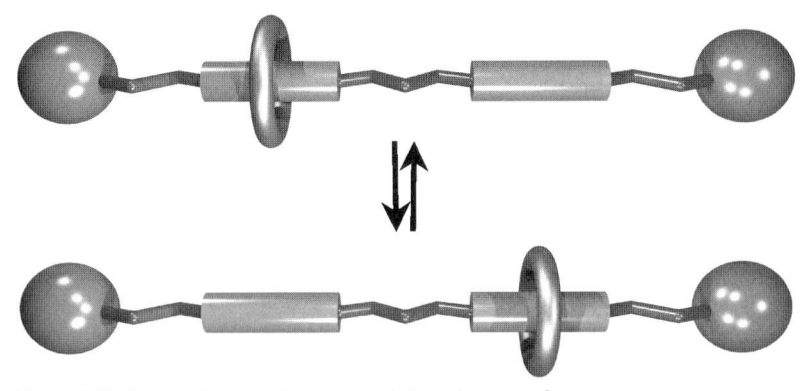

Figure 2.12 Cartoon depicting the two translational isomers of a [2]rotaxane molecular shuttle.

In order actually to control the motions of the components in these molecular shuttles, the 1,2-bis(pyridinium)ethane site was combined with an alkylammonium site which can be turned ON and OFF by protonation and deprotonation [17]. Figure 2.13 shows a molecular shuttle that operates as a bistable switch in dichloromethane solution by the sequential addition of acid and base. When the amine site is unprotonated there is no recognition element to interact with the crown ether and the [2]rotaxane is orange due to the same ICT absorption as found in our ON/OFF [2]

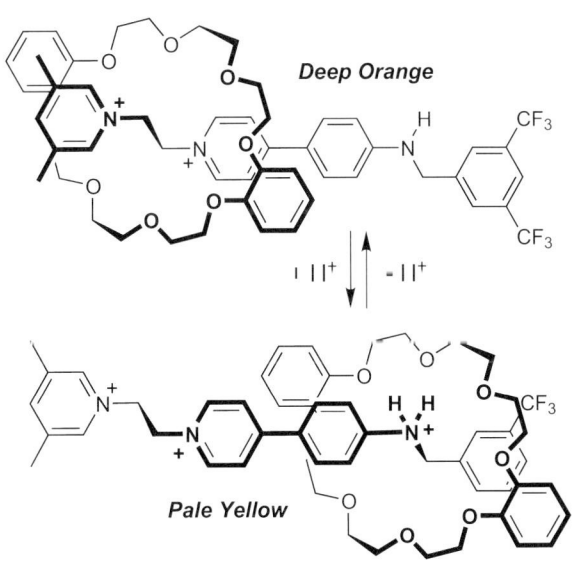

Figure 2.13 A molecular shuttle containing both 1,2-bis(pyridinium)ethane and benzylanilinium recognition sites that can be sensed optically and controlled by the addition of acid and base.

pseudorotaxane system described in Section 2.2.2. Upon protonation, a competition is established between the two sites which is dominated totally by the benzylanilinium recognition site [18] in a nonpolar solvent such as dichloromethane and there is complete elimination of the orange colored form due to quenching of the ICT absorption. This [2]rotaxane molecular shuttle is thus a bistable molecular switch that can be easily controlled by acid–base chemistry and observed by a simple optical read-out.

2.2.4
Flip Switches

We have demonstrated the existence of a new set of positional isomers based on [2]rotaxanes which contain a single recognition site but have different end groups on both the axle and crown ether wheel. The relative positioning of the two interlocked components produces two co-conformations and their reorientation is reminiscent of a mechanical "flip switch" (Figure 2.14) [19].

^{1}H NMR spectroscopy was used to determine that the populations of the two co-conformations were dependent upon the relative degrees of π-stacking between axle and wheel. Since solvent polarity is known to have a profound influence on intramolecular face-to-face π-stacking between aromatic rings [20], we studied a system with two different planar pyridinium groups of different surface area and demonstrated that the ratio of co-conformational isomers could be tuned by solvent polarity (Figure 2.15). The ratio of isomers showed an increase in π-stacking of the naphtho group with the 4,4′-bipyridinium group in more polar solvents. The X-ray structure of one of these [2]rotaxanes containing an unsymmetrical crown ether with both benzo and naphtho aromatic units, BN24C8, is shown in Figure 2.16.

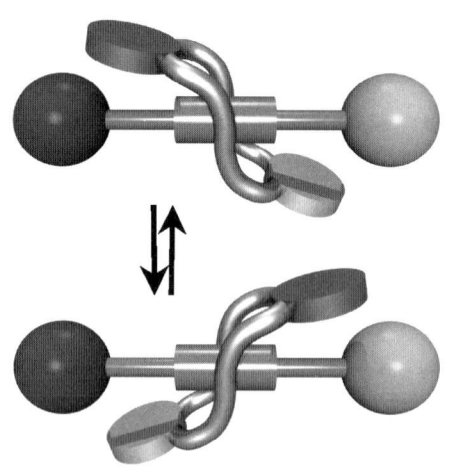

Figure 2.14 Cartoon illustrating the two co-conformations of a [2]rotaxane molecular flip-switch.

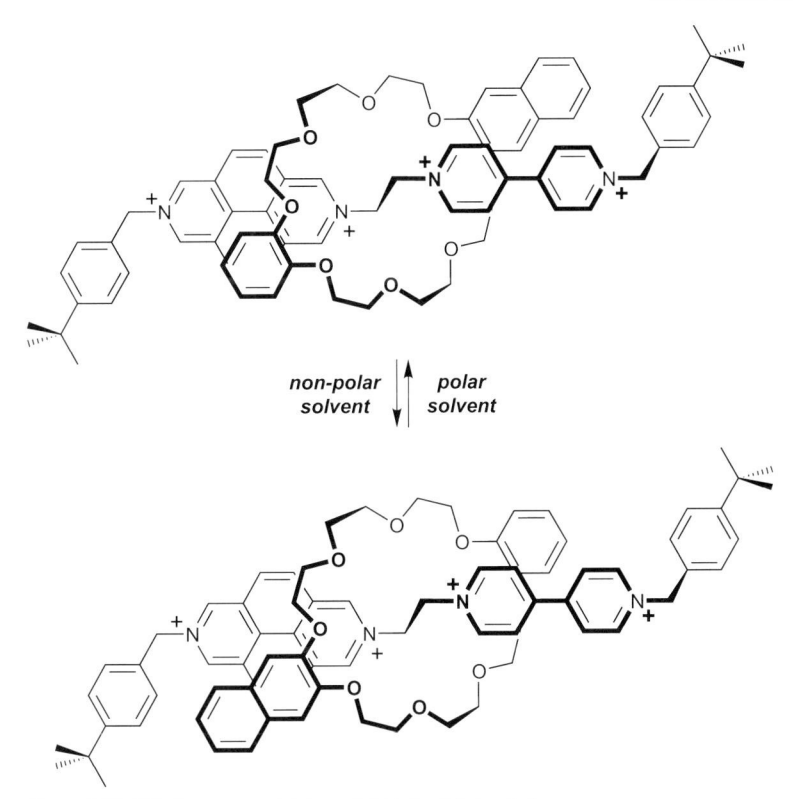

Figure 2.15 [2]Rotaxanes containing a 1,2-bis(pyridinium)ethane recognition site in which both the axle and wheel have different end groups can exist in two co-conformations that depend on the orientation of the components. The preference for a particular co-conformation is dependent upon π-stacking interactions and can be controlled by solvent polarity.

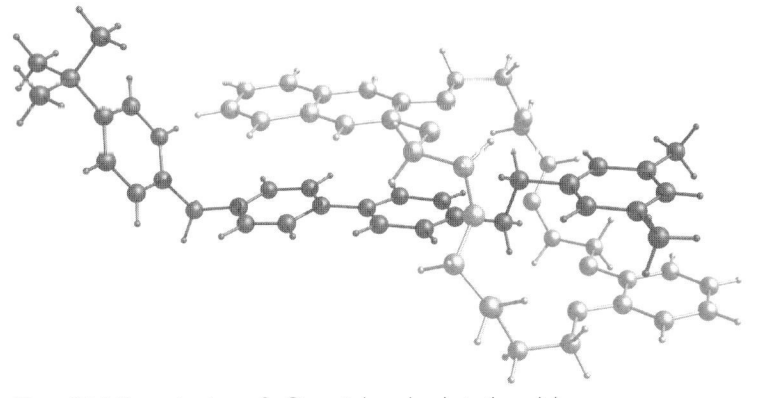

Figure 2.16 X-ray structure of a flip-switch molecule in the solid state. The co-conformation is that identified as the major one in solution with the larger naphtho ring π-stacked over the 4,4′-bipyridninium unit.

This new type of molecular switch with a single recognition site is just in its infancy but shows good potential. In particular, it is a very compact system and might be a good candidate for inclusion in solid-state devices in which large-amplitude changes in structure are more difficult to envision. Preliminary work in our laboratory has shown that the switching can be observed optically and the barrier to "flipping" controlled by inclusion of appropriate substituents on the axle and wheel [17].

2.3
Interlocked Molecules and Ligands

2.3.1
[2]Pseudorotaxanes as Ligands

Almost any metal complex with a single open coordination site is bulky enough to be used as an effective cap and form a metal-based [2]rotaxane. Our initial study involved using the axle $1g^{2+}$ and DB24C8 to pre-form the [2]pseudorotaxane ligand followed by capping with the palladated pincer fragment $\{Pd[C_6H_3(CH_2SPh)_2]\}^+$ [10]. The ^{1}H NMR spectrum of the reaction mixture showed quantitative formation of the metal-capped [2]rotaxane which could be isolated from solution by crystallization; an X-ray structure verified the interlocked nature of the product. Subsequently, it was shown that the metal-containing caps could be larger, more sophisticated units such as the porphyrin complex [Ru(CO)(TTP)] [21] or simple anionic metal fragments such as $[MBr_3]_2$ (M = Mn, Co) [6]. The X-ray structures of the PdS_2- and $CoBr_3$-capped complexes are show in Figure 2.17.

2.3.2
[2]Rotaxanes as Ligands

One of the shortcomings of a self-assembly strategy for metal ion incorporation is that conditions for formation of the metal–ligand bonds must be compatible with maintaining the weaker noncovalent interactions between axle and wheel. A simple modification that circumvented this problem was replacement of one of the monodentate pyridine donors of $1g^{2+}$ with a multidentate terpyridine group to give a new axle 3^{3+}. This gave rise to a series of permanently interlocked [2]rotaxanes containing 24C8, DB24C8 and DN24C8. Figure 2.18 shows the X-ray structure of the new interlocked ligand $[(3) \subset (DN24C8)]^{3+}$ ready for coordination via the tridentate terpy site [22].

We then studied the coordination chemistry of $[(3) \subset (24C8)]^{3+}$, $[(3) \subset (DB24C8)]^{3+}$ and $[(3) \subset (DN24C8)]^{3+}$ with Fe(II) [22] and Ru(II) [23]. Since the ligands are already [2]rotaxanes. the conditions under which coordination to a metal center is conducted do not effect the integrity of the mechanical linkage. For example, the Ru(II) complexes were prepared at reflux in polar solvents with no decomposition

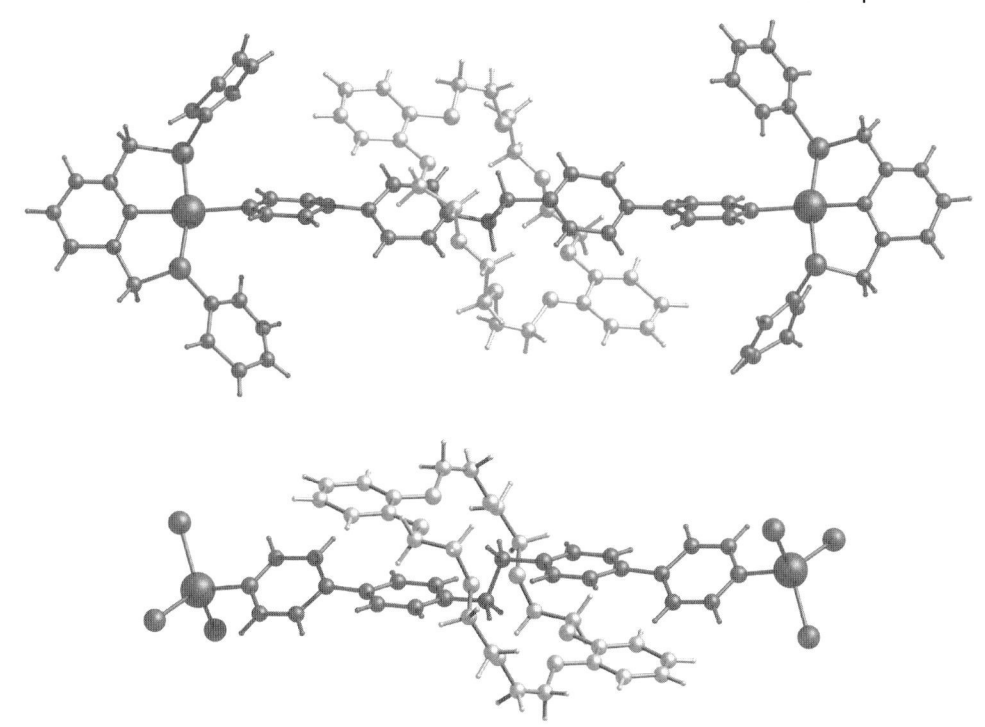

Figure 2.17 X-ray crystal structures of the two metal-based [2] rotaxanes derived from a combination of **1g**$^{2+}$ and DB24C8 capped by (top) the palladated pincer fragments [Pd$(C_6H_3(CH_2SPh)_2)$]$^+$ and (bottom) the anionic fragments CoBr$_3$.

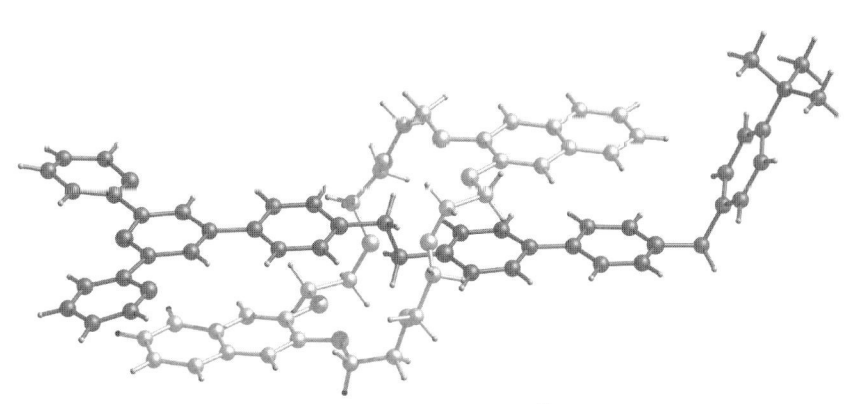

Figure 2.18 X-ray crystal structure of the [2]rotaxane ligand **3**$^{3+}$ featuring a terpyridine chelating axle, capped with a *tert*-butylbenzyl group and containing DB24C8.

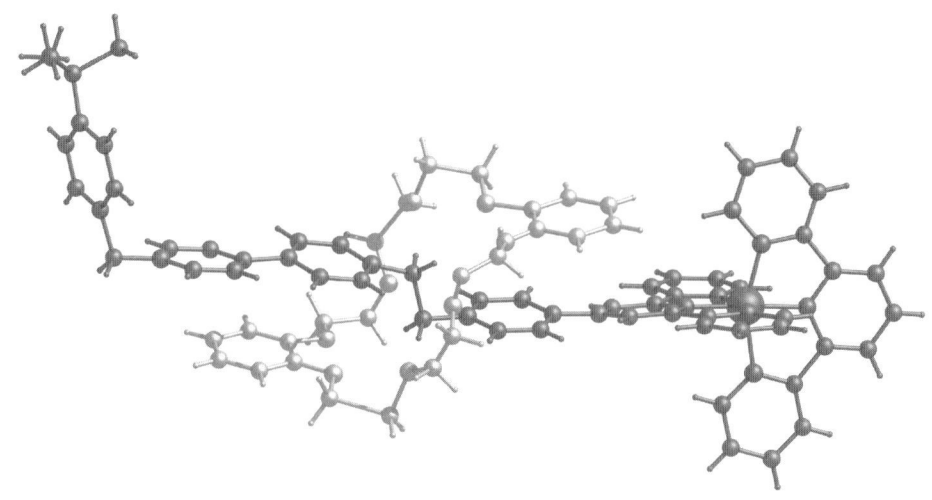

Figure 2.19 X-ray crystal structure of the Ru(II) complex of ligand **3**$^{3+}$ formed with an equivalent of [Ru(terpy)]$^{2+}$.

of the [2]rotaxane ligands. Figure 2.19 shows an X-ray structure of the Ru(II) complex cation {(terpy) Ru[(**3**) ⊂ (DB24C8)]}$^{5+}$ containing a chelating [2]rotaxane ligand.

A detailed investigation of the Ru(II) complexes showed some unique absorption and fluorescence properties that were dependent on the nature of the crown ether [23]. This synthetic approach has the potential to expand greatly the conditions under which metal complexes with mechanical linkages can be prepared and could produce a wide variety of [2]rotaxane metal complexes with unique electronic, magnetic or photo-physical properties [24].

2.4
Materials from Interlocked Molecules

2.4.1
Metal–Organic Rotaxane Frameworks (MORFs)

Although a great deal of information about the fundamental properties of mechanically interlocked molecules has been derived from solution studies [25,28], there still need to be methods for imposing order in these systems. Some ideas that have been studied are (i) attachment to surfaces (ii) tethering between electrodes (iii) incorporation into organic polymers [31] or dendrimers [9] and (iv) assembly into the repeating framework of a crystalline lattice The ultimate goal of this work is to produce materials that contain functional components that are addressable and controllable. We are interested in the solid state and in particular the formation of crystalline materials we have termed metal–organic rotaxane frameworks (MORFs) [33].

The use of simple [2]rotaxanes as linkers in MORF materials should provide a blueprint for the eventual inclusion of molecular machines into three-dimensional frameworks. For example, one defining attribute of a conventional metal–organic framework (MOF) material is the ability to modify the linker unit by organic synthesis [33]. With MORFs, an additional degree of flexibility is available as the rotaxane linker can be modified by retaining the axle unit but exchanging the wheel component. This *supramolecular* modification can potentially be used to tune the internal properties of the material [34].

The initial challenges that we have taken up and describe here are (1) the design of appropriate ligands, (2) the development of suitable synthetic routes and protocols for self-assembly and crystallization and (3) specific methods for materials characterization.

2.4.2
One-dimensional MORFs

The evolution from a [2]rotaxane capped with metal fragments to a 1D MORF was straightforward. Mixing **1g**$^{2+}$ with an excess of DB24C8 in MeCN resulted in a solution which contained an equivalent of the [2]pseudorotaxane ligand. Diffusion of a solution of $[(\mathbf{1g}) \subset (DB24C8)]^{2+}$ (as the BF_4 salt) into an MeCN solution of $[M(H_2O)_6][BF_4]_2$ (M = Co, Zn) resulted in isolation of crystalline material with formula $\{Co(H_2O)_2(MeCN)_2(\mathbf{1g}) \subset (DB24C8))][BF_4]_4 \cdot (MeCN)_2(H_2O)_2\}_x$ (yields: 71% Co, 92% Zn).

The top structure in Figure 2.20 shows that the use of Co(II) ions in the presence of a coordinating solvent such as MeCN results in an octahedral coordination sphere comprised of two equivalents of $[(\mathbf{1g}) \subset (DB24C8)]^{2+}$, two MeCN molecules and two water molecules with each set of ligands having a *trans* orientation. The result is a coordination polymer in which every linker is a [2]rotaxane.

The linearity of the framework is due to the fact that **1g**$^{2+}$ must adopt an *anti* conformation at the central ethylene unit when threaded through DB24C8. This structural feature combined with a *trans* geometry at the metal ion produces a linear 1D MORF with a Co···Co distance of 22.1 Å [35]. This polymer can be viewed as a metal ligand "wire" in which the crown ethers act as a protective coating somewhat analogous to the use of long-chain hydrocarbons attached to phosphine ligands to surround metal–polyacetylene linkages [36]. This MORF also contains infinite channels parallel to the polymer chains which are filled with anions and solvent. The bottom structure in Figure 2.20 shows several strands of the framework. It is likely that the girth of this rotaxane ligand contributes both to the parallel organization of the chains and the channels between them. As evidence for this, we have crystallized two 1D polymers utilizing only **1g**$^{2+}$ as the linker. In both cases, we observed a classic herringbone pattern; there was no parallel alignment of chains and neither compound contained void channels for solvent occlusion [35].

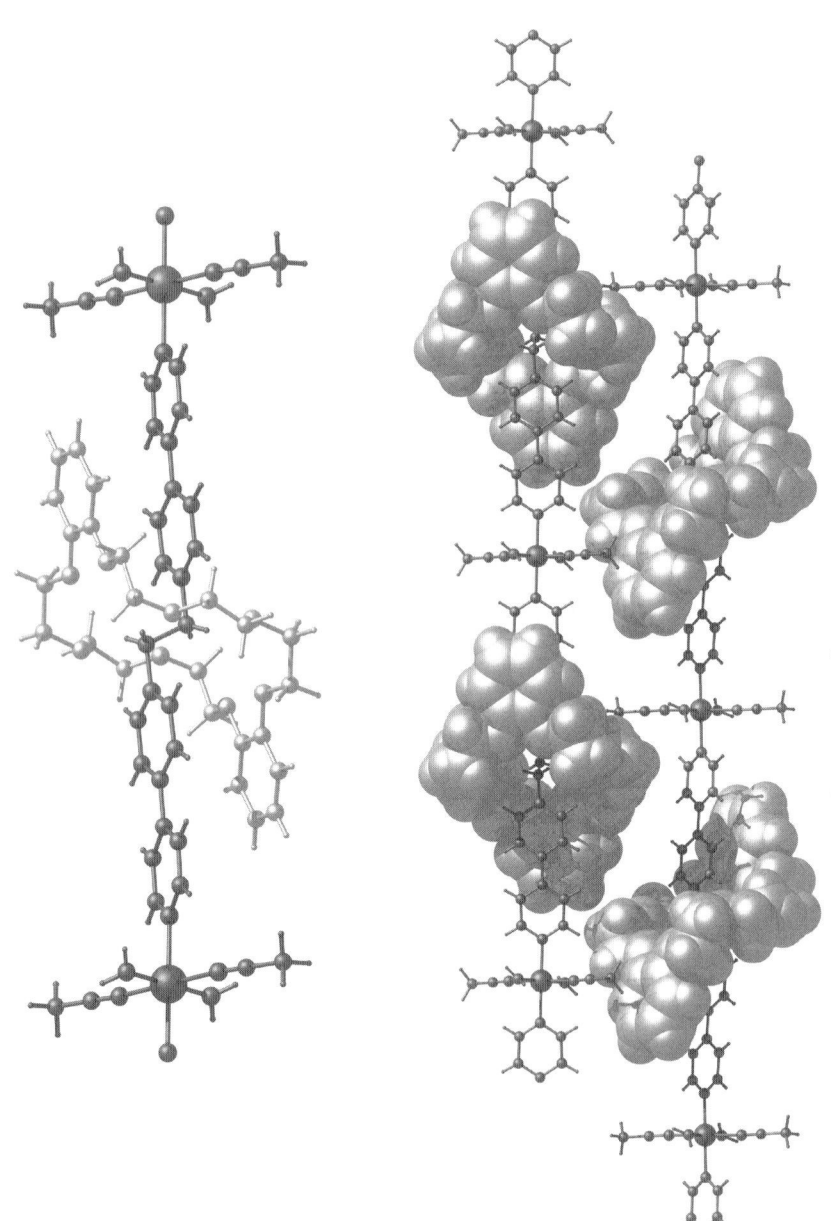

Figure 2.20 X-ray crystal structure a 1D MORF with Co(II) ion nodes and [(**1g**) ⊂ (DB24C8)]$^{2+}$ linkers showing (top) a single repeating unit and (bottom) two parallel strands.

2.4.3
Two-dimensional MORFs

Since the linear 1D MORFs contained an octahedral metal ion in which the ancillary ligands were solvent molecules, we reasoned that it should be possible to induce higher orders of dimensionality by employing a greater amount of $[(\mathbf{1g}) \subset (DB24C8)]^{2+}$ in a noncoordinating solvent. The reaction of two equivalents of $\mathbf{1g}^{2+}$ (as the BF_4 salt) with four equivalents of DB24C8 and one equivalent of $[M(H_2O)_6][BF_4]_2$ (M = Cu, Cd, Ni) in $MeNO_2$ produced X-ray quality crystalline material (average yields ~80%) [35]. The top structure in Figure 2.21 shows how the use of these metal ions in a non-coordinating solvent allows for a an octahedral coordination geometry comprising four $[(\mathbf{1g}) \subset (DB24C8)]^{2+}$ linkers in a square-planar arrangement, along with one water molecule and one coordinated BF_4 anion. The bottom structure shows how propagation of these units results in a 2D MORF with square nets and a formula $\{[Cd(H_2O)(BF_4)(\mathbf{1g}) \subset (DB24C8))_2][BF_4]_5(MeNO_2)_{15}\}_x$. The sides of the square are defined by Cd···Cd distances of 22.2 Å. The interlayer spacings are 12.0 and 10.0 Å with the layers stacked in a pattern that gives rise to an open framework material with large infinite channels lined with DB24C8 crown ethers. The channels are filled with anions and solvents; there are 15 molecules of $MeNO_2$ per Cd(II) ion. Calculations estimate that the accessible void space occupied by anions and solvent is 50%; 38% for solvent only [35].

2.4.4
Three-dimensional MORFs

Regardless of the metal to ligand ratio employed, a two-dimensional square net was the highest order MORF that could be attained using $[(\mathbf{1g}) \subset (DB24C8)]^{2+}$ as a bridging ligand. We ascribed the failure to obtain 3D frameworks to the hindrance involved in trying to place six sterically demanding ligands around a single transition metal ion. In order to avoid crowding at the metal center, we made two changes: (1) we synthesised the bis-N-oxide analogue of $\mathbf{1g}^{2+}$, a new axle $\mathbf{4}^{2+}$, which allowed the formation of a new linker $[(\mathbf{4}) \subset (DB24C8)]^{2+}$, and (2) we employed larger lanthanide metal ions.

Three equivalents of $\mathbf{4}^{2+}$ (as the OTf salt) were reacted with nine equivalents of DB24C8 and one equivalent of $[M(OTf)_3]$ (M = Sm, Eu, Gd, Tb) in MeCN. X-ray quality crystalline material was produced in moderate yield (average yields ~50%). The top structure in Figure 2.22 shows that the use of Ln(III) ions as nodes results in an eight-coordinate metal center with a square anti-prismatic geometry comprised of six $[(\mathbf{4}) \subset (DB24C8)]^{2+}$ linkers, one water molecule and one coordinated triflate anion [37]. The bottom structure shows how propagation of these units results in 3D MORFs with formula $\{[M(H_2O)(OTf)(\mathbf{4}) \subset (DB24C8))_3][Cl][OTf]_7 \cdot (2MeCN)\}_x$ (M = Sm, Eu, Gd, Tb) in which every linker is a [2]rotaxane. The edges of the "cube" are defined by Sm···Sm distances of ~23.5 Å. Although the internal cavity of this 3D framework has a volume of ~10 000 Å [3], this apparently void space is filled by the single interpenetration of a parallel net. This is an obvious side-effect of employing

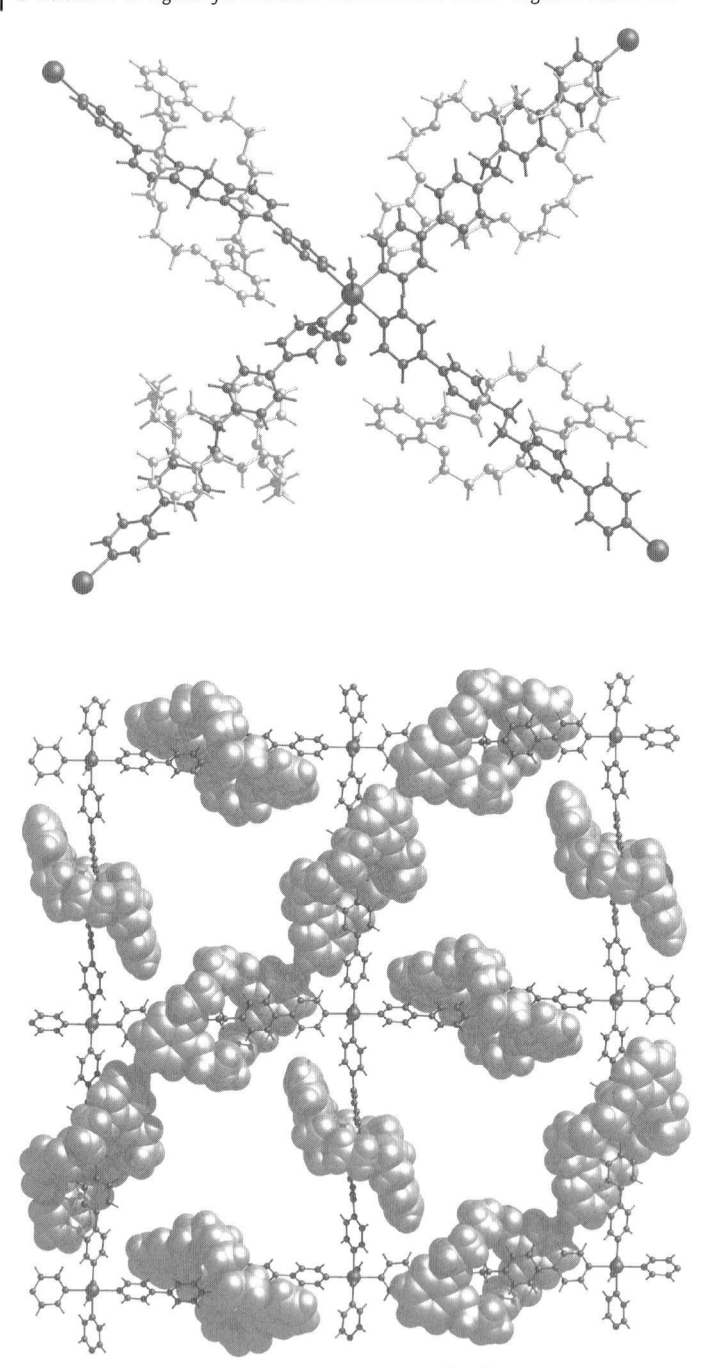

Figure 2.21 X-ray crystal structure a 2D MORF with Cd(II) ion nodes and [(**1g**) ⊂ (DB24C8)]$^{2+}$ linkers showing (top) the coordination sphere around a single metal ion and (bottom) a portion of the 2D square net.

2.4 Materials from Interlocked Molecules | **53**

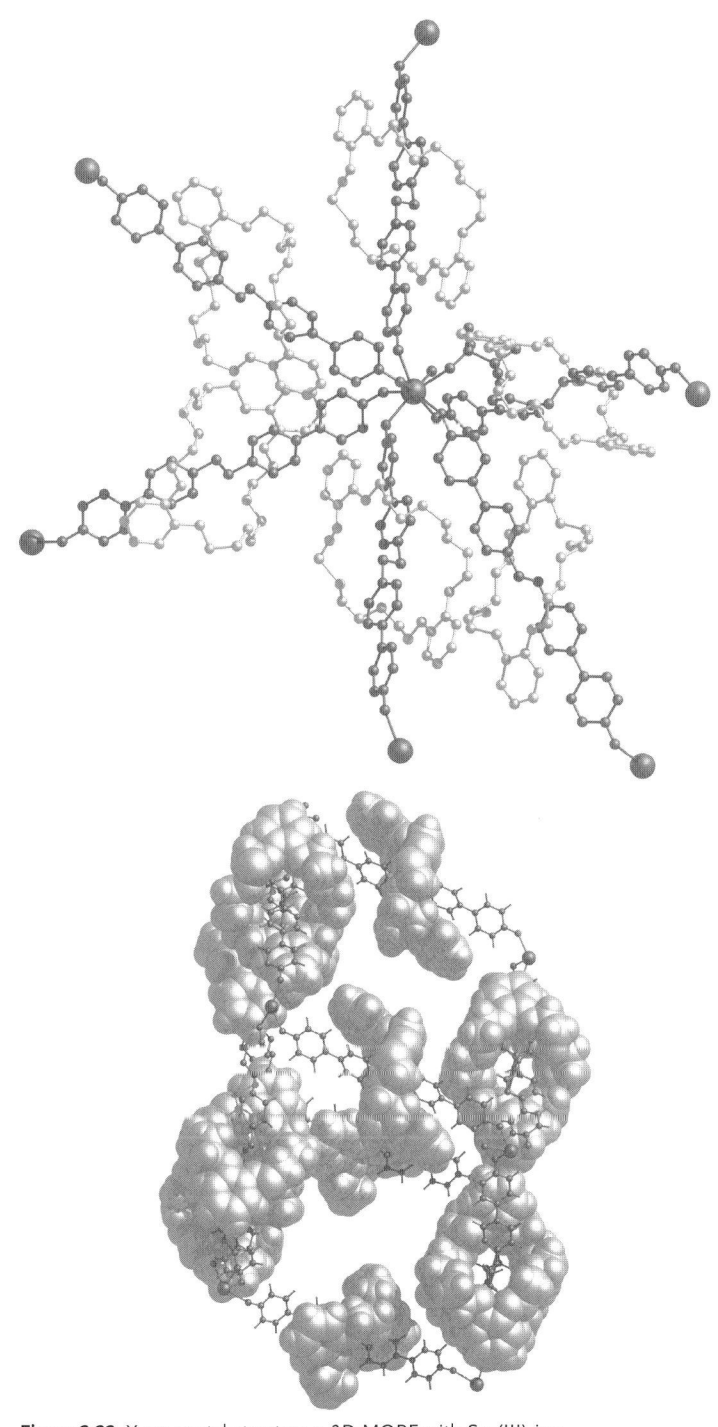

Figure 2.22 X-ray crystal structure a 3D MORF with Sm(III) ion nodes and [(**4**) ⊂ (DB24C8)]$^{2+}$ linkers showing (top) the coordination sphere around a single metal ion and (bottom) a portion of the α-Po-type framework.

the longer axles and increasing the metal–metal distance. This creates an expanded cavity with larger "windows" which can now accommodate the girth of the [2]rotaxane ligand and allow interpenetration to occur.

An isomorphous series of 3D MORFs (Sm, Eu, Gd, Tb) based on eight-coordinate metal centers was possible, but changing to a smaller lanthanide ion, Yb(III) (2.40 vs. 2.51 Å, for Tb), yielded a completely different MORF. A material with formula {[Yb(OTf)(**4**) ⊂ (DB24C8))$_3$][Cl][OTf]$_7$}$_x$ was isolated in moderate yield using the same diffusion procedure as used to produce the other 3D MORFs. In this case, a unique seven-coordinate pentagonal bipyramidal geometry was adopted by the smaller Yb(III) center. Five N-oxide-based rotaxanes occupy the five equatorial sites of the pentagonal plane with a sixth rotaxane and a single triflate ion positioned in the two axial sites, as shown in the top structure in Figure 2.23. Upon reducing the size of the Ln(III) node, a single water molecule is removed and the coordination number decreased from eight to seven. This subtle change has a dramatic effect on the nature of the resulting MORF structure [37]. Since the novel seven-coordinate geometry at Yb(III) contains a pentagonal equatorial plane and close-packed tiling in two dimensions with pentagons is impossible, it was interesting to see how this dilemma was circumvented. What occurs is that each N-oxide ligand "bends" at the Yb−O−N linkage, so a (3/4,5) net is formed, which solves the tiling problem by utilizing a combination of alternating triangles and squares rather than pentagons. Until very recently, this two-dimensional pattern was unknown in chemical systems [38]. The bottom structure in Figure 2.23 shows the tiling in the pentagonal plane. This network propagates one step further into a full 3D MORF by pillaring to alternating layers utilizing the sixth [(**4**) ⊂ (DB24C8)]$^{2+}$ ligand in the apical position. This results in a previously unknown chemical topology which takes the form of a (3/4/6,6) six-connected net comprised of triangles, squares and hexagons. The square openings are used for interpenetration as this is not possible through the more crowded triangular cavities.

2.4.5
Controlling the Dimensionality of a MORF

Attempts to construct a 3D MORF with [(**4**) ⊂ (DB24C8)]$^{2+}$ using smaller transition metal ions such as Cd(II) did yield a polyrotaxane but not the desired 3D framework. Instead, a 2D network was created with only one of the directions utilizing [(**4**) ⊂ (DB24C8)]$^{2+}$ and the other simply employing a "naked" **4**$^{2+}$ as a linker with no crown ether wheel [39]. The bottom structure in Figure 2.24 shows the basic coordination sphere around the Cd(II) center.

The X-ray crystal structure of the MORF material shows that the solid has the formula {[Cd(**4**) ⊂ (DB24C8))(**4**)(OTf)$_2$][OTf]$_4$(MeNO$_2$)$_4$}$_n$. The Cd(II) metal centers adopt an octahedral geometry with three different pairs of ligands in an all-*trans* arrangement. Two rotaxane ligands and two "naked" axles define a square plane while two triflate anions are in the axial positions. This is similar to the Cd(II) MORF prepared using the pyridine-based rotaxane ligand [(**1g**) ⊂ (DB24C8)]$^{2+}$. In that structure, the cavities of the grid were aligned to

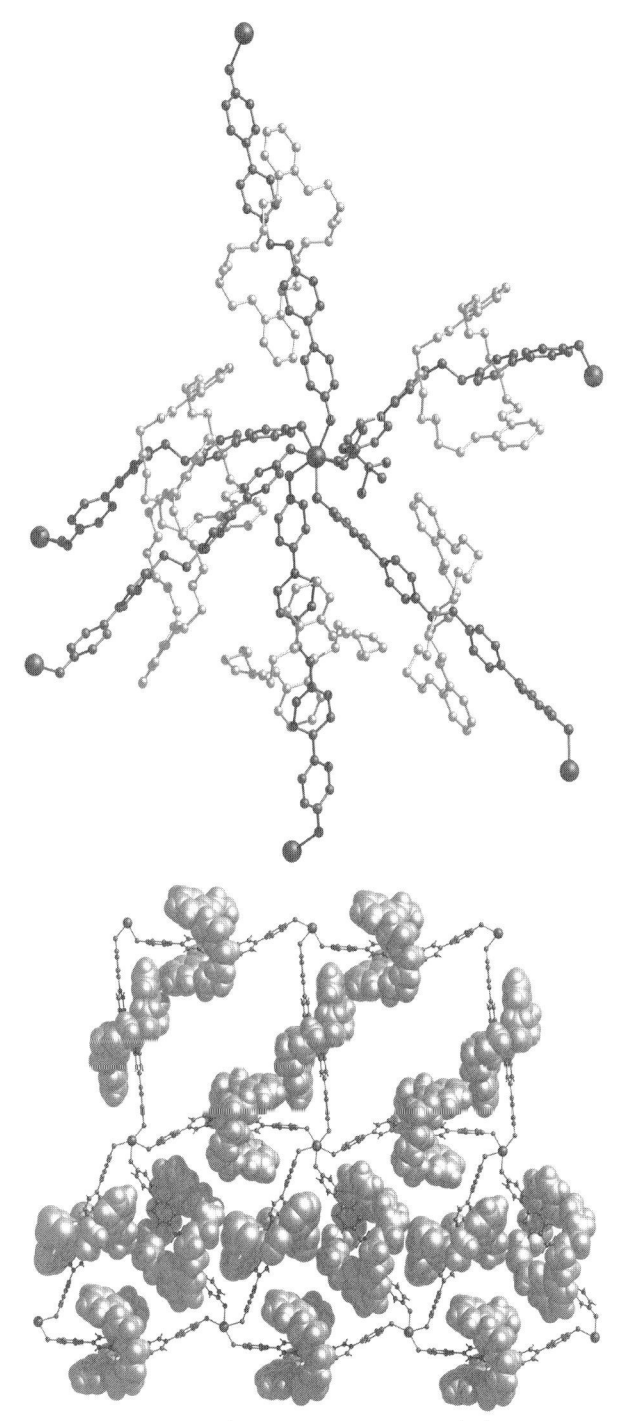

Figure 2.23 X-ray crystal structure a 3D MORF with Yb(III) ion nodes and [(**4**) ⊂ (DB24C8)]$^{2+}$ linkers showing (top) the unique 2D network comprised of square and triangular units.

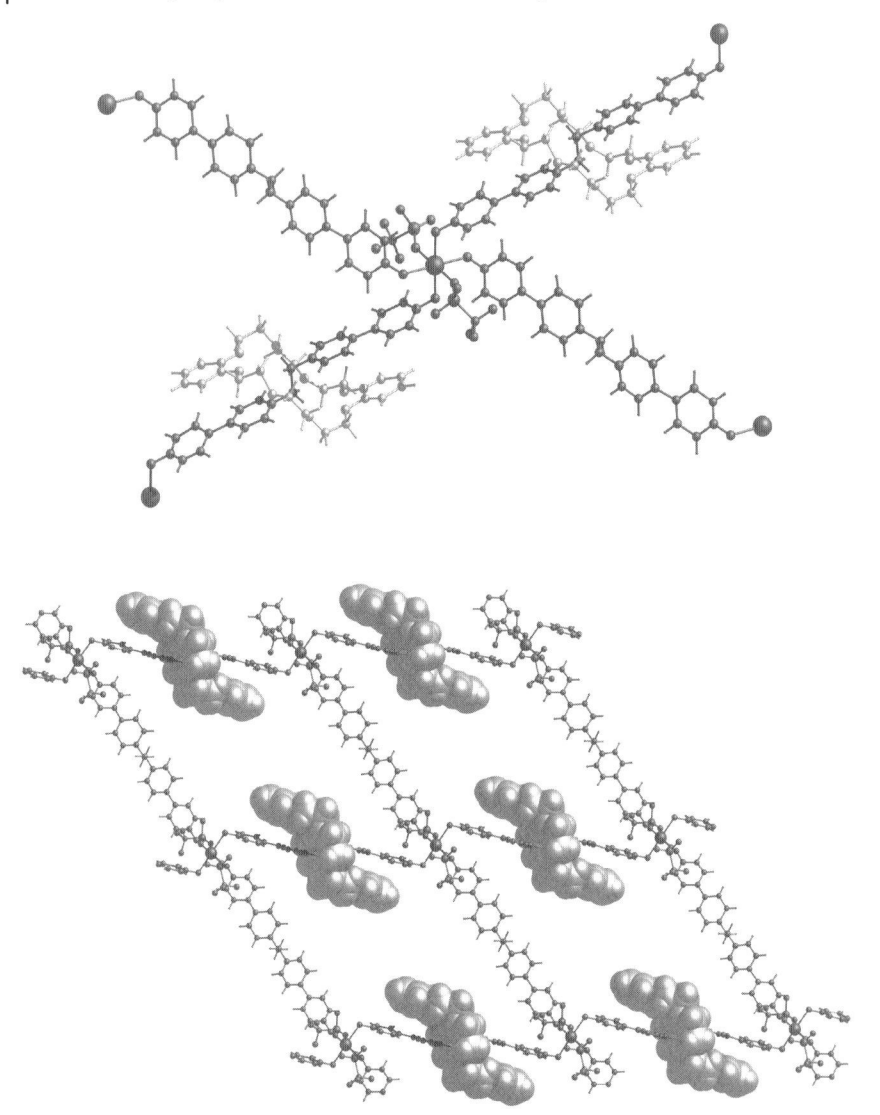

Figure 2.24 X-ray crystal structure of the pillared 1D MORF with Cd(II) ion nodes with [(**4**) ⊂ (DB24C8)]$^{2+}$ rotaxane linkers and naked **4**$^{2+}$ pillars showing (top) the coordination sphere around a single Cd(II) ion and (bottom) a portion of the polar 2D network.

produce large channels filled with solvent [35]. However, in this new structure, each cavity is filled by crown ethers from the layers above and below. The network is actually reminiscent of the 1D MORF structure based on Co(II) or Zn(II) and [(**1g**) ⊂ (DB24C8)]$^{2+}$ as it can be thought of as a pillared 1D MORF. That is, the two *trans*-oriented [(**4**) ⊂ (DB24C8)]$^{2+}$ linkers coordinate to the Cd(II) center in one dimension while the two 4^{2+} molecules pillar the polyrotaxane strands in the second dimension (see the bottom structure in Figure 2.24).

Finally, a property of this new MORF that is intriguing is its crystallization in the noncentrosymmetric space group $P1$. All the crown ether wheels are oriented in the same fashion along each strand of the grid. This is significant as it may be possible to replace DB24C8 with substituted crown ethers and thus orient all the wheel dipoles in the same direction by virtue of their entrapment on the metal–ligand grid. This may have potential as a novel method for creating NLO or similar materials that require ordering of functional groups by taking advantage of this supramolecular modification.

2.4.6
Frameworks Using Hydrogen Bonding

Since we had success in using metal ions as nodes to build MORFs, we looked at the possibility of using hydrogen bonding for the systematic preparation of solid-state materials which contain mechanical linkages. In this vein, we have shown that H^+ can be used in place of a metal ion to form polymeric systems with formula $\{[(H_21g) \subset (DB24C8)(DB24C8)][OTf]_4\}_x$ via hydrogen bonding [40]. In a single-stranded 1D MORF structure, the diprotonated [2]pseudorotaxane $[(H_21g) \subset (DB24C8)]^{4+}$ acts as the H-bond donor while a second equivalent of DB24C8 acts as the H-bond acceptor. The bottom structure in Figure 2.25 shows this H-bonded pair, and the bottom structure shows how this motif extends into a polymeric structure with parallel strands.

2.5
Properties of MORFs: Potential as Functional Materials

2.5.1
Robust Frameworks

All of the MORF materials prepared to date show the same basic stability. They are highly crystalline materials that occlude solvent to some degree and each loses some portion of the trapped solvent rapidly upon removal from the mother liquor at room temperature. Thermogravimetric analysis showed that all residual solvent was removed after heating to $\sim 100°C$. Each MORF studied then showed a stable phase until $\sim 225–250°C$, at which point decomposition of the metal–ligand framework was indicated by loss of DB24C8. It should be noted that the interlocked crown ether, although originally held in place by weak noncovalent bonds, can only be removed by breaking a covalent bond in the metal–ligand backbone. Thus, the loss of DB24C8 is actually a sensitive and unambiguous detection of framework breakdown.

In the case of the 3D MORFs, powder XRD patterns of the stable desolvated phase were consistent with retention of the 3D framework observed in the

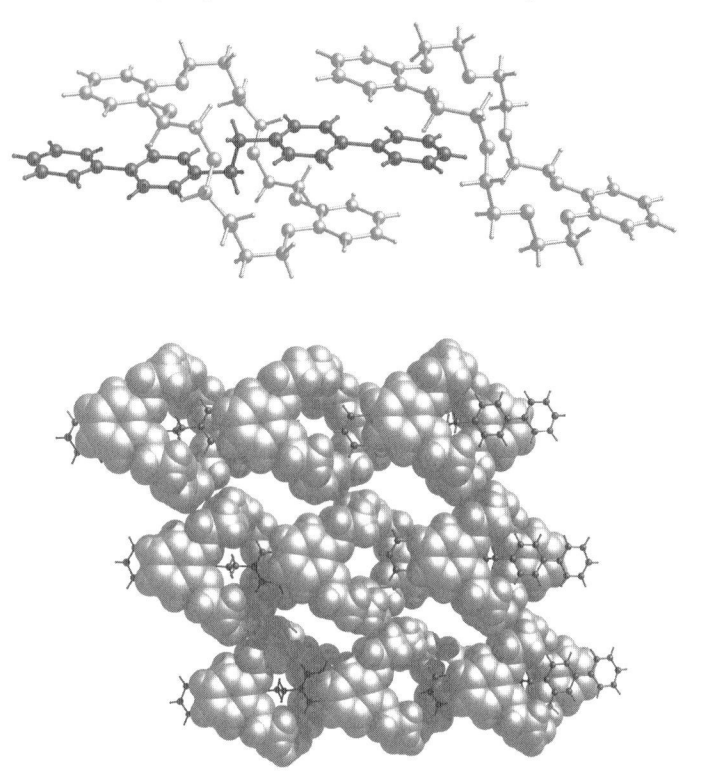

Figure 2.25 X-ray crystal structure of the 1D MORF with alternating [(**1g**) ⊂ (DB24C8)]$^{2+}$ units and DB24C8 held together by hydrogen bonding showing (top) the repeating unit and (bottom) three parallel strands.

single-crystal structures. Although the 1D and 2D materials do not retain their original lattices, the polyrotaxane frameworks remain intact, as evidenced by the observation of a stable phase over a ∼150°C range prior to loss of crown ether. We can confidently conclude that these new materials are at least as stable as basic coordination polymers and there is nothing inherently unstable about a MORF structure.

For this chemistry to evolve, we must be able to prepare truly robust MORFs. This is an issue that can be addressed by utilizing stronger metal–ligand interactions and metal clusters as nodes following the well-documented evolution of conventional MOF materials. We have already shown that rotaxane ligands such as [(**3**) ⊂ (DB24C8)]$^{3+}$, with a terpy chelator, will allow metal complexation under extreme synthetic conditions. This type of approach should therefore greatly extend the synthetic conditions under which MORF assembly can be conducted and allow the preparation of materials with increased stability.

2.5.2
Porosity and Internal Properties

The challenge of creating porous MOF materials for a variety of applications such as gas storage and catalysis is ongoing and some tremendous progress has been made [41]. One of the contributions that MORF materials can make to this area is fine tuning the internal properties of porous materials. For a conventional MOF, the properties of the internal cavities or channels are dictated by the chemical structure of the organic linkers which define the great majority of the internal surface area. In a MORF, this internal surface is primarily related to the nature of the cyclic component and not the linking backbone. So, when a robust and porous MORF can be created, the axle and metal nodes will dictate the shape of the framework but the cyclic wheel will define the internal surface chemistry. It should then be a facile supramolecular event to replace, for example, hydrophobic groups on the exterior of one wheel with hydrophilic groups of another and thereby change the internal property of the material in a predictable fashion without altering the framework of the MORF.

2.5.3
Dynamics and Controllable Motion in the Solid State

One of the ultimate goals of MORF chemistry is to create solid-state materials that contain arrays of ordered molecular machines based on mechanically interlocked species. We have already created the molecules and observed their properties in solution and learned how to create basic MORF structures. The next step in the development of these materials will be to prepare robust, crystalline materials with mechanically switchable components [42]. The ultimate result could be bulk materials with individual components that can be individually addressed by simple external signals (chemical, electrochemical, photochemical, etc.), thus combining the chemistry of molecular machines and the properties of solid-state materials.

References

1 Arico, F., Badjic, J.D., Cantrill, S.J., Flood, A.H., Leung, K.C.-F., Liu, Y. and Stoddart, J.F. (2005) *Top. Curr. Chem.*, **249**, 203, and references therein.
2 Fyfe, M.C.T., Stoddart, J.F. and Fraser, J. (1997) *Acc. Chem. Res.*, **30**, 393.
3 Loeb, S.J. and Wisner, J.A. (1998) *Angew. Chem. Int. Ed.*, **37**, 2838.
4 (a) Loeb, S.J., Tiburcio, J., Vella, S.J. and Wisner, J.A. (2006) *Org. Biomol. Chem.*, **4**, 667. (b) Loeb, S.J. and Wisner, J.A. (1998) *Chem. Commun.*, 2757.
5 (a) Ashton, P.R., Chrystal, E.J.T., Glink, P.T., Menzer, S., Schiavo, C., Spencer, N., Stoddart, J.F., Tasker, P.A., White, A.J.P. and Williams, D.J. (1996) *Chem. Eur. J.*, **2**, 709. (b) Ashton, P.R., Bartsch, R.A., Cantrill, S.J., Hanes, R.E. Jr., Hickingbottom, S.K., Lowe, J.N., Preece, J.A., Stoddart, J.F., Talanov, V.S. and

Wang, Z.-H. (1999) *Tetrahedron Lett.*, **40**, 3661.
6 Davidson, G.J.E., Loeb, S.J., Parekh, N.A. and Wisner, J.A. (2001) *Dalton Trans.*, 3135, and references therein.
7 Georges, N., Loeb, S.J., Tiburcio, J. and Wisner, J.A. (2004) *Org. Biomol. Chem.*, **2**, 2751.
8 Huang, F. and Gibson, H.W. (2005) *Prog. Polym. Sci.*, **30**, 982 and references therein.
9 Lee, J.W. and Kim, K. (2003) *Top. Curr. Chem.*, **228**, 111, and references therein.
10 Loeb, S.J. and Wisner, J.A. (1998) *Chem. Commun.*, 2757.
11 Loeb, S.J. and Tramontozzi, D.A. (2005) *Org. Biomol. Chem.*, **3**, 1393.
12 Hubbard, A.L., Davidson, G.J.E., Patel, R.H., Wisner, J.A. and Loeb, S.J. (2004) *Chem. Commun.*, 138.
13 Balzani, V., Credi, A., Raymo, F.M. and Stoddart, J.F. (2000) *Angew. Chem. Int. Ed.*, **39**, 3348.
14 Vella, S.J., Tiburcio, J., Gauld, J.W. and Loeb, S.J. (2006) *Org. Lett.*, **8**, 3421.
15 Loeb, S.J. and Wisner, J.A. (2000) *Chem. Commun.*, 845.
16 Loeb, S.J. and Wisner, J.A. (2000) *Chem. Commun.*, 1939.
17 Vella, S.J. (2006) PhD Thesis, University of Windsor, Windsor, ON.
18 Vella, S.J., Tiburcio, J. and Loeb, S.J. (2005) *Org. Lett.*, **7**, 4923.
19 Loeb, S.J., Tiburcio, J. and Vella, S.J. (2006) *Chem. Commun.*, 1598.
20 Hunter, C.A. (2004) *Angew. Chem., Int. Ed.*, **43**, 5310.
21 Chichak, K., Walsh, M.C. and Branda, N.R. (2000) *Chem. Commun.*, 847.
22 Davidson, G.J.E. and Loeb, S.J. (2003) *Dalton Trans.*, 4319.
23 Davidson, G.J.E., Loeb, S.J., Passaniti, P., Silvi, S. and Credi, A. (2006) *Chem. Eur. J.*, **12**, 3233.
24 Bonnet, S., Collin, J.-P., Koizumi, M., Mobian, P. and Sauvage, J.-P. (2006) *Adv. Mater.*, **18**, 1239, and references therein.
25 Pease, A.R., Jeppesen, J.O., Stoddart, J.F., Luo, Y., Collier, C.P. and Heath, J.R. (2001) *Acc. Chem. Res.*, **34**, 433.
26 Collin, J.-P., Dietrich-Buchecker, C., Gaviña, P., Jiminez-Molero, M.C. and Sauvage, J.-P. (2001) *Acc. Chem. Res.*, **34**, 477.
27 Shipway, A.N. and Willner, I. (2001) *Acc. Chem. Res.*, **34**, 421.
28 Ballardini, R., Balzani, V., Credi, A., Gandolfi, M.T. and Venturi, M. (2001) *Acc. Chem. Res.*, **34**, 445, and references therein.
29 Tseng, H.-R., Wu, D., Fang, N.X., Zhang, X. and Stoddart, J.F., (2004) *ChemPhysChem* **5**, 111 and references therein.
30 Yu, H., Luo, Y., Beverly, K., Stoddart, J.F., Tseng, H.-R. and Heath, J.R. (2003) *Angew. Chem. Int. Ed.*, **42**, 5706.
31 Gibson, H.W., Nagvekar, D.S., Yamaguchi, M., Bhattacharjee, S., Wang, H., Vergne, M.J. and Hercules, D.M. (2004) *Macromolecules*, **37**, 7514, and references therein.
32 Yaghi, O.M., O'Keefe, M., Ockwing, N.W., Chae, H.K., Eddaoudi, M. and Kim, J. (2003) *Nature*, **423**, 705, and references therein.
33 Loeb, S.J. and Wisner, J.A. (2005) *Chem. Commun.*, 1511. The term MORF is used in this article to designate a sub-class of MOF as defined by Yaghi; See Rowsell, J.L.C. and Yaghi, O.M. (2004) *Microporous Mesoporous Mater.*, **73**, 3.
34 The preparation of these materials pre-dates those described herein. For an excellent review of this chemical system, see. Kim, K. (2002) *Chem. Rev.*, **31**, 96
35 Davidson, G.J.E. and Loeb, S.J. (2003) *Angew. Chem. Int. Ed.*, **42**, 74.
36 Stahl, J., Bohling, J.C., Bauer, E.B., Peters, T.B., Mohr, W., Martín-Alvarez, J.M., Hampel, F. and Gladysz, J.A. (2002) *Angew. Chem. Int. Ed.*, **41**, 1872, and references therein.
37 Hoffart, D.J. and Loeb, S.J. (2005) *Angew. Chem. Int. Ed.*, **117**, 901.
38 Li, J.-R., Bu, X.-H., Zhang, R.-H. (2004) *Inorg. Chem.*, **43**, 237 and references therein.
39 Hoffart, D.J. and Loeb, S.J. (2007) *Supramol. Chem.*, **19**, 89.

40 Tiburcio, J., Davidson, G.J.E. and Loeb, S.J. (2002) *Chem. Commun.*, 1282.

41 Mueller, U., Schubert, M., Teich, F., Puetter, H., Schierle-Arndt, K. and Pastre, J. (2006) *J. Mater. Chem.*, **16**, 626.

42 For a review on creating molecular machines in the crystalline state, see. Khuong, T.-A.V., Nuñez, J.E., Godinez, C.E. and Garicia-Garibay, M.A. (2006) *Acc. Chem. Res.*, **39**, 413.

3
Strategic Anion Templation for the Assembly of Interlocked Structures
Michał J. Chmielewski and Paul D. Beer

3.1
Introduction

Anions are usually perceived as elusive species, difficult to trap and thus presenting a particular challenge to supramolecular chemists aiming at the design of efficient anion receptors [1]. Low charge densities (in comparison with isoelectronic cations) and high solvation energies are traditionally blamed for their sometimes disappointingly weak interactions with synthetic receptors. This, together with their weakly pronounced coordination preferences, is why anions may have been overlooked as potential templates.

However, an increasing body of evidence now seems to indicate [2] that although these limitations certainly exist, they are of a less fundamental nature then previously thought and may be overcome by improved receptor design. Continuing progress in the construction of anion receptors has resulted in the development of simple, organic, hydrogen bond donating units that exhibit strong affinities towards anions. Furthermore, these affinities are solvent dependent and sometimes it is enough to use a less competitive solvent to achieve the desired strength of binding. Strong binding in turn is a prerequisite for the use of anions as templates in the supramolecular assembly processes.

Over the past few years, numerous, often serendipitous, discoveries of anion templating phenomena have demonstrated that the potential of using anions as templates was underestimated [2]. In this chapter, we wish to illustrate that by systematic and rational research this potential can be developed into a highly effective strategy, which compares favorably with previous approaches based on cationic and neutral templating agents. More specifically, the use of anion recognition for the formation of pseudorotaxanes, rotaxanes and catenanes will be described, with emphasis placed on recent work of our group on the use of halide anions as templates in the formation of new architectures with sensory properties [3].

Organic Nanostructures. Edited by Jerry L. Atwood and Jonathan W. Steed
Copyright © 2008 WILEY-VCH Verlag GmbH & Co. KGaA, Weinheim
ISBN: 978-3-527-31836-0

Our motivations for carrying out this work were twofold. First, we wanted to explore the scope and limitations of anion templation as a novel method for the construction of interpenetrated structures. Second, such interlocked structures, after template removal, may act as hosts for anionic guests by virtue of their unique, topologically constrained three-dimensional cavities with orthogonal alignment of binding units. We hoped that a high degree of preorganization in the receptors' three-dimensional cleft and complete encapsulation of anions would lead to strong association and high selectivity in anion binding. Indeed, the only literature precedent, a pyrrole amide-based catenane by Sessler, Vögtle and coworkers [4], exhibits binding characteristics that compares favorably with a macrocyclic control. Unfortunately, this catenane was obtained in just 4% yield and attempts to improve the yield by anion templation were unsuccessful.

Three-dimensional cavities of rotaxanes and catenanes appear even more appealing in the context of difficulties afflicting two-dimensional receptors [5]. Acyclic clefts and macrocycles, which dominate the current landscape of anion receptors, often suffer from poor selectivities. It is becoming more and more apparent that, despite synthetic difficulties, they need to be elaborated into three-dimensional receptors able to embrace anions fully, in order to overcome this issue. Traditional approaches towards this goal may be exemplified by the construction of cryptands [6] or lariat-type compounds [7]. Alternatively, one may thread a one-dimensional, acyclic receptor through the annulus of a two-dimensional, macrocyclic receptor, thus creating a three-dimensional cavity for anion recognition. Such a design has a number of potential advantages, such as the aforementioned orthogonal disposition of the two binding units, which is difficult to achieve by other means, lower risk of slow exchange kinetics owing to less rigid, mechanically bound structure than that of covalently linked polycyclic cage compounds and a number of potential mechanisms of signal transduction, based on anion-induced change in co-conformation or in mutual interactions between thread and macrocycle. Thus we believe that rotaxanes and catenanes have huge potential in anion recognition and sensing.

3.2
Precedents of Anion-directed Formation of Interwoven Architectures

Two key types of self-assembly have been employed in the construction of interwoven molecules (Figure 3.1). In the first, a discrete interweaving template [8] is used to direct complex formation. In this case, the template is not a part of any component and may be removed from the interlocked system following synthesis. The archetypical example of such an approach is provided by the seminal work of Sauvage and coworkers, who used Cu(I) metal cations to control the formation of a wide range of interlocked species, including the [2]catenane illustrated [9]. Alternatively, the assembly may be driven by direct interaction between the components leading to the reversible formation of a precursor complex, which is then "trapped" by covalent bond formation to give a permanently interlocked system. This strategy, dubbed by

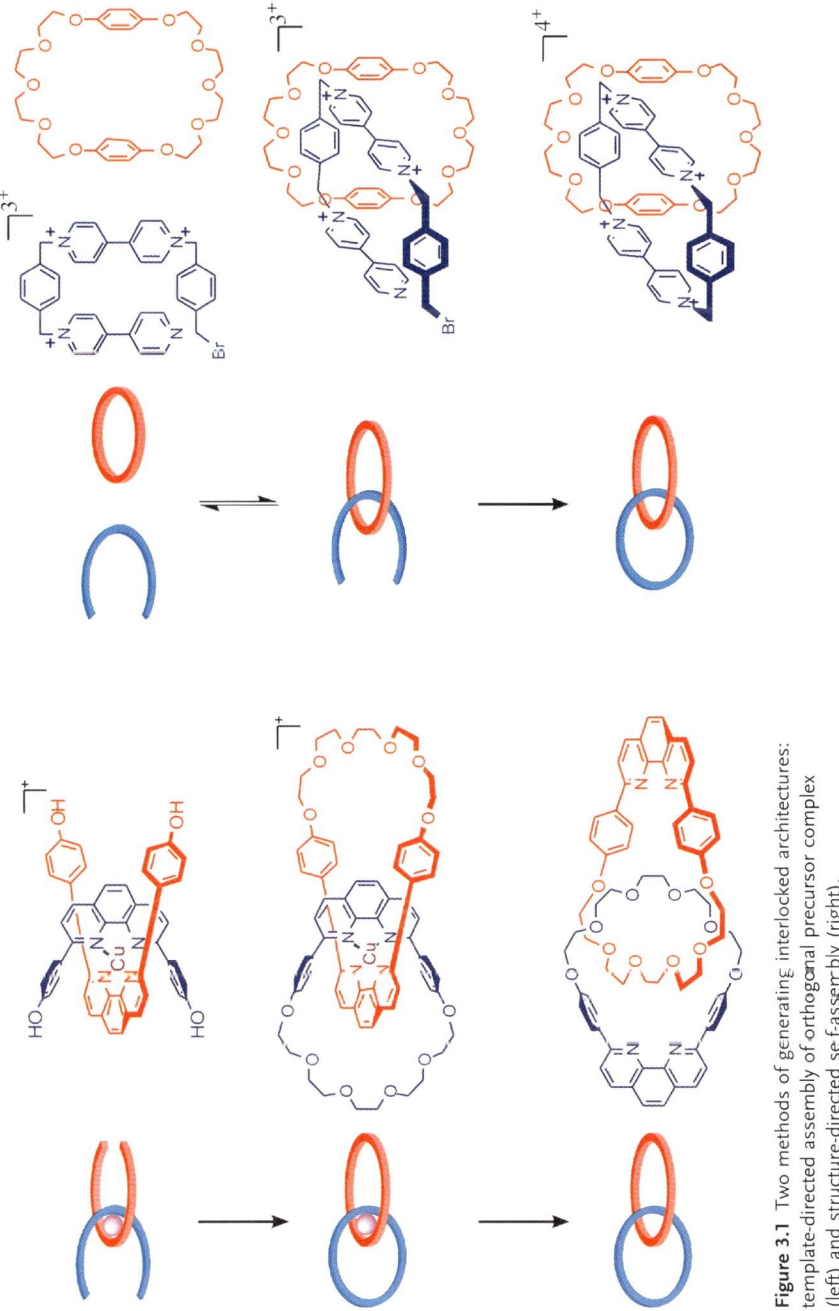

3.2 Precedents of Anion-directed Formation of Interwoven Architectures | **65**

Figure 3.1 Two methods of generating interlocked architectures: template-directed assembly of orthogonal precursor complex (left) and structure-directed self-assembly (right).

Stoddart "structure-driven self-assembly" [10], is exemplified by his catenane synthesis [11], guided by attractive π–π interactions.

All previous applications of anion recognition in the synthesis of interlocked molecules fall into the second category. In 1999, Vögtle and coworkers introduced a new threading method in which anion binding of a bulky phenolate (a stopper precursor) by a tetraamide macrocycle is used to assemble a complex termed a "wheeled nucleophile" or "semi-rotaxane" [12,13]. This complexed phenoxide anion reacts with a stoppered electrophile through the annulus of the macrocycle (making an axle inside the wheel) and thus a rotaxane is produced (Scheme 3.1) . Yields of up to a spectacular 95% have been obtained by this method, although more commonly yields are below 50%. The lack of any anion binding sites in the axle means that these rotaxanes were not designed as potential anion receptors.

One year later, this elegant anion recognition-directed threading methodology was employed by Smith and coworkers in the synthesis of ion-pair binding rotaxanes [14]. They used macrobicyclic wheels comprising a cation binding macrocycle bridged by an anion binding cleft (Scheme 3.2). The anion binding isophthalamide site played its major role during the synthesis of these systems using Vogtle's methodology, but was also shown to bind anions after the rotaxane assembly [15]. The cation binding crown ether macrocycle allowed for cation binding-induced modulation of the rotaxane's dynamics [16].

Further investigation of this template motif by Schalley and coworkers revealed certain limitations of this methodology, namely that application of less bulky phenolates gave disappointingly low yields of rotaxanes [17]. This was traced back to the double role of phenolate oxygen, which acts both as a nucleophile and the assembly-directing anionic agent at the same time. To circumvent this problem, they designed a modified thread in which the only role of the phenolate anion was to direct the assembly of the pseudorotaxane precursors (Scheme 3.3) [18]. This assembly is subsequently locked by covalent attachment of two stoppers, which occurs at amino groups remote from the recognition site. Interestingly, the phenolate anion encapsulated inside the rotaxane does not react even with very powerful alkylating agents such as methyl iodide.

Such protection of a thread by a macrocycle was also used by Smith and co-workers to stabilize near-infrared dyes, squaraines, in the form of rotaxanes (Scheme 3.4) [19]. These examples are also notable because of the role of anion recognition in their synthesis. More precisely, the macrocyclic wheel associates around the squaraine thread, which acts as a template by virtue of its negatively charged oxygen atoms.

The above examples leave no doubt that anion recognition may be used advantageously to direct the assembly of intertwined structures. However, in all of the above methods, negatively charged groups are covalently attached parts of the assembling structures and furthermore none of these methods have been extended to the synthesis of catenanes thus far. To uncover the full potential of anion templation in supramolecular synthesis, the use of discrete anions as interpenetrating templates needs to be developed in analogy with what has been achieved with cationic templates by Sauvage and others [20]. We set ourselves the challenge to achieve this goal.

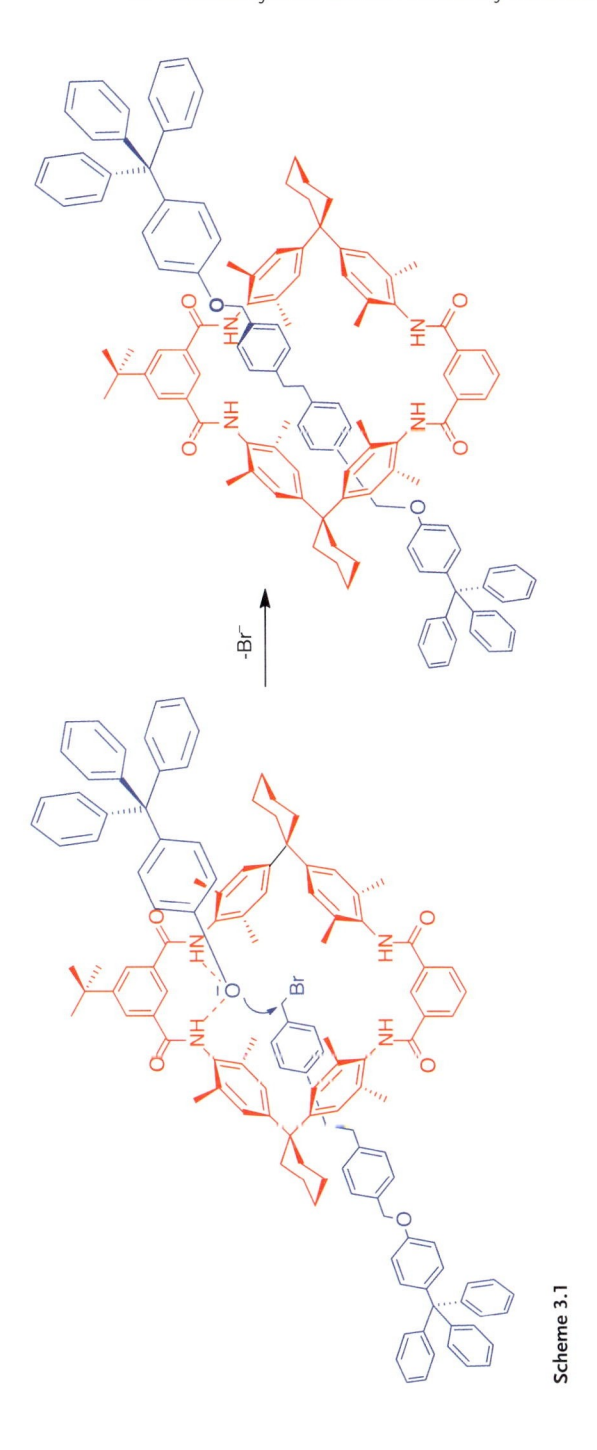

3.2 Precedents of Anion-directed Formation of Interwoven Architectures | 67

Scheme 3.1

Scheme 3.2

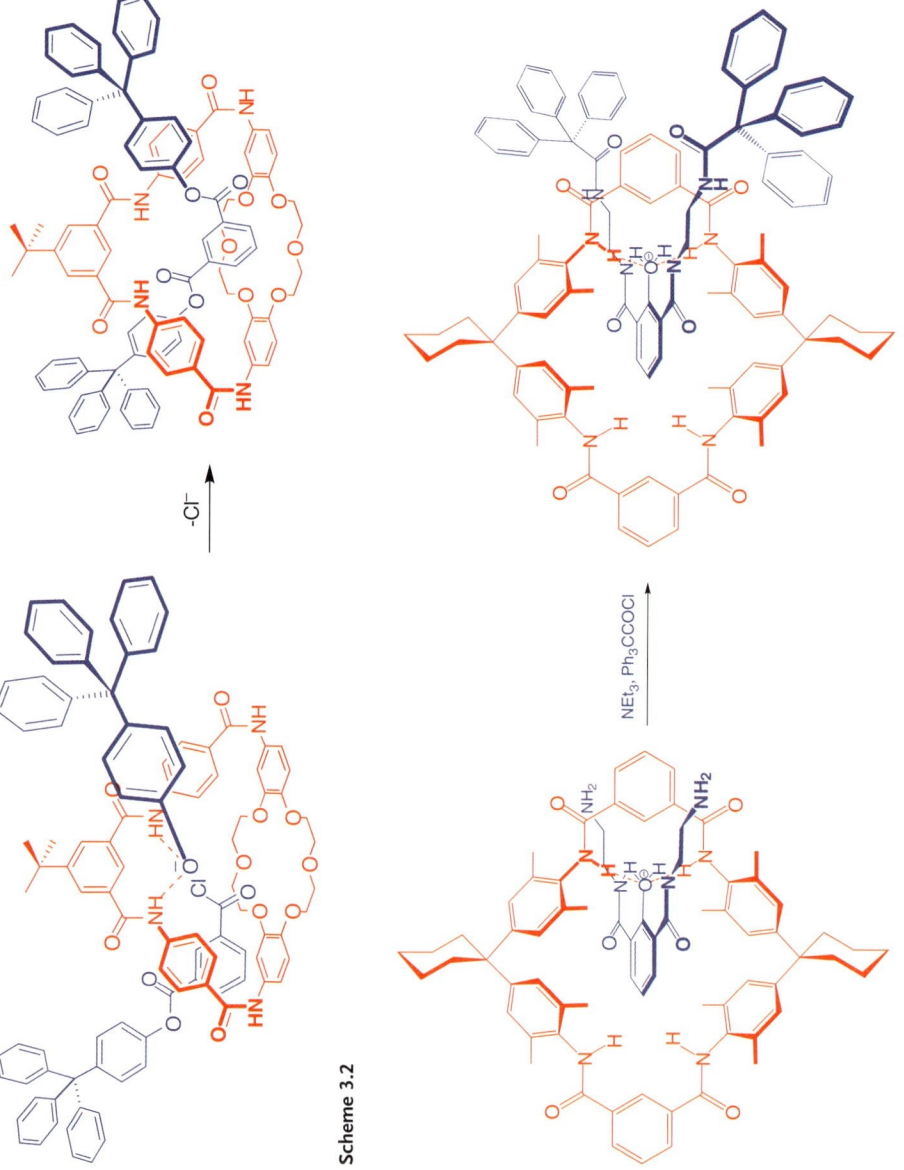

Scheme 3.3

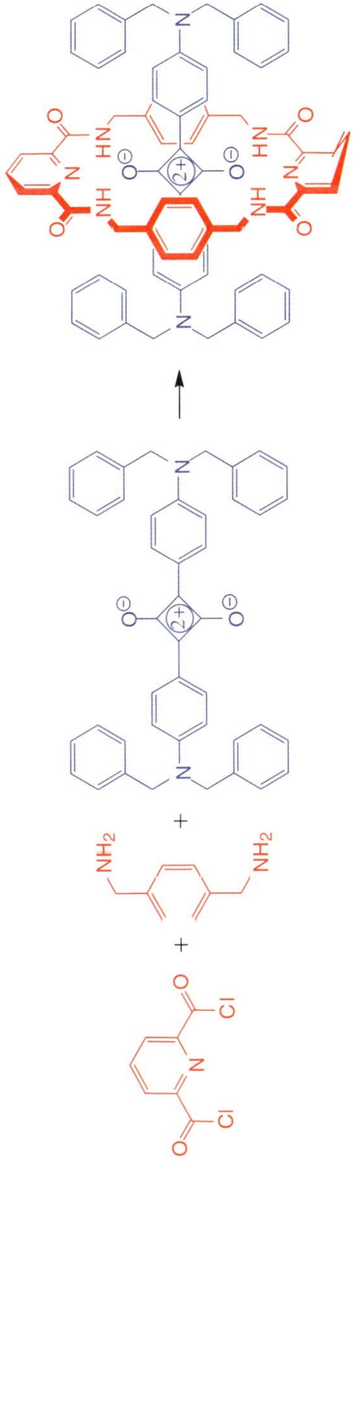

3.2 Precedents of Anion-directed Formation of Interwoven Architectures | 69

Scheme 3.4

3.3
Design of a General Anion Templation Motif

The basic motif in the cation templated syntheses of intertwined structures is an orthogonal assembly of two U-shaped ligands held together by a spherical metal cation. With the aim of producing an analogous motif but with reversed polarity, we focused on simple spherical halide anions.

A key requirement for an interwoven template is that it brings together two components and directs them in a more or less perpendicular manner. Of course, it is topology, not geometry, that matters here, so in principle any nonplanar arrangement should be equally good. However, from the practical point of view it is far more desirable that a perpendicular arrangement is formed, as it minimizes the probability of unwanted reactions such as macrocyclization. Sauvage achieved perpendicular arrangements of phenanthroline ligands using the copper(I) cation, which has a strong preference for tetrahedral coordination and this feature of the metal cation template was often mentioned as indispensable for the success of catenane synthesis. This calls into question the idea of using halide anions for the same purpose, as they have no marked geometric preferences and are therefore certainly unable to compel geometric relationships between organic building blocks in a manner close to that achieved by transition metal cations. However, the same effect can be achieved by steric repulsion between ligands, thus leaving only one requirement for the potential template – its ability to interact simultaneously with two or more ligands strongly enough to overcome the unfavorable entropy of association and mutual repulsions of ligands.

The simple isophthalamide cleft receptors introduced by Crabtree were deemed to be good analogs of phenanthroline ligands because they are planar, U-shaped and bind halide anions strongly in nonpolar organic solvents (Figure 3.2) [21]. Unfortunately, however, their halide complexes exhibit exclusive 1 : 1 stoichiometry. To overcome this problem, a new pyridinium-based ligand **2** was designed, in which a positive charge was introduced to the aromatic ring, while the anion binding site was left unaltered [22]. This new ligand exists as a tightly associated ion pair with halide counterions, with an association constant $K_{ass} > 10^5 \, M^{-1}$ estimated for the equilibrium:

$$\mathbf{2}^+PF_6^- + TBA^+Cl^- \rightleftharpoons \mathbf{2}^+Cl^- + TBA^+PF_6^-$$

by ^{1}H NMR titration in acetone-d_6. The chloride anion is strongly held within the binding cleft of this receptor by a combination of electrostatic and hydrogen bonding interactions and importantly its coordination sphere is still unsaturated: the anion presents an empty meridian orthogonal to the pyridinium cation, which is available for further complexation to a second hydrogen bond donating ligand. This desired association indeed occurred and the isopthalamide compound **1** and pyridinium chloride ion pair $\mathbf{2}^+Cl^-$ produced a 1 : 1 association constant of $100 \, M^{-1}$ in acetone-d_6, whereas no association was observed for weakly coordinating anions such as hexafluorophosphate (Figure 3.2).

It is worth noting, that under the conditions used, the above process is exclusively observed as heterodimerization. This is due to each pyridinium receptor associating

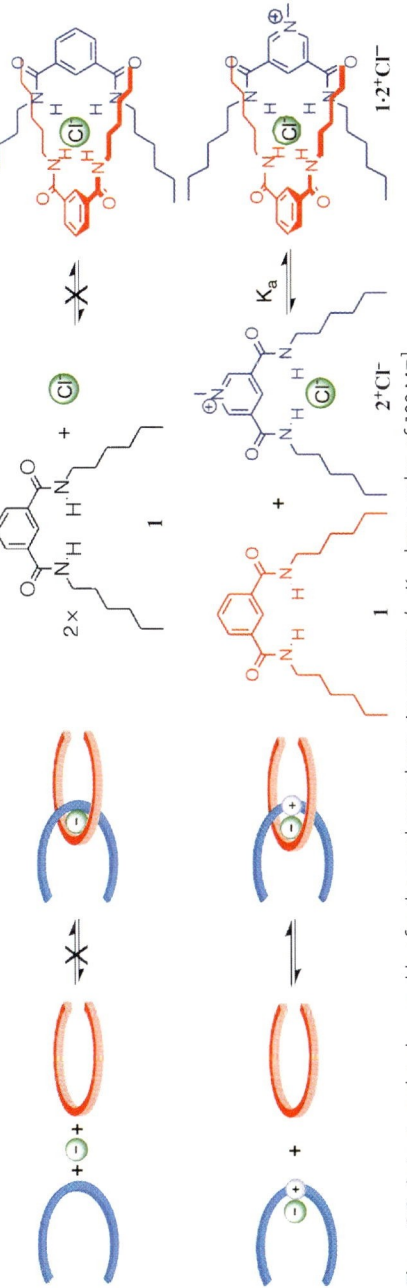

Figure 3.2 Anion-templated assembly of orthogonal complexes. In acetone-d_6, K_{ass} has a value of $100\ M^{-1}$.

strongly with its own chloride, so that the anions are not shared, whereas, as already discussed, the isophthalamide itself has no propensity for the formation of 2 : 1 complexes.

Encouraged by this result, we elaborated the second component of the self-assembly, isophthalamide cleft receptor in such a way so as to strengthen the heterodimeric complex by additional interactions (Figure 3.3) [23]. Thus the sequential introduction of an electron-rich hydroquinone moiety (**3**) and ether oxygens (**4**) led to incremental enhancements in the anion-templated orthogonal complex association constants, as measured by ^{1}H NMR titrations in dichloromethane-d_2. The hydroquinone functionality allows π-donor–π-acceptor interactions to occur with the pyridinium cation unit, with such interactions being detectable by UV/visible spectroscopy. The extra ether oxygens then form weak, charge-assisted C–H···O hydrogen bonds with the *N*-methyl group. However, these interactions are secondary to the anion templation event, as no orthogonal complex formation could be detected in the presence of noncoordinating anions, such as hexafluorophosphate.

3.4
Anion-templated Interpenetration

All the above structural elements were combined in the structure of the macrocyclic receptor **5**, designed to demonstrate anion-templated threading [22]. As was hoped, the chloride salt of pyridinium receptor **2**$^+$Cl$^-$ threaded through the annulus of the macrocycle driven by the affinity of the halide anion for the isophthalamide cleft. This is evidenced by characteristic changes in ^{1}H NMR spectra of both components observed upon their mixing. First, large downfield shifts in the macrocycle isophthalyl and amide protons (C) are seen, indicative of anion binding within macrocycle's cleft. At the same time, upfield shifts in the pyridinium aromatic and amide protons (A) suggest loosened interactions between pyridinium cleft and anion. These two observations are consistent with both threading and competition between receptors for chloride binding, with the second explanation being disfavored by much stronger affinity of charged pyridinium receptor to the anion. The definite proof of threading comes from an upfield shift of the macrocycle hydroquinone proton signals (D) and aromatic C–H protons of the thread caused by π-stacking interactions. Additionally, weak hydrogen bonding interactions between the pyridinium *N*-methyl protons and the ether oxygen atoms are manifested by a small downfield shift of the methyl group signal (B) (Figure 3.4). Thus, ^{1}H NMR spectroscopy confirms the presence of all primary and secondary stabilizing interactions incorporated into our design.

Quantitative ^{1}H NMR titration experiments carried out in acetone-d_6 revealed that the strength of the pseudorotaxane assembly is critically dependent upon the nature of the templating anion. Thus, when chloride is the thread counterion, the pseudorotaxane has an association constant of 2400 M^{-1}, but for bromide, iodide and hexafluorophosphate the constants are lower, being 700, 65 and 35 M^{-1}, respectively. This reflects the relative complexation abilities of these anions.

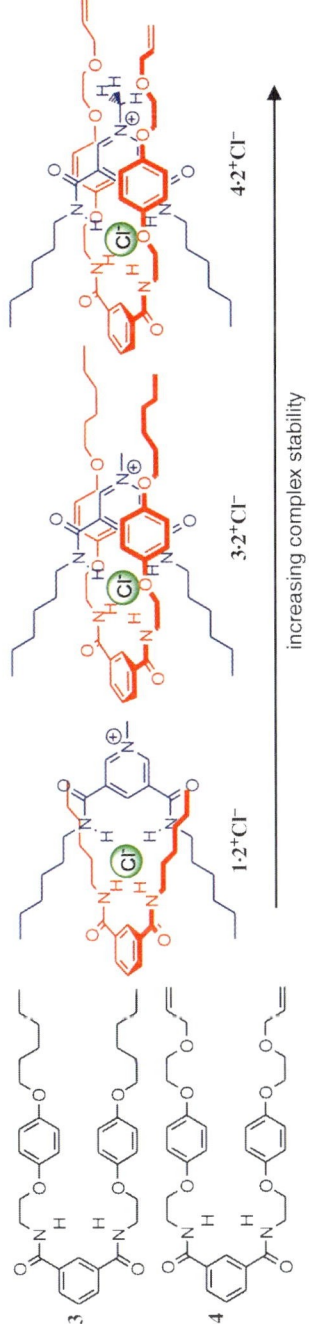

Figure 3.3 Increasing the strength of orthogonal complex formation using secondary interactions.

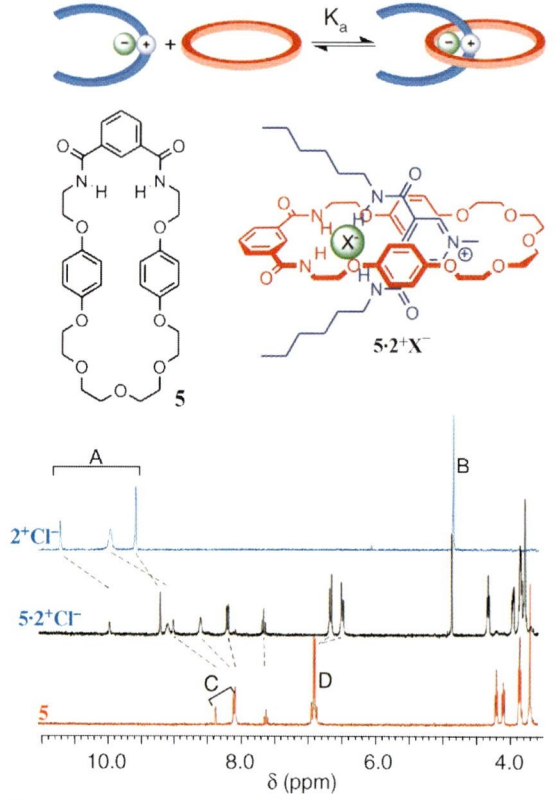

Figure 3.4 Formation of anion-templated [2]pseudorotaxane. General scheme (top), macrocycle **5** and pseudorotaxane **5·2⁺Cl⁻** (middle) and ^{1}H NMR spectroscopic evidence for pseudorotaxane formation (bottom). For spectrum labeling, see text.

Single-crystal X-ray analysis further confirmed the interlocked nature of the assembly and the presence of all the secondary stabilizing interactions inferred from the ^{1}H NMR spectra (Figure 3.6). Furthermore, it gives a detailed picture of the anion binding cavity. The amide hydrogen bond donors are shown to form an approximately tetrahedral coordination sphere around the spherical chloride anion, resulting from an orthogonal arrangement of the organic components. Additionally, the anion forms two short C—H···Cl hydrogen bonds, one with each aromatic ring.

3.5
Probing the Scope of the New Methodology

The large amount of detail required in the design of the first anion-templated pseudorotaxane provoked a question about the versatility of this templation strategy.

3.5 Probing the Scope of the New Methodology

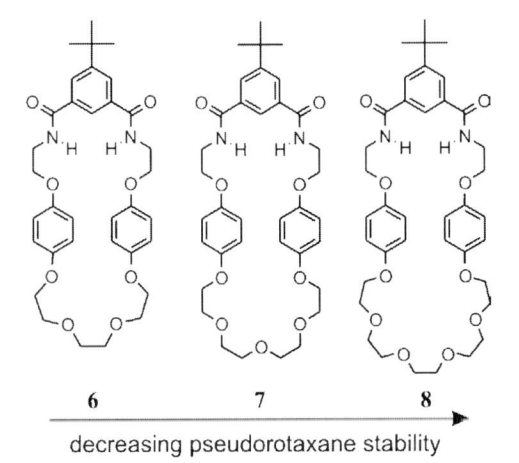

Figure 3.5 Variation of [2]pseudorotaxane stability with macrocycle ring size.

To address this question, we independently varied (i) the structure of the macrocycle, (ii) the thread and (iii) the nature of the anion [24].

The length of the macrocyclic polyether chain was found to affect significantly the stability of the [2]pseudorotaxane assemblies with the pyridinium chloride thread, which decreased with increasing ring size: **6 > 7 > 8** (Figures 3.5 and 3.6). As the anion binding ability of the macrocycles remains unaltered in this series, this trend was explained in terms of increasing entropic cost of association resulting from the increasing flexibility of the polyether chains and decreasing second-sphere stabilization between complexes.

ITC investigations on **6·2$^+$Cl$^-$** and **7·2$^+$Cl$^-$** in 1,2-dichloroethane revealed that although in both cases the association was enthalpy driven and opposed by entropy, the entropic cost of association was much higher for **7·2$^+$Cl$^-$** than for **6·2$^+$Cl$^-$**.

Subtle differences in second sphere stabilization were found in the X-ray structures of pseudorotaxanes **6·2$^+$Cl$^-$**, **5·2$^+$Cl$^-$** (which differs from **7·2$^+$Cl$^-$** with *t*-Bu group only) and **8·2$^+$Cl$^-$** (Figure 3.6). The smallest macrocycle **6·2$^+$Cl$^-$** is apparently

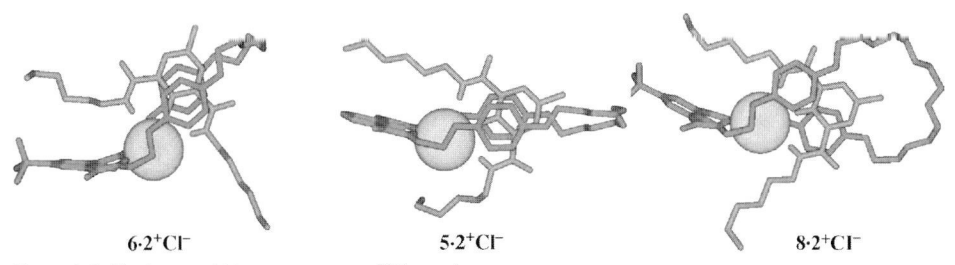

Figure 3.6 Single-crystal X-ray structures of [2]pseudorotaxanes **6·2$^+$Cl$^-$**, **5·2$^+$Cl$^-$** and **8·2$^+$Cl$^-$**. All hydrogens except those in the primary anion coordination sphere have been omitted for clarity. Chloride is represented as a space-filling sphere.

Figure 3.7 Alternative threads for pseudorotaxane assembly.

too small to encircle the thread and, as a result, the pyridinium ring of the thread lies out of the plane of the polyether ring. This allows for a short bifurcated hydrogen bond between the aromatic CH proton from the position 2 of the pyridinium ring and two ether oxygen atoms, which is not present in the crystal structures of the other two pseudorotaxanes. Another interesting feature was found in the crystal structure of the largest macrocycle $8\cdot2^+Cl^-$. The π-stacking interaction between the pyridinium cation and one of the hydroquinones is greatly diminished in this complex, with the distance between ring centroids being 4.266 Å, versus 3.540–3.728 Å in other two cases. This observation suggests that **8** is too large to accommodate the pyridinium chloride ion pair comfortably.

The above structure–affinity investigations were subsequently extended to other types of cationic threads based on the nicotinamide, imidazolium, benzimidazolium and guanidinium moieties (Figure 3.7) [25]. All these threads, as chloride salts, form interpenetrated complexes with macrocyclic receptors **6–8**, illustrating the versatility of this anion templating methodology. However, the stability of these complexes was lower and decreased across the above series, as a result of the decreasing role of second-sphere coordination effects. This is not surprising given that these macrocycles were deliberately optimized for N-methylpyridinium-based threads. As a corollary, a much less pronounced effect of macrocycle size on pseudorotaxane stability was observed in these cases (Figure 3.8).

The key role of anion binding in all the above assembly processes is evidenced by the observation that the stability of pseudorotaxanes increases with the hydrogen bond accepting ability of anions in the series $PF_6^- < I^- < Br^- < Cl^-$ [7].

Thus the stability of the resulting pseudorotaxane systems is heavily dependent on the nature of the components, in particular the nature of the templating anion, but also on the size of macrocycle, propensities for secondary stabilizing interactions and

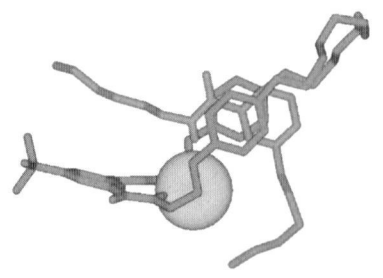

Figure 3.8 Single-crystal X-ray structure of $7\cdot9^+Cl^-$.

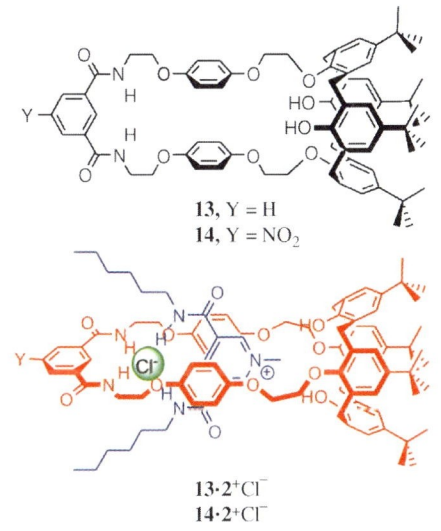

Figure 3.9 Calix[4]arene macrocycles for [2]pseudorotaxane formation.

the strength of thread-anion pairing. Importantly, however, in the presence of chloride anion all the above cationic threads do form pseudorotaxanes, demonstrating structural versatility of the methodology. This encouraged us to divert even further from the original design of macrocycle.

First, the crown ether portion of **6–8** was replaced with a calix[4]arene unit, as in **13** (Figure 3.9) [26]. This macrobicycle also undergoes [2]pseudorotaxane formation on treatment with thread 2^+Cl^-, although the association constant is much reduced (170 M^{-1} in acetone-d_6) compared with the simpler systems, possibly due to steric constraints. It is possible to enhance pseudorotaxane stability by improving the anion binding properties of the macrocycle. Thus the introduction of a 5-nitro group into isophthalamide moiety (as in **14**), which increases the hydrogen bond acidity of the amide donors, also increases the observed pseudorotaxane association constant (240 M^{-1} in acetone-d_6).

Second, the anion binding isophthalamide fragment of the macrocycle was exchanged for rhenium(I) bipyridylbisamide unit, which may be thought of as an expanded version of isophthalamide moiety with two aromatic CH hydrogen bond donors instead of just one (Figure 3.10) [27].

Its major advantage is the potential to provide an optical signal for anion binding [28] and, when the anion is paired to a suitable thread, for pseudorotaxane formation. Macrocycle **15** was shown by ^{1}H NMR titrations in acetone-d_6 to bind anions much more strongly than isophthalamide macrocycles and, accordingly, formed strong, interpenetrated complexes with the halide salts of pyridiniumdiamide, pyridinium-nicotinamide, benzimidazolium and guanidinium cations in the same solvent. The single-crystal X-ray structure determination of the macrocycle-pyridinium chloride complex (Figure 3.11) shows the interpenetrated nature of the assembly and binding

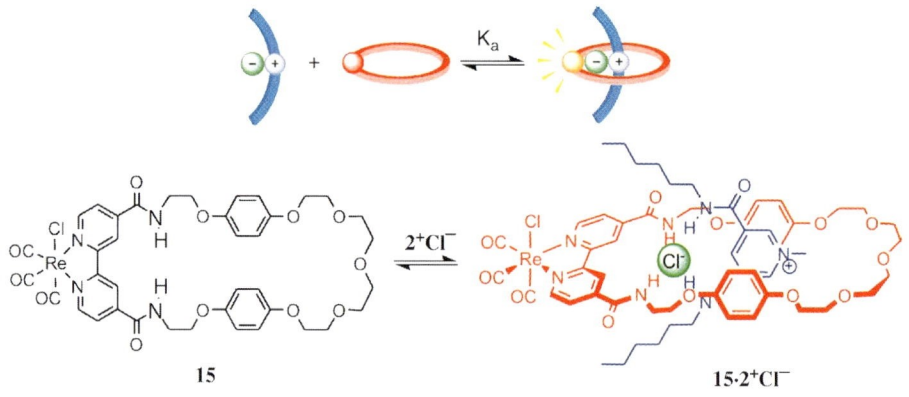

Figure 3.10 Luminescence sensing of [2]pseudorotaxane assembly using a rhenium-containing macrocycle. General schematic (top) and formation of pseudorotaxane **15·2⁺Cl⁻** (bottom).

of the chloride anion by seven hydrogen bonds: four with amide NH protons, two with 3,3′-bipyridyl CH protons and one with 4-pyridinium CH proton. Interestingly, it is the only crystal structure available thus far in which hydroquinone rings and pyridinium ring are not offset but lie directly one above the other. This is probably simple geometric consequence of larger bite size of bipyridyl moiety in comparison to isophthalamide.

As was hoped, the addition of **2⁺Cl⁻** to the macrocycle was signaled through an enhancement in the rhenium–bipyridine ^{3}MLCT emissive response. This was exploited for anion sensing in a permanently locked system described later.

The successful formation of such a variety of pseudorotaxanes using anion templation demonstrated structural robustness of this methodology and encouraged its application in the synthesis of permanently interlocked derivatives.

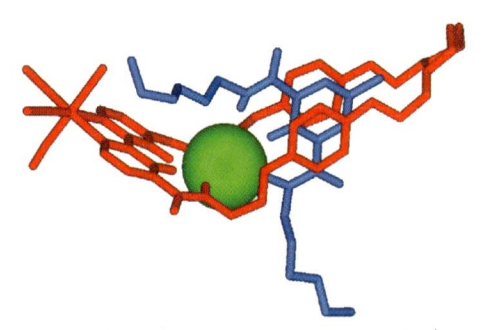

Figure 3.11 Single-crystal X-ray structure of **15·2⁺Cl⁻**. All hydrogens except those in the primary anion coordination sphere have been omitted for clarity. Chloride is represented as a space-filling sphere.

3.6
Anion-templated Synthesis of Rotaxanes

Having demonstrated the feasibility of using anions to direct the assembly of orthogonal complexes and pseudorotaxanes, it was hoped that the same methodology would be readily applicable to the synthesis of permanently interlocked rotaxane and catenane species. However, as mentioned above, discrete anionic templates have never been used for this purpose before and some practical problems were encountered on the way to this goal. One of them was the limited availability of suitably large stoppering groups. An inherent feature of our methodology is that a macrocycle has to be large enough to accommodate both thread and anion. As a corollary, we needed particularly bulky stoppers to prevent the macrocycle from slipping off. No such suitable building blocks are commercially available and an assortment of easy to make synthons in the literature is very restricted. Initially, the well-known amino-functionalized tetraphenylmethane-type stoppers appended with *tert*-butyl groups for added bulkiness were used.

Second, our choice of synthetic reactions was limited by the requirement of their compatibility with anion recognition. For example, reactions involving anions as substrates (nucleophiles) or products (nucleofuges) should be avoided in view of possible competition with the template. Furthermore, reactions taking place in nonpolar solvents are preferred, because such conditions maximize the strength of anion binding.

The above considerations made the "stoppering" route to rotaxanes problematic and prompted us to switch to a "clipping" methodology (Figure 3.12). This was achieved using a ring-closing olefin metathesis reaction, which is well known to give high yields in macrocyclizations, works very well in nonpolar solvents and has excellent functional group tolerability (depending on the catalyst) [20,29].

In a first system, two tetraphenyl stoppers were covalently attached to a pyridinium chloride threading unit making up a thread component 16^+Cl^- (Figure 3.12). The second component was the above-described charge neutral macrocycle precursor 4, terminating with allyl groups capable of undergoing ring-closing metathesis reaction. By virtue of the chloride anion template the two components associate strongly in non-competitive solvents and RCM reaction with Grubbs' catalyst in dichloromethane led to the expected [2]rotaxane product 17^+Cl^- in 47% yield [23]. No rotaxane formation was detected with analogous bromide, iodide and hexafluorophosphate pyridinium salts, indicating the critical templating role of the chloride anion. It is noteworthy that in this case the second-sphere coordination interactions which support the primary chloride recognition process also facilitate the ring closure process around the ion pair thread by directing the allyl-terminated arms into close proximity.

Single-crystal X-ray structural analysis confirmed the interpenetrated nature of the product and gave valuable insight into the binding mode of the anion inside the rotaxane cavity (Figure 3.13). It turned out that the two stoppers are not just innocent spectators of the anion binding, but actually donate two $C-H\cdots Cl$ hydrogen bonds from phenyl rings adjacent to amide groups. Thus, the anion binding cleft of the

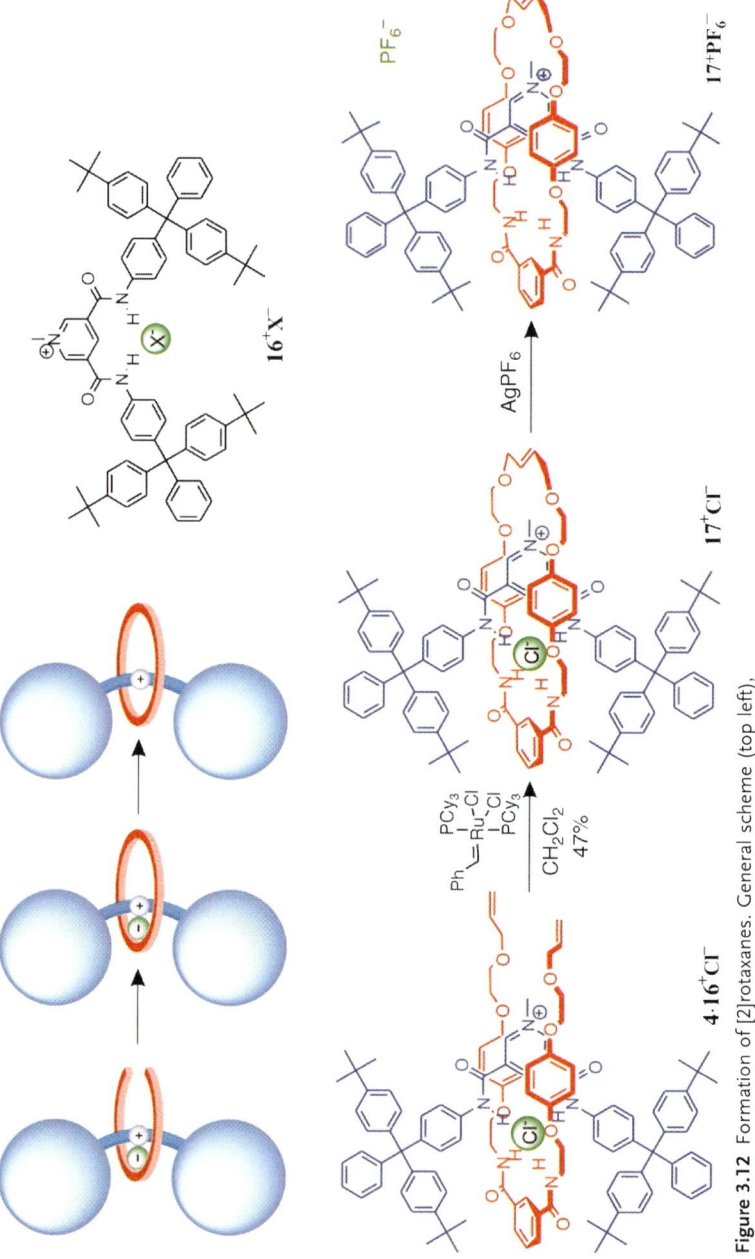

Figure 3.12 Formation of [2]rotaxanes. General scheme (top left), stoppered thread **16**$^+$Cl$^-$ (top right) and formation of rotaxanes **17**$^+$Cl$^-$ and **17**$^+$PF$_6^-$ (bottom).

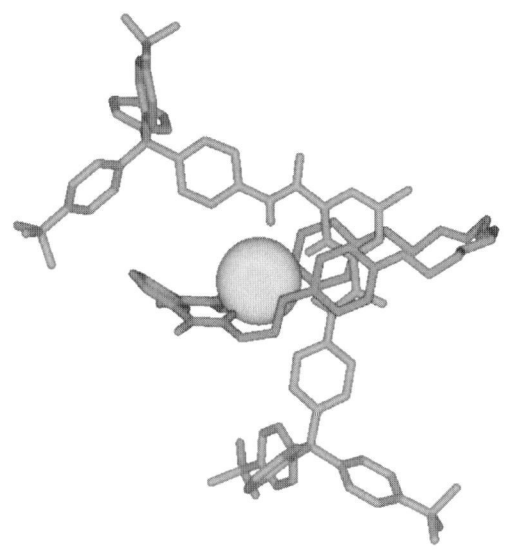

Figure 3.13 Single-crystal X-ray structure of 17^+Cl^-. All hydrogens except those in the primary anion coordination sphere have been omitted for clarity. Chloride is represented as a space-filling sphere.

thread is defined by five, not three, almost exactly coplanar hydrogen atoms. The anion's meridian is occupied by the macrocycle's isophthalamide cleft, which donates two strong hydrogen bonds from the amide groups. The third potential hydrogen bond donor, the internal aromatic proton of the isophthalamide unit, forms only a very weak bond with the anion, so that the coordination number of chloride in this complex may be thought of as seven. The structure also provides evidence for the existence of the second sphere π-stacking and $N^+-CH_3 \cdots O$ hydrogen bonding interactions "designed into" this system.

Anion exchange of the chloride template for the noncoordinating hexafluorophosphate anion allowed us, for the first time, to compare the anion binding properties of the free and interlocked components. Satisfyingly, the [2]rotaxane $17^+PF_6^-$ binds anions strongly in very competitive protic solvent mixture such as CH_3OH : $CDCl_3 = 1:1$ with remarkable reversal of selectivity with respect to the pyridinium thread (the macrocycle itself does not bind anions in this solvent system). Thus, whereas the thread binds anions according to their hydrogen bond accepting ability $[K_{(Cl^-)} = 125\ M^{-1}, K_{(H_2PO_4^-)} = 260, K_{11(AcO^-)} = 22\,000\ M^{-1}, K_{12(AcO^-)} = 140\ M^{-1}]$, the rotaxane has a notable preference for chloride ($K = 1130\ M^{-1}$) over dihydrogenphosphate ($K = 300\ M^{-1}$) and acetate ($K_{11} = 100\ M^{-1}, K_{12} = 40\ M^{-1}$). This is postulated to be the result of a unique hydrogen bonding pocket of the rotaxane, formed by orthogonal clefts of the thread and macrocycle and its high degree of complementarity to the guest chloride anion. The complexation of larger anions would result in the significant, unfavorable distortion of the binding cavity thus reducing complex stability.

18⁺Cl⁻, Y=I, (43%)
19⁺Cl⁻, Y=NO₂, (60%)

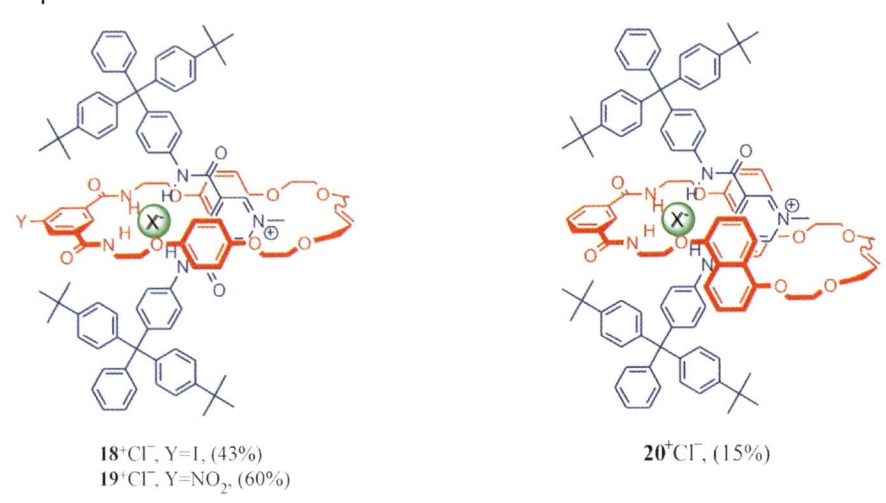

20⁺Cl⁻, (15%)

Figure 3.14 Anion-templated [2]rotaxanes. The percentages in parentheses correspond to the yields of the [2]rotaxane formation step from the acyclic precursors.

This synthetic route to [2]rotaxanes has been used to generate a number of species similar to **17** (Figure 3.14) [30]. Again, the process is highly dependent on the nature of the templating anion and, interestingly, on the anion binding properties of macrocycle. The rotaxane yield was increased to 60% for the nitro-substituted rotaxane **19**, as a result of the precursor orthogonal complex assembly being more stable owing to increased amide hydrogen bond acidity. The structural tolerance of the method towards changes in the macrocycle structure was demonstrated by the synthesis of naphthyl-containing [2]rotaxane **20**. The yield was lower in this case (15%), perhaps due to the increase in macrocycle size and flexibility. Importantly, the interlocked nature of these systems could withstand the removal of the chloride anion template, which allowed the study of their anion binding properties. The introduction of the electron withdrawing nitro or iodo functionalities (**19**⁺PF₆⁻ and **18**⁺PF₆⁻) leads to an enhancement in anion binding, which was mirrored in the corresponding macrocycles. These results indicate that it is possible to fine tune the binding properties of the interlocked species by small structural changes of their components.

3.7
Anion-templated Synthesis of Catenanes

In a major development of this methodology, the first example of the use of anion templation in the synthesis of catenanes was demonstrated [31]. The strategy employed is shown in Figure 3.15. A chloride anion, as a part of tight ion pair, promotes the initial formation of a [2]pseudorotaxane and a subsequent clipping reaction afforded the [2]catenane structure.

Figure 3.15 Formation of anion-templated catenanes from [2]pseudorotaxane precursor. General scheme (top left), structure of thread precursor **21** (top right) and formation of [2]-catenane **22** and [3]-catenane **23** (bottom). Yields for the catenane formation process are given in parentheses.

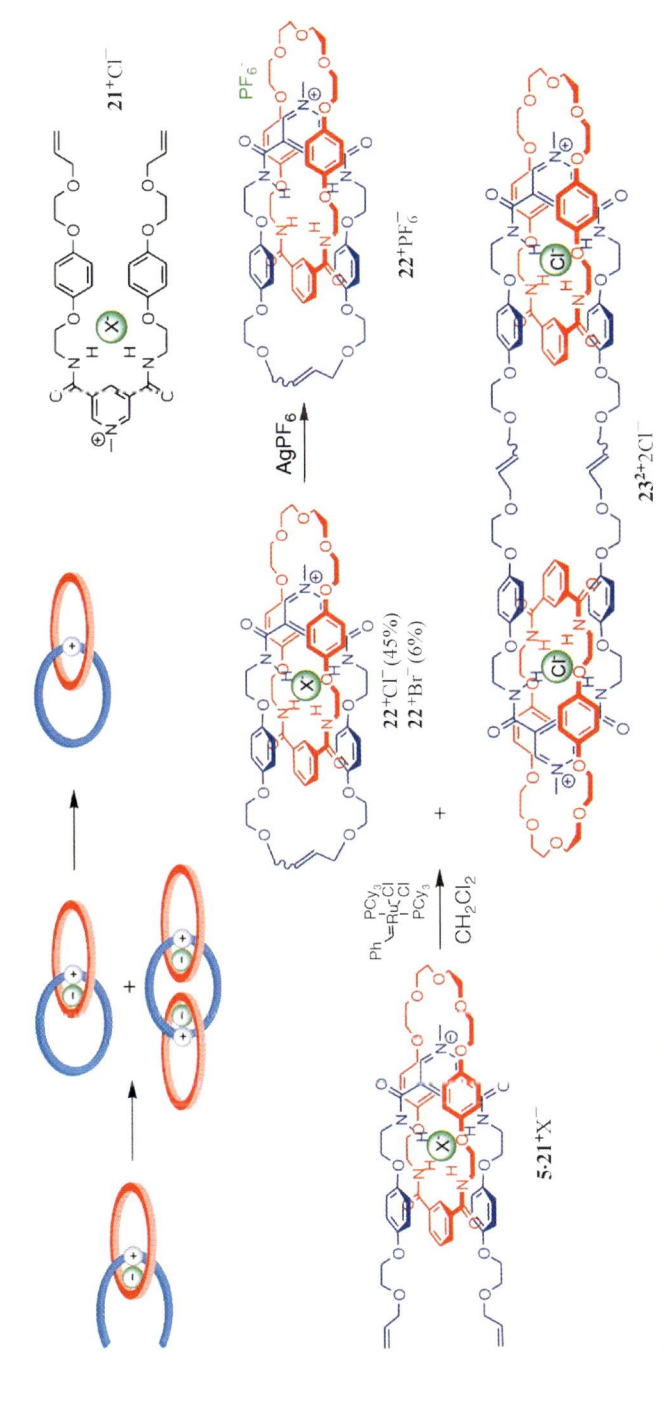

Mixing macrocycle **5** and pyridinium chloride allyl functionalized derivative 21^+Cl^- in dichloromethane followed by addition of Grubbs' catalyst afforded the [2]catanene 22^+Cl^- in 45% yield and a [3]catenane, $23^{2+}2Cl^-$, in <5% yield (Figure 3.15). It is noteworthy that analogous RCM reactions of macrocycle **5** with the corresponding bromide pyridinium component gave the desired [2]catenane 22^+Br^- in only 6% yield and no catenanes were isolated from RCM reactions of the macrocycle with iodide or hexafluorophosphate pyridinium derivatives. As with the [2]rotaxane synthesis discussed previously, this again highlights the crucial role the chloride ion template plays whereby threading of the pyridinium cation 21^+ is driven by recognition of its chloride counterion by the macrocycle.

The interlocked nature of the product was established both from NMR spectroscopic experiments and single-crystal X-ray analysis (Figure 3.16). It was clear from this structure that chloride is essential in controlling the orientation of the two interpenetrated components and that the secondary π-stacking and hydrogen bonding interactions involving the pyridinium function are present. The anion is held within the catenane cavity by six hydrogen bonds, four with amide groups and two with aromatic protons located between amide arms. The pyridinium cation is sandwiched between two hydroquinone rings, whereas the isophthalamide moiety forms a less perfect stacking with just one hydroquinone moiety.

The removal of the chloride anion template was achieved by addition of silver hexafluorophosphate to produce the [2]catenane$^+$ PF_6^- salt. Quantitative ^1H NMR binding studies in methanol-d_4:chloroform-d_3 1 : 1 mixture revealed that the pyridinium macrocyclic precursor $21^+PF_6^-$ displays a strong affinity for acetate and dihydrogen phosphate and only binds chloride weakly, whereas the catenane $22^+PF_6^-$ exhibits a reverse binding trend: $Cl^- > H_2PO_4^- > AcO^-$. Table 3.1 shows that chloride anion binding is significantly enhanced upon catenane formation, whereas the binding of the oxoanions is weakened. In a similar fashion to the [2]rotaxane binding studies discussed previously, the removal of the templating anion creates a unique topologically defined hydrogen bond donating pocket which is highly selective for chloride anions.

Furthermore, the assembly process is tolerant of major changes in the neutral macrocyclic components, with for example the [2]catenane 24^+Cl^- being formed in 29% yield from reaction of calix[4]arene macrocycle **13** and 21^+Cl^- (Figure 3.17) [17].

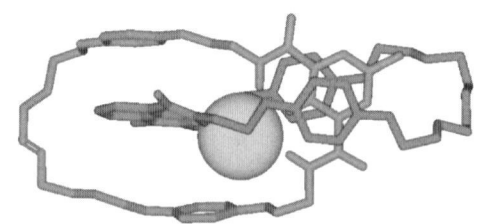

Figure 3.16 Single-crystal X-ray structure of 22^+Cl^-. All hydrogens except those in the primary anion coordination sphere have been omitted for clarity. Chloride is represented as a space-filling sphere.

Table 3.1 Comparison of anion binding properties of macrocycle precursor $21^+PF_6^-$ and catenane $22^+PF_6^-$. Units M^{-1}, solvent 1:1 $CD_3OD-CDCl_3$, errors <10%.

	Cl^-	$H_2PO_4^-$	AcO^-
$21^+PF_6^-$	$K_{11} = 230$	$K_{11} = 1360$ $K_{12} = 370$	$K_{11} = 1500$ $K_{12} = 345$
$22^+PF_6^-$	$K_{11} = 730$	$K_{11} = 480$ $K_{12} = 520$	$K_{11} = 230$

As before, this process was reliant on the presence of a suitable anion template and exchange of this template to give the hexafluorophosphate salt $24^+PF_6^-$ also proved possible.

As mentioned in the Introduction, the development of our anion templation methodology was inspired by Sauvage's elegant synthesis of a [2]catenane by two simultaneous macrocyclizations performed on an orthogonal precursor complex assembled around a copper(I) cation. Thus, the ultimate challenge for our newly developed methodology was to emulate this "double clipping" synthesis. This has recently been achieved with the assembly of two identical acyclic pyridinium precursors around a single chloride anion template followed by a double ring closing metathesis reaction, which yielded the doubly charged catenane **26** in excellent yield (Figure 3.18) [32].

In this experiment, a pyridinium-based macrocyclic precursor 25^+X^-, featuring extended polyether chains, was used. Addition of TBA^+Cl^- to $25^+PF_6^-$ in $CDCl_3$ led to the observation by ^1H NMR spectroscopy of the coexistence of 1:1 and 1:2 host–guest complexes, indicating the presence of an interwoven assembled species. Further evidence of the orthogonal assembly was provided by the upfield shifts of the hydroquinone protons, due to the familiar effect of favorable π–π stacking interactions. Mixing an equimolar solution of 25^+Cl^- and $25^+PF_6^-$ in dichloromethane, followed by double macrocyclization using Grubbs' catalyst, afforded the dicationic [2]catenane **26** in the exceptionally high yield of 78% (Figure 3.18). When the hexafluorophosphate salt $25^+PF_6^-$ was so treated, the [2]catenane product $26^{2+}2PF_6^-$ also formed, but in much lower yield (16%). This demonstrates the role of secondary interactions which, in the absence of chloride template and with

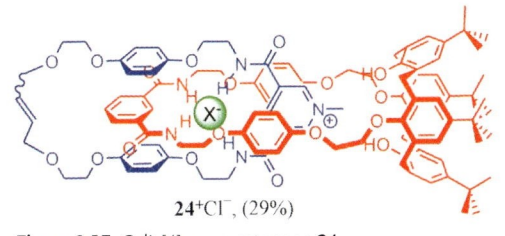

Figure 3.17 Calix[4]arene catenane **24**.

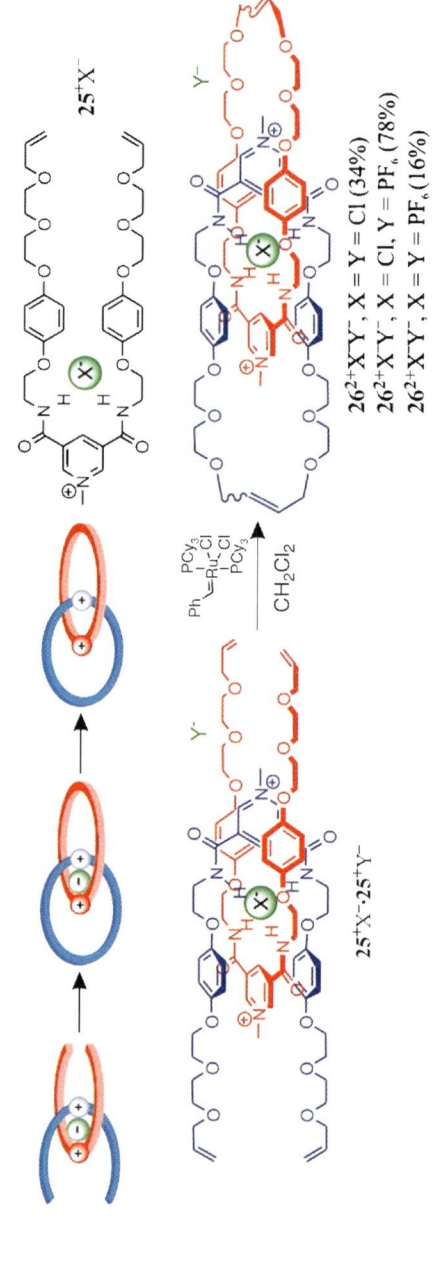

Figure 3.18 "Double clipping" route to [2]catenane. General schematic (top left), precursor molecule **25** (top right) and formation of [2]catenanes **26** (bottom).

$26^{2+}XY^-$, X = Y = Cl (34%)
$26^{2+}XY^-$, X = Cl, Y = PF$_6$ (78%)
$26^{2+}XY^-$, X = Y = PF$_6$ (16%)

possible aid from weak hexafluorophosphate binding, direct the assembly of the orthogonal complex. Of course, the presence of a templating chloride anion substantially enhances the efficacy of the reaction. The 2 : 1 receptor : template ratio is essential; when the chloride precursor 25^+Cl^- was treated with Grubbs' catalyst, the yield of catenane product was lower (34%) due to competition from 1 : 1 binding mode, which favors simple macrocyclization.

Single-crystal X-ray structural analysis of $26^{2+}Cl^-PF_6^-$ confirms the interlocked nature of the product and reveals its self-complementary structure – the two macrocycles bind each other more efficiently than in the previous unsymmetrical catenane due to the presence of a second, positively charged, pyridinium ring instead of neutral isophthalamide moiety. This permits stronger π-stacking interactions, additional hydrogen bonding with ether oxygens and also strengthens anion binding. These effects explain the extraordinarily high yield obtained in this synthesis (even without optimization).

There is an excellent match between the guest chloride anion and the host cavity. The anion is coordinated by six hydrogen bonds in a distorted octahedral manner and, as shown by space-filling model, almost completely surrounded by the catenane molecule (Figure 3.19).

Anion exchange with $AgPF_6$ gave $26^{2+}2PF_6^-$ whose anion binding properties were investigated by ^1H NMR spectroscopic titration in $CDCl_3$–acetone-d_6 (1 : 1). Analysis of the titration data obtained upon the addition of chloride, bromide and acetate produced association constants with a major 1 : 1 host:guest binding stoichiometry and a minor 1 : 2 binding component. The association constants reveal a remarkable selectivity for chloride ($K_{11} = 9240\,M^{-1}$, $K_{12} = 160\,M^{-1}$) over bromide ($K_{11} = 790\,M^{-1}$, $K_{12} = 40\,M^{-1}$) and acetate ($K_{11} = 420\,M^{-1}$, $K_{12} = 40\,M^{-1}$).

It is noteworthy that this "double clipping" catenane synthesis is much shorter than the previous one, owing to the symmetrical structure of the product; the two wheels in previous catenane, although very similar, required independent synthesis. More generally, it is a great advantage of catenanes over rotaxanes that the sometimes laborious synthesis of stoppers and threads is no longer required. This advantage grows in importance with the increasing size of wheels, which requires larger and larger stoppers.

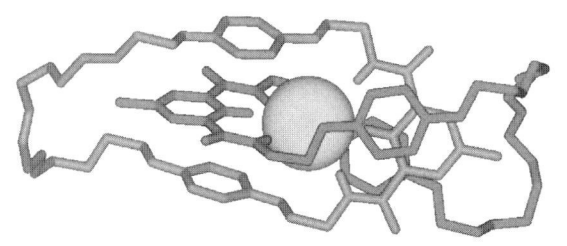

Figure 3.19 Single-crystal X-ray structure of $26^+Cl^-PF_6^-$. All hydrogens except those in the primary anion coordination sphere have been omitted for clarity. Chloride is represented as a space-filling sphere.

3.8
Functional Properties of Anion-templated Interlocked Systems

Thus far it has been shown that the unique interlocked binding domain topology lends anion-templated rotaxanes and catenanes interesting anion binding characteristics distinct from their "parent" species. The observed increase in binding strength coupled with reversal of selectivity makes these interlocked structures of great interest in the molecular sensing arena. In order to apply these receptors to anion sensing, it is necessary to provide some means of signal transduction and amplification. The above-described advances in templation methodology have made the synthesis of such sophisticated structures, equipped with reporter groups able to signal the recognition event by electrochemical or spectroscopic means, possible. Another challenge on the way to practical applications is the attachment of molecular sensors to a solid surface, which is a prerequisite for robust device fabrication. Recent progress towards these goals is outlined below.

The first photo-active anion-sensing rotaxane was based on a luminescent rhenium(I) bipyridyl motif being incorporated into macrocyclic wheel (Figure 3.20), as in the previously described pseudorotaxane $15 \cdot 2^+Cl^-$ [33]. The chloride derivative 27^+Cl^- was prepared in 21% yield via the now established ring clipping of the neutral rhenium(I) bipyridyl-containing precursor **29** around the pyridinium chloride thread 28^+Cl^-; the bulky calix[4]arene stopper groups were necessary to prevent dethreading of the larger macrocycle. As before, replacement of the chloride template with hexafluorophosphate gave a [2]rotaxane $27^+PF_6^-$ which contained not only a three-dimensional anion binding domain but also a luminescent transition metal bipyridyl center able to sense optically the anion binding event. Thus the addition of TBA anion salts to a solution of $27^+PF_6^-$ in acetone induced an enhancement in the ^{3}MLCT emission band intensity of the rotaxane receptor. Titration experiments in acetone demonstrated that the rotaxane selectively bound hydrogensulfate ($K_a > 10^6$ M^{-1}) over nitrate and chloride, which contrasts with the properties of the macrocycle **30**, which was selective for chloride ($K_a = 8.7 \times 10^4$ M^{-1}). Receptors selective for hydrogensulfate are rare, because hydrogensulfate anion is a poor hydrogen bond acceptor. Importantly, an anion templation approach may therefore be used to synthesize molecular sensors for anions different from template.

The above rotaxane sensor utilizes a common sensing mechanism based on electronic communication between anion and reporter group. However, rotaxane- and catenane-based receptors offer some potential means of signal transduction that are unique to interlocked structures, based on the mutual relationships between the mechanically bound subunits. For example, anion binding may amplify/reduce the interactions between thread and macrocycle and, as a consequence, alter spectroscopic or electrochemical properties of the rotaxane. Alternatively, anions may induce co-conformational change, such as shuttling of the macrocycle along the thread, which could also translate into an observable signal. Although basic mechanisms underlying the signal generation in the above examples are well developed and routinely used to study, for example, molecular switches or machine-like behavior of interlocked molecules, their application to molecular

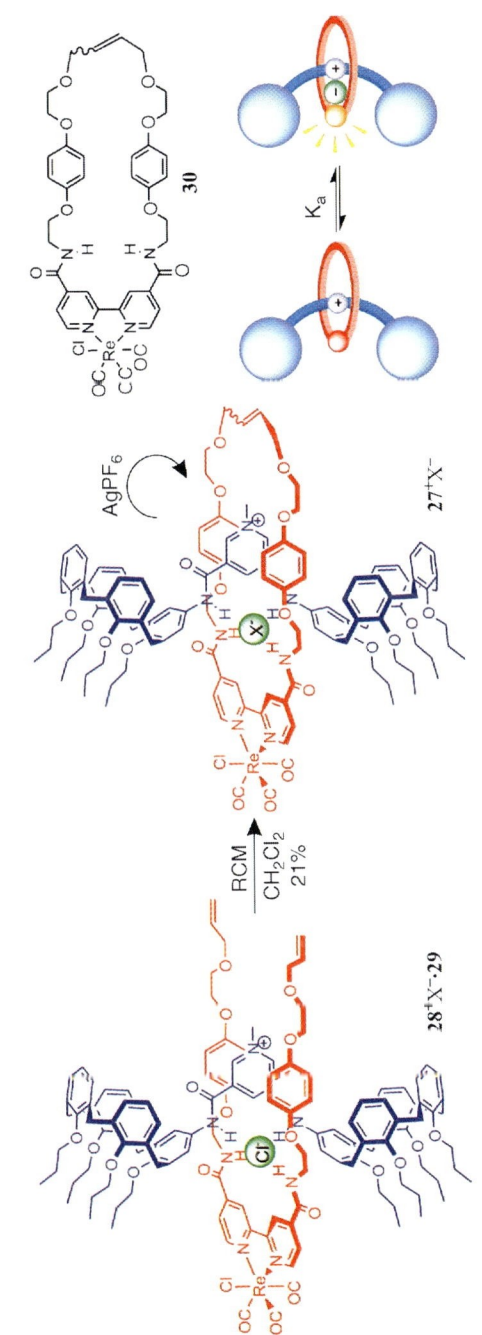

Figure 3.20 [2]Rotaxane anion sensor. Formation of luminescent rotaxanes 27^+Cl^- and $27^+PF_6^-$ from the precursors 28^+Cl^- and 29 by anion-templated Grubbs' ring closing metathesis (left), luminescent macrocycle 30 (top right) and schematic for rotaxane anion sensing (bottom right).

sensing is underexplored, due to the lack of guest binding cavities in previously described catenanes and rotaxanes.

As a prototype sensing system illustrating this paradigm, we set about constructing a pseudorotaxane system with a through-space communication between thread and macrocycle components, which may be influenced by anion binding. The mechanism used to accomplish this was photoinduced energy transfer between a rhenium (I) bipyridyl sensitizer incorporated within the macrocycle **31** and a luminescent lanthanide complex appended to one terminus of the benzimidazolium threads **32** (Figure 3.21) [34]. Addition of threads containing no lanthanide center, or a lanthanide center not suitable for energy transfer such as gadolinium (**32a**), to the rhenium macrocycle **31** resulted in an enhancement of ^{3}MLCT luminescent emission, as observed in **15** and **27**$^+$PF$_6^-$ above. For threads containing a suitable lanthanide metal such as neodymium or ytterbium, however, no such enhancement was observed on pseudorotaxane formation; indeed, for the neodymium thread **32c** a significant quenching of the rhenium-centered luminescence was observed. Furthermore, the evolution of new near-infrared (NIR) emission bands consistent with lanthanide metal emission could be detected. This observation is consistent with the proposed energy transfer between the ^{3}MLCT excited state of the rhenium(I) bipyridyl center and the lanthanide complex. As such an energy transfer process is highly dependent on the distance between the two metal centers, the appearance of NIR luminescence indicates the proximity between the stopper and macrocycle and hence pseudorotaxane formation, which is in turn anion dependent. Thus the same principle may be used for anion sensing or, for example, to monitor anion-induced shuttling of the macrocycle along the thread in a prototype molecular machine-like device based on an anion recognition process.

The confinement of interlocked anion receptors at electrode surfaces should allow the harnessing of their specific binding behavior in electrochemical sensing materials. This possibility was probed by the formation of self-assembled monolayers (SAMs) of redox-active bis-ferrocene functionalized pseudorotaxane **33·34**$^+$Cl$^-$ at a gold surface; the transformation results in a rotaxane with the gold electrode effectively acting as a stopper (Figure 3.22) [35]. The presence of two different redox-active centers on the thread and on the macrocycle allowed for independent monitoring of their presence on the electrode surface. The replacement of the chloride template with hexafluorophosphate proved possible without disrupting the interlocked nature of the surface assembled rotaxanes. The anion binding properties of this redox-active rotaxane-SAM could be probed using electrochemical methods, examining the perturbations in the two redox waves of the system on the addition of various analytes; proximal anion binding should be accompanied by a cathodic shift due to electrostatic stabilization of the oxidized ferrocene unit. In acetonitrile solutions the ferrocene unit of the rotaxane macrocycle was shown to demonstrate a selective voltammetric response to chloride ($\Delta E \approx 40$ mV), even in the presence of a 100-fold excess of competing anion such as dihydrogenphosphate. This contrasts sharply with the solution responses of the free thread **34**$^+$PF$_6^-$ and macrocycle **33**, which demonstrate small cathodic shifts in the presence of halides and basic oxyanions, except for **33** with

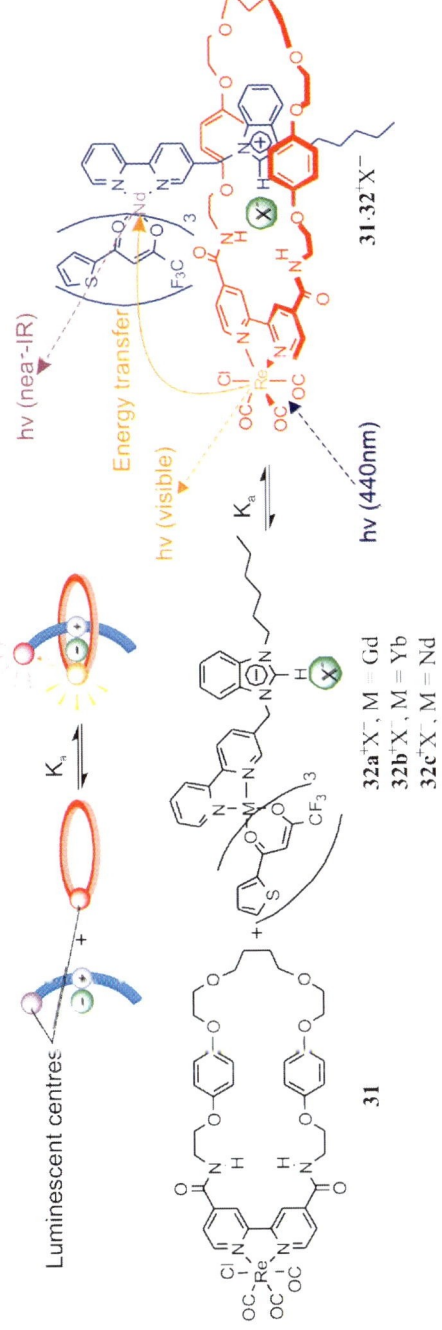

3.8 Functional Properties of Anion-templated Interlocked Systems | 91

Figure 3.21 Sensing of [2]pseudorotaxane formation through Förster energy transfer. General schematic (top), luminescent macrocycle and thread components (bottom) and pseudorotaxane **31·32⁺X⁻** (right).

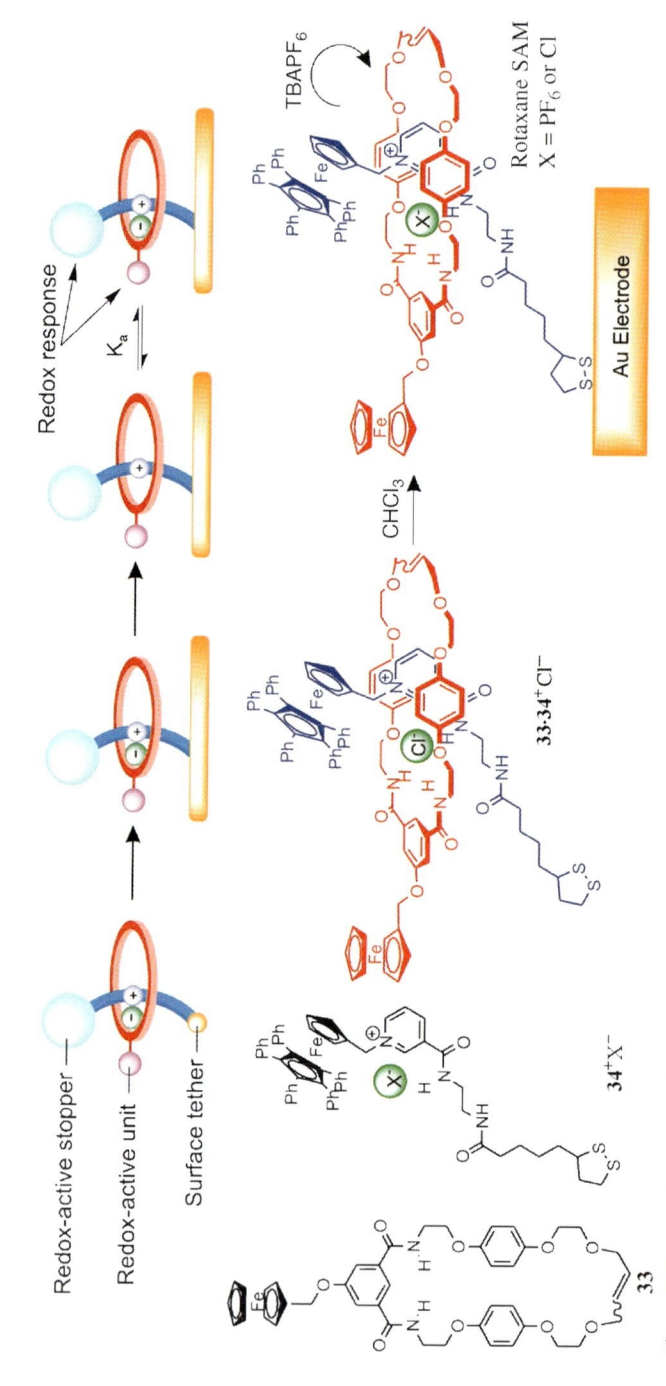

92 | *3 Strategic Anion Templation for the Assembly of Interlocked Structures*

Figure 3.22 Anion-templated SAM rotaxanes for chloride sensing. General schematic (top) and rotaxane SAM components and formation (bottom).

dihydrogenphosphate ($\Delta E \approx 45$ mV) and hydrogensulfate ($\Delta E \approx 15$ mV). This therefore provides another example of the change in selectivity induced by the mutual interpenetration of two components and further demonstrates the ability of this templation methodology to give sophisticated surface-confined species with promising selective electrochemical recognition properties.

To date, the general anion templation synthetic strategy described in this chapter has been exploited in two main areas. The first concerns the removal of the anion template from permanently interlocked molecules, which gives anion receptors demonstrating binding properties dependent upon the presence of a unique three-dimensional hydrogen bond-donating pocket. By including optical and electrochemical readout functionalities and by attaching these derivatives to surfaces, such anion-templated structures have begun to be used for sensory purposes. Second, anion-templated molecular motion in the form of threading has been signaled by luminescence spectroscopic means. This opens the door for molecular machine like devices based on anion recognition processes.

3.9
Summary and Outlook

A comprehensive anionic alternative to existing cation interweaving templation approaches has been developed and exploited in the synthesis of numerous interpenetrated systems. These range from reversible assemblies such as orthogonal complexes and pseudorotaxanes to permanently interlocked compounds such as rotaxanes and catenanes. Removal of the anion template from these permanently interlocked systems leads to novel receptor and sensory behaviors defined by the mutual interpenetration of the two components. These observations underline the general applicability and scope of this novel templation strategy.

This anion templation methodology is, however, currently very much in its infancy; still only a handful of interlocked structures have been obtained and even some threads that have been shown to form pseudorotaxanes are awaiting to be exploited in the construction of permanently interlocked systems. The spectacular progress described in this chapter has been made possible with the exclusive use of halide anions as templates, therefore the application of the whole spectrum of other, more strongly coordinating and structurally complex anions may be expected to open up even more exciting avenues in the field and address current limitations of the methodology. For example, the extension of the currently available threading motifs to uncharged systems would significantly increase its scope.

As shown recently, the application of reversible palladium ligation as the ring-closing reaction in Sauvage's copper-templated catenane synthesis increased the yield from below 30% to quantitative [36]. Certainly, anion-templated synthesis would also benefit from such thermodynamic control. The introduction of reversible reactions compatible with anion supramolecular chemistry may bring a step change in the field.

Further progress in anion-templated rotaxane synthesis is expected with the development of stoppering methods compatible with anion recognition and with broadening of the selection of readily available stoppers of sufficiently large size and various functional groups. Most desirable in this context are functional stoppers such as the electroactive pentaphenylferrocene group from **34**, or photoactive lanthanide complexes from **32**, granting additional degrees of functionality to the resulting rotaxane.

The goal is ultimately to create increasingly sophisticated device-like structures to complement the impressive array of molecular machines already furnished [37] through other templation and self-assembly strategies. We also continue to be fascinated by the potential for these anion-templated architectures to be used for highly selective sensory devices for ionic substrates and we aim to continue the development of such systems. As with the emergence of cationic interweaving templation over two decades ago, however, a huge catalogue of anion-templated structures has yet to be exploited.

References

1 (a) Atwood, J.L., Holman, K.T. and Steed, J.W. (1996) Laying traps for elusive prey: recent advances in the non-covalent binding of anions. *Chem. Commun.*, 1401–1407. (b) for a general introduction, see Sessler, J.L., Gale, P.A. and Cho, W.-S. (2006) *Anion Receptor Chemistry* Royal Society of Chemistry (c) Bianchi, A., Bowman-James, K. and García-España, E. (eds) (1997) *Supramolecular Chemistry of Anions*, Wiley-VCH, New York.

2 Vilar, R. (2003) Anion-templated synthesis. *Angew. Chem. Int. Ed.*, **42**, 1460–1477.

3 Previous accounts: (a) Vickers, M.S. and Beer, P.D. (2007) Anion templated assembly of mechanically interlocked structures. *Chem. Soc. Rev.*, **36**, 211–225. (b) Beer, P.D., Sambrook, M.R. and Curiel, D. (2006) Anion templated assembly of interpenetrated and interlocked structures. *Chem. Commun.*, 2105–2117. (c) Lankshear, M.D. and Beer, P.D. (2006) Strategic anion templation. *Coord Chem. Rev.*, **250**, 3142–3160.

4 Andrievsky, A., Ahuis, F., Sessler, J.L., Vögtle, F., Gudat, D. and Moini, M. (1998) Bipyrrole-based[2]catenane: a new type of anion receptor. *J. Am. Chem. Soc.*, **120**, 9712–9713.

5 For example, Chmielewski, M.J. and Jurczak, J. (2005) Anion recognition by neutral macrocyclic amides *Chem. Eur. J.*, **11**, 6080–6094.

6 (a) Kang, S.O., Llinares, J.M., Powell, D., VanderVelde, D. and Bowman-James, K. (2003) New polyamide cryptand for anion binding. *J. Am. Chem. Soc.*, **125**, 10152–10153. (b) Bisson, A.P., Lynch, V.M., Monahan, M.C. and Anslyn, E.V. (1997) Recognition of anions through NH-π hydrogen bonds in a bicyclic cyclophane – selectivity for nitrate. *Angew. Chem. Int. Ed. Engl.*, **36**, 2340–2342. (c) Schmidtchen, F.P. (1977) Einschluss von Anionen in makrotricyclische quartare Ammoniumsalze. *Angew. Chem.*, **89**, 751–752. (d) Graf, E. and Lehn, J.M. (1976) Anion cryptates: highly stable and selective macrotricyclic anion inclusion complexes. *J. Am. Chem. Soc.*, **98**, 6403–6405. (e) Park, C.H. and Simmons, H.E. (1968) *J. Am. Chem. Soc.*, **90**, 2431–2432.

7 Sasaki, S., Mizuno, M., Naemura, K. and Tobe, Y. (2000) Synthesis and anion-selective complexation of cyclophane-based cyclic thioureas. *J. Org. Chem.*, **65**, 275–283.

8 Anderson, S., Anderson, H.L. and Sanders, J.K.M. (1993) Expanding roles for templates in synthesis. *Acc. Chem. Res.*, **26**, 469–475.

9 Sauvage, J.-P. (1990) Interlacing molecular threads on transition metals: catenands, catenates and knots. *Acc. Chem. Res.*, **23**, 319–327.

10 Kohnke, F.H., Mathias, J.P. and Stoddart, J. (1989) Structure-directed synthesis of new organic materials. *Angew Chem. Int. Ed. Engl. Adv. Mater.*, **28**, 1103–1110.

11 Ashton, P.R., Goodnow, T.T., Kaifer, A.E., Reddington, M.V., Slawin, A.M.Z., Spencer, N., Stoddart, J.F., Vicent, Ch. and Williams, D.J. (1989) A [2]catenane made to order. *Angew. Chem. Int. Ed.*, **28**, 1396–1399.

12 Hübner, G.M., Gläser, J., Seel, Ch. and Vögtle, F. (1999) High-yielding rotaxane synthesis with an anion template. *Angew. Chem. Int. Ed.*, **38**, 383–386.

13 Seel, Ch. and Vögtle, F. (2000) Templates, "wheeled reagents" and a new route to rotaxanes by anion complexation: the trapping method. *Chem. Eur. J.*, **6**, 21–24.

14 Shukla, R., Deetz, M.J. and Smith, B.D. (2000) [2]Rotaxane with a cation binding wheel. *Chem. Commun.*, 2397–2398.

15 Mahoney, J.M., Shukla, R., Marshall, R.A., Beatty, A.M., Zajicek, J. and Smith, B. (2002) Template conversion of a crown ether-containing macrobicycle into [2]rotaxanes. *J. Org. Chem.*, **67**, 1436–1440.

16 Deetz, M.J., Shukla, R. and Smith, B.D. (2002) Recognition-directed assembly of salt-binding [2]rotaxanes, *Tetrahedron*, **58**, 799–805.

17 Schalley, Ch.A., Silva, G., Nising, C.F. and Linnartz, P. (2002) Analysis and improvement of an anion-templated rotaxane synthesis. *Helv. Chim. Acta*, **85**, 1578–1596.

18 Ghosh, P., Mermagen, O. and Schalley, Ch.A. (2002) Novel template effect for the preparation of [2]rotaxanes with functionalised centre pieces. *Chem. Commun.*, 2628–2629.

19 Arunkumar, E., Forbes, Ch.C., Noll, B.C. and Smith, B.D. (2005) Squaraine-derived rotaxanes: sterically protected fluorescent near-IR dyes. *J. Am. Chem. Soc.*, **127**, 3288–3289.

20 Mohr, B., Weck, M., Sauvage, J.-P. and Grubbs, R.H. (1997) High-yield synthesis of [2]catenanes by intramolecular ring-closing metathesis. *Angew. Chem. Int. Ed.*, **36**, 1308–1310.

21 Kavallieratos, K., Bertao, C.M. and Crabtree, R.H. (1999) Hydrogen bonding in anion recognition: a family of versatile, nonpreorganized neutral and acyclic receptors. *J. Org. Chem.*, **64**, 1675–1683.

22 Wisner, J.A., Beer, P.D. and Drew, M.G.B. (2001) A demonstration of anion templation and selectivity in pseudorotaxane formation. *Angew. Chem. Int. Ed.*, **40**, 3606–3609.

23 Wisner, J.A., Beer, P.D., Drew, M.G.B. and Sambrook, M.R. (2002) Anion-templated rotaxane formation. *J. Am. Chem. Soc.*, **124**, 12469–12476.

24 Sambrook, M.R., Beer, P.D., Wisner, J.A., Paul, R.L., Cowley, A.R., Szemes, F. and Drew, M.G.B. (2005) Anion-templated assembly of pseudorotaxanes: importance of anion template, strength of ion-pair thread association and macrocycle ring size. *J. Am. Chem. Soc.*, **127**, 2292–2302.

25 Wisner, J.A., Beer, P.D., Berry, N.G. and Tomapatanaget, B. (2002) Anion recognition as a method for templating pseudorotaxane formation. *Proc. Natl. Acad. Sci. USA*, **99**, 4983–4986.

26 Lankshear, M.D., Evans, N.H., Bayly, S.R. and Beer, P.D. (2007) Anion-templated calix[4]arene-based pseudorotaxanes and catenanes. *Chem. Eur. J.*, **13**, 3861–3870.

27 Curiel, D., Beer, P.D., Paul, R.L., Cowley, A.R., Sambrook, M.R. and Szemes, F. (2004) Halide anion directed assembly of luminescent pseudorotaxanes. *Chem. Commun.*, 1162–1163.

28 Beer, P.D., Timoshenko, V., Maestri, M., Passaniti, P. and Balzani, V. (1999) Anion recognition and luminescent sensing by new ruthenium(II) and rhenium(I) bipyridyl calyx[4]diquinone receptors. *Chem. Commun.*, 1755–1756.

29 Grubbs, R.H., Miller, S.J. and Fu, G.C. (1995) Ring-closing metathesis and related processes in organic synthesis. *Acc. Chem. Res.*, **28**, 446–452.

30 Sambrook, M.R., Beer, P.D., Lankshear, M.D., Ludlow, R.F. and Wisner, J.A. (2006) Anion-templated assembly of [2] rotaxanes. *Org. Biol. Chem.*, **4**, 1529–1538.

31 Sambrook, M.R., Beer, P.D., Wisner, J.A., Paul, R.L. and Cowley, A.R. (2004) Anion-templated assembly of a [2]catenane. *J. Am. Chem. Soc.*, **126**, 15364–15365.

32 Ng, K.-Y., Cowley, A.R. and Beer, P.D. (2006) Anion templated double cyclization assembly of a chloride selective [2]catenane. *Chem. Commun.*, 3676–3678.

33 Curiel, D. and Beer, P.D. (2005) Anion directed synthesis of a hydrogensulfate selective luminescent rotaxane. *Chem. Commun.*, 1909–1911.

34 Sambrook, M.R., Curiel, D., Hayes, E.J., Beer, P.D., Pope, S.J. and Faulkner, S. (2006) Sensitized near infrared emission from lanthanides via anion-templated assembly of d-f heteronuclear [2] pseudorotaxanes. *New J. Chem.*, **30**, 1133–1136.

35 Bayly, S.R., Gray, T.M., Chmielewski, M.J., Davis, J.J. and Beer, P.D. (2007) Anion templated surface assembly of a redox-active sensory rotaxane. *Chem. Commun.*, **2234–2236**.

36 Dietrich-Buchecker, Ch., Colasson, B., Fujita, M., Hori, A., Geum, N., Sakamoto, S., Yamaguchi, K. and Sauvage, J.-P. (2003) Quantitative formation of [2]catenanes using copper(I) and palladium(II) as templating and assembling centers: the entwining route and the threading approach. *J. Am. Chem. Soc.*, **125**, 5717–5725.

37 (a) Kay, E.R., Leigh, D.A. and Zerbetto, F. (2007) Synthetic molecular motors and mechanical machines. *Angew. Chem. Int. Ed.*, **46**, 72–191. (b) Sauvage, J.-P. and Dietrich-Buchecker, C. (eds) (1999) *Molecular Catenanes, Rotaxanes and Knots: a Journey Through the World of Molecular Topology*, Wiley-VCH, Weinheim.

4
Synthetic Nanotubes from Calixarenes
Dmitry M. Rudkevich[†] *and Voltaire G. Organo*

4.1
Introduction

Synthetic nanotubes represent a novel type of molecular containers. In supramolecular chemistry, they are still overshadowed by more popular cavitands, carcerands and self-assembling capsules [1,2]. They also have not received much attention compared with the similarly shaped carbon nanotubes [3] and biologically relevant ion channels [4]. At the same time, synthetic nanotubes possess unique dimensions and topology and, as a consequence, different and interesting complexation properties. They also offer a variety of applications in chemistry, nanotechnology and medicine. One important feature of nanotubes is the ability to align multiple guest species in one dimension (1D), which is useful for ion and molecular transport, nanowires and information flow. Other potential applications include using nanotubes as reaction vessels and molecular cylinders for separation and storage.

In recent years, a number of general reviews have appeared describing approaches towards the preparation and characterization of organic nanotubes [5]. This chapter will focus on nanotubes that are based on calixarenes. Calixarenes, cyclic oligomers of phenols and aldehydes, play a special role in molecular recognition [6]. They have yielded a great number of excellent receptors for ions and neutral molecules. In particular, calixarenes appear to be useful in the design of molecular containers [1,2]. We will discuss the synthesis of calixarene-based nanotubes and their emerging host–guest properties, including encapsulation, the guest dynamics and exchange and potential applications.

Research on synthetic nanotubes has been inspired, in many ways, by recent successes with ion channels on the one hand and single-walled carbon nanotubes (SWNTs) on the other. As valuable supplements, organic synthesis offers robust and well-defined tubular structures, with a wide variety of sizes. It also helps to control the nanotube length. Through smart molecular design, it is also possible to prepare stable host–guest complexes. All these features are not easy to achieve for ion channels and SWNTs.

Organic Nanostructures. Edited by Jerry L. Atwood and Jonathan W. Steed
Copyright © 2008 WILEY-VCH Verlag GmbH & Co. KGaA, Weinheim
ISBN: 978-3-527-31836-0

Another issue is characterization. The functions of synthetic ion channels are commonly assessed by electrophysiological planar-bilayer voltage-clamp techniques, fluorimetric assays on liposomes and heteronuclear and solid-state NMR spectroscopy. Transmission electron microscopy is used to study the location of molecules inside SWNTs in the solid state. Solution studies with SWNTs are a great challenge because of their poor solubility. As well-defined organic structures, synthetic nanotubes overcome these difficulties. Their complexes can be prepared and handled by standard organic chemistry protocols and studied by conventional organic spectroscopies.

4.2
Early Calixarene Nanotubes

Early calix[4]arene-based nanotubes were reported by Shinkai and coworkers [7]. Taking advantage of the dynamic behavior of metal cation complexes with *1,3-alternate* calix[4]arenes, they connected several such calixarenes to form nanotubes **1**–**4** (Figure 4.1). Conceptually, nanotubes **1**–**4** would allow small metal cations to tunnel through its π-basic interior.

Complexation experiments with $Ag^+CF_3SO_3^-$ revealed the presence of a 1:1 Ag^+ ion complex with calix[4]tube **2**. Analysis of the variable-temperature ^1H NMR spectrum of the complex suggested that the Ag^+ ion is delocalized between two calixarenes.

The authors proposed that the metal cation oscillates between metal-binding sites in calix[4]tubes in two possible modes: intracalixarene metal tunneling and intercalixarene metal hopping (Figure 4.2). This dynamic behavior, however, was not observed in the complexation with calix[4]tube **1**. Instead, a mixture of free tube **1** and the 1:2 **1** Ag^+ complex was found in the ^1H NMR spectra. It was suggested that the *para* substituents used to connect the two calixarenes interfered with the cation–π interactions, thus suppressing the metal tunneling. In this case, the Ag^+ ions were said to be localized at the edges of the tube, interacting with the calixarene rings and the propyloxy-oxygen groups through cation–π and electrostatic $O \cdots Ag^+$ interactions, respectively.

Similarly, there was no evidence of metal tunneling in complexation studies with calix[4]tube **3** and no data were reported for longer tube **4**. A 1:1 mixture of **3** and $Ag^+CF_3SO_3^-$ yielded three different species: free **3**, **3** Ag^+ and **3** $(Ag^+)_2$ in a 1:2:1 ratio. This result implies that Ag^+ is bound to **3** according to simple probability. The lack of metal tunneling was suspected to be the result of several structural features of **3**. These include the *para* substitution of phenyl groups in the calixarene units, the increased distance between two calixarene units relative to other structures and the nonionophoric bridges connecting the calixarene units.

While Shinkai's nanotubes were only 2–4 nm long and of molecular weight up to 1500 Da, it should be possible to prepare much longer structures utilizing the same calixarene precursors. These pioneering studies, initiated in the early 1990s, triggered intense research on calixarene-based tubular (nano)structures for metal ion complexation, tunneling and transport [8–12].

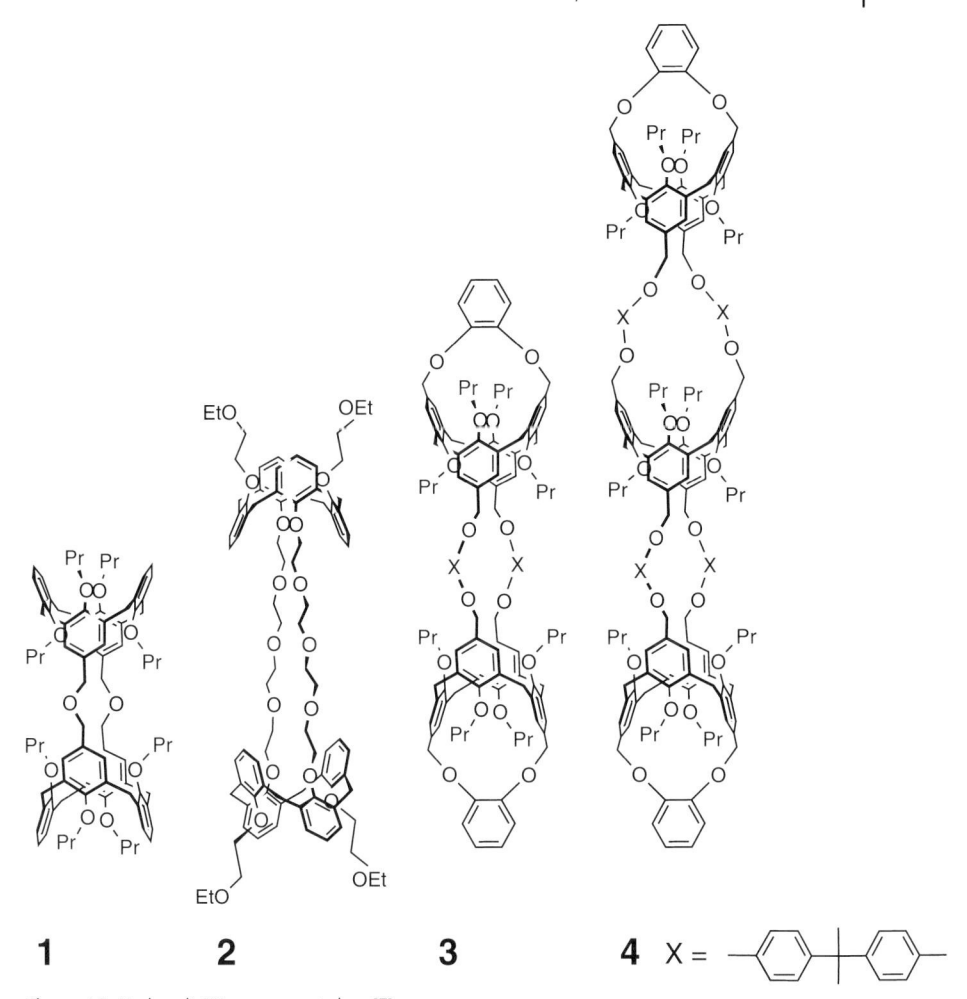

Figure 4.1 Early calix[4]arene nanotubes [7].

4.3
Metal Ion Complexes with Calixarene Nanotubes

Kim and coworkers reported multiply connected *1,3-alternate* calix[4]arene tubes **5** (Figure 4.3), in which the terminal calixarene units were capped with crown ethers [11]. In this design, however, K^+ or Cs^+ cations were bound in tubes **5** ($m = 1, 2$) at the end-calixcrown "stoppers" and metal shuttling was not observed. The X-ray crystal structure of the biscalix[4]crown **5** ($n = 1$, $m = 1$) with K^+ ions revealed that electrostatic interactions between the oxygen donor atoms of the crown ether ring and the metal cation play a major role in entrapping the metal ion whereas the cation–π interaction plays a minor role.

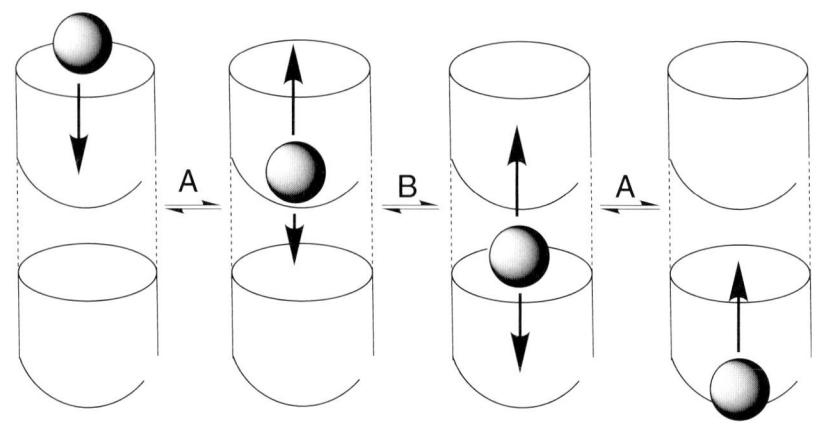

Figure 4.2 Proposed intracalixarene metal tunneling (route A) and intercalixarene metal hopping (route B) [7].

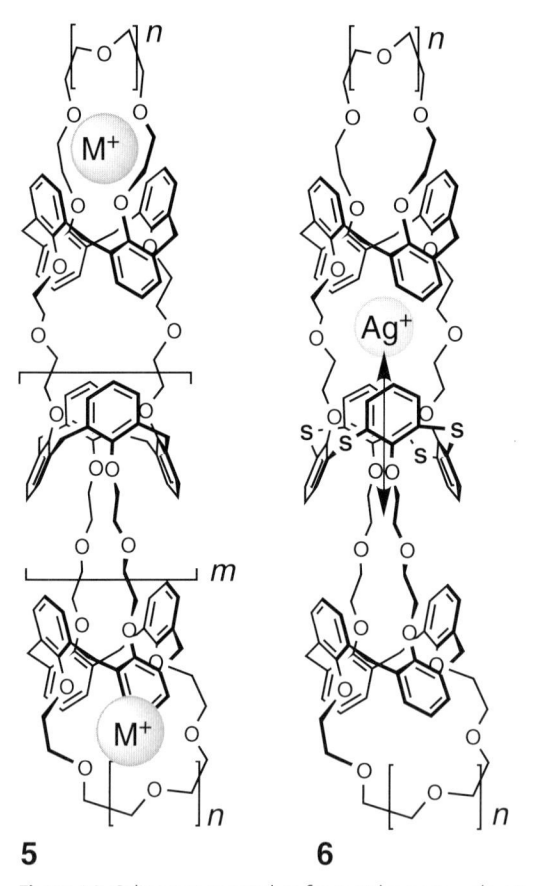

Figure 4.3 Calixcrown nanotubes for metal ions complexation and tunneling [11,12].

The apparent problem with the calixarene nanotubes was the lack of strong cation–π interactions within the interior. Monitoring trapped cationic guests by conventional NMR spectroscopy in these nanotubes was also difficult.

A structural analog of nanotubes **5** was recently prepared which possesses a calixarene unit with higher affinity to metal cations [12]. Nanotubes **6** ($n = 1, 2$) contained thiacalix[4]arene in the middle and the Ag^+ cation was found to be entraped in this central unit in a 1 : 1 fashion (Figure 4.3). In addition to the calixarene aromatic rings, the sulfur atoms provided supplementary coordination sites for transition metal ions. Variable-temperature ^1H NMR spectroscopy revealed that the Ag^+ oscillates through the thiacalixarene and cation–π interactions were important in this case. With some further modifications, it should be possible to synthesize polymeric analogs of tube **6**, inside which Ag^+ ions can freely shuttle [12].

4.4
Nanotubes for NO_x Gases

Synthetic nanotubes have recently been introduced that possess much more pronounced cation–π features. These are based on reversible chemistry between calix[4] arenes and NO_2/N_2O_4 gases [13,14]. NO_2 is paramagnetic and exists in equilibrium with its dimer N_2O_4. N_2O_4 disproportionates to ionic $NO^+NO_3^-$ while interacting with aromatic compounds. It was found that tetrakis-O-alkylated calix[4]arenes, e.g. **7**, react with NO_2/N_2O_4 to form very stable ($K_{assoc} \gg 10^6\,M^{-1}$), charge-transfer calix-nitrosonium (NO^+) complexes **8**. In these, NO^+ cations are strongly encapsulated within the π-electron-rich calix[4]arene tunnel (Figure 4.4). This phenomenon was used in the design of calixarene nanotubes.

In nanotubes **9–12**, *1,3-alternate* calix[4]arenes were rigidly connected from both sides of their rims with pairs of diethylene glycol linkers (Figures 4.5 and 4.6) [15–17]. In this conformation, two pairs of phenolic oxygens are oriented in opposite directions, providing diverse means to enhance modularly the tube length. The nanotubes possess defined inner tunnels 6 Å in diameter and approximately 15, 25, 35 and 45 Å in length for **9, 10, 11** and **12**, respectively. Tubes **11** and **12** have molecular

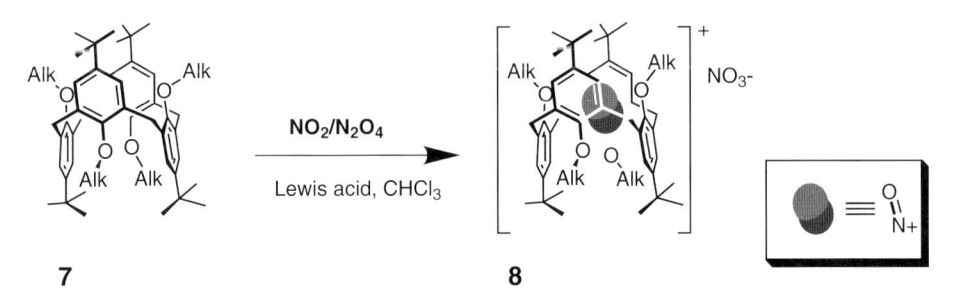

Figure 4.4 Simple calix[4]arenes and their supramolecular interactions with NO_2/N_2O_4 gases; generation of nitrosonium complexes [13].

9   **10**   **11**   **12**

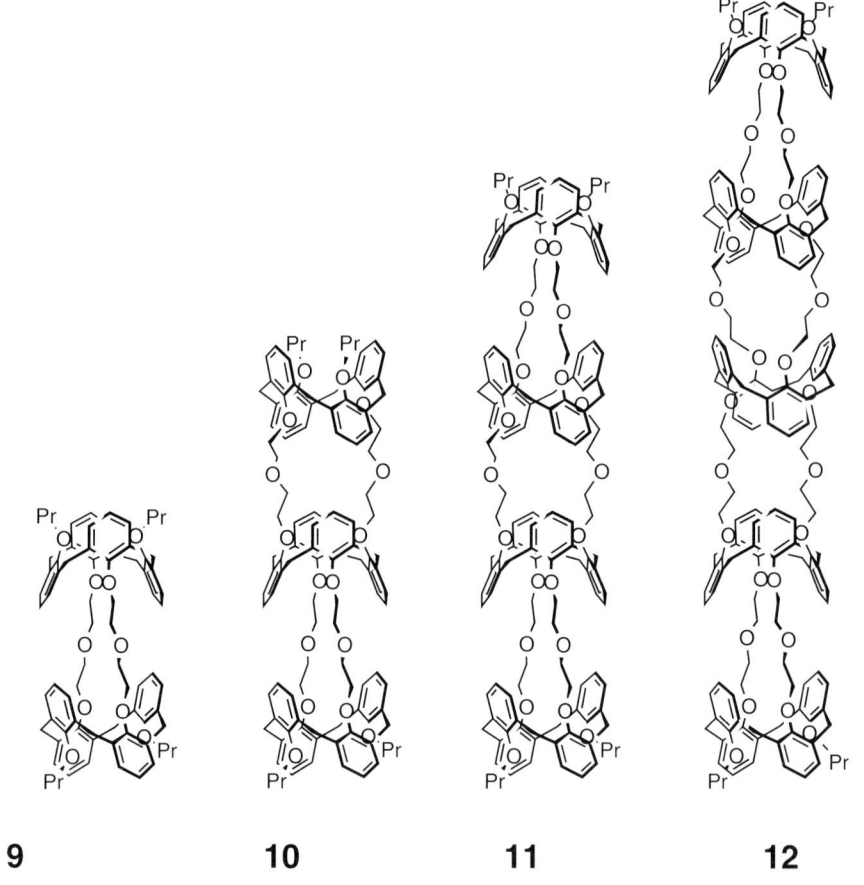

Figure 4.5 Synthetic nanotubes for entrapment of NO_x gases [15–17].

weights of ∼2.3 and 2.8 kDa, respectively. These features place them among the largest nonpolymeric, synthetic molecular containers known to date [1].

The synthesis was based on a straightforward strategy, which incorporated reliable Williamson-type alkylations and provided yields as high as 70–80% (Scheme 4.1). For example, the synthesis of trimeric tube **10** was accomplished in 70% yield by the coupling of tetratosylate **13** with two equivalents of diol **14** in boiling THF with NaH as a base. Tube **11**, which contains four linked calixarenes, was prepared by reaction of biscalixarene diol **15** with ditosylate **16** in 64% yield using NaH and K_2CO_3 in THF. Finally, reaction of two equivalents of diol **16** with tetratosyate **13** under the same conditions afforded pentameric nanotube **12** in a remarkable 82% yield [17].

Addition of excess NO_2/N_2O_4 to nanotubes **9–12** in $(CHCl_2)_2$ in the presence of stabilizing Lewis acids ($SnCl_4$ or $BF_3 \cdot Et_2O$) resulted in quantitative formation of nitrosonium complexes $9 \cdot (NO^+)_2$–$12 \cdot (NO^+)_5$ (Figure 4.7). Similar complexes formed when nanotubes **9–12** were mixed with a commercially available nitrosonium salt,

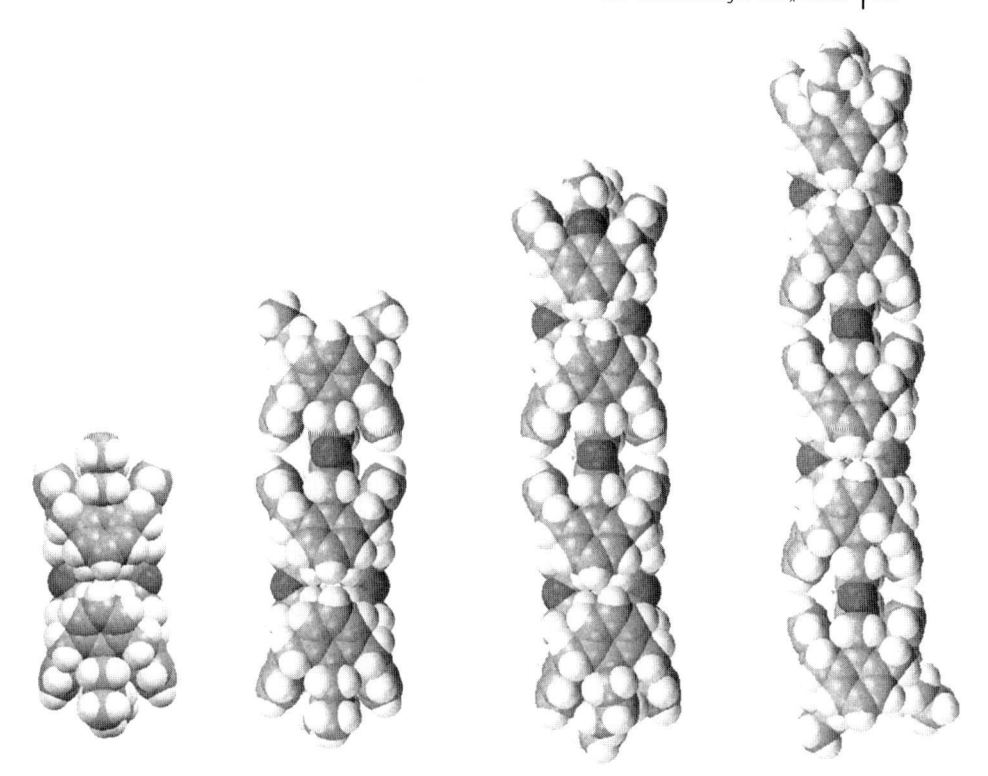

Figure 4.6 Molecular models of nanotubes **9–12**.

$NO^+SbF_6^-$, in $(CHCl_2)_2$. Complexes $9 \cdot (NO^+)_2$–$12 \cdot (NO^+)_5$ were identified by absorption, IR and ^1H NMR spectroscopy. Of particular importance is the characteristic deep purple color. The broad charge-transfer bands responsible for this were observed at $\lambda_{max} \approx 550$ nm in the absorption spectra of all these nanotubes. The charge transfer occurs only when NO^+ guests are tightly entrapped inside the calixarene cavities. Accordingly, the filling process can be monitored visually.

Upon stepwise addition of NO_2/N_2O_4 or NO^+SbF_6 in $(CDCl_2)_2$, the ^1H NMR signals of empty tubes **9–12** and complexes $9 \cdot (NO^+)_2$–$12 \cdot (NO^+)_5$ can be seen separately and in slow exchange. The guests presence and location inside nanotubes $9 \cdot (NO^+)_2$–$12 \cdot (NO^+)_5$ was deduced from conventional ^1H NMR, COSY and NOESY experiments. Chemical shifts of the $Ar-O-CH_2$ and CH_2-O-CH_2 protons and, to lesser extent, the aromatic protons are very sensitive to the encapsulation. In addition to the charge transfer, strong cation–dipole interactions between the calixarene oxygen atoms and the entrapped NO^+ take place. Through ^1H NMR titration experiments and molecular modeling, the stoichiometry of the nanotube complexes was unambigously established: they possess one NO^+ guest per calixarene unit.

FTIR spectra allowed unique information to be obtained on the bonding of multiple NO^+ species inside tubes $9 \cdot (NO^+)_2$–$12 \cdot (NO^+)_5$ in solution [17]. From the literature, nitrosonium salts NO^+Y^- ($Y^- = BF_4^-$, PF_6^-, AsF_6^-) showed a single

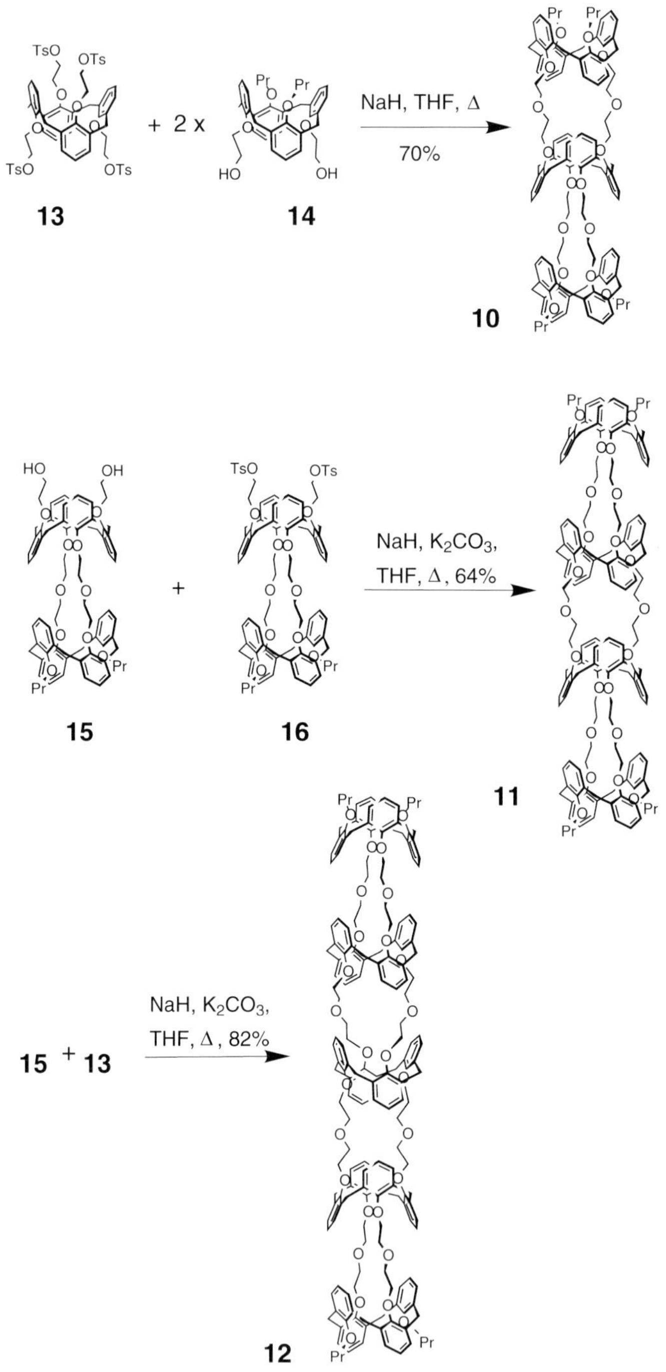

Scheme 4.1 Preparation of calix[4]arene-based nanotubes [17].

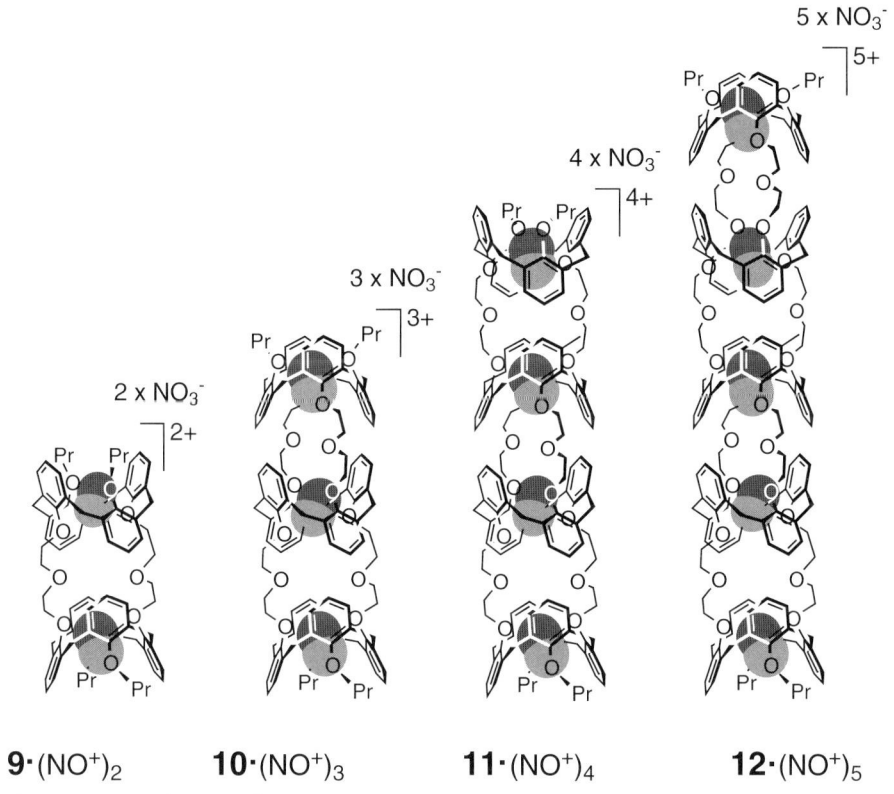

Figure 4.7 Nanotubes **9–12** disproportionate NO_2/N_2O_4 gases and entrap multiple NO^+ cations.

stretching band at $\nu(NO^+) = 2270\,cm^{-1}$ in CH_3NO_2 solutions and the stretching frequency of the neutral diatomic NO gas was $\nu(NO) = 1876\,cm^{-1}$. In calixarene–NO^+ complexes **8**, the NO^+ band significantly shifted ($\Delta\nu = 312\,cm^{-1}$) to the lower energies, compared to free NO^+ cation and appeared at $\nu(NO^+) = 1958\,cm^{-1}$ in $(CHCl_2)_2$. This is due to strong electron donor–acceptor interactions between the encapsulated NO^+ and π-electron-rich aromatic walls of the calixarene. Dimeric complex **9** $(NO^+)_2$ also exhibited similar shifts for the NO^+ guests at $\nu(NO^+) = 1958\,cm^{-1}$. At the same time, longer tubes **10**·$(NO^+)_3$–**12**·$(NO^+)_5$ clearly showed two absorption bands at $\nu(NO^+) = 1958$ and $1940\,cm^{-1}$ in $(CHCl_2)_2$. For trimeric tube **10**·$(NO^+)_3$, these two bands have a comparable intensity, whereas in longer tubes **11**·$(NO^+)_4$ and **12**·$(NO^+)_5$ the band at $\nu(NO^+) = 1940\,cm^{-1}$ dominates. This band was assigned to the NO^+ guest(s), which are situated in the middle of the tubes. Apparently, they are somewhat more strongly bound to the nanotubes walls.

One possible explanation may be a participation of the glycol CH_2OCH_2 oxygens in the complexation process. The FTIR data strongly suggest the *anti-gauche* conformational transition of these CH_2OCH_2 chains upon complexation, so their oxygens appear in close proximity to the complexed NO^+ and thus contribute to dipole–cation interactions [17]. Cooperativity through allosteric effects is also possible. It may be the result of multiple guests aligning in one dimension, which brings an order that cannot be achieved for shorter complexes.

Filled nanotubes $9 \cdot (NO^+)_2$–$12 \cdot (NO^+)_5$ were stable in dry solution at room temperature for hours, but could readily dissociate upon addition of H_2O or alcohol, quantitatively reproducing free **9–12**. The process, however, is not reversible: the released NO^+ species are now converted to nitrous acid and complexes $9 \cdot (NO^+)_2$–$12 \cdot (NO^+)_5$ cannot be regenerated.

However, it was further found that 18-crown-6 could reversibly remove the encapsulated NO^+ species. It is known that crown ethers form stable complexes with NO^+. When ~4 equiv. of 18-crown-6 were added to $(CDCl_2)_2$ solutions of $9 \cdot (NO^+)_2$ and $10 \cdot (NO^+)_3$, empty nanotubes **9** and **10**, respectively, regenerated within minutes and the deep-purple color disappeared. Interestingly, further addition of $SnCl_4$ to the same solutions fully restores complexes $9 \cdot (NO^+)_2$ and $10 \cdot (NO^+)_3$. Apparently, an excess $SnCl_4$ replaces NO^+ from the crown ether moiety and the latter goes back to the calixarene units of the nanotubes. This observation is important, since in this case foreign species can be replaced and returned back without decomposition or changing the solution polarity.

Modeling suggests that NO^+ can enter and leave the nanotube either through its ends or the middle gates between the calixarene modules. The approach and exit through the ends appears to be less hindered. The middle gates between the calixarene units become narrower due to the glycol conformational changes from *anti* to *gauche* upon complexation. The encapsulated NO^+ species should also avoid electrostatic repulsions with each other. Most probably, the tube filling and release occurs through the guest tunneling along the interior.

In the solid state, longer nanotubes **10** (but not tubes **9**) pack head-to-tail, in straight rows, resulting in infinitely long cylinders (Figure 4.8) [16]. The neighboring nanocylinders are aligned parallel to each other. In each nanocylinder, molecules **10** are twisted by 90° relative to each other and the Ar—O—Pr propyl groups effectively occupy the voids between the adjacent molecules. In such an arrangement, the intermolecular distance between two neighboring tubes in the nanocylinder is ~6 Å. The nanocylinders are separated from each other by ~9 Å. This supramolecular order comes with the tube length and is without precedent for conventional, shorter calixarenes. The unique linear nanostructures maximize their intermolecular van der Waals interactions in the crystal through the overall shape simplification. Such a unique arrangement resembles that of SWNT bundling.

Among possible applications of nanotubes **9–12** are nanowires and also optical sensors for NO_x. Chemical fixation of NO_x is also of great interest. Indeed, the tubes can be used for molecular storage of active nitrosonium and act as size–shape-selective nitrosating reagents [14,15]. In addition, they can be used for generating NO gas.

In a one-electron reduction scheme involving the calixarene-NO^+ complexes **8**, $9 \cdot (NO^+)_2$ and $10 \cdot (NO^+)_3$ and simple hydroquinone, NO was smoothly released and free calixarenes **7**, **9** and **10** were quantitatively regenerated (see, for example,

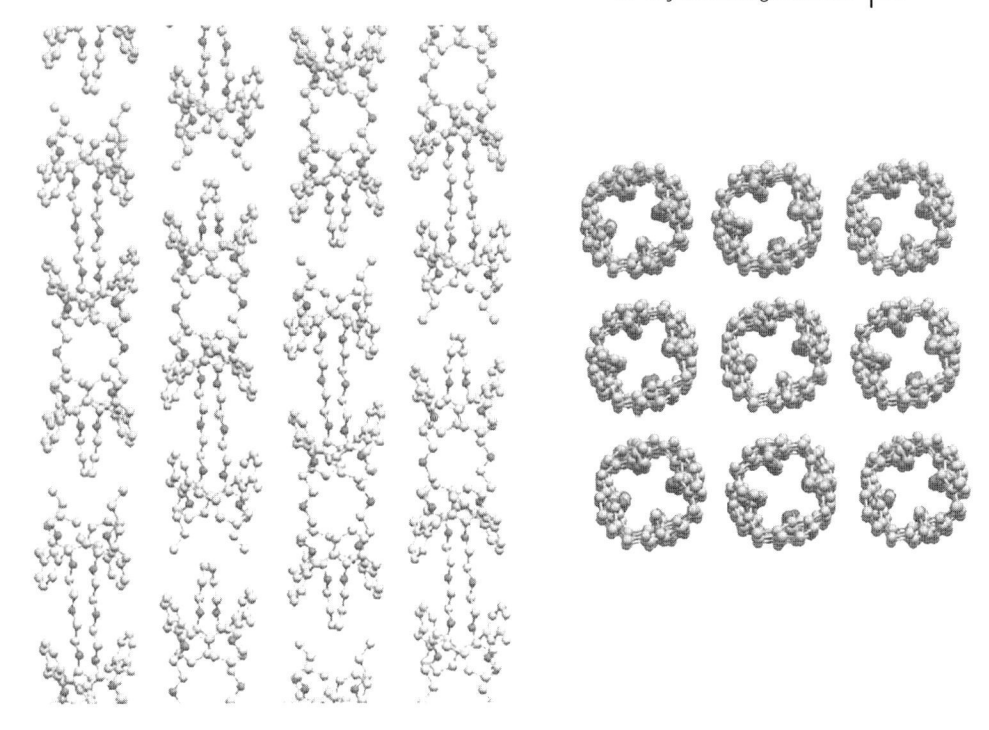

Figure 4.8 Solid-state packing of nanotube **10**, side and top views (from the X-ray structure) [16].

Figure 4.9) [18]. In detail, when a ∼20-fold excess of hydroquinone was added to the $(CDCl_2)_2$ solutions of nitrosonium filled nanotubes **9**·$(NO^+)_2$ and **10**·$(NO^+)_3$, the color changed from deep purple to yellow. The ^1H NMR spectrum clearly showed the quantitative regeneration of the empty nanotubes **9** and **10**. The NO release could be visually detected and identified by UV spectrophotometry. The use of calixarene nanotubes, capable of storing *multiple* NO^+ species, could potentially lead to interesting NO-releasing materials with a high gas capacity.

4.5
Self-assembling Structures

Solid-state self-assembling nanotubes were recently published that are based on calixarenes [19–21]. Although their stability and host–guest behavior in solution still remain to be investigated, it should be possible to use preorganized calixarene cavities for molecular encapsulation, separation and storage. It must be remembered, however, that self-assembling nanotubes are stable only under specific, rather mild conditions, which may not be suitable for some applications. Another important but still unresolved issue is control over their length.

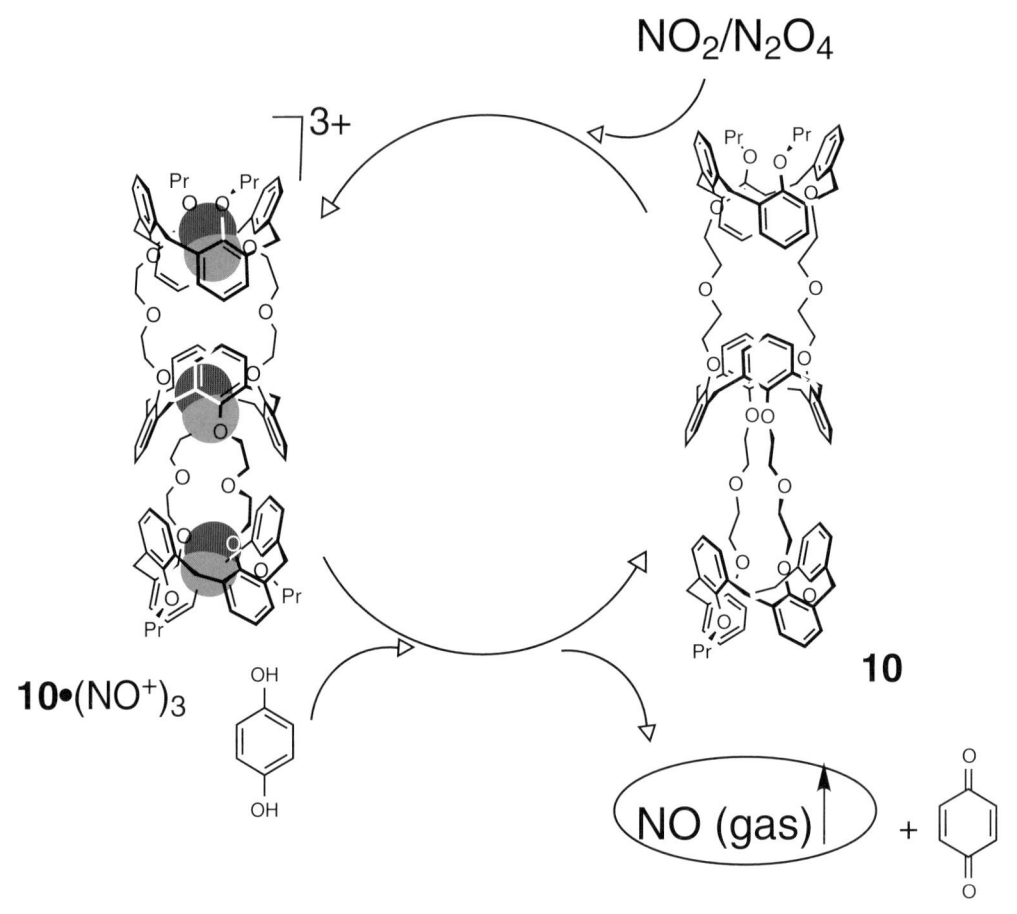

Figure 4.9 Calixarene nanotubes can be used for generation of NO gas [18].

4.6
Conclusions and Outlook

Synthetic nanotubes are promising molecular containers. Their geometric features and actual nanodimensions clearly place them in a unique position compared with conventional molecular containers. Described here, calixarene-based nanotubes simultaneously entrap multiple guests in a 1D fashion. The guest exchange mechanism is also different, since nanotubes are open from both ends and do not require dissociation [22]. This leads to interesting host–guest dynamics and opens the door to such applications as 1D ion mobility for transport and nanowires, inner-space reactions with subsequent product release and also high-capacity porous materials for molecular separation and storage. This also establishes an internal order that cannot be achieved for conventional encapsulation complexes and even may influence the binding strength.

In contrast to SWNTs and ion channels, conventional organic spectroscopy can be used to study the complexation processes and monitor guest behavior within the interior.

Among the future goals will be the synthesis of even more sophisticated nanotubes. There is a need to achieve higher kinetic and thermodynamic stabilities of the encapsulation complexes. It also remains to be seen how the encapsulated guests interact or even react with each other and the nanotube walls and whether their properties in a confined environment are different from those in bulk.

Acknowledgment

The National Science Foundation is acknowledged for financial support.

References

1 (a) Cram, D.J. and Cram, J.M. (1994) *Container Molecules and their Guests*, Royal Society of Chemistry. (b) Rudkevich, D.M. (2002) *Bull. Chem. Soc. Jpn.*, **75**, 393–413. (c) Rudkevich, D.M. (2005) *Functional Artificial Receptors* (eds T. Shrader and A.D. Hamilton), Wiley-VCH, Weinheim, 257–298.

2 (a) Purse, B.W. and Rebek, J. Jr. (2005) *Proc. Natl. Acad. Sci. USA*, **102**, 10777–10782. (b) Jasat, A. and Sherman, J.C. (1999) *Chem. Rev.*, **99**, 931–967. (c) Warmuth, R. and Yoon, J. (2001) *Acc. Chem. Res.*, **34**, 95–105. (d) Collet, A., Dutasta, J.-P., Lozach, B. and Canceill, J. (1993) *Top. Curr. Chem.*, **165**, 103–129. (e) Rebek, J. Jr. (2005) *Angew. Chem. Int. Ed.*, **44**, 2068–2078. (f) Hof, F., Craig, S.L., Nuckolls, C. and Rebek, J. Jr. (2002) *Angew. Chem. Int. Ed.*, **41**, 1488–1508

3 Britz, D.A. and Khlobystov, A.N. (2006) *Chem. Soc. Rev.*, **35**, 637–659.

4 (a) Sisson, A.L., Shah, M.R., Bhosale, S. and Matile, S. (2006) *Chem. Soc. Rev.*, **35**, 1269–1286. (b) Sakai, N., Mareda, J. and Matile, S. (2005) *Acc. Chem. Res.*, **38**, 79–87.

(c) Matile, S., Som, A. and Sorde, N. (2004) *Tetrahedron*, **60**, 6405–6435.

5 (a) Bong, D.T., Clark, T.D., Granja, J.R. and Ghadiri, M.R. (2001) *Angew. Chem. Int. Ed.*, **40**, 988–1011. (b) Pasini, D. and Ricci, M. (2007) *Curr. Org. Synth.*, **4**, 59–80. (c) Baklouti, L., Harrowfield, J., Pulpoka, B. and Vicens, J. (2006) *Mini-Rev. Org. Chem.*, **3**, 356–386.

6 (a) Asfari, Z., Böhmer, V., Harrowfield, J. and Vicens, J. (eds) (2001) *Calixarene 2001*, Kluwer. Dordrecht (b) Gutsche, C.D. (1998) *Calixarenes Revisited*, Royal Society of Chemistry, Cambridge.

7 (a) Ikeda, A. and Shinkai, S. (1994) *J. Chem. Soc., Chem. Commun.*, 2375–2376. (b) Ikeda, A., Kawaguchi, M. and Shinkai, S. (1997) *Anal. Quim. Int. Ed.*, **93**, 408–414.

8 (a) Khomich, E., Kasparov, M., Vatsouro, I., Shokova, E. and Kovalev, V. (2006) *Org. Biomol. Chem.*, **4**, 1555–1560. (b) Gac, S.L., Zeng, X., Reinaud, O. and Jabin, I. (2005) *J. Org. Chem.*, **70**, 1204–1210.

9 (a) Matthews, S.E., Schmitt, P., Felix, V., Drew, M.G.B. and Beer, P.D. (2002) *J. Am. Chem. Soc.*, **124**, 1341–1353. (b)

Matthews, S.E., Felix, V., Drew, M.G.B. and Beer, P.D. (2003) *Org. Biomol. Chem.*, **1**, 1232–1239. (c) Webber, P.R.A., Cowley, A., Drew, M.G.B. and Beer, P.D. (2003) *Chem. Eur. J.*, **9**, 2439–2446 (d) Matthews, S.E., Felix, V., Drew, M.G.B. and Beer, P.D. (2001) *New J. Chem.*, **25**, 13551358.

10 Perez-Adelmar, J.-A., Abraham, H., Sanchez, C., Rissanen, K., Prados, P. and de Mendoza, J. (1996) *Angew. Chem. Int. Ed.*, **35**, 1009–1011.

11 (a) Kim, S.K., Sim, W., Vicens, J. and Kim, J.S. (2003) *Tetrahedron Lett.*, **44**, 805–809. (b) Kim, S.K., Vicens, J., Park, K.-M., Lee, S.S. and Kim, J.S. (2003) *Tetrahedron Lett.*, **44**, 993–997.

12 Kim, S.K., Lee, J.K., Lee, S.H., Lim, M.S., Lee, S.W., Sim, W. and Kim, J.S. (2004) *J. Org. Chem.*, **69**, 2877–2880.

13 (a) Zyryanov, G.V., Kang, Y. and Rudkevich, D.M. (2003) *J. Am. Chem. Soc.*, **125**, 2997–3007. (b) Kang, Y. and Rudkevich, D.M. (2004) *Tetrahedron*, **60**, 11219–11225.

14 Kang, Y., Zyryanov, G.V. and Rudkevich, D.M. (2005) *Chem. Eur. J.*, **11**, 1924–1932.

15 (a) Zyryanov, G.V. and Rudkevich, D.M. (2004) *J. Am. Chem. Soc.*, **126**, 4264–4270. (b) Sgarlata, V., Organo, V.G. and Rudkevich, D.M. (2005) *Chem. Commun.*, 5630–5632.

16 Organo, V.G., Leontiev, A.V., Sgarlata, V., Dias, H.V.R. and Rudkevich, D.M. (2005) *Angew. Chem. Int. Ed.*, **44**, 3043–3047.

17 Organo, V.G., Sgarlata, V., Firouzbakht, F. and Rudkevich, D.M. (2007) *Chem. Eur. J.*, **13**, 4014–4023.

18 Wanigasekara, E., Leontiev, A.V., Organo, V.G. and Rudkevich, D.M. (2007) *Eur. J. Org. Chem.*, 2254–2256.

19 Dalgarno, S.J., Cave, G.W.V. and Atwood, J.L. (2006) *Angew. Chem. Int. Ed.*, **45**, 570–574.

20 Mansikkamaki, H., Busi, S., Nissinen, M., Ahman, A. and Rissanen, K. (2006) *Chem. Eur. J.*, **12**, 4289–4296.

21 (a) Hong, B.H., Lee, J.Y., Lee, C.-W., Kim, K.C., Bae, S.C. and Kim, K.S. (2001) *J. Am. Chem. Soc.*, **123**, 10748–10749. (b) Hong, B.H,. Bae, S.C., Lee, C.-W., Jeong, S. and Kim, K.S. (2001) *Science*, **294**, 348–351. (c) Kim, K.S., Suh, S.B., Kim, J.C., Hong, B.H., Lee, E.C., Yun, S., Tarakeshwar, P., Lee, J.Y., Kim, Y., Ihm, H., Kim, H.G., Lee, J.W., Kim, J.K., Lee, H.M., Kim, D., Cui, C., Youn, S.J., Chung, H.Y., Choi, H.S., Lee, C.-W., Cho, S.J., Jeong, S. and Cho, J.-H. (2002) *J. Am. Chem. Soc.*, **124**, 14268–14279.

22 Palmer, L.C. and Rebek, J. Jr. (2004) *Org. Biomol. Chem.*, **2**, 3051–3059.

5
Molecular Gels – Nanostructured Soft Materials
David K. Smith

5.1
Introduction to Molecular Gels

It seems increasingly likely that *nanochemistry* will underpin manufacturing technology in the 21st century, with the ability to control structure and morphology at the nanometer level allowing the synthesis of a new generation of smart materials [1]. As such, the use of organic building blocks, which can *self-assemble* into nanoscale architectures, offers a simple approach to the spontaneous generation of nanomaterials. Of particular interest in this regard has been the rapid development of *gel-phase materials* constructed via the hierarchical assembly of molecular building blocks; indeed, over the past 15 years, there has been an explosion of publications dealing with the self-assembly of organic molecules into nanostructured gels in a variety of solvents [2]. In 2006, an excellent book, edited by Terech and Weiss, was published dealing with the topic in a comprehensive manner [3]. This chapter makes no attempt to reproduce a comprehensive coverage of the topic, but instead intends to provide a "primer" which will act as an accessible overview and summarize the following issues:

- What is a molecular gel?
- How can molecular gels be prepared?
- How can molecular gels be studied?
- What kind of molecules assemble into molecular gels and why?
- What are the applications, both real and potential, of molecular gels?

By answering these questions, it is hoped that this chapter will provide a useful overview for the non-specialist reader, in addition to offering new insights for those well acquainted with the field.

Gels are a well-known colloidal state of matter and are familiar to us in everyday life (hair gel, toothpaste, cleaning products, air fresheners, contact lenses, foodstuffs, etc.). However, perhaps surprisingly, an accurate definition of the gel state remains

---

Organic Nanostructures. Edited by Jerry L. Atwood and Jonathan W. Steed
Copyright © 2008 WILEY-VCH Verlag GmbH & Co. KGaA, Weinheim
ISBN: 978-3-527-31836-0

5 Molecular Gels – Nanostructured Soft Materials

somewhat elusive. In the simplest terms, Dorothy Jordan Lloyd's definition from the 1920s (paraphrased as "if it looks like 'Jell-O' – it must be a gel") is easy to use on a macroscopic level [4]. However, more useful in terms of visualizing the structure of a gel on a molecular level was her additional statement "... they must be built up from two components, one of which is a liquid at the temperature under consideration and the other of which, the gelling substance proper ... is a solid". Indeed, it is usually argued that in gels, a "solid-like" network is able to "immobilize" the flow of bulk "liquid-like" solvent [5].

Most of the gels familiar from everyday life are constructed from polymers and as such are not strictly "molecular" gels. In polymer gels, crosslinking between polymer chains (either covalent or noncovalent) gives rise to an entangled network, which constitutes the "solid-like" phase. The "liquid-like" solvent is primarily located in pores within the crosslinked network, in addition to solvating the polymer chains. It is worth noting that on the *molecular level*, the solvent is mobile within the gel-phase network [4]; flow of the bulk solvent is only prevented as a consequence of capillary forces and some solvent–gelator interactions. This leads to one of most striking visual properties of gels – when placed in an inverted vial, liquid is unable to flow out of the material under the force of gravity, in spite of the high liquid content (Figure 5.1). Indeed, this simple test is one of the best ways of rapidly identifying a new gelator in addition to performing basic characterization (see Section 5.3.1).

It is relatively straightforward to see how entangled polymer chains can yield gel-phase materials, but it is more difficult to understand how low molecular weight molecules can achieve gelation – so-called molecular gels. The concept of low molecular weight compounds causing viscosity modification is not new – molecular gels based on simple organic building blocks (e.g. fatty acid salts) were first recognized in the 19th century [6]. However, the recent academic interest has

GEL

Figure 5.1 Photograph of a typical gel.

coincided with the ready availability of imaging techniques able to visualize objects on the nanometer length scale. Electron microscopy in particular has allowed researchers to observe gel-phase materials on the nanoscale. In turn, this has enhanced the understanding of the way in which low molecular weight systems can achieve gelation and improved the ability of researchers to study the effect of molecular structural change on the gelation process.

It has recently become well understood that self-assembly processes underpin the formation of molecular gel-phase materials. Intermolecular interactions, such as *hydrogen bonding, π–π stacking, van der Waals interactions* and *solvophobicity* are able to encourage the assembly of molecular building blocks in a one-dimensional manner. Such a process can be considered as a *supramolecular polymerization* [7], i.e. the polymers formed are held together solely by noncovalent interactions between building blocks. These supramolecular polymers may then be able to form entangled networks and hence give rise to gelation. The process of gelation is often described as *hierarchical* and depends on a number of steps:

1. Molecule interacts with adjacent molecule to form a dimer.
2. Dimers interact to form oligomers.
3. Oligomers extend into supramolecular polymer *fibrils*, which have approximately the same width as the molecular building block (e.g. 1–2 nm wide).
4. Fibrils bundle together to form *fibers* (often ca. 20–50 nm wide).
5. Fibers interact to form an *interconnected sample spanning network*. These fiber–fiber interactions are one of the least well-characterized and understood aspects of molecular gels.
6. Sample spanning network "immobilizes" solvent.

It should be noted that in different cases, the levels of hierarchical assembly may be expressed slightly differently; however, some degree of hierarchy is always assumed – two examples from the literature are illustrated in Figure 5.2. One of the most fascinating features of molecular gels is that very small amounts of gelator often achieve complete solvent "immobilization" and this demonstrates the power of hierarchical assembly. Typically, molecular gels contain <2% w/v gelator. Molecules which achieve gelation at concentrations <1% w/v are sometimes described as "*supergelators*".

There is clearly a relationship between gelation and *crystallization*, as in both cases a solid-like component assembles in a liquid-like phase as a consequence of multiple noncovalent interactions. However, in gelation, the solid-like network is compatible with the solvent and remains solvated and hence does not fully phase-separate. As such, gelation can be considered to be a competition between solubilization and phase separation. Furthermore, in gelation, the assembly process is *directional* – i.e. unlike a crystallization process, aggregation does not occur in three dimensions; in gels the growth of the supramolecular polymer is usually one dimensional.

Perhaps one of the key advantages of materials based on low molecular weight building blocks is their *reversibility*. Unlike a covalently crosslinked polymer network, molecular gels can be broken down into their individual molecular building blocks

(A)

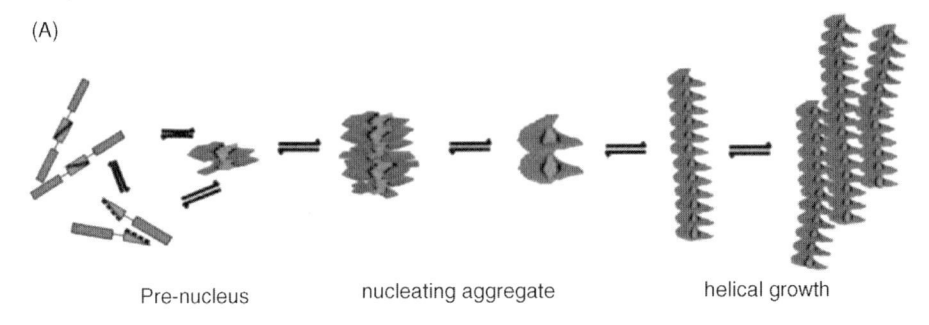

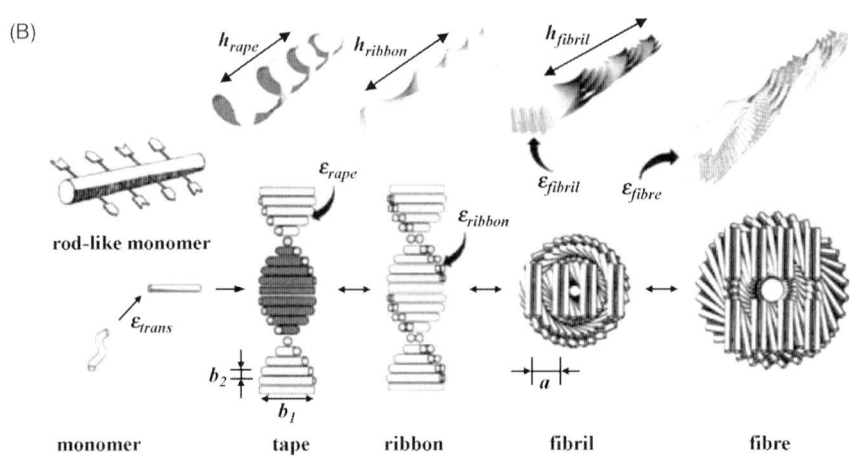

Pre-nucleus   nucleating aggregate   helical growth

(B)

Figure 5.2 Hierarchical assembly of gel-phase materials as illustrated for synthetic systems developed by (A) Meijer's group and (B) Aggeli and coworkers. Ref. [8] adapted with permission of Science and Ref. [9] adapted with permission of National Academy of Sciences.

and this makes molecular gels highly tunable, responsive materials. For example, raising the *temperature* has a profound effect on the assembly process due to the entropy term ($\Delta G = \Delta H - T\Delta S$) and encourages disassembly of the gel into the less ordered "sol" state. Molecular gel formation is also *concentration* dependent; at low concentration, the interaction between molecules is unable to yield supramolecular polymers. Such effects are now well understood in supramolecular polymer chemistry [7] and can be applied to gel-phase materials.

5.2
Preparation of Molecular Gels

Molecular gels are often formed by *heating* the solid low molecular mass gelator in an appropriate solvent, with *sonication* sometimes being applied to the sample [10].

This process results in the solubilization of the gelator. In the isotropic solution, the gelator molecules have a relatively low degree of aggregation – insufficient to support a sample spanning network. On cooling, preferably at a controlled rate, the sol transforms into a gel. In general terms, the higher the concentration of gelator used, the higher is the temperature of the gel–sol boundary. On some occasions, a gel forms instantly on dispersing a gelator in the appropriate liquid [11]. This would imply that in this case, the activation energy required for solubilization and subsequent molecular assembly is lower than that in those cases where a heat–cool cycle (or sonication) is required for gelation. It is essential that the method of gel preparation is accurately reported, as gel properties can vary depending on the precise details of preparation (e.g. rapid cooling versus slow cooling, heating versus sonication).

One key recent study provided a detailed insight into a one-dimensional aggregation process, in which molecules aggregated into dimers, which formed helical fibrils, that then assembled further into fibers and bundles (Figure 5.2A) [8]. It was concluded that the basic model of aggregation described above to form gels was accurate, but that solvent molecules played an unexpected explicit role in forming an ordered shell capable of rigidifying the aggregates and guiding them towards assembly into bundles/gels. This analysis is consistent with the observation that gelation is highly (and sometimes unpredictably) solvent dependent [12]. Understanding these nucleation and assembly mechanisms in more detail will hopefully enable the important goal of extending *rational synthesis into the nanoscale regime* in a more controllable way.

5.3
Analysis of Molecular Gels

Perhaps one of the most interesting features of gels is the ability to gain an understanding of the connection between *molecular structure, nanoscale self-assembled morphology* and *macroscopic materials properties*. In the following sections, we discuss the ways in which gels can be analyzed across the full range of length scales. It should be noted that a full discussion of each of these techniques is beyond the scope of this chapter, which will instead try to focus on the key points of each method and direct the reader to useful references in the literature.

5.3.1
Macroscopic Behavior – "Table-Top" Rheology

Quick and simple ways of visually assessing the physical (i.e. macroscopic) behavior of the gels are referred to as "table-top" rheology [13] and can be readily carried out in any laboratory without sophisticated equipment. These techniques, discussed in further detail below, are particularly useful for exploring the *gel–sol phase boundary*, as the conversion of the material from a gel to a sol can be readily assessed in a visual manner. There are two simple parameters which are widely used to define the

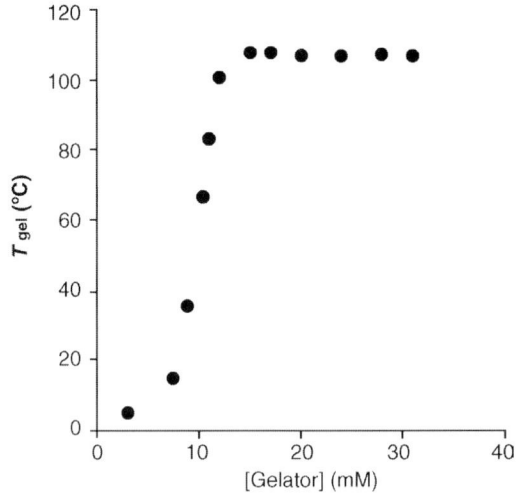

Figure 5.3 Example of a phase diagram for gelation. Reproduced from Ref. [14] with permission of the American Chemical Society.

macroscopic properties of molecular gels:

1. *Minimum gelation concentration* (MGC) – the minimum concentration of gelator required to form a sample-spanning self-supporting gel at a given temperature (usually 25°C).

2. *Gel–sol transition temperature* (T_{gel}) – this value can be determined by monitoring gelation with the use of a high-precision thermoregulated heating–cooling bath. It is concentration dependent and determining T_{gel} values at different concentrations allows the construction of a "phase diagram" (e.g., Figure 5.3).

5.3.1.1 Tube Inversion Methodology

The simplest method of monitoring the gel–sol transition involves inverting a vial of the gel and watching to see whether any flow occurs (for example, the gel shown in Figure 5.1 would satisfy the tube inversion test) [15]. It is essential to use the same sample mass and vial type when performing comparative studies, to ensure that the yield stress exerted on the gel remains constant. This ensures that the gel–sol transition in each case is determined under the same conditions of stress and hence meaningful comparative results can be obtained. However, this limitation means that it can sometimes be difficult to compare results recorded in different laboratories in a quantitative way.

5.3.1.2 Dropping Ball Method

In this method, a small metal ball is placed on the gel and the dropping of the ball through the gel is observed [16]. In the ideal scenario, the ball should be immobile in the gel, but drop rapidly in the sol. In this way, the T_{gel} value at any given concentration of gelator can be determined. In this case, the yield stress is dependent on the density of the ball and its radius. Therefore, the ball must be kept constant across experiments in order to ensure that data can be compared in a meaningful way. The sample tube should also be significantly larger than the ball, otherwise the presence of the nearby walls can affect the motion of the ball (wall effects can never be completely eliminated).

It is possible to analyze the phase diagrams constructed from "table-top" rheology methods in order to estimate thermodynamic parameters associated with gelation. Schrader's relationship can be used to generate plots of ln[gelator] against $1/T_{gel}$, in which the gradient is $-\Delta H/R$, where ΔH can be approximated to the enthalpy associated with the gel–sol transition [17]. This value can therefore provide some insight into the thermodynamics of the interactions between molecular-scale building blocks, although the assumptions inherent in this treatment probably mean that the data are best used to compare the behavior of related gelators. It should also be noted that many gels apparently exhibit a plateau region in T_{gel} above a certain concentration (see Figure 5.3) – corresponding to a point at which the presence of additional gelator does not appear to enhance the thermal macroscopic properties of the gel, presumably because the formation of a sample-spanning network is complete. This behavior cannot be fitted to Schrader's relationship.

5.3.2
Macroscopic Behavior – Rheology

As soft materials, gels can ideally be explored using *rheological methods* [18], although this requires the use of specialist equipment. In typical experiments, the magnitudes and ratios of the elastic (G') and loss (G'') moduli are determined under oscillatory shear. The elastic modulus (G') indicates the ability of a deformed material to regain its shape, whereas the loss modulus (G'') represents the ability of the material to flow under stress. For a gel, the elastic modulus should be independent of oscillatory frequency and G' should exceed G'' by about one order of magnitude. Measurements of viscosity also provide a useful method for comparing gel-phase materials in a quantitative way and generating structure–activity relationships.

Rheology has been performed on a number of molecular gels [19] and it is increasingly clear, from the relationship between strain and stress tensors, that the choice of theoretical model is not straightforward. Molecular gels can be classified rheologically as cellular solids, fractal/colloidal systems or soft glassy materials, depending on their behavior. Evidently, there is the need for continued application of rheological methods to molecular gel materials in order to provide a better insight into the way in which modifying gelator structures on the molecular scale controls the macroscopic rheological properties of the self-assembled material.

5.3.3
Macroscopic Behavior – Differential Scanning Calorimetry

In differential scanning calorimetry (DSC), the difference in the amount of heat required to increase the temperature of a sample and reference is measured as a function of temperature [13]. When the sample of interest undergoes a physical transformation, such as a gel–sol phase transition, more heat will need to flow to it than the reference in order to maintain both at the same temperature. Observing the difference in heat flow associated with this endothermic phase transition permits the measurement of the amount of heat absorbed or released by integration of the DSC trace. This provides a way of directly measuring the phase-change enthalpy ($\Delta H_{gel-sol}$) and can provide an insight into the thermodynamics of the gelator–gelator interactions. DSC should be performed in both heating and cooling modes in order to assess both endothermic and exothermic transitions and assess the thermo-reversibility of the gel–sol phase change.

5.3.4
Nanostructure – Electron Microscopy

Electron microscopy is, perhaps more than any other method, the technique that has opened up the "nanoworld". In the simplest use of electron microscopy to image the nanostructure of gel-phase materials, a gel sample is first allowed to dry on a substrate (either under ambient conditions or *in vacuo*). The sample is coated under vacuum with a thin (ca. 2 nm) metallic layer and then imaged by scanning electron microscopy (SEM). The SEM image obtained in this way therefore depicts a *dried and treated sample*. Usually the network structure of the gel "collapses" on to itself during drying to yield a *xerogel* (if "collapse" does not occur, the structure is referred to as an *aerogel*). It should be noted that changes in the nanostructure other than "collapse" may also occur during drying; however, it is often assumed that such effects are minor.

A range of morphologies are observed for gel-phase materials using SEM (Figure 5.4). In general, transparent gels exhibit nanoscale structuring, whereas opaque gels, which scatter light, have microscale features. Typically assembled *nanofibers* are observed, as might be expected for supramolecular polymers. Other types of "one-dimensional" objects such as *tapes/ribbons* have also been reported. In some cases nanoscale chirality of these objects can be observed – usually in the form of *helicity*. Transmission electron microscopy (TEM) can also be applied to gel imaging, although it is often necessary to apply a heavy metal staining agent to enhance image contrast.

Cryo-electron microscopy techniques are used to try and minimize disruption to the self-assembled network [20]. This method uses a rapid freezing step to prevent thermal motion of the assembled gelator network. In some cases solvent is then sublimed from the sample by freeze-drying (although this may modify the gelator structure, the low temperatures make gelator reorganization less likely). Overall, cryo-electron microscopy leads to significantly less disruption of the gelator network and therefore a typical cryo-EM image shows a more expanded and "solvated" network, in which the solvent pockets can be readily visualized as cavities within the network (Figure 5.5).

In some rare examples, *nonfibrillar gel morphologies* have been observed by electron microscopy methods. For example, platelet-type morphologies have been observed in

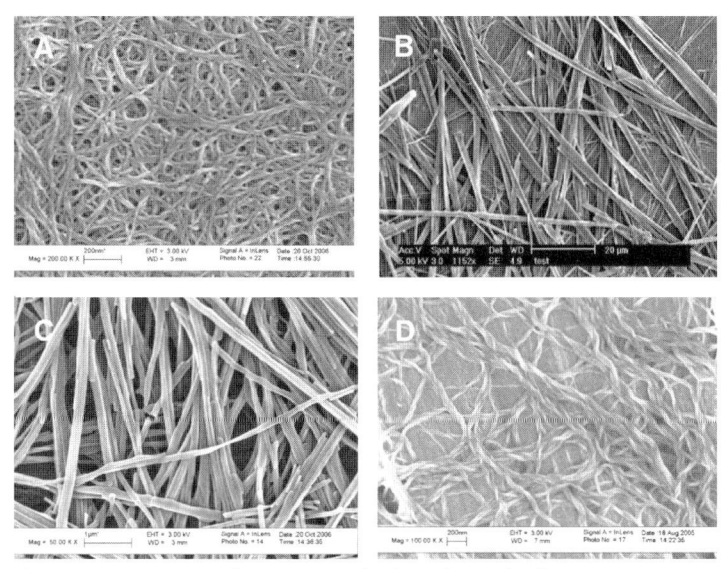

Figure 5.4 SEM images of nanostructured gels synthesized in the author's laboratories showing (A) fibrillar, (B) tape-like, (C) rod-like and (D) helical ribbon morphologies for different gelators.

which interpenetrating networks of platelet-like objects constitute the "solid-like" phase [21]. Figure 5.6 illustrates how a honeycomb network of two-dimensional platelets may lead to the effective "immobilization" of solvent molecules within the nanoscale cavities inherent within the morphology. It has been argued that in these systems, growth of an ordered structure occurs in two dimensions (and is prevented in the third). This would therefore appear to be conceptually related to the more typically observed formation of gel fibers, in which growth is only allowed in one direction and prevented in the other two.

Figure 5.5 Cryo-scanning electron microscopy image of a self-assembled gel. Reproduced from Ref. [12a] with permission from the American Chemical Society.

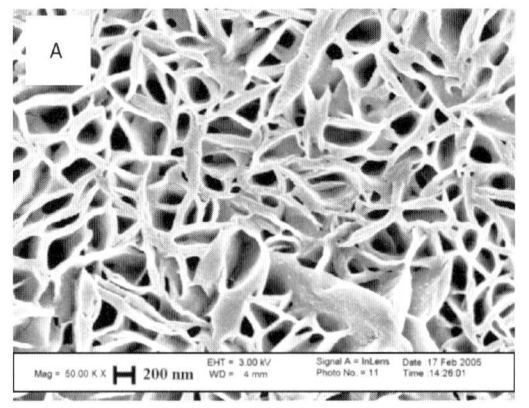

Figure 5.6 SEM image of an unusual nonfibrillar interconnected honeycomb morphology constructed from interlocked platelets underpinning a gel-phase network. Reproduced from Ref. [12d] with permission of Wiley.

5.3.5
Nanostructure – X-Ray Methods

The irregular packing of fibrils and fibers within gel-phase materials means that rather than leading to well-defined diffraction patterns, solvated molecular gels generally scatter X-rays, which can, at low angles of scattering, be fitted to a computer model [22]. For example, the scattering data obtained from a gel can be modeled as a collection of cylinders or tapes. The data can provide useful information about nanoscale dimensions, such as the cylinder diameter. If the model is accurate, this will correspond to the fibril dimensions (e.g. ca. 1–3 nm). This is useful information, as the objects visualized by electron microscopy methods are usually larger fibers and bundles (i.e. not molecular scale) – X-ray scattering analysis therefore provides a way of "visualizing" the *molecular-scale fibrils* present in the gel, which underpin the morphology observed by electron microscopy. X-ray methods can also be used on the dried gels (xerogels), in which case diffraction peaks are generally observed. This can lead to a more precise understanding of the molecular packing within fibrils, although results must be treated with some care, as drying can lead to morphological change [23]. Computer modeling is often an excellent complement to X-ray diffraction (and other) investigations of molecular gels and such approaches can sometimes be used to provide a predictive understanding of fibril/fiber packing [24] and lead to pictorial models such as that illustrated in Figure 5.2B [9].

5.3.6
Molecular Scale Assembly – NMR Methods

NMR spectroscopy provides a useful method to try to understand why, on a molecular level, gelators assemble into fibrillar architectures and to probe the intermolecular interactions between the molecular building blocks which underpin the self-assembly of fibrillar architectures [25]. It should first be noted that NMR in the "rigid" gel-phase is challenging because the low mobility of the molecular building blocks causes broad peaks. As in solid-phase NMR, magic angle spinning (MAS) could, in principle, be employed to obtain NMR spectra of rigid gels [26], although this has not been widely applied to low molecular weight gelation systems. In some rigid gels, unexpected sharp NMR peaks associated with the gelator have been observed; these have been attributed to gelator molecules *not bound* to the nanoscale network. Such peaks can only be observed if the exchange of gelator molecules from bound to unbound forms is kinetically slow [25]. Usually, the concentration of gelator being observed in the NMR spectrum can be quantified by integration with respect to a reference molecule which does not associate with the "solid like" gelator network and has sharp NMR peaks.

In order to facilitate the study of molecular gels, NMR experiments are often performed on soft solids/partial gels or on samples just above the T_{gel} value – in this way, spectra with reasonably sharp peaks can be recorded. Under this regime of concentration/temperature, the hierarchical assembly is at the stage of oligomer formation (rather than fully formed network) and NMR therefore provides an ideal way of monitoring the way in which one molecule interacts with another to form an oligomeric assembled structure. Several typical experiments are performed:

1. *Variable concentration* – NMR titration. NMR spectra are recorded at increasing concentration of gelator. As the concentration increases, the NMR spectra should broaden due to gel formation and loss of gelator mobility. However, the NMR peaks should also shift – particularly those that are involved in noncovalent interactions between the molecules. For example, N–H protons are often observed to shift downfield as concentration increases (Figure 5.7) [27]. This indicates that as the concentration increases, these peaks become increasingly involved in hydrogen bond interactions. Alternatively, aromatic peaks may shift upfield – which would be indicative of the aromatic rings becoming involved in the formation of $\pi-\pi$ interactions.

2. *Variable temperature (VT) experiments.* As the temperature of a sample is increased, the gel should gradually become more mobile and interactions between molecules will weaken. Once again, as these interactions weaken, NMR peaks will shift and this can be assigned to breaking hydrogen bonding and $\pi-\pi$ interactions [28].

3. *NOE experiments* [25,27b,29]. NOE experiments can provide information on the way in which one molecule interacts with another through space and this can help uncover the interactions present within a gel.

4. *Relaxation time experiments* [30]. Relaxation time experiments performed on the broadened peaks at known concentrations can be used to estimate the molecular mobility of the aggregated species (diffusion rates can be obtained from such experiments).

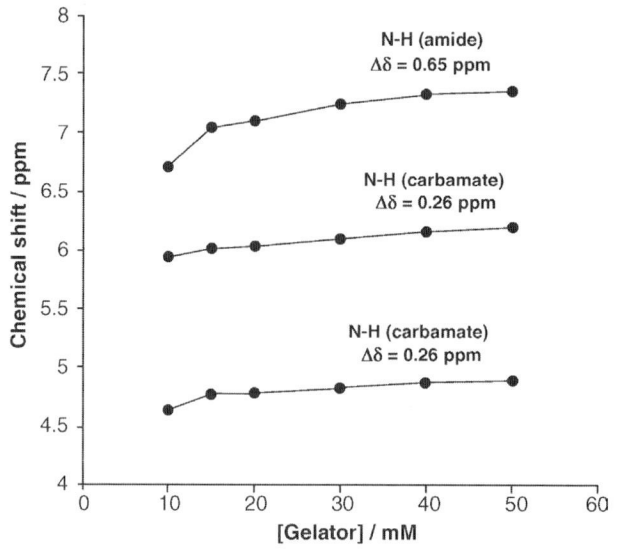

Figure 5.7 Variable concentration NMR experiment indicating downfield shifts of NMR peaks on increasing gelator concentration. Reproduced from Ref. [27c] with permission of the American Chemical Society.

5.3.7
Molecular Scale Assembly – Other Spectroscopic Methods

There are many other spectroscopic methods which can be applied to provide greater insight into the assembly of gel-phase materials. Two illustrative examples are given here:

1. *Infrared (IR) spectroscopy.* This method is very useful for probing hydrogen bond interactions between the molecular building blocks. In particular, O−H, N−H and C=O stretches all show distinctive responses to hydrogen bonding. Van der Waals interactions can also be detected by looking for changes in C−H stretching interactions. Typically, it is necessary to compare IR spectra of the gelator in both the sol and the gel in order to determine the key noncovalent interactions. Variable temperature IR spectra can provide a useful way of probing the response of these interactions to temperature changes [31].

2. *Fluorescence spectroscopy* [32]. Gelators which include fluorophores can exhibit changes in their spectra on aggregation. For example, many fluorophores emit as excimers when in close proximity (such as in a gel fiber) but emit as monomers when present in dilute solution. In one very recent example, excimers were present in the sol, but not in the gel [33]. Fluorescence can also detect smaller changes associated with differences in polarity between the gelator in aggregated and nonaggregated states. Once again, VT fluorescence

can be a useful technique in order to determine which spectral features are responsive to the aggregation process [34].

5.3.8
Chirality in Gels – Circular Dichroism Spectroscopy

Circular dichroism (CD) spectroscopy [35] is an ideal technique which can be used to monitor both molecular and nanoscale chirality [36]. An achiral molecule will exhibit no bands in the CD spectra, whereas a chiral molecule can exhibit a signal (either positive or negative), in the same wavelength region where it has its UV–Vis absorption. In general terms, however, solvated isolated chiral molecules often have relatively small CD signals (unless they have well-organized chromophores inherent within the molecular structure). As individual molecules assemble into a chiral nanostructure, the interaction with polarized light can be significantly enhanced and hence self-assembled nanostructures often exhibit much larger CD bands than their isolated molecular building blocks. The presence of a CD signal therefore provides good evidence for the presence of a chiral nanoscale object. To confirm this assignment, however, it is necessary to carry out a VT experiment or compare the CD spectrum with that observed in a solvent where aggregation does not take place [37]. In VT CD, the CD band should decrease in intensity as the temperature increases, because the nanoscale aggregates break up into their constituent building blocks on heating (Figure 5.8). For well-characterized chromophores, it can sometimes be possible to use CD spectroscopy to predict whether they are packed in a clockwise or an anticlockwise manner within the helical assembly and examples will be discussed later in the chapter.

CD spectroscopy also provides an ideal method for probing the assembly of mixtures of enantiomeric molecular building blocks into gels and can be used to

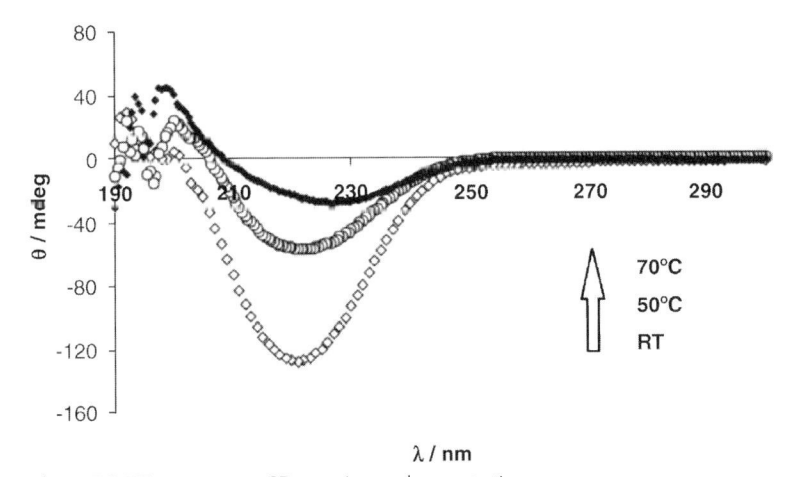

Figure 5.8 VT temperature CD experiment demonstrating disassembly of chiral nanostructure on application of heat. Reproduced from Ref. [37] with permission of the publishers.

distinguish between situations in which:

1. the presence of one enantiomer *disrupts the assembly* of the other and cause a breakdown in chiral order [38];
2. one enantiomer (present in excess) enforces its chirality on to the other enantiomer via a *"majority rules"*-type mechanism – this is well known in supramolecular polymers [39] but not widely explored in gels; or
3. the two enantiomers ignore one another's presence and are able to *"self-sort"* into a proportional mixture of their own homochiral assemblies [37,40].

5.4
Building Blocks for Molecular Gels

Gelation is highly dependent on the choice of solvent – indeed, it is common for researchers to present large tables of data testing gelation behavior across a wide range of solvents. For the purposes of this discussion we will broadly classify gels as either organogels or hydrogels, where *organogels* form in apolar, non-hydrogen-bonding solvents (e.g. hexane, toluene, acetonitrile, food and fuel oils), whereas *hydrogels* form in polar, hydrogen-bonding solvents (e.g. water, methanol, ethanol). The driving forces for gelation in these two classes of solvent are markedly different, as will be discussed below.

The design of a gelator from first principles is a difficult and challenging task. More often than not, gelators are discovered serendipitously. However, there is now sufficient literature available to suggest the following broad conclusions about whether or not a molecule may be expected to gelate:

1. The molecule must be *partly soluble* in the solvent of choice (but not too soluble, otherwise it will simply dissolve).
2. The molecule must be *partly insoluble* in the solvent of choice (but not too insoluble otherwise it will simply precipitate).
3. The molecule must have the potential to form *multiple noncovalent interactions* with itself. Typically, although not exclusively, these might be *hydrogen bonds and/or π–π interactions for organogels* and *hydrophobic and/or donor–acceptor interactions for hydrogels*.
4. *Van der Waals forces* are usually present to support the gelation process and in very rare cases, gelation can occur solely through these very weak interactions.
5. These noncovalent interactions should occur in a *directional manner* (a degree of asymmetry is often advantageous).

The sections that follow highlight the major classes of gelator systems explored in the literature. The examples chosen are highly selective and, for a more comprehensive overview, the interested reader is directed towards relevant review articles at the start of each section.

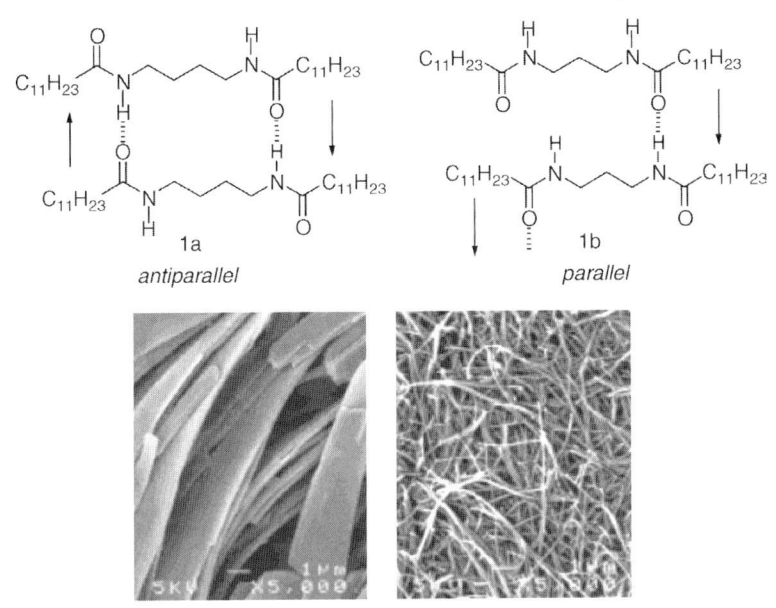

Figure 5.9 Gelators **1a** and **1b** with even and odd spacer chains stack in antiparallel and parallel modes, leading to different gel behavior and nanostructures (SEM images, scale bar = 1 μm). Adapted from Ref. [42a] with permission of the American Chemical Society.

5.4.1
Amides, Ureas, Carbamates (−XCONH− Groups, Hydrogen Bonding)

The −CONH− group has both a hydrogen bond donor (N−H) and a hydrogen bond acceptor (C=O) and, as such, is ideal for the formation of intermolecular hydrogen bonded networks [41]. For example, bisamide compounds **1a** and **1b** (Figure 5.9) form effective gels in aromatic solvents such as mesitylene due to a combination of hydrogen bond and van der Waals interactions [34,42]. Interestingly, the length of the aliphatic spacer chain between the amide groups was observed to exert an *odd/even effect* on gelation. It was argued that the ability of the molecules to assemble in either a parallel or antiparallel manner was dependent on whether the spacer chain had an odd or even number of carbon atoms. Indeed, the gels formed by gelators with odd and even spacer chains had markedly different thermal stabilities and SEM studies indicated that the gels were underpinned by different nanostructures.

Compound **2** (Figure 5.10) is another example of a bisamide gelator [43] and, once again, intermolecular hydrogen bonds and van der Waals forces drive the assembly

Figure 5.10 Compound **2** – a chiral gelator building block.

process. In this case, the molecular building block is chiral. Racemic mixtures of compound **2** with its enantiomer were observed to form much less stable gels than either pure enantiomer, demonstrating that chirality can play an essential role in controlling the molecular packing. This is consistent with the argument that the supramolecular polymers which underpin molecular gels have nanoscale chirality such as helicity (see Section 5.3.8).

In general, if peptidic molecules are to act as hydrogelators, rather than organogelators, they require slightly *higher polarity* in order to ensure sufficient compatibility with the aqueous solvent medium but they must also have sufficiently large hydrophobic surfaces to provide a thermodynamic driving force for assembly. Consequently, many hydrogelators contain unmasked carboxylic acid groups (or amines) to provide polarity and long alkyl chains to provide a hydrophobic surface for assembly. Amphiphilic structures such as **3** and **4** have therefore been employed (Figure 5.11) [44]. These molecules self-assemble as a consequence of hydrophobic interactions (and van der Waals forces) between the long alkyl chains and hydrogen bond interactions between the peptide groups, with the carboxylic acids/amines ensuring compatibility of the "solid-like" assembled network with the "liquid-like" solvent phase.

Ureas can readily replace the peptide groups and bisurea bola-amphiphilic structures also act as highly efficient hydrogelators. Depending on the choice of spacer chain, compounds such as **5** can have minimum gelation concentrations of <0.3% w/v (Figure 5.11) [45]. The gelation ability of many hydrogelators is highly *pH dependent*. Compound **5** forms vesicles at low pH and only assembles into fibrillar gel-phase materials when the pH is increased, a process which optimizes the solubility of the bola-amphiphile.

Compound **6** provides an interesting example of a photoswitchable hydrogel (Figure 5.12) [46]. Only the *trans* form of this compound assembles into nanoscale fibrillar objects and acts as a hydrogelator, whereas the *cis* form of the compound assembles into a microspherical morphology. The *cis* form of the gelator can be

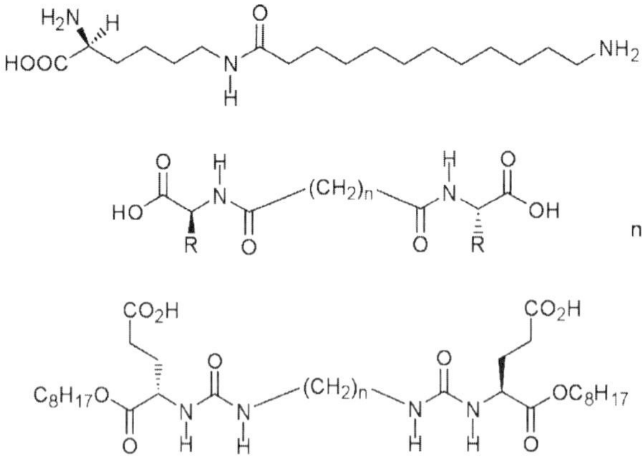

Figure 5.11 Hydrogelators based on peptide (**3** and **4**) and urea (**5**) functional groups.

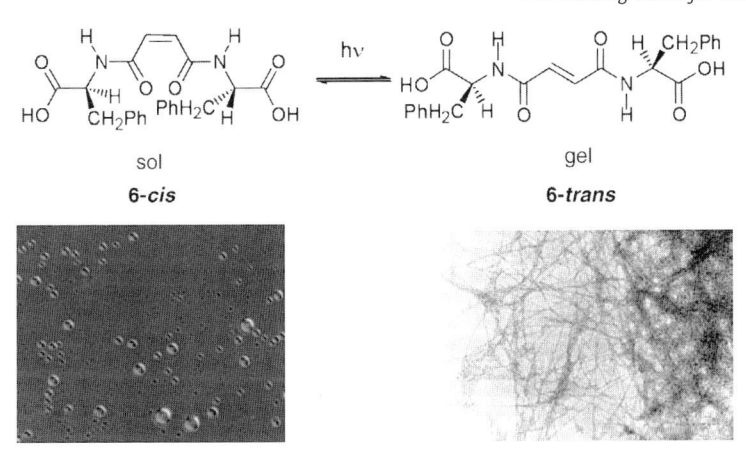

Figure 5.12 Compound **6** – a photoswitchable hydrogelator. Adapted from Ref. [46] with permission of the American Chemical Society.

converted into the *trans* form by photoirradiation and consequently, photoisomerization triggers a morphological transition at the nanoscale level and hence "switches on" gelation. Using *external stimuli* in this way to *trigger* gelation is an important target in controlled release applications such as drug delivery (see Section 5.5.4) and is one of the key ways in which the reversibility of supramolecular gels can be harnessed.

5.4.2
Carbohydrates (Multiple –OH Groups, Hydrogen Bonding)

Like peptides, alcohol groups contain both a *hydrogen bond donor* and an *acceptor*, ideal for setting up intermolecular hydrogen-bonded networks. Carbohydrates are particularly interesting building blocks for gels [47] as this class of molecule has many naturally occurring members which possess multiple –OH groups with subtly different organizations and chiralities, hence providing fertile ground for investigations of *structure–activity relationships* [48]. Normally, for the formation of organogels, some of the sugar alcohol groups will be protected – a completely unprotected sugar would not have the right solubility profile, being too hydrophilic and hence insoluble in organic solvents. Once again, the key appears to be achieving *balanced solubility* combined with the presence of potential self-assembling functional groups.

Chiral molecule **7**, a partly protected version of sorbitol, demonstrates these principles in action and has been extensively studied as a gelator of organic solvents and polymer melts since the first reports, which remarkably appeared as long ago as 1891 (Figure 5.13) [49]. Recently, this molecule has been applied as the matrix gelator in the development of quasi-solid-state dye-sensitized solar cells [50]. A wide range of analogues of this compound have since been investigated as

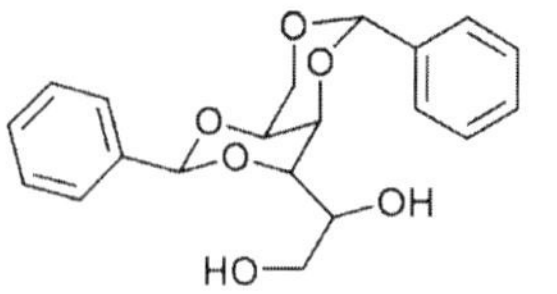

Figure 5.13 Carbohydrate-based organogelator **7**.

organogelators and this has led to an understanding of the role of both intra- and intermolecular hydrogen-bonding pathways on controlling the self-assembly of these molecules [51].

Carbohydrates have also been employed as hydrogelators – for example, gels based on unprotected sorbitol are widely employed in personal care products such as toothpaste and shaving gel. Typically, in hydrogelators, the carbohydrate is fully *deprotected* to provide high compatibility with the aqueous phase and the potential for multiple intermolecular hydrogen bond pathways. Furthermore, the carbohydrate is often functionalized with a significant *hydrophobic* surface in order to provide a *thermodynamic driving force* for assembly [52]. An example of this strategy is provided by amphiphilic compound **8**, which assembles into helical nanoscale fibers (Figure 5.14). Hydrogelators based on sugars have also been constructed with bola-amphiphile type architectures. For example, it was reported that compound **9** remarkably formed gels at concentrations as low as 0.05% w/v (Figure 5.14) [53]. The chromophoric azo dye used as the linker in these gelators experienced an induced CD signal due to its incorporation within a chiral nanoscale self-assembled architecture – the stereochemistry of the glucose units drives the chiral assembly process. The exciton coupling bands in these CD spectra could be assigned to the transition dipoles in the azobenzene units being orientated in a clockwise direction – indicating *right-handed helical growth* of the molecular fibrils. SEM observation of the gel fibers showed that this helicity was also manifested in the fibers observed at the nanometer scale. Transcription of chiral information from the molecular scale to the helicity on the nanoscale was therefore unambiguously demonstrated.

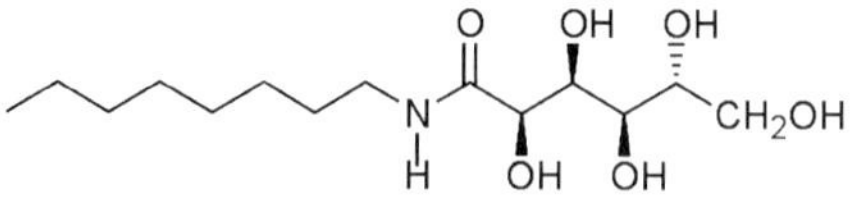

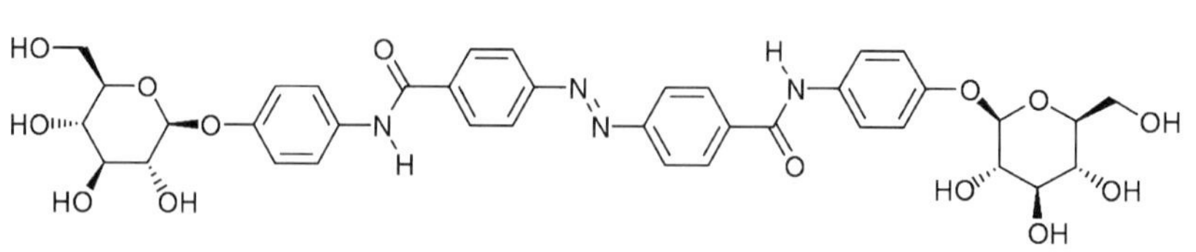

Figure 5.14 Carbohydrate hydrogelators **8** and **9**.

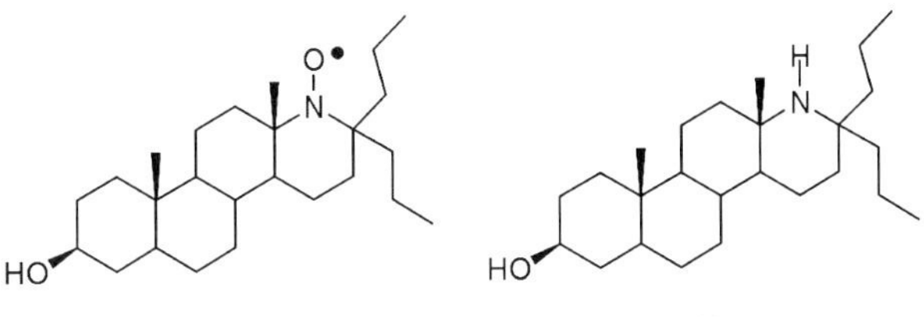

Figure 5.15 Compounds **10** and **11** as simple steroid organogelators.

5.4.3
Steroids/Bile Salts (Hydrophobic Surfaces)

Steroids, such as cholesterol, are naturally occurring lipid molecules which exist in all plants and animals; they have apolar tetracyclic skeletons and relatively planar rigid structures and have been widely explored as gelators [54]. It is well known in biology that steroids can interact with lipid membranes as a consequence of solvophobicity and van der Waals interactions. In gelation, solvophobicity will play the greater role in hydrogel formation whereas van der Waals interactions will be more important in the assembly of organogels. In 1979, it was reported that compounds **10** and **11** could gelate organic solvents and many of the fundamental techniques used to study organogels were developed using these molecules (Figure 5.15) [55]. Interestingly, the free radical on compound **10** meant that this gelator could also be investigated by electron spin resonance (ESR), with this technique even being used to generate a phase diagram.

Steroids modified with aromatic rings can become even more effective gelators due to π–π *stacking*. For example, compound **12** forms gels at concentrations <1% w/v and was described as a "supergelator" (Figure 5.16) [56]. In an elegant study, chiral gelator **13**, a functionalized cholesterol derivative, was investigated by CD spectroscopy (Figure 5.16) [57]. The chromophoric azo dye unit attached to the steroid

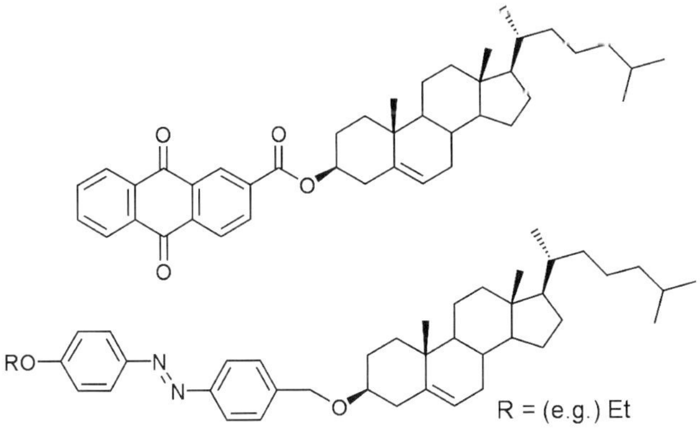

Figure 5.16 Modified steroid gelators **12** and **13**.

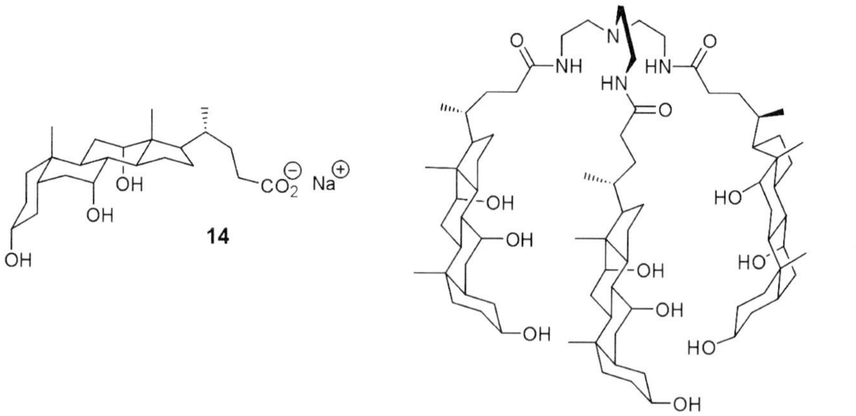

Figure 5.17 Bile acid salt **14** and bile acid derivative **15** both act as hydrogelators.

experienced an induced CD signal due to its incorporation within a nanoscale self-assembled architecture (caused by the presence of the cholesterol, with its high propensity for aggregation). Once again, the directionality of helical stacking within the nanoscale aggregate could be assigned from these CD spectra. In some cases, it was found that the speed of cooling during gel formation could control the 'handedness' of the assembled helical nanostructure – indicating the importance of carefully controlling and reporting the conditions of gel synthesis, as indicated in Section 5.2.

Bile acids, such as **14**, possess a rigid steroid skeleton, but are functionalized with polar −OH groups on the concave α-face and apolar methyl groups on the convex β-face (Figure 5.17). In addition, they contain a carboxylic acid unit which increases water solubility. As such, these molecules have a degree of *facial amphiphilicity* and they might be expected to have higher compatibility with aqueous solvents than simple steroids such as cholesterol – indeed, they are able to act as hydrogelators, with the first observation of this process being made for sodium deoxycholate as long ago as 1913 [58]. As expected for hydrogelators, *hydrophobicity* (between β-faces) drives the assembly process, supported by hydrogen bond interactions. Simple bile acids have also been used to construct more sophisticated gelator architectures. For example, compound **15**, which contains three cholic acid units attached to a triamine scaffold, forms gels at remarkably low concentrations (Figure 5.17) (0.02% w/v) [59]. The formation of hydrophobic "cavities" in the gel was characterized using 8-anilinonaphthalene-1-sulfonic acid (ANS) as a fluorescent probe of polarity.

5.4.4
Nucleobases (Hydrogen Bonding and π–π Stacking)

Nucleobases are substituted heteroaromatic pyrimidines and purines and are primary building blocks in the structures of nucleic acids. They constitute interesting building blocks for gel-phase materials [60] because they can form hydrogen bonding interactions within the plane of the heteroaromatic ring and π–π interactions

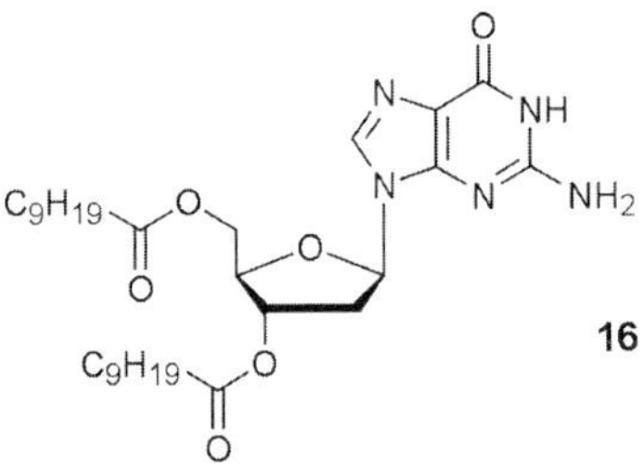

5.4 Building Blocks for Molecular Gels | 131

Figure 5.18 Compound **16** is a nucleobase organogelator.

perpendicular to the plane. This allows these molecules to become involved in extended networks of intermolecular interactions.

Compound **16** is a typical organogelator incorporating a guanosine nucleobase (Figure 5.18) [61]. These molecules formed gel-like liquid crystalline phases in hydrocarbon solvents, but fibrillar structures were not observed – instead, a planar tape-like assembly was indicated, with in-depth X-ray diffraction analysis providing evidence that the tapes contained the hydrogen-bonded guanine residues while the hydrocarbon chains and solvent molecules filled the gap between the tapes. Derivatives of thymidine have also been systematically investigated as organogelators [62].

One of the interesting features of nucleobase derivatives is their ability to form *specific interactions* with their *complementary partner nucleobase* – this can modify the self-assembly process and hence the properties of the gel. For example, thymidine derivative **17** (Figure 5.19) forms opaque gels in organic solvents such as benzene, but on mixing with complementary poly(A) (as a complex solubilized by cationic lipid), the opaque gel became transparent [63]. However, adding non-complementary poly (C) had no effect on the optical properties of the gel. SEM demonstrated that the addition of poly(A) changed the morphology of the gel from a *microscale* platelet-like structure (opaque) to an entangled *nanoscale* fibrillar network (transparent). As expected, the addition of poly(C) caused no perturbation of the microscale platelet-like morphology of compound **17**.

Nucleobases have been used to generate bola-amphiphile-type structures such as **18**, with two thymine units attached to either end of a spacer chain. This molecule assembles into hydrogels in a similar manner to the bola-amphiphiles described in previous sections [64]. Stacking of the nucleobases was indicated by a large hypochromicity in the UV–Vis spectrum. This system was demonstrated to respond

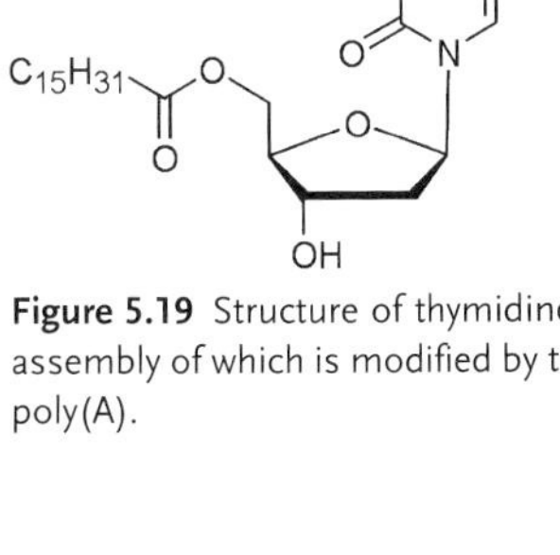

Figure 5.19 Structure of thymidine derivative **17**, the self-assembly of which is modified by the addition of complementary poly(A).

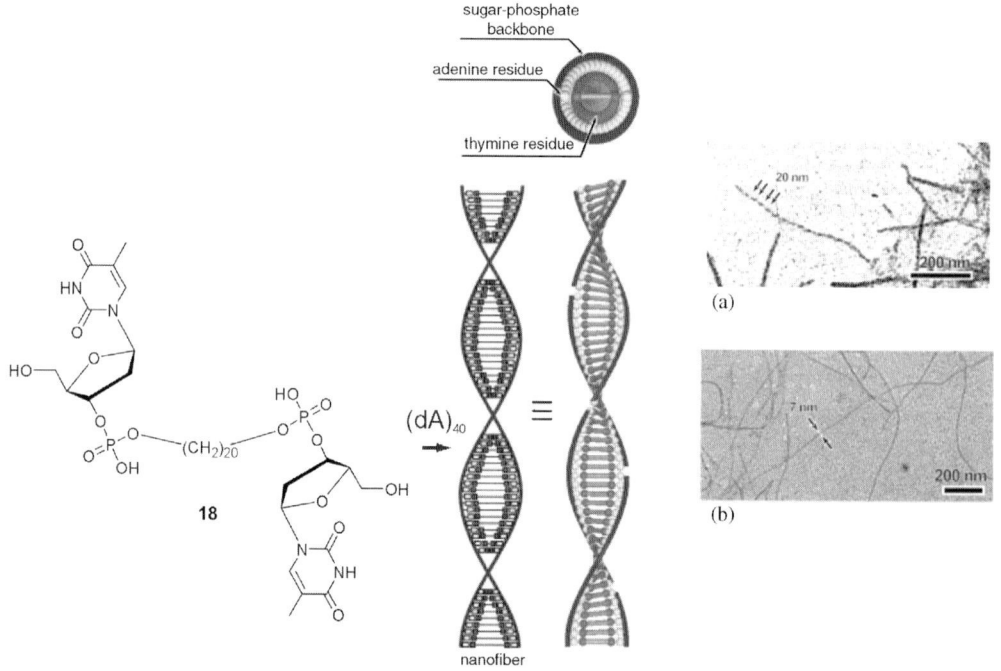

Figure 5.20 Compound **18** based on thymine, self-assembled with oligo(adenine) to yield twisted nanofibers based on complementary A–T interactions. TEM imaging [(a) and (b)] indicated the helical pitch and diameter of these nanoscale assemblies. Adapted from Ref. [65] with permission of Wiley.

to the presence of oligo(adenine) (Figure 5.20) [65]. In the presence of longer strands of oligo(A), helical nanofibers 7–8 nm in width and several hundred nanometers in length were observed. In these nanostructures, the oligo(A) strand had wrapped itself around the fibrillar assembly of compound **18** as a consequence of complementary hydrogen bond interactions between the A and T nucleobases.

Guanosine derivatives are also known to form gels in aqueous solvents as a consequence of G-quartet (**19**) formation (Figure 5.21) – a process which can be templated by Na^+ or K^+ cations (which interact with free lone pairs of the C=O groups pointing into the cavity). These G quartets can assemble into columnar gels, primarily driven by hydrophobic stacking [66].

5.4.5
Long-chain Alkanes (van der Waals Interactions)

Alkyl chains are *widely incorporated* into gelators as they provide a simple way of tuning the solubility of the molecular building blocks – indeed they are near ubiquitous in gelation systems. Alkyl chains can pack together via van der Waals forces and this can enhance gelation by supporting other intermolecular forces, particularly during the formation of organogels. Alkyl chains can also play a role in

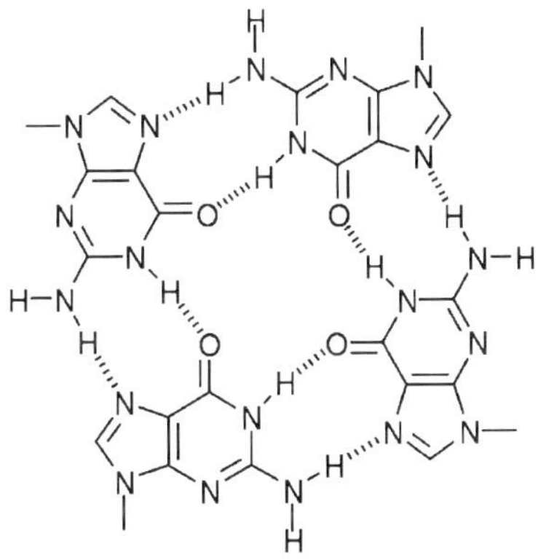

Figure 5.21 G quartet structure, 19.

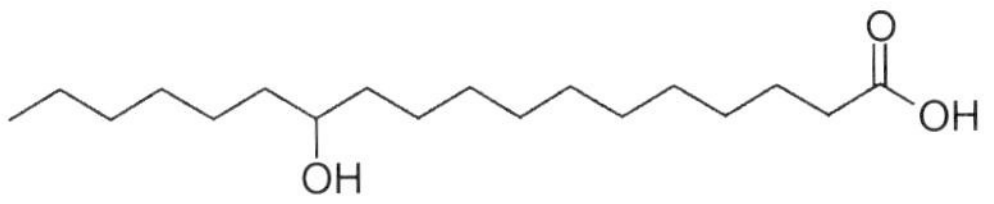

Figure 5.22 Fatty acid **20** (12-hydroxystearic acid, HSA) is capable of gelating a wide range of organic solvents and has widespread industrial use in the form of its lithium salt (Section 5.5.1).

ensuring the compatibility of the molecules with organic oil-type solvents and hence preventing phase separation. In hydrogels, alkyl chains often play a key role in driving the aggregation process as a consequence of the hydrophobic effect.

Fatty acids and their salts (e.g. **20**, Figure 5.22) have been widely employed for their gelation potential [6] and are industrially applied in a range of, e.g., cosmetic products. In 1925, Zsigmondy won the Nobel Prize for his investigations of colloidal systems, including fatty acid salts [67]. These systems are all underpinned by fibrillar self-assembled morphologies.

In one unusual example, van der Waals interactions and packing forces alone are sufficient to drive the gelation process. Simple long-chain alkanes such as n-hexatriacontane [$H(CH_2)_{36}H$] are able to act as gelators in a variety of hydrocarbon liquids, alcohols, halogenated liquids and a silicone oil [21a]. The extended alkyl chains pack together in a lamellar fashion to form a type of platelet microcrystal, which subsequently forms an interpenetrating network of platelets (Figure 5.23). Interestingly, it is known that this type of aggregation process causes problems in oil pipelines and production equipment, where so-called "waxes" of long-chain alkanes are deposited – understanding and prevention of this aggregation process are therefore of commercial importance.

5.4.6
Dendritic Gels

In recent years, *dendritic* (branched) molecules have been explored as gelators [68]. Dendritic systems possess *multiple functional groups* within the branched structure

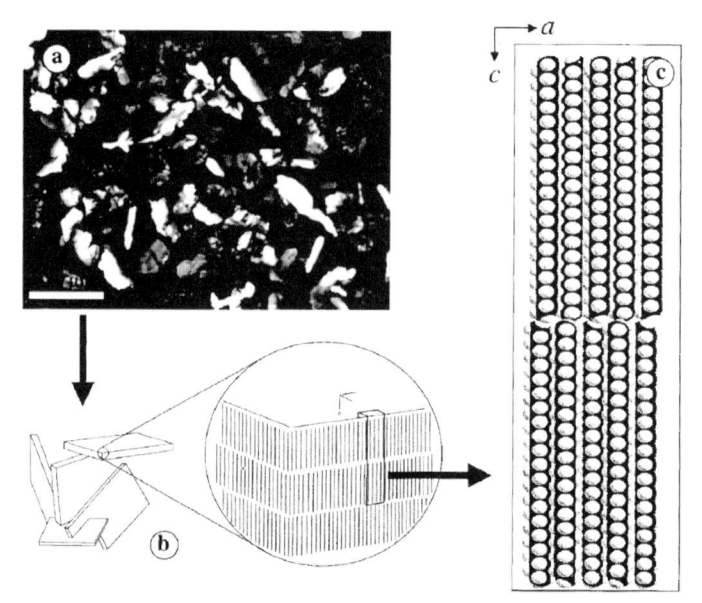

Figure 5.23 (a) Optical micrograph (scale bar 100 μm) of the platelet morphology formed by n-hexatriacontane in 1-octanol; (b) proposed model of platelet formation; (c) proposed molecular packing of long-chain hydrocarbons driven by van der Waals forces. Reproduced from Ref. [21a] with permission of the American Chemical Society.

and as such are capable of forming multiple intermolecular interactions, making their self-assembly [69] and gelation properties of considerable interest.

Compound **21** was one of the first dendritic organogelators [70]. In this system, intermolecular hydrogen bond interactions between the peptide groups at the focal point of the dendron were responsible for the self-assembly process (Figure 5.24).

Figure 5.24 Compound **21**.

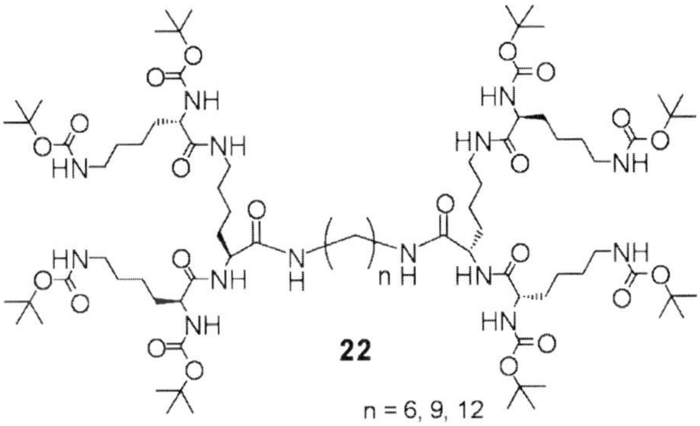

Figure 5.25 Compound 22.

Lower generation dendritic analogues (with less branching) did not lead to gelation, instead giving precipitate. It seems likely that in this case, the dendritic branching plays a steric role in enforcing the directionality of the assembly process and preventing the formation of "three-dimensional" aggregates.

Compounds such as **22**, with dendritic lysine head groups attached to an alkyl spacer chain, have been investigated in detail as gelators (Figure 5.25) [27c]. These molecules assemble predominantly as a consequence of *hydrogen bond interactions* between the dendritic peptide head groups, as demonstrated by NMR titration and variable temperature experiments. In this case, the highest generation system investigated (third generation) was the most effective gelator. It has been argued [27c] that dendritic effects on gelation are a balance between favorable additional intermolecular interactions and unfavorable steric hindrance and entropic penalties. In this case, therefore, it appears that the additional hydrogen bonds possible on introducing more branching more than compensate for the steric and entropic penalty of assembling more bulky and flexible molecular building blocks.

In a key study [37], the gelation of mixtures of these building blocks has been investigated (Figure 5.26). It was reported that in mixtures where the peptide head groups are different, such as those based on *"size"* or *"chirality"*, the dendritic building blocks are able to *self-sort* and form their own independent nanostructures. However, in systems where the dendritic head groups remain the same and only the spacer chain is changed (*"shape"*) the molecular recognition pathways between the peptide head groups become confused, self-assembly is disrupted and the nanoscale assemblies which result are constructed from a mixture of the different building blocks.

Amphiphilic dendritic organogelators [*dendron rod-coil* (DRC) molecules] such as **23** have been investigated as gelators (Figure 5.27) [72]. In this case, structure–activity relationship studies indicated that at least four hydrogen bonding O–H groups were required to support gelation, that the rigid aromatic rod became involved in π–π stacking interactions and that a sufficiently long apolar coil was required. These gelators have been widely exploited and some examples will be given in Section 5.5.5.

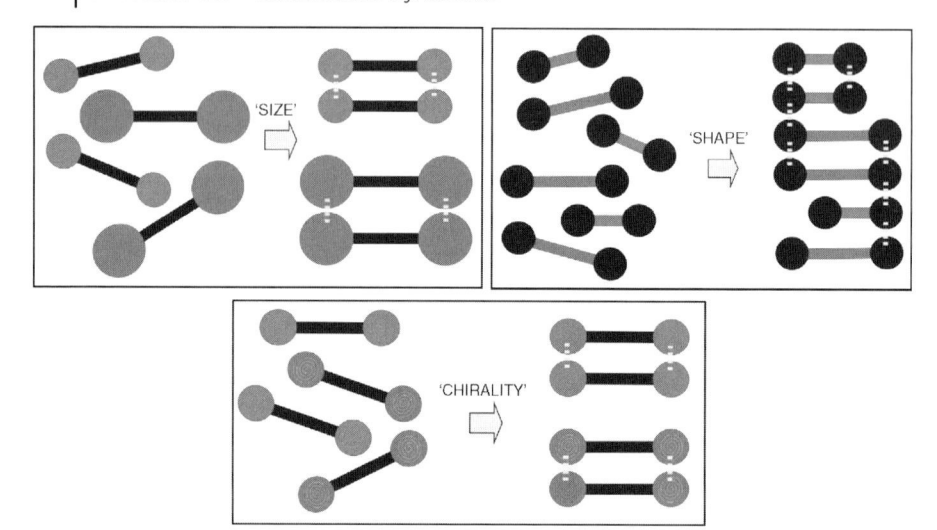

Figure 5.26 Self sorting mixtures – size, shape and chirality. Reproduced from Ref. [37] with permission of Wiley.

Dendritic hydrogelators are also well established. Once again, it is no surprise that some of these structures are bola-amphiphilic in nature, with hydrophobicity driving self-assembly. For example, compounds **24** and **25** form gel-phase materials at concentrations of 1% w/v in water, with fibrous assemblies being visualized by TEM methods (Figure 5.28) [73]. A series of structure–activity relationship experiments were performed and it was reported that there was an *optimum balance* between the length of the spacer chain and the size of the dendritic head group in order for gelation to occur. Indeed, compound **24** will form a gel when the spacer chain contains only eight carbon atoms, whereas compound **25**, which has more extensive polar head groups, requires at least 10 carbon atoms in the spacer chain for gelation to be observed. This hydrophobic/hydrophilic balance reflects the balance in solubility which is required for effective gelation.

Figure 5.27 Dendron rod-coil **23**.

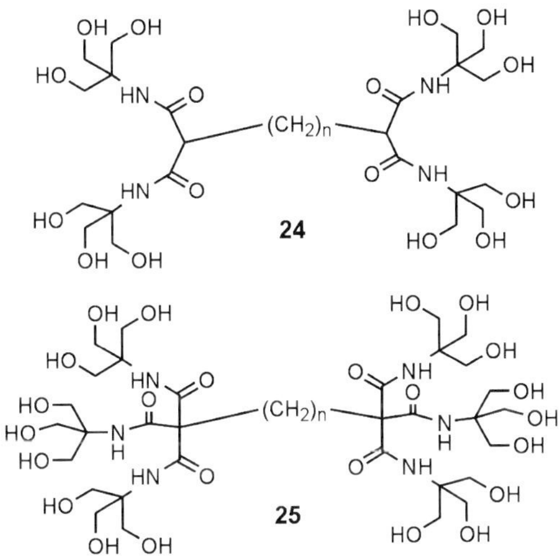

Figure 5.28 Dendritic hydrogelators **24** and **25**.

5.4.7
Two-component Gels

Two-component gelation systems are of rapidly increasing importance [74]. In such systems, two independent components are *both* required to be present in order for gels to form. The two components must first form a *complex*, which subsequently self-assembles into a gel-phase material. This introduces another step into the hierarchical assembly process and consequently provides another level by which the gelation event can be controlled and tuned.

An early example of a two-component gelation process made use of the well-known interaction between barbituric acid (**26**) and pyrimidine (**27**) (Figure 5.29). Appropriate functionalization of these building blocks sterically blocked the formation of a three-dimensional aggregate and encouraged the formation of a linear supramolecular polymer held together by hydrogen bonds [75]. Although this system only formed gels at relatively high concentrations, it demonstrated the principle that two components could act in a synergistic manner to form a gel-phase material.

Extensive studies have been made of a two-component gelation system in which a twin-tailed anionic surfactant, sodium bis(2-ethylhexyl)sulfosuccinate (AOT, **28**) interacts with substituted phenols (**29**) (Figure 5.30) [76]. Gelation is initiated via the formation of a hydrogen bonding interaction (possibly with associated proton transfer) between the phenol and the sulfonate group. Subsequently, the aromatic phenols form a π-stacked architecture. The alkyl chains of the surfactant are then able to contribute van der Waals interactions, in addition to ensuring the compatibility of the fibrillar aggregates with the surrounding solvent. Small-angle X-ray scattering was used to demonstrate that molecular-scale fibrils (diameter ca. 2.1 nm) assemble into fibers (diameter ca. 10 nm), which then aggregate further to yield fiber bundles (diameters 20–100 nm) – a clear example of the *hierarchical assembly process*.

138 | *5 Molecular Gels – Nanostructured Soft Materials*

Figure 5.29 Compounds **26** and **27**.

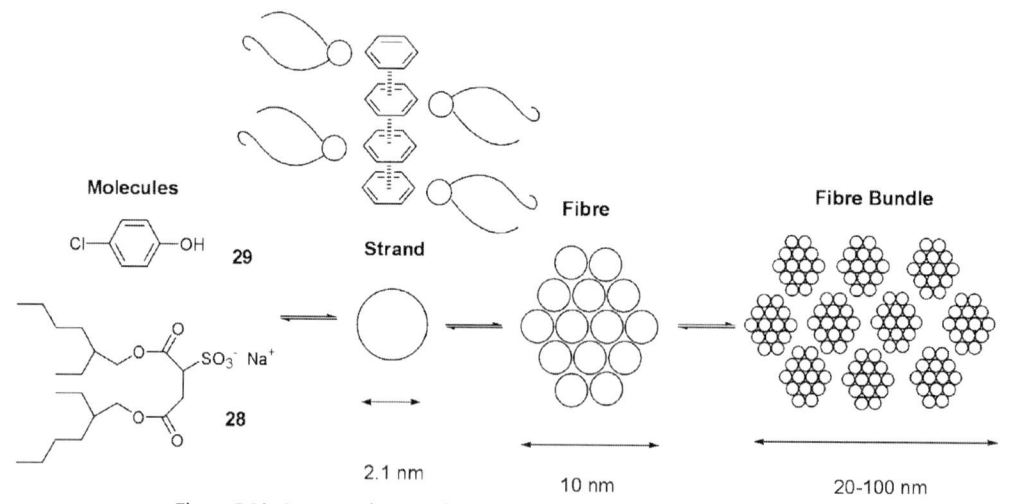

Figure 5.30 Compounds **28** and **29** assemble hierarchically to form strands, fibrils and fibers. Adapted from Ref. [74] with permission of Wiley.

Figure 5.31 Two-component gelation system **30**.

Two-component gels based on dendritic building blocks such as **30** have been reported and fully characterized (Figure 5.31) [12a,14,77]. Intriguingly, it has been demonstrated that the molar ratio of the two components can control the nanoscale morphology into which these molecular-scale building blocks assemble [21d]. A stoichiometric 2 : 1 ratio of dendron to diamine gave rise to well-defined nanofibrillar architectures; however, when the amount of dendron was decreased relative to the diamine (e.g. 1 : 4.5 ratio), the dendron was less able to stabilize this extended fibrillar morphology. Gel-phase materials were still obtained, but it was found that these were underpinned by networks of *micro- or nano-crystalline platelets* (Figure 5.32). It was argued that when sufficient dendron is present, the growth of diamine "crystals" is prevented in two dimensions, giving rise to one-dimensional fibers. However, when the amount of dendron becomes insufficient, it can only prevent growth of diamine "crystals" in one dimension and hence two-dimensional platelets result. Different diamines gave rise to different morphologies (platelet, square, rosette); however, all of them were fundamentally composed of interlocked two-dimensional objects. This demonstrates how two-component gels can provide additional control over nano-fabrication.

Two-component gels have been constructed based on well-known receptors widely used in *molecular recognition*. For example, compound **31** employs a dibenzo-24-

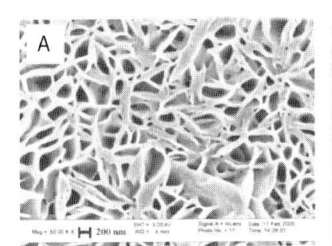

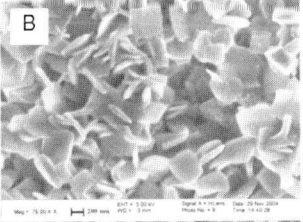

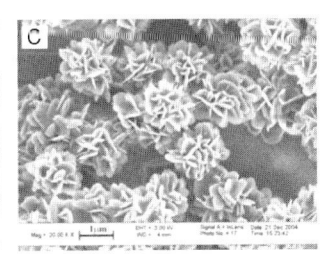

Figure 5.32 SEM images of nanoplatelets formed from different diamines in the case where there is insufficient dendron present to prevent growth of diamine platelet crystals in two dimensions. Adapted from Ref. [21d] with permission of the publishers.

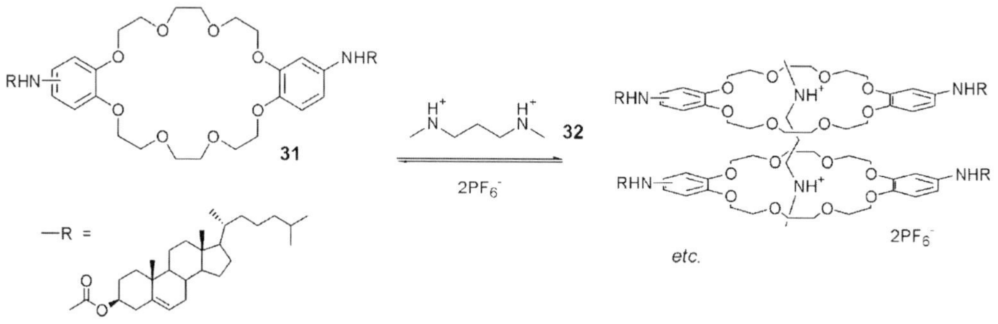

Figure 5.33 Compounds **31** and **32**.

crown-8 building block appended with steroid units [78]. Addition of substrate **32**, capable of binding to the crown ether, was observed to enhance significantly the propensity for gelation (Figure 5.33). It was proposed that molecular recognition induced a conformational change in compound **31** which encouraged the stacking of the steroid units and organization of the hydrogen bonding groups. This indicates how using a two-component approach, gelation can become responsive to specific chemical triggers. Other researchers have also developed gels which respond to the presence of specific ions via a host–guest binding mechanism and such systems have potential in controlled-release applications (Section 5.5.4) [79].

Donor–acceptor π–π interactions have also been employed for the assembly of two-component gels. Bile acid derivative **33**, functionalized with a donor group, formed gels when present in 1:1 stoichiometry with acceptor molecule **34** (Figure 5.34) [80]. UV–Vis spectroscopy indicated the presence of *donor–acceptor interactions*, as the gelation process was accompanied by a substantial increase in the *charge-transfer*

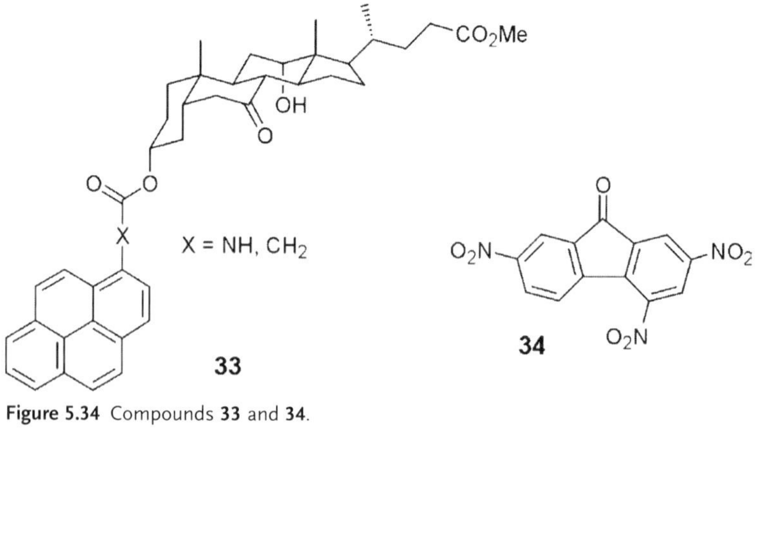

Figure 5.34 Compounds **33** and **34**.

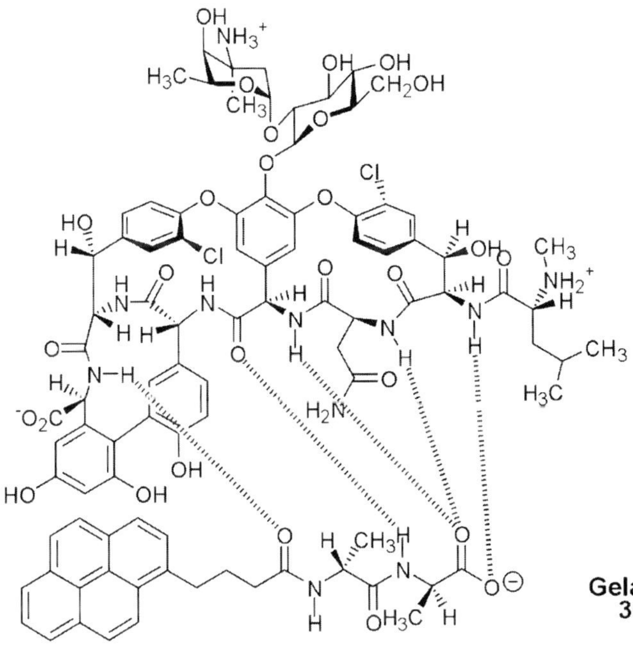

Figure 5.35 Vancomycin **35** triggers the enhanced gelation of compound **36** via host–guest interactions as illustrated.

band. Indeed, whereas the individual components are colorless/pale yellow, the nanostructured assembled gel is brightly colored.

Two-component hydrogels have also been uncovered in recent years. Perhaps one of the most interesting makes use of the medicinally relevant interaction between the antibiotic vancomycin (**35**) and pyrene-functionalized D-Ala,D-Ala derivatives such as **36** (Figure 5.35) [81]. It was demonstrated that the addition of vancomycin to a solution of gelator **36** gave rise to a significant increase in the mechanical strength of the gel – indeed, a dramatic 10^5-fold increase in the storage modulus of the material was reported. SEM investigations demonstrated that vancomycin changed the structure which underpins gelation transforming a self-assembled linear superstructure into a highly crosslinked two-dimensional sheet.

5.5
Applications of Molecular Gels

The endless and subtle synthetic variations which can be made to molecular gelators as illustrated in Section 5.4, means that the resultant materials have vast potential for application in a variety of different areas. The following section aims to provide a brief overview of some current and proposed applications.

5.5.1
Greases and Lubricants

Many *lubricating greases* in everyday commercial use are in fact molecular gels, although perhaps surprisingly, this is not always widely recognized in the recent scientific literature dealing with organogels [82]. In most greases, the "liquid-like" phase or base oil makes up 65–95% of the grease and is composed of hydrocarbons with between 25 and 45 carbon atoms (molecular weights 350–700) and with boiling points between 350 and 500 °C. One of the most commonly applied types of grease, referred to as "lithium grease", is based on a blend of mineral oil and the lithium salt of 12-hydroxystearic acid (HSA) **(20)**. Over 50% of the lithium grease market is based on the use of compound **20** as the thickening agent – this compound is obtained from castor oil, with 65% of the world's supply being produced in India. This grease contains a hydrogen bonding OH group, a long alkyl chain (van der Waals interactions) and a charged head group, all of which will assist in the self-assembly process. The fibrillar nature of the self-assemblies formed by these low molecular weight additives is well characterized in the grease literature [83].

5.5.2
Napalm

Unfortunately, napalm, one of the most notorious inventions in the history of mankind, is a gel-phase material [84]. Napalm was originally a mixture of aluminum salts of naphthenic and palmitic acids in petrol and is a *sticky incendiary gel*. Naphthenic acid refers to the mixture of carboxylic acids obtained from crude oil. Palmitic acid is the most common saturated fatty acid found in plants, $C_{15}H_{31}COOH$. Napalm was used in flamethrowers and bombs by the US and Allied forces to increase the effectiveness of flammable liquids by causing them to adhere to materials. When used against human targets, napalm rapidly deoxygenates the available air in addition to creating large amounts of carbon monoxide, causing suffocation. In some cases, napalm incapacitates and kills its victims very quickly, but victims who suffer second-degree burns from splashed napalm will be in significant amounts of pain. Napalm was used in the Second World War by US troops and has also been used in many subsequent military conflicts. Most infamously, napalm was used in the Vietnam war; however, this later version of napalm used a polymer to achieve gelation rather than using low molecular weight compounds.

5.5.3
Tissue Engineering – Nerve Regrowth Scaffolds

On a more positive note, self-assembled gels have recently been of great interest for their ability to act as *biomaterials*. In particular, they can be considered to provide a form of *nanoscaffolding*, which may have uses in tissue engineering [85]. There have been interesting studies investigating nerve growth through gels within animal model systems [86]. In one study, hamsters were surgically blinded by cutting the

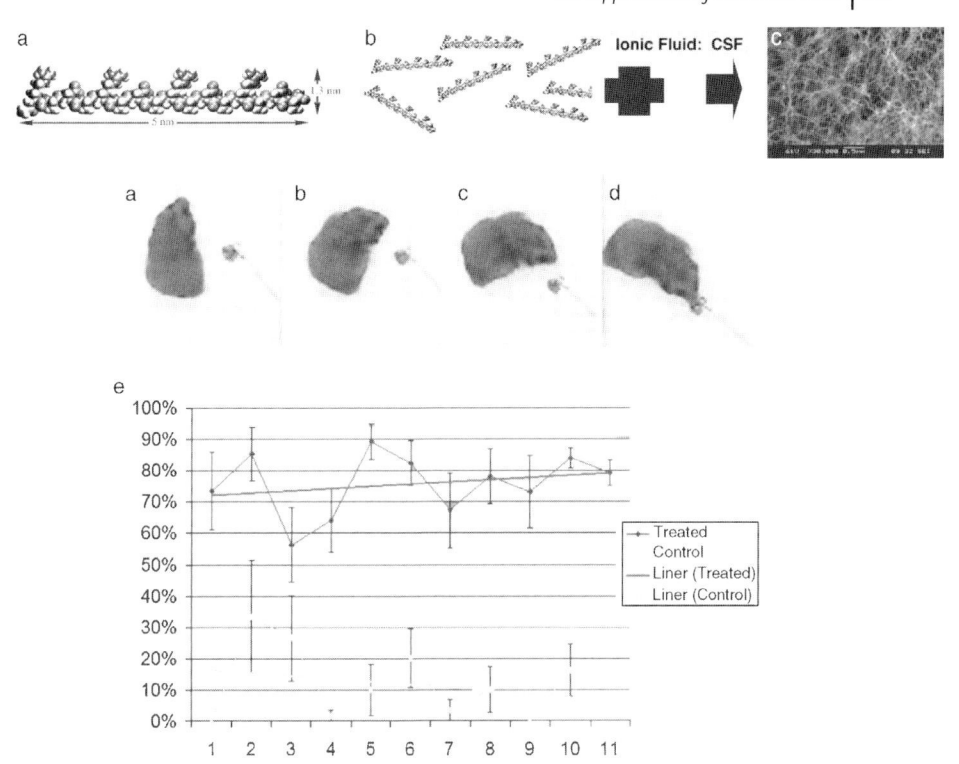

Figure 5.36 Peptide hydrogelator assembles into a fibrillar network as a consequence of hydrophobic and hydrogen bond interactions. When applied to the damaged optic nerve of hamsters with blinding in their right eye, vision is regenerated and the hamster responds to stimulus (a–d). Data (e) indicated that treated hamsters regained ca. 80% of vision, whereas untreated animals regained only 10%. Adapted from Ref. [86] with permission of the National Academy of Sciences of the USA.

optic nerve to one eye. Some of the hamsters then had a self-assembling peptide hydrogel injected in liquid form into the optic nerve – these animals regained ca. 80% of their vision as the nerves regrew through the nanostructured gel matrix (Figure 5.36). Untreated animals remained blind. This demonstrates that this kind of material is compatible with living tissue and has the potential to act as a matrix to encourage the growth of nerve cells.

In a different study, nanofibers have been self-assembled as a result of interactions between synthetic amphiphilic peptide **37** and the polyanionic biopolymer heparin (Figure 5.37) [87]. The nanostructured gel which resulted has been demonstrated to promote the growth of new blood vessels (angiogenesis), a process which is essential in wound healing and will be critical in tissue engineering applications. It is known that heparin binds many of the required angiogenic growth factors. It was argued that immobilization of the heparin as an integral part of the self-assembled nanofiber scaffold should help orient the domains of growth factors, in addition to protecting

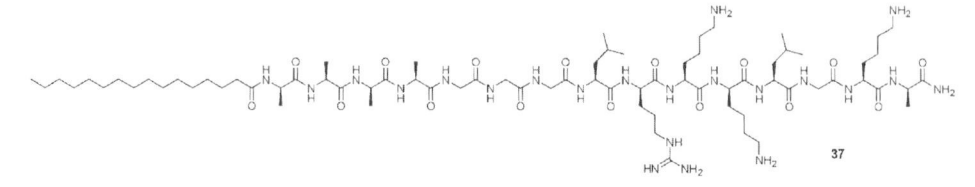

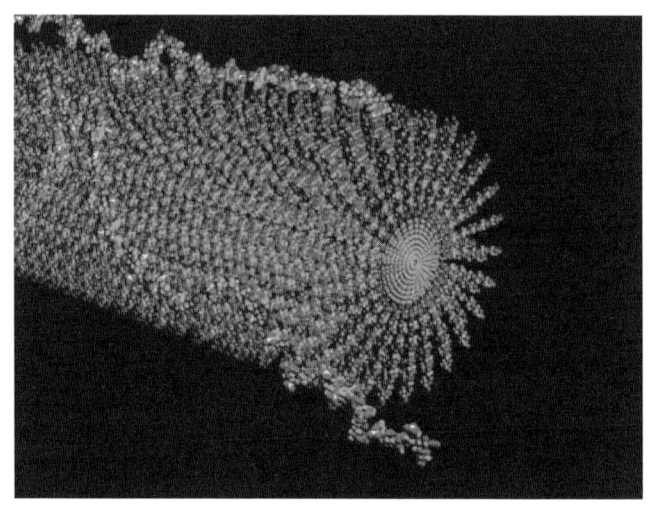

Figure 5.37 Peptide **37** used in tissue engineering studies and a schematic illustration of the way in which heparin can interact with the surface of the self-assembled nano-cylinder. Adapted from Ref. [87] with permission of the American Chemical Society.

them from protease activity – hence promoting angiogenesis – indeed, *in vivo*, these systems stimulated extensive new blood vessel formation. Once again, the fact that the components can self-assemble *in vivo* means that they can easily be delivered by liquid injection.

The fact that molecular hydrogels are based on relatively low molecular weight peptide building blocks means they should be *biocompatible* and furthermore will gradually *degrade in vivo*. Hence, over time, these gels should break down and the degradation products (simple amino acids) are biocompatible. This potential biodegradability is a major advantage of molecular gels over polymeric hydrogels, which are well known in medical applications. Obviously, the long-term stabilities and toxicities of these gels still have to be fully determined, but there is great promise in this area.

5.5.4
Drug Delivery – Responsive Gels

Molecular gels have great potential as drug delivery matrices. Indeed, it is worth noting that a wide range of pharmaceuticals are already *formulated as gels* (normally using polymeric systems) for oral delivery. Gels can also be subcutaneously injected along with a drug in liquid form and then allowed to gelate *in vivo* [88] for sustained release of a drug by allowing slow diffusion of the drug molecule out of the gel. The rate of release could be controlled depending on whether the drug molecule itself has any *specific interactions* with the gelator fibers (which would inhibit release of the drug). Alternatively, responsive molecular gels can achieve rapid ("burst") delivery of encapsulated drugs as a consequence of going through a gel–sol transition. As has been discussed above, molecular gels can be responsive to a wide range of different *stimuli* – temperature, pH, photoirradiation, ionic and molecular triggers, etc., and as such, the design of responsive drug delivery scaffolds is an area of considerable interest [89].

5.5.5
Capturing (Transcribing) Self-assembled Architectures

Although gels are soft materials, it is possible to try and capture the structures into which they assemble in a more rigid form. In this way, the self-assembled organic nanoarchitecture can be considered to act as a *template* for the formation of an inorganic material. One of the first examples of this approach was the use of compound **38** in the synthesis of nanostructured silica (Figure 5.38) [90]. A gel was formed in a mixture of acetic acid, tetraethyl orthosilicate (TEOS) and water. After 10 days, the solidified material was dried and calcined and was shown to consist of tubular silica. The charged quaternary ammonium salt was essential for encouraging the deposition of the silica onto the gel fibers.

Since this initial report, there have been numerous reports of transcription of organic gel morphologies into silica. Particularly eye-catching has been the generation of spiral/helical silica morphologies using different gelator structures (Figure 5.39) [91]. It has been hypothesized that these nanostructured inorganic or hybrid materials may have applications in enantioselective catalysis or separation.

This approach is not limited to the generation of silica-type materials. Dendron rod-coil molecules have been transcribed into CdS nanohelices [92]. These gelators self-assemble into one-dimensional tape-like objects with a uniform width of 10 nm and lengths up to 10 μm. When these dendron rod-coils were present during the

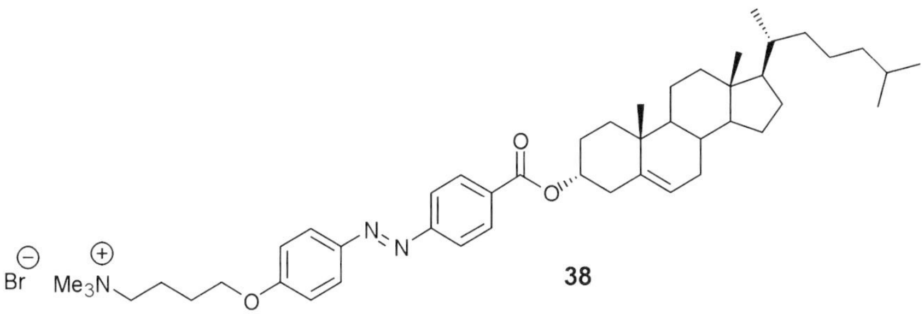

Figure 5.38 Compound **38** m used for transcription into silica.

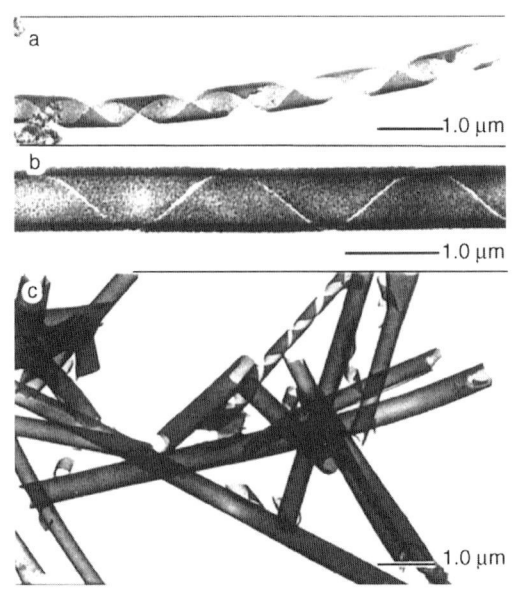

Figure 5.39 Spiral and tubular silica structures created by silica synthesis in the presence of a self-assembled gelator, as imaged by TEM. Reproduced from Ref. [91c] with permission of the American Chemical Society.

synthesis of CdS, they had a direct impact on the morphology of the product – TEM analysis indicated the formation of right- and left-handed helices of CdS, which had been templated on the nanoribbon architecture assembled by the dendron rod-coil. ZnO nanocrystals have also been generated using dendron rod-coils in a similar approach [93]. When these organic–inorganic hybrid assemblies were placed in an electric field, a degree of alignment occurred and these aligned nanocomposites had a lower threshold for lasing behavior than pure ZnO nanocrystals. These results indicate how self-assembly nanofabrication methods can be used to generate *hybrid nanomaterials* with potential photonic applications.

In related studies, gelators have been present during the polymerization of monomers such as styrene or methyl methacrylate. It has been demonstrated that polymerization of these gelated monomers leads to materials which have embedded nanostructures [94]. These *embedded nanostructures* can have a direct impact on the materials properties of the resultant polymer, leading to toughening/hardening effects and, in general terms, this approach may yield new advanced functional materials, in particular if functional gelators can be embedded within the polymer network. Subsequent washing of these nanoembedded polymers can often remove the self-assembled fibers and generate materials that are effectively imprinted by the memory of the nanoscale self-assembled networks [95]. Such *nanoimprinted polymers* have potential applications in catalysis and separation science.

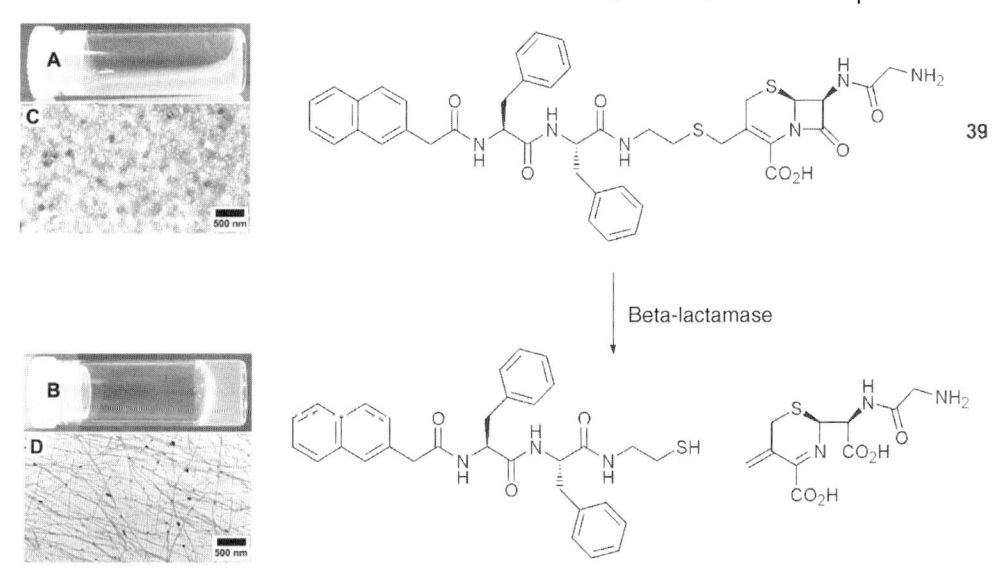

Figure 5.40 Compound **39**, which reacts with β-lactamase enzymes, present in penicillin-resistant bacteria, to generate a more effective gelator. Adapted from Ref. [96] with permission of the American Chemical Society.

5.5.6
Sensory Gels

A recent study has demonstrated the potential of hydrogel materials to act as sensors [96]. Compound **39** is not a gelator in its own right, but can be considered as a latent gelator. This compound includes a β-lactam unit; such structures are important in penicillin-type antibiotic drugs (Figure 5.40). One of the key mechanisms of bacterial resistance to penicillin-type drugs is caused by hydrolysis of the β-lactam ring by β-lactamase enzymes. Compound **39** is therefore also susceptible to β-lactamase enzymes and, in their presence, a chemical conversion takes place which activates the gelator, making it able to form a self-supporting network.

This system, therefore, is capable of *detecting the presence of penicillin-resistant bacteria* by undergoing a sol–gel transition – a clear and easily monitored *sensory response*. It was demonstrated that this system could operate in biological cell lysates and therefore clearly illustrates how gelation can be a useful detection event which can be incorporated into biosensors. The ability to modify the structure of gelators in a precise manner using organic synthesis means that their potential in this field is very high.

5.5.7
Conductive Gels

Given that gel-phase materials are usually underpinned by fibrillar objects, the analogy of these nanoscale structures with *"wires"* is evident. There has therefore

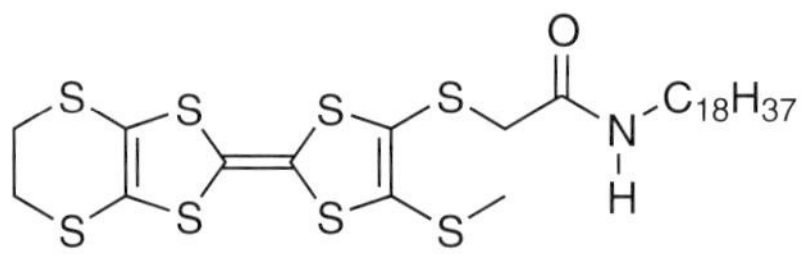

Figure 5.41 Compound **40**.

been considerable interest in the incorporation of conjugated molecular building blocks into gel-phase materials. This can be considered as a bottom-up approach to the construction of well-defined nanoscale wires. Recently, compound **40** has been demonstrated to form gel-phase materials in hexane (Figure 5.41) [97]. Drying the gel on a TEM grid to give the aerogel, followed by doping with iodine vapor and annealing, gave rise to an *electrically conducting thin film*. The hydrogen bond interactions which underpin gelation were essential for organizing the molecular-scale building blocks in order to generate nanowires, which could be visualized by TEM. The self-assembly approach to nanofabrication has considerable potential in the development of *molecular-scale electronics*.

5.6
Conclusions

Gel-phase materials are one of the most intriguing classes of self-assembled material and one of the few systems in which the impact of *synthetic changes made on the molecular level* is clearly evident on the *nanoscale* and furthermore has direct control over *macroscopic materials behavior* visible to the naked eye. This chapter has aimed to provide a "primer" to the area and tried to *highlight key principles* and *emerging themes* in terms of gelator design and applications. There is no doubt that over the coming years, the simplicity of *nanofabrication* using a *self-assembly strategy* will ensure that gel-phase materials are increasingly widely explored. In particular, it seems likely that materials incorporating ever increasing degrees of functional behavior will be developed and for this reason it is predicted that nanochemistry and nanofabrication methods will become dominant themes of new manufacturing industries in the 21st century.

References

1 Whitesides, G.M. (2005) *Small*, **1**, 172–179.
2 For reviews, see (a) Low Molecular Mass Gelators – Design, Self Assembly, Function, F. Fages (ed.), *Top. Curr. Chem.* (2005) p. 256. (b) Terech, P. and Weiss, R.G. (1997) *Chem. Rev.*, **97**, 3133–3159. (c) Gronwald, O., Snip, E. and Shinkai, S. (2002) *Curr. Opin. Colloid Interface Sci.*, **7**, 148–156. (d) Shimizu, T. (2003) *Polym. J.*, **35**, 1–22. (e) Sangeetha, N.M. and Maitra, U. (2005) *Chem. Soc. Rev.*, **34**,

821–836. (f) Smith, D.K. (2007) *Tetrahedron*, **63**, 7283–7284.

3 R.G. Weiss and P. Terech (eds) (2006) *Molecular Gels, Materials with Self-Assembled Fibrillar Networks*, Springer, Dordrecht.

4 Jordan Lloyd, D. (1926) *Colloid Chemistry* (ed. J. Alexander), Chemical Catalog Co., New York, vol. 1, pp. 767–782.

5 (a) Graham, T. (1861) *Philos. Trans. R. Soc. London*, **151**, 183–224. (b) Dean, R.B. (1948) *Modern Colloids*, Van Nostrand, New York, p.2. (c) Hermans, P.H. (1949) *Colloid Science* (ed. H.R. Kruyt), Elsevier, Amsterdam, vol. II, p. 484. (d) Ferry, J.D. (1961) *Viscoelastic Properties of Polymers*, Wiley, New York, p. 391. (e) Flory, P. (1974) *Discuss. Faraday Soc.*, **57**, 7–18. (f) Gelbart, W.M. and Ben-Shaul, A. (1996) *J. Phys. Chem.*, **100**, 13169–13189.

6 (a) Zsigmondy, R. and Batchmann, W. (1912) *Z. Chem. Ind. Kolloide*, **11**, 145–157, this paper indicates that this phenomenon had been recognized since the 19th century. (b) Hatschek, F. (1912) *Z. Chem. Ind. Kolloide*, **11**, 158–165. (c) Hardy, W.B. (1900) *Proc. R. Soc. London*, **66**, 95–109.

7 (a) Brunsveld, L., Folmer, B.J.B., Meijer, E.W. and Sijbesma, R.P. (2001) *Chem. Rev.*, **101**, 4071–4097. (b) Ciferri, A. (ed.) (2005) *Supramolecular Polymers*, Taylor and Francis, Boca Raton, FL.

8 Jonkheijm, P., van der Schoot, P., Schenning, A.P.H.J. and Meijer, E.W. (2006) *Science*, **313**, 80–83.

9 Aggeli, A., Nyrkova, I.A., Bell, M., Harding, R., Carrick, L., McLeish, T.C.B., Semenov, A.N. and Boden, N. (2001) *Proc. Natl. Acad. Sci. USA*, **98**, 11857–11862.

10 Li, Y., Wang, T. and Liu, M. (2007) *Tetrahedron*, **63**, 7468–7473.

11 For an example, see Trivedi, D.R. and Dastidar, P. (2006) *Chem. Mater.*, **18**, 1470–1478.

12 (a) Hirst, A.R. and Smith, D.K. (2004) *Langmuir*, **20**, 10851–10857. (b) Zhu, G. and Dordick, J.S. (2006) *Chem. Mater.*, **18**, 5988–5995.

13 Raghavan, S.R. and Cipriano, B.H. (2006) 'Gel Formation: Phase Diagrams using Tabletop Rheology and Calorimetry', in *Molecular Gels, Materials with Self-Assembled Fibrillar Networks* (eds. R.G. Weiss and P. Terech), Springer, Dordrecht, Chapter 8.

14 Hirst, A.R., Smith, D.K., Feiters, M.C. and Geurts, H.P.M. (2003) *J. Am. Chem. Soc.*, **125**, 9010–9011.

15 Macosko, C.W. (1994) *Rheology: Principles, Measurements and Applications*, VCH, New York.

16 Takahashi, A., Sakai, M., Kato, T. and Polym, J. (1980) **12**, 335–341.

17 Murata, K., Aoki, M., Suzuki, T., Hanada, T., Kawabata, H., Komori, T., Ohseto, F., Ueda, K. and Shinkai, S. (1994) *J. Am. Chem. Soc.*, **116**, 6664–6676.

18 Sollich, P. (2006) 'Soft Glassy Rheology', in *Molecular Gels, Materials with Self-Assembled Fibrillar Networks* (eds. R.G. Weiss and P. Terech), Springer, Dordrecht, Chapter 5.

19 For an excellent recent example, see Terech, P. and Friol, S. (2007) *Tetrahedron*, **63**, 7366–7374.

20 Wade, R.H., Terech, P., Hewat, E.A., Ramasseul, R. and Volino, F. (1986) *J. Colloid Interface Sci.*, **114**, 442–451.

21 (a) Abdallah, D.J., Sirchio, S.A. and Weiss, R.G. (2000) *Langmuir*, **16**, 7558–7561. (b) Ashbaugh, H.S., Radulescu, A., Prud'homme, R.K., Schwahn, D., Richter, D. and Fetters, L.J. (2002) *Macromolecules*, **35**, 7044–7055. (c) Schmidt, R., Adam, F.B., Michel, M., Schmutz, M., Decher, G. and Mésini, P.J. (2003) *Tetrahedron Lett.*, **44**, 3171–3174. (d) Hirst, A.R., Smith, D.K. and Harrington, J.P. (2005) *Chem. Eur. J.*, **11**, 6552–6559.

22 (a) Anne, M. (2006) 'X-Ray Diffraction of Poorly Organized Systems and Molecular Gels', in *Molecular Gels, Materials with Self-Assembled Fibrillar Networks* (eds R.G. Weiss and P. Terech), Springer, Chapter 11. (b) Glatter, O. and Kratky, O. (eds) (1982) *Small*

Angle X-ray Scattering, Academic Press, London.
23 Kumar, D.K., Jose, D.A. and Das, A. (2004) *Langmuir*, **20**, 10413–10418.
24 For selected examples, see (a) Terech, P., Ostuni, E. and Weiss, R.G. (1996) *J. Phys. Chem.*, **100**, 3759–3766. (b) Jeong, Y., Friggeri, A., Akiba, I., Masunaga, H., Sakurai, K., Sakurai, S., Okamoto, S., Inoue, K. and Shinkai, S. (2005) *J. Colloid Interface Sci.*, **283**, 113–122.
25 Escuder, B., Llusar, M. and Miravet, J.F. (2006) *J. Org. Chem.*, **71**, 7747–7752.
26 Schnell, I. (2005) *Curr. Anal. Chem.*, **1**, 3–27.
27 For a general overview of NMR titration methods in molecular recognition, see (a) Fielding, L. (2000) *Tetrahedron*, **56**, 6151–6170. For examples applied to gels, see (b) Schoonbeek, F.S., van Esch, J.H., Hulst, R., Kellogg, R.M. and Feringa, B.L. (2000) *Chem. Eur. J.*, **6**, 2633–2643. (c) Huang, B., Hirst, A.R., Smith, D.K., Castelletto, V. and Hamley, I.W. (2005) *J. Am. Chem. Soc.*, **127**, 7130–7139.
28 Hirst, A.R., Smith, D.K., Feiters, M.C. and Geurts, H.P.M. (2004) *Langmuir*, **20**, 7070–7077.
29 (a) Becerril, J., Burguette, M.I., Escuder, B., Galindo, F., Gavara, R., Miravet, J.F., Luis, S.V. and Peris, G. (2004) *Chem. Eur. J.*, **10**, 3879–3890. (b) Makarević, J., Jokić, M., Raza, Z., Stefanic, Z., Kojic-Prodic, B. and Žinić, M. (2003) *Chem. Eur. J.*, **9**, 557–5580.
30 (a) Capitani, D., Rossi, E., Segre, A.L., Giustini, M. and Luisi, P.L. (1993) *Langmuir*, **9**, 685–689. (b) Capitani, D., Segre, A.L., Dreher, F., Walde, P. and Luisi, P.L. (1996) *J. Phys. Chem.*, **100**, 15211–15217. (c) Duncan, D.C. and Whitten, D.G. (2000) *Langmuir*, **16**, 6445–6452.
31 Suzuki, M., Nanbu, M., Yumoto, M., Shirai, H. and Hanabusa, K. (2005) *New J. Chem.*, **29**, 1439–1444.
32 (a) Terech, P., Furman, I. and Weiss, R.G. (1995) *J. Phys. Chem.*, **99**, 9558–9566. (b) Geiger, C., Stanescu, M., Chen, L. and Whitten, D.G. (1999) *Langmuir*, **15**, 2241–2245. (c) Ajayaghosh, A. and George, S.J. (2001) *J. Am. Chem. Soc.*, **123**, 5148–5149. (d) Ikeda, M., Takeuchi, M. and Shinkai, S. (2003) *Chem. Commun.*, 1354–1355. (e) Sugiyasu, K., Fujita, N. and Shinkai, S. (2004) *Angew. Chem. Int. Ed.*, **43**, 1229–1233. (f) Desvergne, J.-P., Brotin, T., Meerschaut, D., Clavier, G., Placin, F., Pozzo, J.-L. and Bouas-Laurent, H. (2004) *New J. Chem.*, **28**, 234–247.
33 Kamikawa, Y. and Kato, T. (2007) *Langmuir*, **23**, 274–278.
34 Coates, I.A., Hirst, A.R. and Smith, D.K. (2007) *J. Org. Chem.*, **72**, 3937–3940.
35 Berova, N., Nakanishi, K. and Woody, R.W. (1994) *Circular Dichroism: Principles and Applications*, 2nd edn. Wiley-VCH, Weinheim.
36 Mateos-Timoneda, M.A., Crego-Calama, M. and Reinhoudt, D.N. (2004) *Chem. Soc. Rev.*, **33**, 363–372.
37 Hirst, A.R., Castelletto, V., Hamley, I.W. and Smith, D.K. (2007) *Chem. Eur. J.*, **13**, 2180–2188.
38 (a) Hirst, A.R., Smith, D.K., Feiters, M.C. and Geurts, H.P.M. (2004) *Chem. Eur. J.*, **10**, 5901–5910. (b) Fuhrhop, J.-H., Schnieder, P., Rosenberg, J. and Boekema, E. (1987) *J. Am. Chem. Soc.*, **109**, 3387–3390. (c) Fuhrhop, J.-H., Bedurke, T., Hahn, A., Grund, S., Gatzmann, J. and Riederer, M. (1994) *Angew. Chem. Int. Ed. Engl.*, **33**, 350–351. (d) Jokić, M., Makarević, J. and Žinić, M. (1995) *J. Chem. Soc., Chem. Commun.*, 1723–1724. (e) Bhattacharya, S., Acharya, S.N.G. and Raju, A.R. (1996) *Chem. Commun.*, 2101–2102. (f) Mamiya, J.I., Kanie, K., Hiyama, T., Ikeda, T. and Kato, T. (2002) *Chem. Commun.*, 1870–1871. (g) Boettcher, C., Schade, B. and Fuhrhop, J.-H. (2001) *Langmuir*, **17**, 873–877. (h) Becerril, J., Escuder, B., Miravet, J.F., Gavara, R. and Luis, S.V. (2005) *Eur. J. Org. Chem.*, 481–485. (i) Koga, T., Matsuoka, M. and Higashi, N. (2005) *J. Am. Chem. Soc.*, **127**, 17596–17597.
39 (a) van Gestel, J. (2004) *Macromolecules*, **37**, 3894–3898. (b) van Gestel, J., Palmans,

A.R.A., Titulaer, B., Vekemans, J.A.J.M. and Meijer, E.W. (2005) *J. Am. Chem. Soc.*, **127**, 5490–5494.

40 (a) Singh, A., Burke, T.G., Calvert, J.M., Georger, J.H., Herendeen, B., Price, R.R., Schoen, P.E. and Yager, P. (1988) *Chem. Phys. Lipids*, **47**, 135–148. (b) Yamada, N., Sasaki, T., Murata, H. and Kunitake, T. (1989) *Chem. Lett.*, 205–208. (c) Gulik-Krzywicki, T., Fouquey, C. and Lehn, J.-M. (1993) *Proc. Natl. Acad. Sci. USA*, **90**, 163–167. (d) Messmore, B.W., Sukerkar, P.A. and Stupp, S.I. (2005) *J. Am. Chem. Soc.*, **127**, 7992–7993. (e) Terech, P., Rodriguez, V., Barnes, J.D. and McKenna, G.B. (1994) *Langmuir*, **10**, 3406–3418. (f) Spector, M.S., Selkinger, J.V., Singh, A., Rodriguez, J.M., Price, R.R. and Schnur, J.M. (1998) *Langmuir*, **14**, 3493–3500.

41 Fages, F., Vögtle, F. and Žinić, M. (2005) *Top. Curr. Chem.*, **256**, 77–131.

42 (a) Tomioka, K., Sumiyoshi, T., Narui, S., Nagaoka, Y., Iida, A., Miwa, Y., Taga, T., Nakano, M. and Handa, T. (2001) *J. Am. Chem. Soc.*, **123**, 11817–11818. (b) Hanabusa, K., Tanaka, R., Suzuki, M., Kimura, M. and Shirai, H. (1997) *Adv. Mater.*, **9**, 1095–1097; (c) Luo, X., Li, C. and Liang, Y. (2000) *Chem. Commun.*, 2091–2092. (d) van Gorp, J.J., Vekemans, J.A.J.M. and Meijer, E.W. (2002) *J. Am. Chem. Soc.*, **124**, 14759–14769.

43 Hanabusa, K., Yamada, M., Kimura, M. and Shirai, H. (1996) *Angew. Chem., Int. Ed. Engl.*, **35**, 1949–1951.

44 (a) Fuhrhop, J.-H., Spiroski, D. and Boettcher, C. (1993) *J. Am. Chem. Soc.*, **115**, 1600–1601. (b) Franceschi, S., de Viguerie, N., Riviere, M. and Lattes, A. (1999) *New J. Chem.*, **23**, 447–452. (c) D'Aléo, A., Pozzo, J.-L., Fages, F., Schmutz, M., Mieden-Gundert, G., Vögtle, F., Caplar, V. and Žinić, M. (2004) *Chem. Commun.*, 190–191. (d) Nakashima, T., Kimizuka, N. (2002) *Adv. Mater.*, **14**, 1113–1116. (e) Suzuki, M., Yumoto, M., Kimura, M., Shirai, H. and Hanabusa, K. (2003) *Chem. Eur. J.*, **9**, 348–354.

45 Estroff, L.A. and Hamilton, A.D. (2001) *Angew. Chem. Int. Ed.*, **39**, 3447–3450.

46 Frkanec, L., Jokić, M., Makarević, J., Wolsperger, K. and Žinić, M. (2002) *J. Am. Chem. Soc.*, **124**, 9716–9717.

47 Fujita, N. and Shinkai, S. (2006) Design and Function of Low Molecular-Mass Organic Gelator (LMOGs) Bearing Steroid and Sugar Groups, in *Molecular Gels, Materials with Self-Assembled Fibrillar Networks* (eds R.G. Weiss and P. Terech), Springer, Dordrecht, Chapter 15.

48 (a) Kiyonaka, S., Sugiyasu, K., Shinkai, S. and Hamachi, I. (2002) *J. Am. Chem. Soc.*, **124**, 10954–10955. (b) Kiyonaka, S., Shinkai, S. and Hamachi, I. (2003) *Chem. Eur. J.*, **9**, 976–983.

49 (a) Meunier, M.J. (1891) *Ann. Chim. Phys.*, **22**, 412. (b) Yamamoto, S. (1942) *Kogyou Kagaku Zasshi*, **45**, 695. (c) Yamamoto, S. (1943) *J. Chem. Soc. Jpn., Ind. Chem. Sect.*, **46**, 779.

50 Mohmeyer, N., Wang, P., Schmidt, H.-W., Zakeeruddin, S.M. and Grätzel, M. (2004) *J. Mater. Chem.*, **14**, 1905–1909.

51 See, for example, (a) James, T.D., Murata, K., Harada, T., Ueda, K. and Shinkai, S. (1994) *Chem. Lett.*, 273–276. (b) Amanokura, N., Yoza, K., Shinmori, H. and Shinkai, S. (1998) *J. Chem. Soc., Perkin Trans.*, **2**, 2585–2591. (c) Yoza, K., Amanokura, N., Ono, Y., Akao, T., Shinmori, H., Takeuchi, M., Shinkai, S. and Reinhoudt, D.N. (1999) *Chem. Eur. J.*, **5**, 2722–2729. (d) Luboradzki, R., Gronwald, O., Ikeda, M., Shinkai, S. and Reinhoudt, D.N. (2000) *Tetrahedron*, **56**, 9595–9599.

52 (a) Pfannemüller, B. and Welte, W. (1985) *Chem. Phys. Lipids*, **37**, 227–240. (b) Fuhrhop, J.-H., Schnieder, P., Rosenberg, J. and Boekema, E. (1987) *J. Am. Chem. Soc.*, **109**, 3387–3390. (c) Bhattacharya, S. and Acharya, S.N.G. (1999) *Chem. Mater.*, **11**, 3504–3511.

53 Kobayashi, H., Friggeri, A., Koumoto, K., Amaike, M., Shinkai, S. and Reinhoudt, D.N. (2002) *Org. Lett.*, **4**, 1423–1426.

54 Žinić, M., Vögtle, F. and Fages, F. (2005) *Top. Curr. Chem.*, **256**, 39–76.

55 (a) Martin-Borret, O., Ramasseul, R. and Rassat, R. (1979) *Bull. Soc. Chim. Fr.*, **7–8**, II-401. (b) Terech, P., Ramasseul, R. and Volino, F. (1983) *J. Colloid Interface Sci.*, **91**, 280–282. (c) Terech, P. (1985) *J. Colloid Interface Sci.*, **107**, 244–255.

56 (a) Lin, Y.-c. and Weiss, R.G. (1987) *Macromolecules*, **20**, 414–417. (b) Lin, Y.-c., Kachar, B. and Weiss, R.G. (1989) *J. Am. Chem. Soc.*, **111**, 5542–5551. (c) Mukkamala, R. and Weiss, R.G. (1995) *J. Chem. Soc., Chem. Commun.*, 375–376. (d) Mukkamala, R. and Weiss, R.G. (1996) *Langmuir*, **12**, 1474–1482.

57 Murata, K., Aoki, M., Suzuki, T., Harada, T., Kawabata, H., Komori, T., Ohseto, F., Ueda, K. and Shinkai, S. (1994) *J. Am. Chem. Soc.*, **116**, 6664–6676.

58 (a) Schryver, S.B. (1914) *Proc. R. Soc. London, Ser. B*, **87**, 366–374. (b) Sobotka, H. and Czeczowiczka, N. (1958) *J. Colloid Sci.*, **13**, 188–191. (c) Rich, A. and Blow, D.M. (1958) *Nature*, **182**, 423–426. (d) Blow, D.M. and Rich, A. (1960) *J. Am. Chem. Soc.*, **82**, 3566–3571.

59 Maitra, U., Mukhopadhyay, S., Sarkar, A., Rao, P. and Indi, S.S. (2001) *Angew. Chem. Int. Ed.*, **40**, 2281–2283.

60 Araki, K. and Yoshikawa, I. (2005) *Top. Curr. Chem.*, **256**, 133–165.

61 (a) Gottarelli, G., Masiero, S., Mezzina, E., Spada, G.P., Mariani, P. and Recanatini, M. (1998) *Helv. Chim. Acta*, **81**, 2078–2092. (b) Giorgi, T., Grepioni, F., Manet, I., Mariani, P., Masiero, S., Mezzina, E., Pieraccini, S., Saturni, L., Spada, G.P. and Gottarelli, G. (2002) *Chem. Eur. J.*, **8**, 2143–2152. (c) Sato, T., Seko, M., Takasawa, R., Yoshikawa, I. and Araki, K. (2001) *J. Mater. Chem.*, **11**, 3018–3022.

62 Yun, Y.J., Park, S.M. and Kim, B.H. (2003) *Chem. Commun.*, 254–255.

63 Sugiyasu, K., Numata, M., Fujita, N., Park, S.M., Yun, Y.J., Kim, B.H. and Shinkai, S. (2004) *Chem. Commun.*, 1996–1997.

64 (a) Shimizu, T., Iwaura, R., Masuda, M., Hanada, T. and Yase, K. (2001) *J. Am. Chem. Soc.*, **123**, 5947–5955. (b) Shimizu, T. (2002) *Macromol. Rapid Commun.*, **23**, 311–331. (c) Iwaura, R., Yoshida, K., Masuda, M., Yase, K. and Shimizu, T. (2002) *Chem. Mater.*, **14**, 3047–3053.

65 Iwaura, R., Yoshida, K., Masuda, M., Ohnishi-Kameyama, M., Yoshida, M. and Shimizu, T. (2003) *Angew. Chem. Int. Ed.*, **42**, 1009–1012.

66 For a review of G quartets, see (a) Davis, J.T. (2004) *Angew. Chem. Int. Ed.*, **43**, 668–698. For examples, see (b) Ghoussoub, A. and Lehn, J.-M. (2005) *Chem. Commun.*, 5763–5765. (c) Sreenivasachary, N. and Lehn, J.-M. (2005) *Proc. Natl. Acad. Sci. USA*, **102**, 5938–5943.

67 http://nobelprize.org/nobel_prizes/chemistry/laureates/1925/zsigmondy-lecture.pdf.

68 For reviews, see (a) Hirst, A.R. and Smith, D.K. (2005) *Top. Curr. Chem.*, **256**, 237–273. (b) Smith, D.K. (2006) *Adv. Mater.*, **18**, 2773–2778.

69 Smith, D.K., Hirst, A.R., Love, C.S., Hardy, J.G., Brignell, S.V. and Huang, B. (2005) *Prog. Polym. Sci.*, **30**, 220–293.

70 (a) Jang, W.-D., Jiang, D.-L. and Aida, T. (2000) *J. Am. Chem. Soc.*, **122**, 3232–3233. (b) Jang, W.-D. and Aida, T. (2003) *Macromolecules*, **36**, 8461–8469.

71 Love, C.S., Hirst, A.R., Chechik, V., Smith, D.K., Ashworth, I. and Brennan, C. (2004) *Langmuir*, **10**, 6580–6585.

72 (a) Zubarev, E.R., Pralle, M.U., Sone, E.D. and Stupp, S.I. (2001) *J. Am. Chem. Soc.*, **123**, 4105–4106. (b) Zubarev, E.R. and Stupp, S.I. (2002) *J. Am. Chem. Soc.*, **124**, 5762–5773. (c) de Gans, B.J., Wiegand, S., Zubarev, E.R. and Stupp, S.I. (2002) *J. Phys. Chem. B*, **106**, 9730–9736.

73 (a) Newkome, G.R., Baker, G.R., Saunders, M.J., Russo, P.S., Gupta, V.K., Yao, Z.Q., Miller, J.E. and Bouillion, K. (1986) *J. Chem. Soc., Chem. Commun.*, 752–753. (b) Newkome, G.R., Baker, G.R., Arai, S., Saunders, M.J., Russo, P.S., Theriot, K.J., Moorefield, C.N., Rogers, L.E., Miller, J.E., Lieux, T.R., Murray, M.E., Phillips, B. and

Pascal, L. (1990) *J. Am. Chem. Soc.*, **112**, 8458–8465. (c) Newkome, G.R., Moorefield, C.N., Baker, G.R., Behera, R.K., Escamilla, G.H. and Saunders, M.J. (1992) *Angew. Chem. Int. Ed. Engl.*, **31**, 917–919. (d) Yu, K.H., Russo, P.S., Younger, L., Henk, W.G., Hua, D.W., Newkome, G.R. and Baker, G. (1997) *J. Polym. Sci., Polym. Phys. Ed.*, **35**, 2787–2793. (e) Newkome, G.R., Lin, X.F., Yaxiong, C. and Escamilla, G.H. (1993) *J. Org. Chem.*, **58**, 3123–3129.

74 For a review, see Hirst, A.R. and Smith, D.K. (2005) *Chem. Eur. J.*, **11**, 5496–5508.

75 Hanabusa, K., Miki, T., Taguchi, Y., Koyama, T. and Shirai, H. (1993) *J. Chem. Soc., Chem. Commun.*, 1382–1384.

76 (a) Xu, X., Ayyagari, M., Tata, M., John, V.T. and McPherson, G.L. (1993) *J. Phys. Chem.*, **97**, 11350–11353. (b) Tata, M., John, V.T., Waguespack, Y.Y. and McPherson, G.L. (1994) *J. Am Chem. Soc.*, **116**, 9464–9470. (c) Tata, M., John, V.T., Waguespack, Y.Y. and McPherson, G.L. (1994) *J. Phys. Chem.*, **98**, 3809–3817. (d) Simmons, B.A., Taylor, C.E., Landis, F.A., John, V.T., McPherson, G.L., Schwartz, D.K. and Moore, R. (2001) *J. Am. Chem. Soc.*, **123**, 2414–2421. (e) Simmons, B., Li, S.C., John, V.T., McPherson, G.L., Taylor, C., Schwartz, D.K. and Maskos, K. (2002) *Nano Lett.*, **2**, 1037–1042.

77 (a) Partridge, K.S., Smith, D.K., Dykes, G.M. and McGrail, P.T. (2001) *Chem. Commun.*, 319–320. (b) Hirst, A.R., Smith, D.K., Feiters, M.C. and Geurts, H.P.M. (2003) *J. Am. Chem. Soc.*, **125**, 9010–9011. (c) Hirst, A.R. and Smith, D.K. (2004) *Org. Biomol. Chem.*, **2**, 2965–2971. (d) Hardy, J.G., Hirst, A.R., Smith, D.K., Ashworth, I. and Brennan, C. (2005) *Chem. Commun.*, 385–387.

78 Kawano, S., Fujita, N. and Shinkai, S. (2003) *Chem. Commun.*, 1352–1353.

79 (a) Brignell, S.V. and Smith, D.K. (2007) *New J. Chem.*, **31**, 1243–1249. (b) Sohna, J.E.S. and Fages, F. (1997) *Chem. Commun.*, 327–328. (c) Amanokura, N., Kanekiyo, Y., Shinkai, S. and Reinhoudt, D.N. (1999) *J. Chem. Soc., Perkin Trans.*, **2**, 1999–2005. (d) Stanley, C.E., Clarke, N., Anderson, K.M., Elder, J.A., Lenthall, J.T. and Steed, J.W. (2006) *Chem. Commun.*, 3199–3201.

80 (a) Maitra, U., Kumar, P.V., Chandra, N., D'Souza, L.J., Prasanna, M.D. and Raju, A.R. (1999) *Chem. Commun.*, 595–596. (b) Babu, P., Sangeetha, N.M., Vijaykumar, P., Maitra, U., Rissanen, K. and Raju, A.R. (2003) *Chem. Eur. J.*, **9**, 1922–1932.

81 (a) Zhang, Y., Gu, H., Yang, Z. and Xu, B. (2003) *J. Am. Chem. Soc.*, **125**, 13680–13681. (b) Zhang, Y., Yang, Z., Yuan, F., Gu, H., Gao, P. and Xu, B. (2004) *J. Am. Chem. Soc.*, **126**, 15028–15029.

82 For an accessible overview, see Donahue, C.J. (2006) *J. Chem. Educ.*, **83**, 862–869.

83 Dresel, W. and Heckler, R. (2001) Lubricating Greases, in *Lubricants and Lubrications* (eds T. Mang and W. Dresel), Wiley-VCH, Wenheim, Chapter 16.

84 (a) Fieser, L.F., Harris, G.C., Hershberg, E.B., Morgana, M., Novello, F.C. and Putnam, S.T. (1946) *Ind. Eng. Chem.*, **38**, 768–773. (b) Mysels, K.J. (1949) *Ind. Eng. Chem.*, **41**, 1435–1438.

85 (a) Woolfson, D.N. and Ryadnov, M.G. (2006) *Curr. Opin. Chem. Biol.*, **10**, 559–567. (b) Levenberg, S. and Langer, R. (2004) *Curr. Top. Dev. Biol.*, **61**, 113–134.

86 Ellis-Behnke, R.G., Liang, Y.-X., You, S.-W., Tay, D.K.C., Zhang, S., So, K.-F. and Schneider, G.E. (2006) *Proc. Natl. Acad. Sci. USA*, **103**, 5054–5059.

87 Rajangam, K., Behanna, H.A., Hui, M.J., Han, X., Hulvat, J.F., Lomasney, J.W. and Stupp, S.I. (2006) *Nano Lett.*, **6**, 2086–2090.

88 Chitkara, D., Shikanov, A., Kumar, N. and Domb, A.J. (2006) *Macromol. Biosci.*, **6**, 977–990.

89 de Jong, J.J.D., Feringa, B.L. and van Esch, J. (2006) Responsive Molecular Gels, in *Molecular Gels, Materials with Self-Assembled Fibrillar Networks* (eds R.G. Weiss and P. Terech), Springer, Dordrecht, Chapter 26.

90 Ono, Y., Nakashima, K., Sano, M., Kanekiyo, Y., Inoue, K., Hojo, J. and Shinkai, S. (1998) *Chem. Commun.*, 1477–1478.

91 (a) Ono, Y., Nakashima, K., Sano, M., Hojo, J. and Shinkai, S. (1999) *Chem. Lett.*, 1119–1120. (b) Jung, J.H., Ono, Y. and Shinkai, S. (2000) *Angew. Chem. Int. Ed.*, **39**, 1862–1865. (c) Jung, J.H., Kobayashi, H., Masuda, M., Shimizu, T. and Shinkai, S. (2001) *J. Am. Chem. Soc.*, **123**, 8785–8789. (d) Sugiyasu, K., Tamaru, S., Takeuchi, M., Berthier, D., Huc, I., Oda, R. and Shinkai, S. (2002) *Chem. Commun.*, 1212–1213. (e) Kawano, S.-i., Tamaru, S.-i., Fujita, N. and Shinkai, S. (2004) *Chem. Eur. J.*, **10**, 343–351.

92 Sone, E.D., Zubarev, E.R. and Stupp, S.I. (2002) *Angew. Chem. Int. Ed.*, **41**, 1705–1709.

93 Li, L., Beniash, E., Zubarev, E.R., Xiang, W., Rabatic, B.M., Zhang, G. and Stupp, S.I. (2003) *Nature Mater.*, **2**, 689–694.

94 (a) Zubarev, E.R., Pralle, M.U., Sone, E.D. and Stupp, S.I. (2002) *Adv. Mater.*, **14**, 198–203. (b) Stendahl, J.C., Li, L.M., Zubarev, E.R., Chen, Y.-R. and Stupp, S.I. (2002) *Adv. Mater.*, **14**, 1540–1543. (c) Stendahl, J.C., Zubarev, E.R., Arnold, M.S., Hersam, M.C., Sue, H.J. and Stupp, S.I. (2005) *Adv. Funct. Mater.*, **15**, 487–493.

95 (a) Hafkamp, R.J.H., Kokke, B.P.A., Danke, I.M., Geurts, H.P.M., Rowan, A.E., Feiters, M.C. and Nolte, R.J.M. (1997) *Chem. Commun.*, 545–546. (b) Gu, W., Lu, L., Chapman, G.B. and Weiss, R.G. (1997) *Chem. Commun.*, 543–545. (c) Beginn, U., Keinath, S. and Möller, M. (1998) *Macromol. Chem. Phys.*, **199**, 2379–2384. (d) Beginn, U., Sheiko, S. and Möller, M. (2000) *Macromol. Chem. Phys.*, **201**, 1008–1015. (e) Tan, G., Singh, M., He, J., John, V.T. and McPherson, G.L. (2005) *Langmuir*, **21**, 9322–9326.

96 Yang, Z., Ho, P.-L., Liang, G., Chow, K.H., Wang, Q., Cao, Y., Guo, Z. and Xu, B. (2007) *J. Am. Chem. Soc.*, **129**, 266–267.

97 Puigmartí-Luis, J., Laukhin, V., Pérez del Pino, A., Vidal-Gancedo, J., Rovira, C., Laukhina, E. and Amabilino, D.B. (2007) *Angew. Chem. Int. Ed.*, **46**, 238–241.

6
Nanoporous Crystals, Co-crystals, Isomers and Polymorphs from Crystals

Dario Braga, Marco Curzi, Stefano L. Giaffreda, Fabrizia Grepioni, Lucia Maini, Anna Pettersen, and Marco Polito

6.1
Introduction

Molecules have nanometric dimensions, hence crystal engineering, the intelligent and purposeful assembly of molecules in crystalline structures, falls in the burgeoning field of nanochemistry [1–3]. Crystal engineering [4] is at the intersection of supramolecular chemistry [5,6] (molecular aggregates of higher complexity, "supermolecules") with materials chemistry [7–11] (materials made of molecules, i.e. "molecular materials"). Research in supramolecular chemistry has demonstrated that molecules are convenient nanometer-scale building blocks that can be used, in a bottom-up approach, to construct larger aggregates, whether supermolecules or crystalline materials. The controlled assembly and manipulation of three-dimensional nanostructures with well-defined shapes, profiles and functionalities present a significant challenge to nanotechnology. Since chemists know how to synthesize, characterize and exploit molecular aggregation, the molecular approach to functional nanostructures is a natural development of progress in the field.

Making crystals by design is the paradigm of crystal engineering [12,13], the main goal being that of obtaining *collective* properties from the convolution of the physical and chemical properties of the individual building blocks with crystal periodicity and symmetry. This is also the main objective of nanotechnology because the aggregation via *intermolecular* bonds of the component units (molecules or ions) will determine the nature of the *collective* properties and potential applications. Moreover, if "holes", "channels" or "cavities" are present in the crystalline materials because of the way in which the crystalline scaffold is organized, the size of these empty spaces will have, necessarily, nanometric dimensions. The "bottom-up" construction of such materials requires a crystal engineering approach. This will be the focus of this chapter.

One can envisage two main sub-areas of crystal engineering, namely that of coordination frameworks [14–27] or *periodical coordination chemistry* and that of molecular materials [28–39]. This is a practical subdivision, however, and all possible intermediate situations are possible. The possibility of exploiting metal–organic coordination frameworks (MOFs) for practical applications (such as absorption of molecules and reactions in cavities) critically depends on whether the networks contain large empty spaces [40–46] or are close packed because of interpenetration and self-entanglement. [27,47–50] The possibility of a sponge-like behavior to accommodate/release guest molecules is of paramount importance if applications of coordination networks as molecular reservoirs, traps and sensors are sought [24,51–54].

Whereas in periodical coordination chemistry it is useful to focus on the *knots* and *spacers* in order to describe the topology of the network, when dealing with molecular materials what matters most are the characteristics of the component molecules or ions and the type of interactions (van der Waals, hydrogen bonds, π-stacking, ionic interactions, ion pairs, etc.) holding building blocks together [54–72]. The self-assembly of "zeotype" coordination compounds is an extremely active area of research, especially for the synthesis of metal–organic coordination networks, whether finite (triangles, squares, etc.) [73–76] or infinite networks [27,77–83]. The reader is referred to the many relevant books and reviews on this subject [84–96].

This chapter is devoted to the progress made in our research group in the "bottom-up" construction of crystalline aggregates. In particular, we will describe recent results obtained in the construction of (i) nanoporous coordination network crystals able to uptake/release small molecules, (ii) hybrid organic–organometallic and inorganic–organometallic co-crystals and (iii) isomers and polymorphs of crystals obtained from preformed crystalline materials. In the last section we will discuss the dynamic behavior of crystalline systems under the effect of changing temperature.

6.2
Nanoporous Coordination Network Crystals for Uptake/Release of Small Molecules

As mentioned above, the construction of MOFs takes the "lion's share" of the current efforts in the preparation and exploitation of nanoporous materials. Recently, it has been shown that, in addition to traditional solution chemistry or more forcing hydrothermal conditions, coordination framework structures can also be easily prepared by mechanical methods, i.e. by simple grinding of the component units in the air.

Mechanochemical methods pertain to the domain of organic solid-state chemistry. Pioneering work has been carried out by the groups of Rastogi, Etter, Curtin and Paul [97–103]. More recently, there has been an upsurge of interest in solventless methods as it has become apparent that these methods could be used successfully to prepare supramolecular aggregates, crystalline materials, co-crystals and new crystal forms of a same species (polymorphs and solvates). Remarkable results have also been obtained by Steed, Batten, Raston and coworkers [104–107].

We have explored the preparation of 1-D network structures by co-grinding of transition metal salts and organic dinitrogen base that could be used as *divergent* ligands. In particular *trans*-1,4-diaminocyclohexane, $H_2NC_6H_{10}NH_2$ (dace)

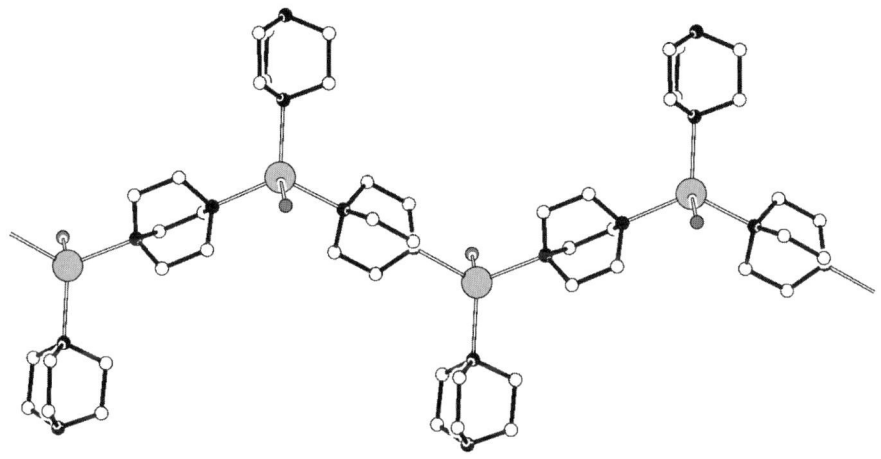

Figure 6.1 The coordination network in Ag[N(CH$_2$CH$_2$)$_3$N]$_2$[CH$_3$COO]·5H$_2$O. Note the chain of Ag$^{(+)}$–[N(CH$_2$CH$_2$)$_3$N]–Ag$^{(+)}$–[N(CH$_2$CH$_2$)$_3$N]–Ag$^{(+)}$ with each silver cation carrying an extra pendant [N(CH$_2$CH$_2$)$_3$N] ligand and a coordinated water molecule in tetrahedral coordination geometry. H atoms not shown for clarity.

and 1,4-diazabicyclo[2.2.2]octane, N(CH$_2$CH$_2$)$_3$N (dabco), coordination polymer [40–46,108,109], Ag[N(CH$_2$CH$_2$)$_3$N]$_2$[CH$_3$COO]·5H$_2$O, has been prepared by co-grinding of silver acetate and dabco in a 1 : 2 ratio (Figure 6.1).

When ZnCl$_2$ is used instead of AgCOOCH$_3$ in the equimolar reaction with dabco, different products are obtained from the solution and solid-state reactions. Figure 6.2 shows that the structure of Zn[N(CH$_2$CH$_2$)$_3$N]Cl$_2$ is based on a one-dimensional coordination network constituted of alternating [N(CH$_2$CH$_2$)$_3$N] and ZnCl$_2$ units, joined by Zn—N bonds.

More recently, we have applied the same procedure to the solid-state co-grinding of AgCOOCH$_3$ and dace in a 1 : 1 ratio, resulting in the preparation of a crystalline powder tentatively formulated as Ag[dace][CH$_3$COO]·xH$_2$O [110]. We have ascertained the formation of different isomers of the same coordination network depending on the preparation and crystallization conditions. The relationship between supramolecular isomerism and network topology has been discussed [6]. Crystallization of the same compound from anhydrous MeOH yields two types of products depending on the solvent evaporation conditions: crystals of Ag[dace][CH$_3$COO][MeOH]·0.5H$_2$O are obtained by crystallization under an argon flow, whereas slow evaporation in the air results in crystals of Ag[dace][CH$_3$COO]·3H$_2$O. Structural analysis shows that both of these compounds contain two isomeric forms of the coordination network {Ag[dace]$^+$}$_\infty$. If the same reaction between AgCOOCH$_3$ and dace is carried out directly in MeOH–water solution, a third crystalline material is obtained, namely the tetrahydrate Ag[dace][CH$_3$COO]·4H$_2$O. In all cases, correspondence between bulk powder and single crystals was ascertained by comparing computed and observed powder diffractograms. In terms of chemical composition,

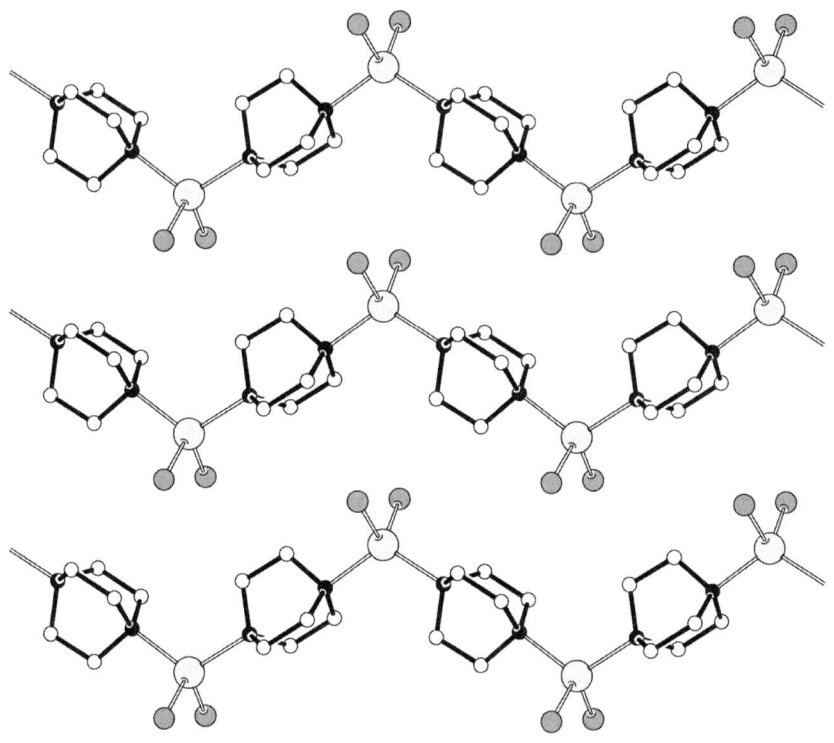

Figure 6.2 The one-dimensional coordination network present in crystals of Zn[N(CH$_2$CH$_2$)$_3$N]Cl$_2$. H atoms not shown for clarity.

the three compounds differ only in the degree and nature of solvation. The differences in topology are, however, much more dramatic and the three compounds must be regarded as isomers of the same basic coordination network. The crystal structure of Ag[dace][CH$_3$COO][MeOH]·0.5H$_2$O is constituted of a two-dimensional coordination network (Figure 6.3) formed by the divergent bidentate dace ligand and two silver atoms, which are joined together by an Ag···Ag bond of length 3.323(1) Å and are asymmetrically bridged by two methanol molecules. There is a close structural relationship between the coordination networks in the Ag[dace][CH$_3$COO][MeOH]·0.5H$_2$O and in the trihydrated compound. This latter structure is built around a zig-zag chain Ag$^{(+)}$···[dace]···Ag$^{(+)}$···[dace]···Ag$^{(+)}$ units as shown in Figure 6.3. The Ag atom is coordinated in a linear fashion. A projection perpendicular to the dace planes shows how the zig-zag chains extend in parallel fashion. The Ag$^{(+)}$···[dace]···Ag$^{(+)}$···[dace]···Ag$^{(+)}$ chains are bridged together via hydrogen bonds involving the N–H donors, the water molecules and the acetate anions. The tetrahydrated species Ag[dace][CH$_3$COO]·4H$_2$O, contains an isomeric form of the coordination networks present in the two former compounds. In the trihydrated compound, two ligands are in a cisoid relative orientation with respect to the silver atom, whereas in the tetrahydrated compound the two ligands adopt a transoid conformation. This is made possible by the different orientation of the N

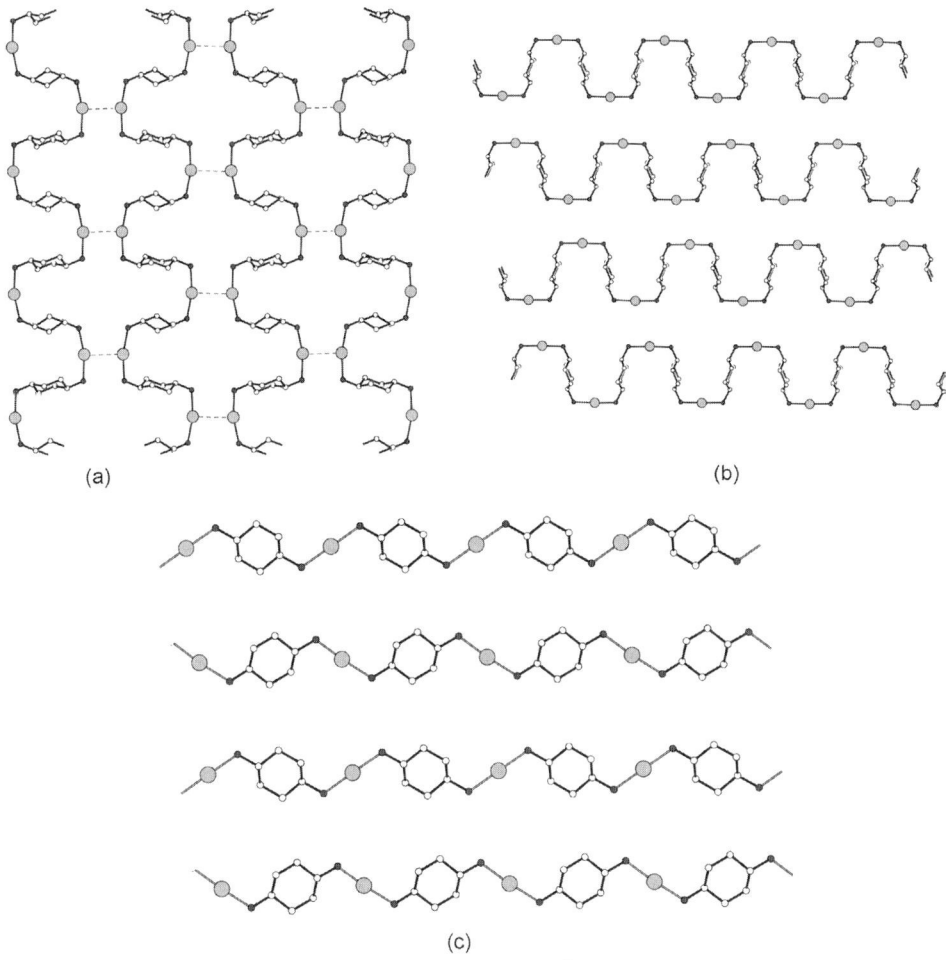

Figure 6.3 Comparison between the isomeric $\{Ag[H_2NC_6H_{10}NH_2]^+\}_\infty$ chains in (a) Ag[dace][CH$_3$COO][MeOH]·0.5H$_2$O, (b) Ag[dace][CH$_3$COO]·3H$_2$O and (c) Ag[dace][CH$_3$COO]·4H$_2$O.

atom lone pairs in the dace ligand. The acetate anions form a hydrated network and interact with the base and the water molecules.

In a further study of the mechanochemical utilization of dace, we have reported that the compound [CuCl$_2$(dace)]$_\infty$ can be obtained by thermal treatment of the hydrated compound [CuCl$_2$(dace)(H$_2$O)]$_\infty$, which is prepared by *kneading* of solid CuCl$_2$ and dace in the presence of a small amount of water [111]. The structure of [CuCl$_2$(dace)]$_\infty$ is not known, since it is insoluble in most organic solvents, which does not permit the growth of single crystals of X-ray quality. However, the DMSO adduct [CuCl$_2$(dace)(DMSO)]$_\infty$ has been fully characterized by single-crystal X-ray diffraction, which gives some insight into the structure of [CuCl$_2$(dace)]$_\infty$. The DMSO adduct can also be easily obtained by kneading solid CuCl$_2$ and dace in the

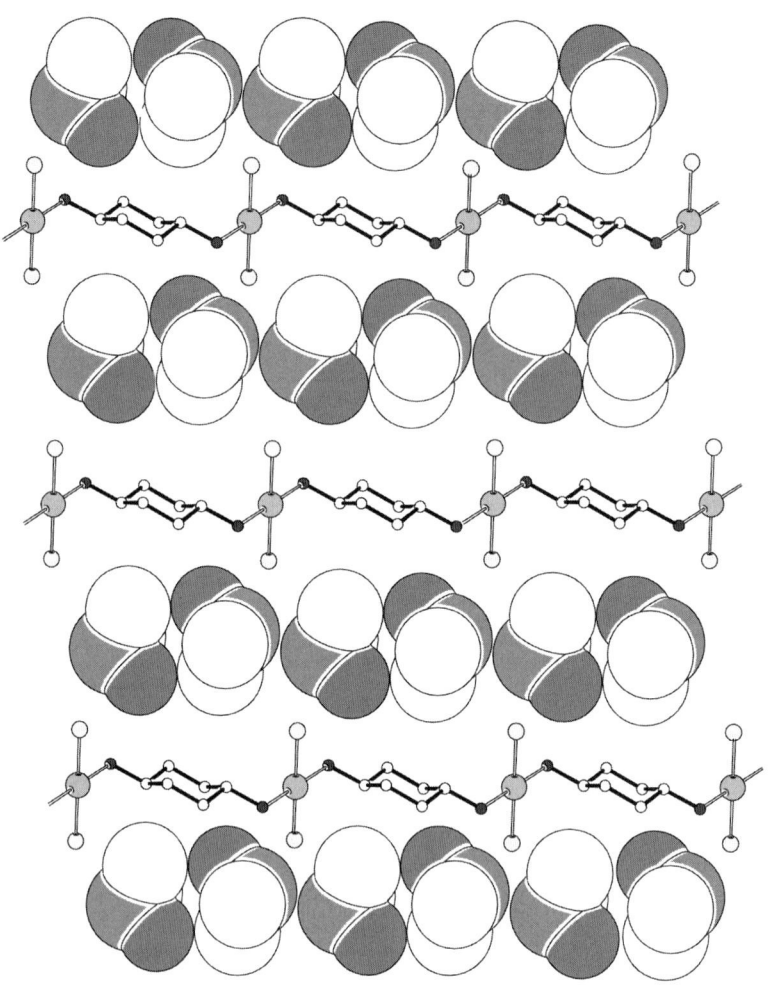

Figure 6.4 A perspective view of the packing of [CuCl$_2$(dace)(DMSO)]$_\infty$, showing the layers formed by parallel 1D networks of alternating CuCl$_2$ and dace units. H atoms omitted for clarity.

presence of a small amount of DMSO. This compound is formed of 1-D coordination networks, constituted of an alternate sequence of CuCl$_2$ and dace ligands (Figure 6.4). Parallel 1-D CuCl$_2$-dace networks form layers and between the layers, the co-crystallized DMSO is intercalated.

The compound [CuCl$_2$(dace)(DMSO)]$_\infty$ is interesting in view of its behavior upon thermal treatment. When [CuCl$_2$(dace)(DMSO)]$_\infty$ is heated to 130°C it converts to [CuCl$_2$(dace)]$_\infty$, as is easily ascertained by comparing X-ray diffraction powder diffractograms. From the structure of [CuCl$_2$(dace)(DMSO)]$_\infty$ and from the knowledge of its thermal behavior, it is possible to infer that the structure of [CuCl$_2$(dace)]$_\infty$ is based on the stacking sequence of layers as in [CuCl$_2$(dace)(DMSO)]$_\infty$, but "squeezed" at a shorter inter-layer separation as a consequence of DMSO removal.

When a guest molecule enters between the layers, the spacing between the CuCl$_2$–dace chains is expanded and the layers are shifted back in position. A series of small molecules can be taken up/released depending on the preparation method, i.e. kneading, suspension in the liquid guest or kneading followed by suspension. The last approach is the most productive; when suspended in the desired liquid guest the [CuCl$_2$(dace)]$_\infty$ only takes up relatively small molecules (DMSO, acetone, water, methanol, etc.) whereas in kneading other guest molecules are also taken up. However, if [CuCl$_2$(dace)]$_\infty$ is first kneaded with a small amount of the desired liquid and then left stirring in the same liquid for 12 h, partial or complete filling of the compound is observed, independently on the guest molecule.

6.3
Hybrid Organic–organometallic and Inorganic-organometallic Co-crystals

Co-crystals are heteromolecular crystals, i.e. molecular crystals constituted of two or more molecular components that are solid at room temperature (to distinguish co-crystals from solvates). The existence of co-crystals is well known, but only recently have they begun to be considered in debates concerning molecular materials mainly because of the claims of their potential interest as pharmaceutical novelties [27].

The formation of co-crystals is possible only when the heteromolecular assembly is more stable than the homomolecular crystals. For this reason, co-crystals are often formed with hydrogen-bonded systems, where the intermolecular hydrogen bonds may afford the additional stabilization required for the co-crystals to form. In fact, the most common way to prepare heteromolecular hydrogen bonded crystals is by mixing (whether in solution of by solid-state grinding) a hydrogen bonding "donor" (usually an acidic species) and a hydrogen bond "acceptor" (usually a base). The mixing of an acid and a base may imply proton transfer from the acid to the base, i.e. protonation, when the pK_a and pK_b values are sufficiently close.

We have exploited the mechanical mixing of solid reactants to prepare novel molecular crystals containing ferrocenyl moieties. It is useful to remember that reactions involving solid reactants or occurring between solids and gases avoid the recovery, storage and disposal of solvents, hence they are of interest in the field of "green chemistry", where environmentally friendly processes are actively sought [112]. Furthermore, solventless reactions often lead to very pure products and reduce the formation of solvate species. In spite of these investigations, solid-state processes involving organometallic systems have only recently begun in a systematic way [113–118].

Manual grinding of the ferrocenyldicarboxylic acid complex [Fe(η^5-C$_5$H$_4$COOH)$_2$] with solid nitrogen-containing bases, namely 1,4-diazabicyclo[2.2.2]octane, 1,4-phenylenediamine, piperazine, *trans*-1,4-cyclohexanediamine and guanidinium carbonate, generates quantitatively the corresponding organic–organometallic adducts (Figure 6.5) [119,120]. The case of the adduct [HC$_6$N$_2$H$_{12}$][Fe(η^5-C$_5$H$_4$COOH)(η^5-C$_5$H$_4$COO)] is particularly noteworthy, because the same product can be obtained in three different ways: (i) by reaction of solid [Fe(η^5-C$_5$H$_4$COOH)$_2$] with vapor of 1,4-diazabicyclo[2.2.2]octane (which possesses a small but significant vapor

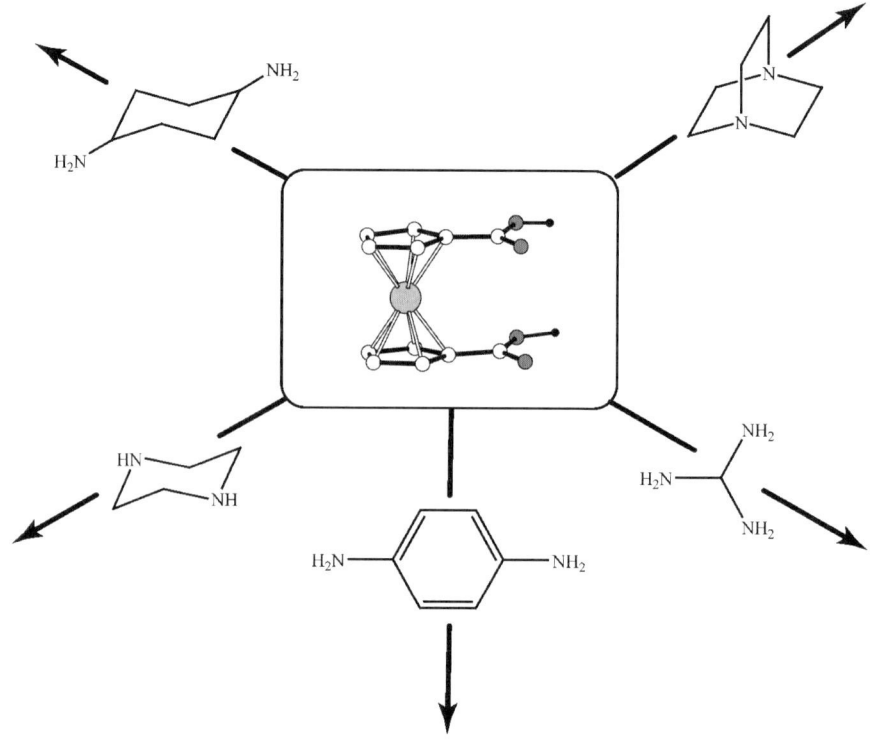

Figure 6.5 Grinding of the organometallic complex [Fe(η^5-C$_5$H$_4$COOH)$_2$] as a solid polycrystalline material with the solid bases 1,4-diazabicyclo[2.2.2]octane, guanidinium cation, 1,4-phenylenediamine, piperazine and trans-1,4-cyclohexanediamine generates quantitatively the corresponding adducts [HC$_6$H$_{12}$N$_2$][Fe(η^5-C$_5$H$_4$COOH)(η^5-C$_5$H$_4$COO)], [C(NH$_2$)$_3$]$_2$[Fe(η^5-C$_5$H$_4$COO)$_2$]·2H$_2$O, [HC$_6$H$_8$N$_2$][Fe(η^5-C$_5$H$_4$COOH)(η^5-C$_5$H$_4$COO)], [H$_2$C$_4$H$_{10}$N$_2$][Fe(η^5-C$_5$H$_4$COO)$_2$] and [H$_2$C$_6$H$_{14}$N$_2$][Fe(η^5-C$_5$H$_4$COO)$_2$]·2H$_2$O.

pressure), (ii) by reaction of solid [Fe(η^5-C$_5$H$_4$COOH)$_2$] with solid 1,4-diazabicyclo[2.2.2]octane, i.e. by co-grinding of the two crystalline powders, and (iii) by reaction in MeOH solution of the two reactants. It is also interesting to note that the base can be removed by mild treatment, regenerating the structure of the starting dicarboxylic acid.

Bis-substituted pyridine/pyrimidine ferrocenyl complexes have also been obtained by a mechanically induced Suzuki coupling reaction [121] in the solid state starting from the complex ferrocene-1,1′-diboronic acid, [Fe(η^5-C$_5$H$_4$-B(OH)$_2$)$_2$] [122]. The ligand [Fe(η^5-C$_5$H$_4$-1-C$_5$H$_4$N)$_2$] (**1**), obtained by both solution and solid-state methods, was then used to prepare a whole family of hetero-bimetallic metallomacrocycles by reaction with AgNO$_3$, Cd(NO$_3$)$_2$, Cu(CH$_3$COO)$_2$, Zn(CH$_3$COO)$_2$ and ZnCl$_2$; the complexes [Fe(η^5-C$_5$H$_4$-1-C$_5$H$_4$N)$_2$]$_2$Ag$_2$(NO$_3$)$_2$·1.5H$_2$O, [Fe(η^5-C$_5$H$_4$-1-C$_5$H$_4$N)$_2$]$_2$Cu$_2$(CH$_3$COO)$_4$·3H$_2$O, [Fe(η^5-C$_5$H$_4$-1-C$_5$H$_4$N)$_2$]$_2$Cd$_2$(NO$_3$)$_4$·CH$_3$OH·0.5C$_6$H$_6$, [Fe(η^5-C$_5$H$_4$-1-C$_5$H$_4$N)$_2$]$_2$Zn$_2$(CH$_3$COO)$_4$ and [Fe(η^5-C$_5$H$_4$-1-C$_5$H$_4$N)$_2$]$_2$Zn$_2$Cl$_4$ were obtained (examples are shown in Figure 6.6) [121]. Reaction of mechanochemically

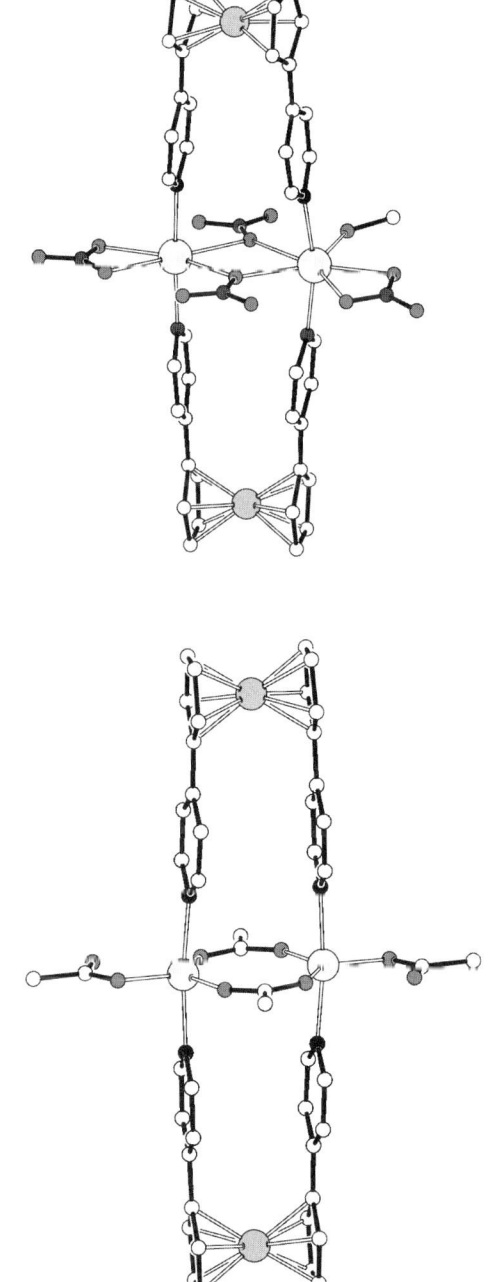

6.3 Hybrid Organic–organometallic and Inorganic-organometallic Co-crystals | 163

Figure 6.6 The metallomacrocycles produced by reaction of [Fe(η^5-C$_5$H$_4$-1-C$_5$H$_4$N)$_2$] and the salts Zn(CH$_3$COO)$_2$ (left) and Cd(NO$_3$)$_2$ (right). The starting material was obtained by a Suzuki coupling reaction in the solid state starting from the complex ferrocene-1,1′-diboronic acid [Fe(η^5-C$_5$H$_4$-B(OH)$_2$)$_2$].

prepared [Fe(η^5-C$_5$H$_4$-1-C$_5$H$_4$N)$_2$] with the ferrocenyl dicarboxylic acid complex [Fe(η^5-C$_5$H$_4$COOH)$_2$] led to the formation of the supramolecular adduct [Fe(η^5-C$_5$H$_4$-1-C$_5$H$_4$N)$_2$][Fe(η^5-C$_5$H$_4$COOH)$_2$] [121].

More recently, the same building block has been used in the preparation of supramolecular metallomacrocycles with dicarboxylic acids of variable aliphatic chain length [123].

The supramolecular macrocyclic adducts of general formula {[Fe(η^5-C$_5$H$_4$-C$_5$H$_4$N)$_2$]·[HOOC(CH$_2$)$_n$COOH]}$_2$ with $n=4$ (adipic acid), $n=6$ (suberic acid), $n=7$ (azelaic acid) and $n=8$ (sebacic acid) have been obtained quantitatively by *kneading* powdered samples of the crystalline organometallic and organic reactants with drops of MeOH (for $n=4, 6, 7$) and by direct crystallization from MeOH for $n=8$ (sebacic), whereas the adduct with $n=5$ (pimelic) represents an isomeric open-chain alternative to the macrocycle. All complexes, with the exception of **1**·pimelic(5), share a common structural feature, namely the formation of supramolecular macrocycles constituted of two organometallic and two organic units linked in large tetramolecular units by O–H···N hydrogen bonds between the –COOH groups of the dicarboxylic acids and the N atom of the ferrocenyl complex. Figure 6.7 shows the structures of **1**·adipic(4), **1**·suberic(6), **1**·azelaic(7) and **1**·sebacic(8). It can be appreciated how the even–odd alternation of carbon atoms in the organic spacers is accommodated by the twist of the cyclopentadienyl-pyridyl groups and by the eclipsed or staggered juxtaposition of the organic moieties

The compounds depicted in Figure 6.7 allow speculation on the *semantics* of these crystal structures: being finite aggregates they cannot be considered as co-crystals but rather as supramolecular systems. On the other hand, they possess a chemical identity and a crystal structure on their own as separate molecules, therefore their solid-state assembly (via *grinding*, incidentally) into the aggregates falls under the broad definition of co-crystals provided above.

The cationic bisamide [Co(η^5-C$_5$H$_4$CONHC$_5$H$_4$NH)$_2$] has been used in the formation of a co-crystal with [Fe(η^5-C$_5$H$_4$COOH)$_2$] [124]. In crystalline [Co(η^5-C$_5$H$_4$CONHC$_5$H$_4$N)$_2$][Fe(η^5-C$_5$H$_4$COOH)$_2$][PF$_6$] the two moieties are linked by an O–H···N hydrogen bond forming a sort of dimer (Figure 6.8) reminiscent of that observed in crystalline [Co(η^5-C$_5$H$_4$CONHC$_5$H$_4$NH) (η^5-C$_5$H$_4$COO)] [PF$_6$].

The zwitterion sandwich complex [Co^{III}(η^5-C$_5$H$_4$COOH)(η^5-C$_5$H$_4$COO)] [86], thanks to its amphoteric behavior, undergoes reversible gas–solid reactions with the hydrated vapors of a variety of acids (e.g. HCl [125], CF$_3$COOH, CCl$_3$COOH, CHF$_2$COOH, HBF$_4$ and HCOOH) [126–128] and bases (e.g. NH$_3$, NMe$_3$ and NH$_2$Me) [125] and also solid–solid reactions with crystalline alkali metal and ammonium salts of formula MX (M=K$^+$, Rb$^+$, Cs$^+$, NH$_4^+$; X=Cl$^-$, Br$^-$, I$^-$, PF$_6^-$) [126]. These products could also be obtained by solution methods, as discussed earlier.

Similar behavior is shown towards other volatile acids. Exposure of the zwitterion to vapors of CF$_3$COOH and HBF$_4$, for instance, quantitatively produces the corresponding salts of the cation [Co(η^5-C$_5$H$_4$COOH)$_2$]$^+$, viz. [Co(η^5-C$_5$H$_4$COOH)$_2$][CF$_3$COO] and [Co(η^5-C$_5$H$_4$COOH)$_2$][BF$_4$]. Exposure of the solid zwitterion to vapor of CHF$_2$COOH quantitatively produces the corresponding salts of the cation [Co(η^5-C$_5$H$_4$COOH)$_2$][CHF$_2$COO].

adipic (4)

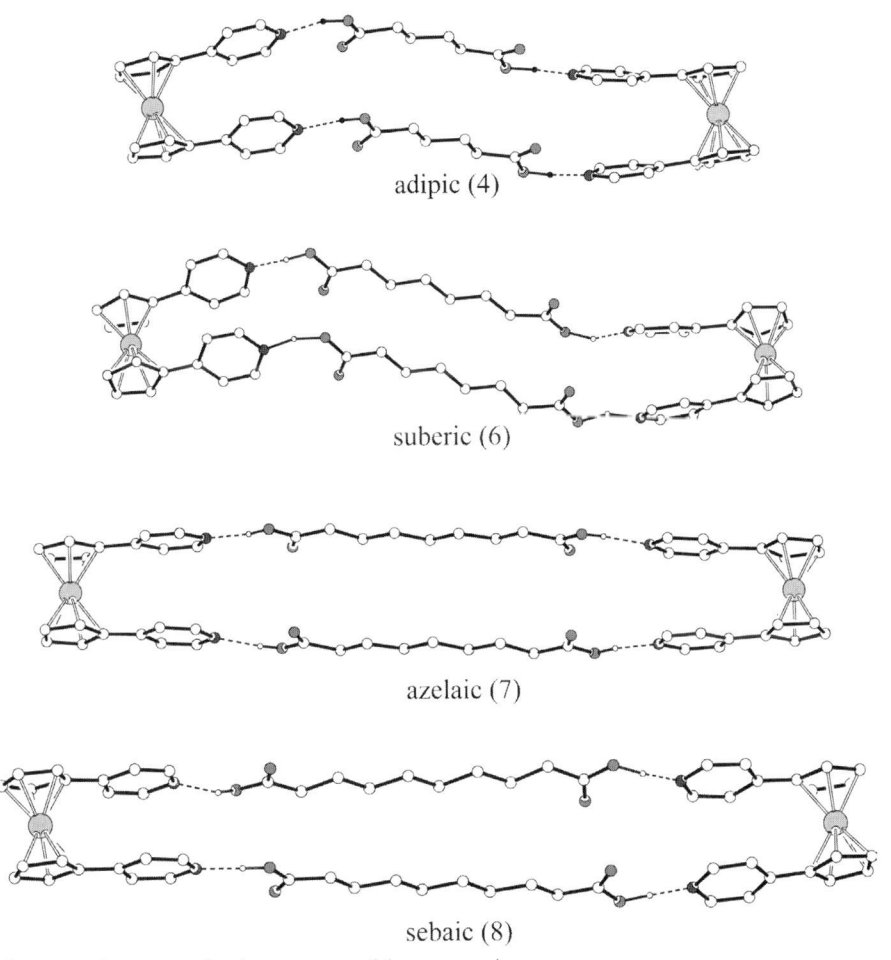

suberic (6)

azelaic (7)

sebaic (8)

Figure 6.7 The supramolecular structures of the macrocycles 1·adipic(4), 1·suberic(6), 1·azelaic(7) and 1·sebacic(8) showing the hydrogen bond links between the two outer organometallic molecules and the inner organic spacers.

The zwitterion also reversibly absorbs formic acid from humid vapors, forming selectively a 1 : 1 co-crystal, $[Co(\eta^5\text{-}C_5H_4COOH)(\eta^5\text{-}C_5H_4COO)][HCOOH]$ (Figure 6.9), from which the starting material can be fully recovered by mild thermal treatment. Contrary to the other compounds of this class, no proton transfer from the adsorbed acid to the organometallic moiety has been observed (Figure 6.9). Hence the reaction between $[Co(\eta^5\text{-}C_5H_4COOH)(\eta^5\text{-}C_5H_4COO)]$ (solid) and HCOOH (vapor) would be more appropriately described as a special kind of solvation rather than as a heterogeneous acid–base reaction, as also confirmed by ^{13}C CP/MAS NMR spectroscopy [128].

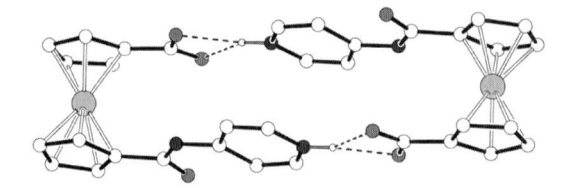

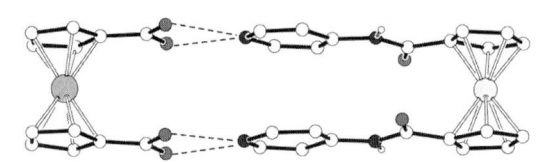

Figure 6.8 (Top) the dimers of [Co^{III}(η^5-C$_5$H$_4$CONHC$_5$H$_4$NH)(η^5-C$_5$H$_4$COO)][PF$_6$] formed via bifurcate N−H$^{(+)}$···O$^{(−)}$ hydrogen-bonding interactions [N···O 2.84(2) and 2.88(2) Å]. (Bottom) the ferrocenedicarboxylic acid molecule and the diamido molecule [Co^{III}(η^5-C$_5$H$_4$CONHC$_5$H$_4$NH)$_2$] in [Co^{III}(η^5-C$_5$H$_4$CONHC$_5$H$_4$N)$_2$][Fe(η^5-C$_5$H$_4$COOH)$_2$][PF$_6$] are linked via O−H···N hydrogen bonds [N···O 2.52(3) Å].

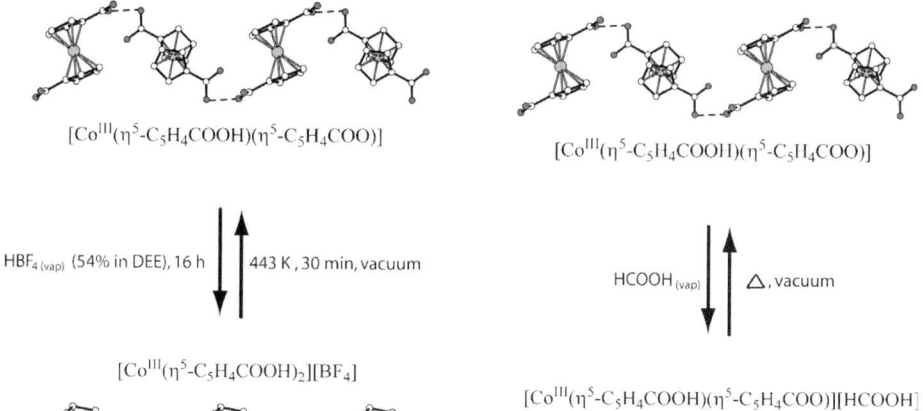

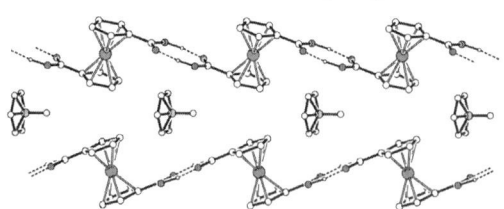

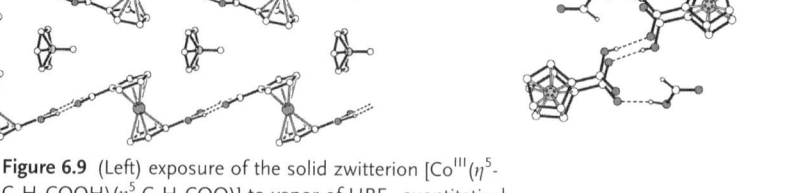

[Co^{III}(η^5-C$_5$H$_4$COOH)(η^5-C$_5$H$_4$COO)]

[Co^{III}(η^5-C$_5$H$_4$COOH)(η^5-C$_5$H$_4$COO)]

HBF$_4$ (vap) (54% in DEE), 16 h ⇅ 443 K, 30 min, vacuum

HCOOH$_{(vap)}$ ⇅ Δ, vacuum

[Co^{III}(η^5-C$_5$H$_4$COOH)$_2$][BF$_4$]

[Co^{III}(η^5-C$_5$H$_4$COOH)(η^5-C$_5$H$_4$COO)][HCOOH]

Figure 6.9 (Left) exposure of the solid zwitterion [Co^{III}(η^5-C$_5$H$_4$COOH)(η^5-C$_5$H$_4$COO)] to vapor of HBF$_4$ quantitatively produces the corresponding salt [Co(η^5-C$_5$H$_4$COOH)$_2$][BF$_4$], whereas (right) exposure to vapor of HCOOH yields the co-crystalline material [Co(η^5-C$_5$H$_4$COOH)(η^5-C$_5$H$_4$COO)][HCOOH]. Both products revert back to the solid zwitterion after mild thermal treatment.

6.4
Crystal Isomers and Crystal Polymorphs

The phenomenon of polymorphism, viz. the existence of more than one crystal structure for a given compound, is well known and has been widely studied [129,130]. However, the structural, thermodynamic and kinetic factors associated with the nucleation and crystallization of molecular compounds are not yet fully understood. The experimental investigation of crystal polymorphism is still mainly based on a systematic, and sometimes tedious, exploration of all possible crystallization and interconversion conditions [129–131] ("polymorph screening"), while theoretical polymorph prediction is still embryonic [132,133]. The screening of different crystal forms of a compound is not only an academic challenge but also is becoming one of the most important goals in the pharmaceutical industries, since the majority of drugs are administered as solids and solid-state properties significantly influence the bioavailability and stability of the final product. When two or more polymorphs occur, a full characterization of these forms and of the relationship among the different solid phases should be studied, which is best achieved by using complementary techniques such as X-ray diffraction and differential scanning calorimetry combined with IR, Raman and solid-state NMR (SSNMR) spectroscopy.

In the context of this chapter, it is worth discussing a classical case of crystal polymorphism, namely that of the compound 3,4-dicarboxypyridine, also known as cinchomeronic acid [134].

3,4-Dicarboxypyridine is one of the six isomers of the acid pyridinedicarboxylic. All these isomers are widely utilized in the construction of coordination networks, since their metal coordination modes make possible different architectures [135–137]. Furthermore, some of these isomers are biologically active and the acid has been studied for its ability to promote the growth of radishes [138,139].

The structural relationship between the two crystal forms of cinchomeronic acid has been investigated by single-crystal X-ray diffraction, IR, Raman and solid-state NMR spectroscopy, showing that the two polymorphs form a monotropic system, with the orthorhombic form I being the thermodynamically stable form, whereas the monoclinic form II is unstable. In both forms cinchomeronic acid crystallizes as a zwitterion and decomposes before melting. The crystal structure and spectroscopic analysis indicate that the difference in stability can be ascribed to the strength of the hydrogen bonding patterns established by the protonated N atom and the carboxylic/carboxylate O atoms.

Although cinchomeronic acid has been known for almost a century [139] and the presence of two forms has been reported in the PDF-2 [140] database since 1971, no scientific report seems to mention the existence of these two polymorphs. The crystal structure of form **I** (according to the name in PDF-2) was described by Takusagawa et al. in 1973 [141,142].

The molecule is present as a zwitterion both in the solid state and in solution [143], with one acid hydrogen on the ring nitrogen. The presence of different kinds of acceptor and donor groups for hydrogen bonding makes it a worthy candidate for a study of the competition among of different supramolecular synthons that can be formed [144]. We have fully characterized both forms **I** and **II** by single-crystal X-ray diffraction and spectroscopic methods. Both forms were initially obtained as concomitant [145] polymorphs from an ethanol–water solution.

In a further study, we investigated the behavior of the different crystal forms of barbituric acid ($C_4H_4N_2O_3$) towards basic vapors [146]. The two known crystalline polymorphic forms of barbituric acid ($C_4H_4N_2O_3$) (**I** and **II**) [147,148] and the dihydrate form [($C_4H_4N_2O_3$)·$2H_2O$] [149,150] have been reacted with vapors of ammonia, methylamine and dimethylamine and the crystalline products investigated by means of single-crystal and powder X-ray diffraction, thermogravimetric analysis, differential scanning calorimetry and infrared spectroscopy. It has been shown that forms **I** and **II** and the dihydrate form of barbituric acid react with ammonia leading to the same crystalline ammonium barbiturate salt, $NH_4(C_4H_3N_2O_3)$ (**1a**), whereas the gas–solid reactions of form **II** with methylamine and dimethylamine yield the corresponding crystalline salts, $CH_3NH_3(C_4H_3N_2O_3)$ (**1b**) and $(CH_3)_2NH_2(C_4H_3N_2O_3)$ (**1c**). Thermal desorption of the bases at ca. 200°C leads to formation of a new crystal form of barbituric acid, form **III**, as confirmed by ^1H NMR spectroscopy and by its chemical behavior. Dehydration of the dihydrate form has also been investigated, showing that it releases water to yield exclusively crystals of form **II**.

Polymorphism within the barbiturate family is widespread and well studied also in view of the pharmaceutical interest [151–154].

The three crystal forms are easily prepared separately and could thus be used independently in the reaction with gases. A similar experiment has been carried out before by Stowell's group [155] in the course of the investigation of the reactivity of indomethacin amorphous and crystal forms with ammonia. In the case of barbituric acid, the main idea was not only to check whether different crystal forms of the same acidic molecular species reacted in a different way with vapors of the same base but also whether the reverse reactions, namely ammonia desorption, would lead to one of the known crystal form or to the formation of a new polymorph [129,130].

In the first instance, we have discovered that the dihydrate form exclusively transforms upon dehydration into form **II** of barbituric acid. Second, we have shown that all three forms, whether as anhydrous crystals or as the dihydrated form, react in the same way, yielding the same anhydrous barbiturate ammonium salt, **1a**. This analogy in behavior may be taken as an indication that the gas–solid reactions require destruction and reconstruction of the crystal phases and are, therefore, not selective. Form **II** has beeen used in subsequent reactions with vapors of ammonia, methylamine and dimethylamine. In all cases it has been possible to ascertain the rapid formation of the corresponding anhydrous salts in the form of polycrystalline powders. The structures of the salts **1a**, **1b** and **1c** have been investigated by X-ray diffraction experiments on single crystals grown by carrying out the acid–base reaction in solution. In addition to ascertaining the detailed structural features of the salts, the data were used to calculate theoretical powder diffraction patterns for comparison. In all cases, the structures determined by single-crystal X-ray diffraction experiments have been shown to coincide with those of the bulk products.

A related situation has been observed on reacting solid malonic acid, HOOC(CH$_2$)COOH, with solid N(CH$_2$CH$_2$)$_3$N in a 1:2 molar ratio [156]. Two different crystal forms of the salt [HN(CH$_2$CH$_2$)$_3$NH][OOC(CH$_2$)COOH]$_2$ are obtained, depending on the preparation technique (grinding or solution) and crystallization rate. Form **I**, containing monohydrogenmalonate anions forming conventional intramolecular O−H···O hydrogen bonds and interionic N−H···O hydrogen bonds, is obtained by solid-state co-grinding or by rapid crystallization, whereas form **II**, containing *both* intermolecular and intramolecular O−H···O hydrogen bonds, is obtained by slow crystallization (Figure 6.10). Forms **I** and **II** do not interconvert, and form **I** undergoes an order–disorder phase transition on cooling. One can envisage the two crystalline forms as *hydrogen bond isomers* of the same *solid supermolecule*.

Proton transfer along a hydrogen bond poses an interesting question about polymorph definition. In fact, proton mobility along a hydrogen bond [say from

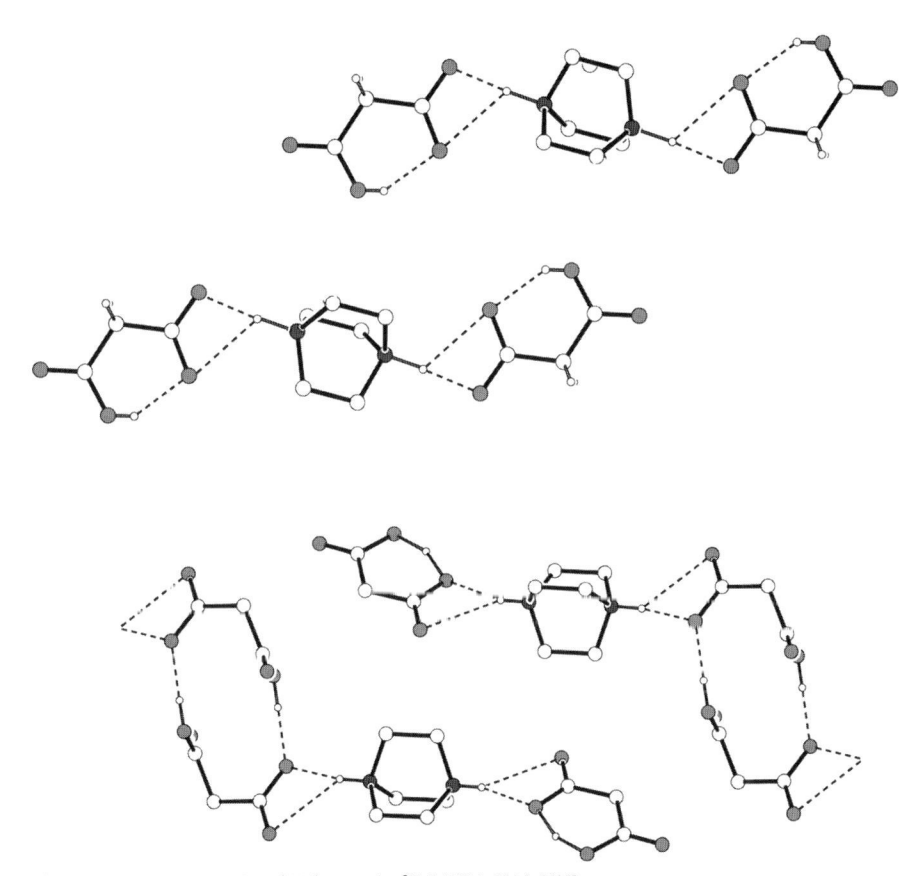

Figure 6.10 Forms **I** (top) and **II** (bottom) of [HN(CH$_2$CH$_2$)$_3$NH][OOC(CH$_2$)COOH]$_2$ and the hydrogen-bonded anion···cation chains present in their crystals. Form **I** is obtained by solid-state co-grinding or by rapid crystallization, whereas form **II** is obtained by slow crystallization.

O−H···N to $^{(-)}$O···H−N$^{(+)}$] may not be associated with a phase transition, even though it implies the formal transformation of a molecular crystal into a molecular salt. This situation has been observed, for instance, for the proton migration along an O−H···O bond in a co-crystal of urea–phosphoric acid (1:1) as a function of temperature [157]. Wiechert and Mootz, on the other hand, have isolated two co-crystals of pyridine and formic acid: in the 1:1 co-crystal proton transfer from formic acid to pyridine does not take place, whereas in the 1:4 co-crystal N−H$^{(+)}$···O$^{(-)}$ interactions are present [158]. Examples of this kind are rare, but serve to stress how the phenomenon of polymorphism can be, at times, full of ambiguity.

An intriguing case of interconversion between unsolvate and solvate crystals is observed when N(CH$_2$CH$_2$)$_3$N is reacted with maleic acid, HOOC(HC=CH)COOH. The initial product is the anhydrous salt [HN(CH$_2$CH$_2$)$_3$N][OOC(HC=CH)COOH], which contains chains of $^{(+)}$N−H···N$^{(+)}$-bonded [HN(CH$_2$CH$_2$)$_3$N]$^+$ cations and "isolated" [OOC(HC=CH)COOH]$^-$ anions [159]. On exposure to humidity, the anhydrous salt converts within a few hours into the hydrated form [HN(CH$_2$CH$_2$)$_3$N][OOC(HC=CH)COOH]·0.25H$_2$O, which contains more conventional "charge-assisted" $^{(+)}$N−H···O$^{(-)}$ hydrogen bonds between the anion and cation (Figure 6.11). This latter form can also be obtained by co-grinding.

6.5
Dynamic Crystals – Motions in the Nano-world

In a further exploration of the use of mechanochemical methods to prepare hydrogen-bonded nanometric adducts, we have used crown ethers to capture alkali metal cations and the ammonium cation in extended hydrogen-bonded networks formed by hydrogensulfate and dihydrogenphosphate anions [160]. Crown ether complexes have been the subject of a large number of studies because of the interest in ion recognition, complexation and self-assembly processes [6,161–164]. The hydrogensulfate and dihydrogenphosphate salts, on the other hand, have found applications in a number of devices such as H$_2$ and H$_2$O sensors, fuel and steam cells and high-energy density batteries [165–168]. Manual co-grinding of solid 18-crown-6 and solid [NH$_4$][HSO$_4$] in air leads to the formation of the crown ether complex 18-crown-6·[NH$_4$][HSO$_4$]·2H$_2$O, the water molecules being taken up from ambient humidity during grinding. In the complex the ammonium cation is trapped via O$_{crown}$···H−N hydrogen bonds by the crown ethers, whereas on the exposed side it interacts with the hydrogensulfate anion (Figure 6.12). The sulfate anion and the water molecules also interact via hydrogen bonding, forming a ribbon that is sandwiched between 18-crown-6·[NH$_4$]$^+$ units. Hydrogen bonds are also observed between water molecules and oxygen atoms in the crown ether. Analogous behavior is shown by the potassium complex 18-crown-6·K[HSO$_4$]·2H$_2$O, which converts into 18-crown-6·K[HSO$_4$] on dehydration and which undergoes, on further heating, enantiotropic solid–solid transitions associated with the onset of a solid-state dynamic process [169].

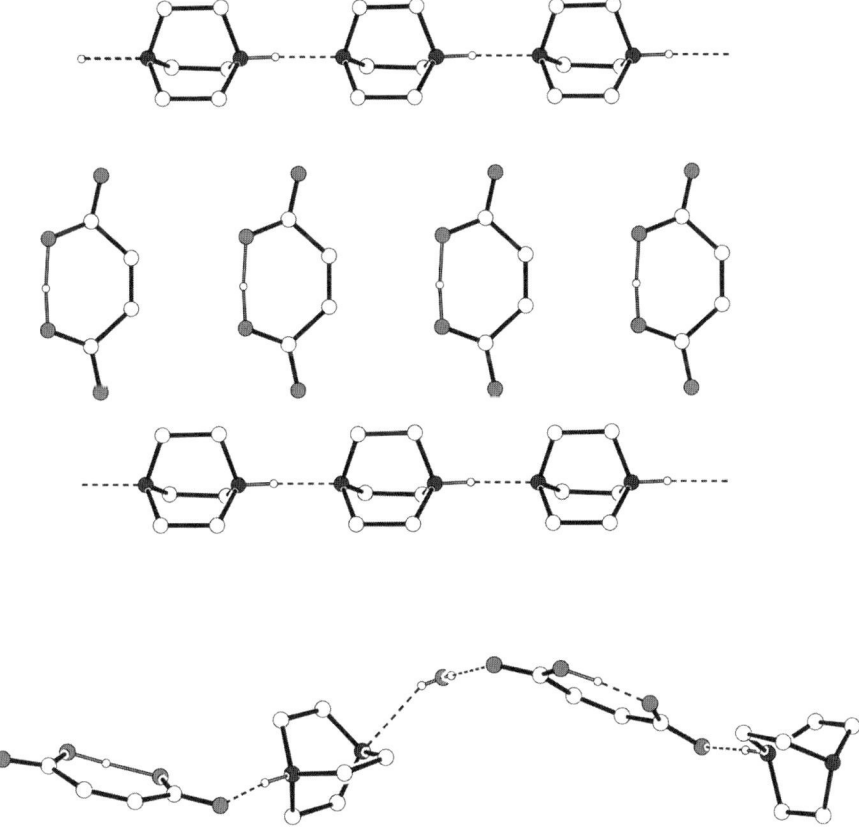

Figure 6.11 Views of the packing and hydrogen bonding in the anhydrous salt [HN(CH$_2$CH$_2$)$_3$N] [OOC(HC=CH)COOH] (top) and of the hydrated salt [HN(CH$_2$CH$_2$)$_3$N][OOC(HC=CH)COOH]H$_2$O$_{0.25}$ (bottom).

The crown ether 15-crown-5 is a liquid in room temperature, so when it is kneaded instead of ground, with ammonium hydrogensulfate, a similar reaction as for 18-crown-6 takes place. The product, (15-crown-5)$_3$·[NH$_4$]$_2$[HSO$_4$]$_3$·H$_2$O, also fully structurally determined by single-crystal X-ray analysis, is also obtained when the reaction takes place in solution. (15-Crown-5)$_3$·[NH$_4$]$_2$[HSO$_4$]$_3$·H$_2$O is reminiscent of 18-crown-6 because of the formation of hydrogen-bonded ribbons intercalated between the crown ether layers. The difference between the two adducts is, however, the two different types of interactions between the ammonium cation and the crown ether that is present in the 15-crown-5 adduct. One ammonium cation is sandwiched between two crown ether units, whereas the other is linked to the hydrogensulfate anion by N−H···O hydrogen bonds. In the 15-crown-5 adduct a hydrogen-bonded [H$_3$O]$^+$ ion is also needed to neutralize the overall charge.

More recently, this chemistry has been extended to encompass the investigation of the behavior under thermal treatment of the transition metals complexes 15-crown-5·

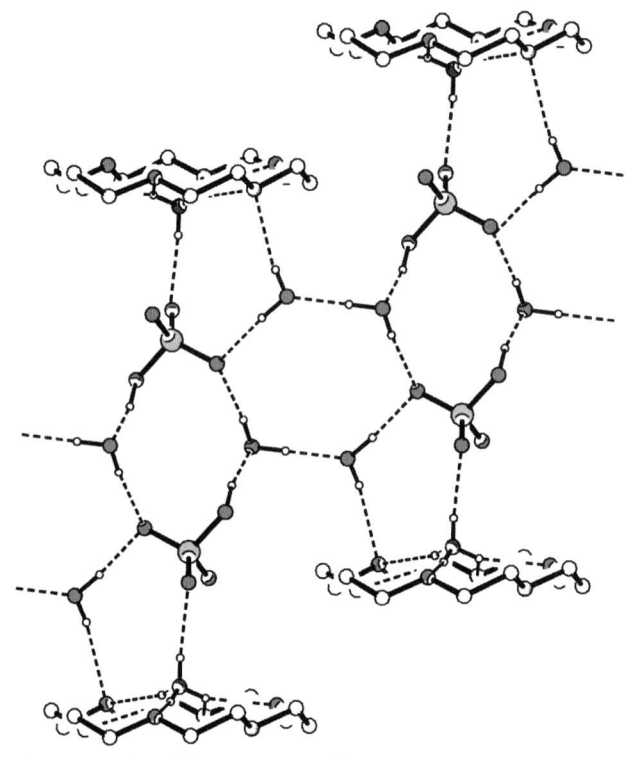

Figure 6.12 The solid-state structure of 18-crown-6·[NH$_4$][HSO$_4$]· 2H$_2$O. Note how the crown ether molecules interact via O$_{crown}$···H−N hydrogen bonds with the ammonium cations, which, in turn, form hydrogen bonds with the hydrogensulfate anions. Two water molecules (per formula unit) contribute to the formation of the central hydrogen-bonded ribbon.

[Mn(H$_2$O)$_2$(HSO$_4$)$_2$], 15-crown-5·[Cd(H$_2$O)$_2$(HSO$_4$)$_2$], 15-crown-5·[(H$_7$O$_3$)(HSO$_4$)$_2$]· H$_2$O, 18-crown-6·[Pb(HSO$_4$)$_2$] and 18-crown-6·[(H$_5$O$_2$)(HSO$_4$)], obtained by mechanical mixing or from solution crystallization. It has been shown that the Mo and Cd complexes undergo reversible water release–uptake processes, accompanied by a change in coordination at the metal centers [170].

6.6
Conclusions

In this chapter, we have illustrated, by means of examples coming mainly from our own work, that the basic ideas of crystal engineering belong also to the expanding field of nanochemistry. The possibility of rationally designing and exploring experimentally the construction of nanoporous materials is a valid example. Many research groups world-wide are applying crystal design principles to construct materials for applications such as sensors, traps and reservoirs. Another area, which we have not touched upon, is that related to nanocomputing, i.e. the use of self-assembly processes to construct addressable nanoarrays based on molecules.

Dealing with molecules, ligands and ions naturally "brings in" the nanometer scale to the scientific strategy, if for no other reason because these chemical entities have nanometric dimensions. Under this premise, the construction of a co-crystal, i.e. the heteromolecular assembly of two or more component units denoted by their own chemical and physical identities, is a nanochemical reaction. The nanochemical reaction may be "mediated" by the formation of hydrogen bonds. Polymorphism, on the other hand, may be seen as nanochemical isomerism, viz. the possibility of different arrangements of the same nano-sized building block in space.

Acknowledgments

We acknowledge financial support from MUR and by the University of Bologna. We thank the Swedish Research Council (VR) and PolyCrystalLine s.r.l. for postdoctoral grants to A.P. and M.C., respectively.

References

1 Ball, P. (1996) *Nature*, **381**, 648.
2 Bryce, M. (1999) *J. Mater. Chem.*, **9**, xi.
3 Ozin, G. and Arsenault, A. (2005) *Nanochemistry: a Chemistry Approach to Nanomaterials*, RSC Publishing, Cambridge, U.K.
4 Braga, D., Grepioni, F. and Orpen, A.G., (1999) Crystal Engineering: from Molecules and Crystals to Materials. Proceedings of the NATO Advanced Study Institute, held 12–23 May 1999, Erice, Italy. *NATO Sci. Ser.*, *Ser.* 538.
5 Lehn, J.M. (1995) *Supramolecular Chemistry: Concepts and Perspectives*, Wiley-VCH, Weinheim, Germany.
6 Steed, J.W. and Atwood, J.L. (2000) *Supramolecular Chemistry: a Concise Introduction*, Wiley, Chichester, U.K.
7 Lehn, J.-M., Singh, A. and Fouquey, C. (1996) *Polym. Prepr. Am. Chem. Soc. Div. Polym. Chem.*, **37**, 476.
8 Wilson, C.C. (2000) *Single Crystal Neutron Diffraction from Molecular Materials*, World Scientific, Singapore.
9 Coronado, E., Galan-Mascaros, J.R., Gimenez-Saiz, C., Gomez-Garcia, C.J. and Ruiz-Perez, C. (2003) *Eur. J. Inorg. Chem*, 2290.
10 Gatteschi, D. and Sorace, L. (2001) *J. Solid State Chem.*, **159**, 253.
11 Alberola, A., Coronado, E., Galan-Mascaros, J.R., Gimenez-Saiz, C., Gomez-Garcia, C.J. and Romero, F.M. (2003) *Synth. Met.*, **133**, 509.
12 Braga, D. and Grepioni, F. (2006) *Making Crystals by Design*, Wiley-VCH, Weinheim.
13 Desiraju, G. (1989) *Crystal Engeneering: The Design of Organic Solids*, Elsevier, Amsterdam.
14 Blake, A.J., Champness, N.R., Hubberstey, P., Li, W.-S., Withersby, M.A. and Schroder, M. (1999) *Coord. Chem. Rev.*, **183**, 117.
15 Batten, S.R., Hoskins, B.F. and Robson, R. (2000) *Chem. Eur. J.*, **6**, 156.
16 Rather, B. and Zaworotko, M.J. (2003) *Chem. Commun.*, 830.
17 Moulton, B., Abourahma, H., Bradner, M.W., Lu, J., McManus, G.J. and

Zaworotko, M.J. (2003) *Chem. Commun.*, 1342.
18 Fujita, M. (1998) *Chem. Soc. Rev.*, **27**, 417.
19 Olenyuk, B., Fechtenkotter, A. and Stang, P.J. (1998) *J. Chem. Soc., Dalton Trans.*, 1707.
20 Pan, L., Sander, M.B., Huang, X., Li, J., Smith, M., Bittner, E., Bockrath, B. and Johnson, J.K. (2004) *J. Am. Chem. Soc.*, **126**, 1308.
21 Ferey, G., Latroche, M., Serre, C., Millange, F., Loiseau, T. and Percheron-Guegan, A. (2003) *Chem. Commun.*, 2976.
22 Cotton, F.A., Lin, C. and Murillo, C.A. (2001) *J. Chem. Soc., Dalton Trans.*, 499.
23 Cotton, F.A., Lin, C. and Murillo, C.A. (2001) *Chem. Commun.*, 11.
24 Mori, W. and Takamizawa, S. (2000) *J. Solid State Chem.*, **152**, 120.
25 Carlucci, L., Ciani, G., Proserpio, D.M. and Rizzato, S. (2002) *CrystEngComm*, **4**, 121.
26 Lu, J., Moulton, B., Zaworotko, M.J. and Bourne, S.A. (2001) *Chem. Commun.*, 861.
27 Moulton, B. and Zaworotko, M.J. (2001) *Chem. Rev.*, **101**, 1629.
28 Bruce, D.W. and O'Hare, D. (1992) *Inorganic Materials*, Wiley, New York.
29 Long, N.J. (1995) *Angew. Chem. Int. Ed. Engl.*, **34**, 21.
30 Kanis, D.R., Ratner, M.A. and Marks, T.J. (1994) *Chem. Rev.*, **94**, 195.
31 Tucker, M.S., Khan, I., Fuchs-Young, R., Price, S., Steininger, T.L., Greene, G., Wainer, B.H. and Rosner, M.R. (1993) *Brain Res.*, **631**, 65.
32 Gatteschi, D. (1994) *Adv. Mater.*, **6**, 635.
33 Ward, M.D. (2003) *Science*, **300**, 1104.
34 Miller, J.S. (2003) *Angew. Chem. Int. Ed.*, **42**, 27.
35 Vos, T.E., Liao, Y., Shum, W.W., Her, J.H., Stephens, P.W., Reiff, W.M. and Miller, J.S. (2004) *J. Am. Chem. Soc.*, **126**, 11630.
36 Vickers, E.B., Giles, I.D. and Miller, J.S. (2005) *Chem. Mater.*, **17**, 1667.
37 Toma, L., Lescouezec, R., Vaissermann, J., Delgado, F.S., Ruiz-Perez, C., Carrasco, R., Cano, J., Lloret, F. and Julve, M. (2004) *Chem. Eur. J.*, **10**, 6130.
38 Anderson, K.M., Kuendig, E.P., Norman, N.C., Orpen, A.G., Pardoe, J.A.J., Smith, D.W. and Timms, P.L. (2001) *Acta Crystallogr., Sect. E: Struct. Rep. Online*, **E57**, m419.
39 Miller, J.S. and Epstein, A.J. (1995) *Chem. Eng. News*, **73**, 30.
40 Rosi, N.L., Eddaoudi, M., Kim, J., O'Keeffe, M. and Yaghi, O.M. (2002) *Angew. Chem. Int. Ed.*, **41**, 284.
41 Yaghi, O.M., Li, H., Davis, C., Richardson, D. and Groy, T.L. (1998) *Acc. Chem. Res.*, **31**, 474.
42 Li, H., Eddaoudi, M., O'Keeffe, M. and Yaghi, O.M. (1999) *Nature*, **402**, 276.
43 Eddaoudi, M., Kim, J., Rosi, N., Vodak, D., Wachter, J., O'Keeffe, M. and Yaghi, O.M. (2002) *Science*, **295**, 469.
44 Rosi, N.L., Eddaoudi, M., Kim, J., O'Keeffe, M. and Yaghi, O.M. (2002) *CrystEngComm*, **4**, 401.
45 Rosi, N.L., Eckert, J., Eddaoudi, M., Vodak, D.T., Kim, J., O'Keeffe, M. and Yaghi, O.M. (2003) *Science*, **300**, 1127.
46 Braun, M.E., Steffek, C.D., Kim, J., Rasmussen, P.G. and Yaghi, O.M. (2001) *Chem. Commun.*, 2532.
47 Batten, S.R. and Robson, R. (1998) *Angew. Chem. Int. Ed.*, **37**, 1461.
48 Carlucci, L., Ciani, G. and Proserpio, D.M. (2003) *CrystEngComm*, **5**, 269.
49 Carlucci, L., Ciani, G. and Proserpio, D.M. (2003) *Coord. Chem. Rev.*, **246**, 247.
50 Barnett, S.A. and Champness, N.R. (2003) *Coord. Chem. Rev.*, **246**, 145.
51 Benmelouka, M., Messaoudi, S., Furet, E., Gautier, R., Le Fur, E. and Pivan, J.Y. (2003) *J. Phys. Chem. A*, **107**, 4122.
52 Laliberte, D., Maris, T. and Wuest, J.D. (2004) *J. Org. Chem.*, **69**, 1776.
53 Saied, O., Maris, T. and Wuest, J.D. (2003) *J. Am. Chem. Soc.*, **125**, 14956.
54 Biradha, K. (2003) *CrystEngComm*, **5**, 374.
55 Nangia, A. (2002) *CrystEngComm*, **4**, 93.
56 Hosseini, M.W. and De Cian, A. (1998) *Chem. Commun.*, 727.

57 Braga, D. and Grepioni, F. (1996) *Chem. Commun.*, 571.
58 Aakeroy, C.B. and Seddon, K.R. (1993) *Chem. Soc. Rev.*, **22**, 397.
59 Roesky, H.W. and Andruh, M. (2003) *Coord. Chem. Rev.*, **236**, 91.
60 Beatty, A.M. (2003) *Coord. Chem. Rev.*, **246**, 131.
61 Bruton, E.A., Brammer, L., Pigge, F.C., Aakeroey, C.B. and Leinen, D.S. (2003) *New J. Chem.*, **27**, 1084.
62 Braga, D., Grepioni, F., Biradha, K., Pedireddi, V.R. and Desiraju, G.R. (1995) *J. Am. Chem. Soc.*, **117**, 3156.
63 Zaworotko, M.J. (1994) *Chem. Soc. Rev.*, **23**, 283.
64 Calhorda, M.J. (2000) *Chem. Commun.*, 801.
65 Braga, D., Maini, L., Polito, M. and Grepioni, F. (2004) *Struct. Bonding*, **111**, 1.
66 Yang, X.J., Wu, B., Sun, W.H. and Janiak, C. (2003) *Inorg. Chim. Acta*, **343**, 366.
67 George, S., Nangia, A., Bagieu-Beucher, M., Masse, R. and Nicoud, J.-F. (2003) *New J. Chem.*, **27**, 568.
68 Nishio, M. (2004) *CrystEngComm*, **6**, 130.
69 Desiraju, G.R. and Hulliger, J. (2002) *Curr. Opin. Solid State Mater. Sci.*, **6**, 107.
70 Brammer, L., Burgard, M.D., Eddleston, M.D., Rodger, C.S., Rath, N.P. and Adams, H. (2002) *CrystEngComm*, **4**, 239.
71 Brammer, L. (2003) *J. Chem. Soc., Dalton Trans.*, 3145.
72 Burrows, A.D. (2004) in: *Crystal Engineering Using Multiple Hydrogen Bonds*, (ed. D.M.P. Mingos), Vol. 108 p. 55, Springer-Verlag, Berlin and Heidelberg.
73 Philp, D. and Stoddart, J.F. (1996) *Angew. Chem. Int. Ed. Engl.*, **35**, 1155.
74 Holliday, B.J., Arnold, F.P. and Mirkin, C.A. (2003) *J. Phys. Chem. A*, **107**, 2737.
75 Leininger, S., Olenyuk, B. and Stang, P.J. (2000) *Chem. Rev.*, **100**, 853.
76 Fujita, M. (2000) in: Molecular Paneling Through Metal-Directed Self-Assembly, (eds M. Fujita), Vol. 96, 177.
77 Saalfrank, R.W., Maid, H., Hampel, F. and Peters, K. (1999) *Eur. J. Inorg. Chem.*, 1859.
78 Robson, R. (2000) *J. Chem. Soc., Dalton Trans.*, 3735.
79 Eddaoudi, M., Moler, D.B., Li, H., Chen, B., Reineke, T.M., O'Keeffe, M. and Yaghi, O.M. (2001) *Acc. Chem. Res.*, **34**, 319.
80 Evans, O.R. and Lin, W. (2002) *Acc. Chem. Res.*, **35**, 511.
81 Chen, B., Eddaoudi, M., Hyde, S.T., O'Keeffe, M. and Yaghi, O.M. (2001) *Science*, **291**, 1021.
82 Biradha, K. and Fujita, M. (2002) *Angew. Chem. Int. Ed.*, **41**, 3392.
83 Desiraju, G.R. (2001) *Nature*, **412**, 397.
84 Tiekink, E.R.T. and Vittal, J.J. (2005) *Frontiers in Crystal Engineering*, Wiley, New York.
85 Braga, D., Grepioni, F. and Desiraju, G.R. (1998) *Chem. Rev.*, **98**, 1375.
86 Braga, D. (2000) *J. Chem. Soc., Dalton Trans.*, 3705.
87 Braga, D., Desiraju, G.R., Miller, J.S., Orpen, A.G. and Price, S.L. (2002) *CrystEngComm*, **4**, 500.
88 Hollingsworth, M.D. (2002) *Science*, **295**, 2410.
89 Oh, M., Carpenter, G.B. and Sweigart, D.A. (2002) *Organometallics*, **21**, 1290.
90 Oh, M., Carpenter, G.B. and Sweigart, D.A. (2001) *Angew. Chem. Int. Ed.*, **40**, 3191.
91 Oh, M., Carpenter, G.B. and Sweigart, D.A. (2002) *Angew. Chem. Int. Ed.*, **41**, 3650.
92 Oh, M., Carpenter, G.B. and Sweigart, D.A. (2002) *Chem. Commun.*, 2168.
93 Oh, M., Carpenter, G.B. and Sweigart, D.A. (2003) *Organometallics*, **22**, 2364.
94 Oh, M., Carpenter, G.B. and Sweigart, D.A. (2003) *Angew. Chem. Int. Ed.*, **42**, 2026.
95 Oh, M., Carpenter, G.B. and Sweigart, D.A. (2004) *Acc. Chem. Res.*, **37**, 1.
96 Braga, D., Chen, S., Filson, H., Maini, L., Netherton, M.R., Patrick, B.O., Scheffer, J.R., Scott, C. and Xia, W. (2004) *J. Am. Chem. Soc.*, **126**, 3511.
97 Paul, I.C. and Curtin, D.Y. (1973) *Acc. Chem. Res.*, **6**, 217.

98 Chiang, C.C., Lin, C.T., Wang, H.J., Curtin, D.Y. and Paul, I.C. (1977) *J. Am. Chem. Soc.*, **99**, 6303.
99 Desiraju, G.R., Paul, I.C. and Curtin, D.Y. (1977) *J. Am. Chem. Soc.*, **99**, 1594.
100 Patil, A.O., Pennington, W.T., Desiraju, G.R., Curtin, D.Y. and Paul, I.C. (1986) *Mol. Cryst. Liq. Cryst.*, **134**, 279.
101 Etter, M.C. (1991) *J. Phys. Chem.*, **95**, 4601.
102 Ojala, W.H. and Etter, M.C. (1992) *J. Am. Chem. Soc.*, **114**, 10288.
103 Etter, M.C., Reutzel, S.M. and Choo, C.G. (1993) *J. Am. Chem. Soc.*, **115**, 4411.
104 Nichols, P.J., Raston, C.L. and Steed, J.W. (2001) *Chem. Commun.*, 1062.
105 Belcher, W.J., Longstaff, C.A., Neckenig, M.R. and Steed, J.W. (2002) *Chem. Commun.*, 1602.
106 Batten, S.R. (2001) *Curr. Opin. Solid State Mater. Sci.*, **5**, 107.
107 Sun, H.-L., Gao, S., Ma, B.-Q., Su, G. and Batten, S.R. (2005) *Cryst. Growth Des.*, **5**, 269.
108 Stark, J.L., Rheingold, A.L. and Maatta, E.A. (1995) *J. Chem. Soc., Chem. Commun.*, 1165.
109 Eddaoudi, M., Kim, J., Rosi, N., Vodak, D., Wachter, J., O'Keeffe, M. and Yaghi Omar, M. (2002) *Science*, **295**, 469.
110 Braga, D., Curzi, M., Grepioni, F. and Polito, M. (2005) *Chem. Commun.*, 2915.
111 Braga, D., Curzi, M., Johansson, A., Polito, M., Rubini, K. and Grepioni, F. (2006) *Angew. Chem. Int. Ed.*, **45**, 142.
112 Anastas, P. and Warner, J. (1998) *Green Chemistry: Theory and Practice*, Oxford University Press, Oxford.
113 Braga, D., Maini, L., Grepioni, F., Elschenbroich, C., Paganelli, F. and Schiemann, O. (2001) *Organometallics*, **20**, 1875.
114 Tanaka, K. and Toda, F. (2000) *Chem. Rev.*, **100**, 1025.
115 Tanaka, K. (2003) *Solvent-Free Organic Synthesis*, Wiley-VCH, Weinheim, Germany.
116 Rotheberg, G., Downie, A.P., Raston, C.L. and Scott, J.L. (2001) *J. Am. Chem. Soc.*, **123**, 8701.
117 Toda, F. (2002) *CrystEngComm*, **4**, 215.
118 Nassimbeni, L.R. (2003) *Acc. Chem. Res.*, **36**, 631.
119 Braga, D., Maini, L., Polito, M., Mirolo, L. and Grepioni, F. (2003) *Chem. Eur. J.*, **9**, 4362.
120 Braga, D., Maini, L., Polito, M., Mirolo, L. and Grepioni, F. (2002) *Chem. Commun.*, 2960.
121 Braga, D., Polito, M., D'Addario, D., Tagliavini, E., Proserpio, D.M., Grepioni, F. and Steed, J.W. (2003) *Organometallics*, **22**, 4532.
122 Braga, D., Polito, M., Bracaccini, M., D'Addario, D., Tagliavini, E., Sturba, L. and Grepioni, F. (2003) *Organometallics*, **22**, 2142.
123 Braga, D., Giaffreda, S.L. and Grepioni, F. (2006) *Chem. Commun.*, 3877.
124 Braga, D., Polito, M. and Grepioni, F. (2004) *Cryst. Grow. Des.*, **4**, 769.
125 Braga, D., Cojazzi, G., Emiliani, D., Maini, L. and Grepioni, F. (2001) *Chem. Commun.*, 2272.
126 Braga, D., Cojazzi, G., Emiliani, D., Maini, L. and Grepioni, F. (2002) *Organometallics*, **21**, 1315.
127 Braga, D., Maini, L., Mazzotti, M., Rubini, K. and Grepioni, F. (2003) *CrystEngComm*, **5**, 154.
128 Braga, D., Maini, L., Mazzotti, M., Rubini, K., Masic, A., Gobetto, R. and Grepioni, F. (2002) *Chem. Commun.*, 2296.
129 Bernstein, J. (2002) *Polymorphism in Molecular Crystals*, Oxford University Press, Oxford.
130 Brittain, H.G. (1999) *Drugs Pharm. Sci.*, **95**, 227.
131 Hilfiker, R. (2006) *Polymorphism*, Wiley-VCH, Weinheim.
132 Day, G.M., Motherwell, W.D.S., Ammon, H.L., Boerrigter, S.X.M., Della Valle, R.G., Venuti, E., Dzyabchenko, A. Dunitz, J.D., Schweizer, B., van Eijck, B.P., Erk, P., Facelli, J.C., Bazterra, V.E., Ferraro, M.B., Hofmann, D.W.M., Leusen, F.J.J., Liang, C., Pantelides, C.C., Karamertzanis, P.G., Price, S.L., Lewis, T.C., Nowell, H., Torrisi, A., Scheraga, H.A., Arnautova, Y.A.,

Schmidt, M.U. and Verwer, P. (2005) *Acta Crystallogr., Sect. B: Struct. Sci.*, **B61**, 511.

133 Datta, S. and Grant David, J.W. (2004) *Nat. Rev. Drug Discov.*, **3**, 42.

134 Michele, L.M.C.F.P.T. and Dario Braga, R.C.R.G. (2007) *Chem. Eur. J.*, **13**, 1222.

135 Tong, M.-L., Wang, J., Hu, S. and Batten, S.R. (2005) *Inorg. Chem. Commun.*, **8**, 48.

136 Tong, M.-L., Wang, J. and Hu, S. (2005) *J. Solid State Chem.*, **178**, 1518.

137 Senevirathna, M.K.I., Pitigala, P.K.D.D.P., Perera, V.P.S. and Tennakone, K. (2005) *Langmuir*, **21**, 2997.

138 Nishitani, H., Nishitsuji, K., Okumura, K. and Taguchi, H. (1996) *Phytochemistry*, **42**, 1.

139 Taguchi, H., Maeda, M., Nishitani, H., Okumura, K., Shimabayashi, Y. and Iwai, K. (1992) *Biosci. Biotechnol. Biochem.*, **56**, 1921.

140 PDF-2 Release (2005) *International Centre for Diffraction Data*, Newtown Square, PA, U.S.A.

141 Griffiths, P.J.F. (1963) *Acta Crystallogr.*, **16**, 1074.

142 Takusagawa, F., Hirotsu, K. and Shimada, A. (1973) *Bull. Chem. Soc. Jpn.*, **46**, 2669.

143 Harmon, K.M. and Shaw, K.E. (1999) *J. Mol. Struct.*, **513**, 219.

144 Vishweshwar, P., Nangia, A. and Lynch, V.M. (2002) *J. Org. Chem.*, **67**, 556.

145 Bernstein, J., Davey, R.J. and Henck, J.-O. (1999) *Angew. Chem. Int. Ed.*, **38**, 3441.

146 Braga, D., Cadoni, M., Grepioni, F., Maini, L. and Rubini, K. (2006) *CrystEngComm*, **8**, 756.

147 Lewis, T.C., Tocher, D.A. and Price, S.L. (2004) *Cryst. Growth Des.*, **4**, 979.

148 Bolton, W. (1963) *Acta Crystallogr.*, **16**, 166.

149 Jeffrey, G.A., Ghose, S. and Warwicker, J.O. (1961) *Acta Crystallogr.*, **14**, 881.

150 Al-Karaghouli, A.R., Abdul-Wahab, B., Ajaj, E. and Al-Asaff, S. (1977) *Acta Crystallogr., Sect. B: Struct. Sci.*, **B33**, 1655.

151 Lewis, T.C., Tocher, D.A. and Price, S.L. (2005) *Cryst. Grow. Des.*, **5**, 983.

152 Cleverley, B. and Williams, P.P. (1959) *Tetrahedron*, **7**, 277.

153 Caillet, J. and Claverie, P. (1980) *Acta Crystallogr., Sect. B: Struct. Sci.*, **B36**, 2642.

154 Bojarski, J.T., Mokrosz, J.L., Barton, H.J. and Paluchowska, M.H. (1985) *Adv. Heterocycl. Chem.*, **38**, 229.

155 Chen, X., Morris, K.R., Griesser, U.J., Byrn, S.R. and Stowell, J.G. (2002) *J. Am. Chem. Soc.*, **124**, 15012.

156 Braga, D. and Maini, L. (2004) *Chem. Commun.*, 976.

157 Wilson, C.C. (2001) *Acta Crystallogr., Sect. B: Struct. Sci.*, **B57**, 435.

158 Wiechert, D. and Mootz, D. (1999) *Angew. Chem. Int. Ed.*, **38**, 1974.

159 Braga, D., Rubini, K. and Maini, L. (2004) *CrystEngComm*, **6**, 236.

160 Braga, D., Curzi, M., Lusi, M. and Grepioni, F. (2005) *CrystEngComm*, **7**, 276.

161 Calleja, M., Mason, S.A., Prince, P.D., Steed, J.W. and Wilkinson, C. (2003) *New J. Chem.*, **27**, 28.

162 Junk, P.C., McCool, B.J., Moubaraki, B., Murray, K.S., Spiccia, L., Cashion, J.D., Steed, J.W. (2002) *J. Chem. Soc., Dalton Trans.*, 1024.

163 Atwood, J.L., Holman, K.T. and Steed, J.W. (1996) *Chem Commun.*, 1401.

164 Steed, J.W. (2001) *Coord. Chem. Rev.*, **215**, 171.

165 Lipkowski, J., Baranowski, B. and Lunden, A. (1993) *Pol. J. Chem.*, **67**, 1867.

166 Haile, S.M. (1999) *Mater. Res. Soc. Symp. Proc.*, **547**, 315.

167 Haile, S.M., Boysen, D.A., Chisholm, C.R.I. and Merle, R.B. (2001) *Nature*, **410**, 910.

168 Chisholm, M.H. and Hollandsworth, C.B. (2005) *Multiple Bonds Between Metal Atoms*, 3rd edn. p. 203, Springer-Verlag, Berlin and Heidelberg.

169 Braga, D., Gandolfi, M., Lusi, M., Paolucci, D., Polito, M., Rubini, K. and Grepioni, F. (2007) *Chem. Eur. J.*, **13**, 5249.

170 Braga, D., Gandolfi, M., Lusi, M., Polito, M., Rubini, K. and Grepioni, F. (2007) *Cryst. Grow. Des.*, **7** (5), 919–924.

7
Supramolecular Architectures Based On Organometallic Half-sandwich Complexes

Thomas B. Rauchfuss and Kay Severin

7.1
Introduction

Cationic half-sandwich reagents have evolved as versatile building blocks that permit controlled growth of stable functional nanostructures. The key half-sandwich modules of interest are cations with the formula $[(\pi\text{-ligand})M]^{n+}$. The most popular derivatizing agents are the dications $[Cp^*M]^{2+}$ (M = Rh, Ir). These half-sandwiches are facial and tritopic, meaning that they have three mutually adjacent points of attachment. The facial arrangement of the sites encourages the formation of molecular structures, whereas meridional tritopic Lewis acids would favor the formation of polymers. In terms of stabilizing nanostructures, the advantages of the half-sandwich components containing cyclopentadienyl ligands are several:

- The large organic ligand partially protects the exterior of the nanostructure with a lipophilic sheath that enhances the solubility of its derivatives, compensating for the low solubility that is often characteristic of large aggregates. Furthermore, bulky C_5Me_5 ligands inhibit polymerization by facilitating intramolecular condensations, giving entropically favored closed structures [1].

- The cyclopentadienyl ligands are virtually inert. This important characteristic is perhaps best demonstrated by the syntheses of $[(C_5R_5)M(CN)_3]^-$ (M = Co, Rh, Ir) and $[(C_5R_5)Ru(CN)_3]^{2-}$, which are generated by treatment of the precursor half-sandwich halides with excess cyanide salt [2].

- The Cp* ligand facilitates characterization of complicated ensembles because it provides a conveniently intense, simple label for ^1H and ^{13}C NMR analysis of complex mixtures and the mass spectra of Cp*M-derived species are relatively simple.

- Cyclopentadienyl ligands can be obtained commercially with diverse substitution patterns; for example, C_5Me_4H and C_5Me_4Et can be readily attached to Rh(III) and Ir(III) to modify usefully the solubility, spectroscopy, crystallizability and geometry of the resulting ensembles.

Organic Nanostructures. Edited by Jerry L. Atwood and Jonathan W. Steed
Copyright © 2008 WILEY-VCH Verlag GmbH & Co. KGaA, Weinheim
ISBN: 978-3-527-31836-0

[(π-ligand)M]$^{2+}$

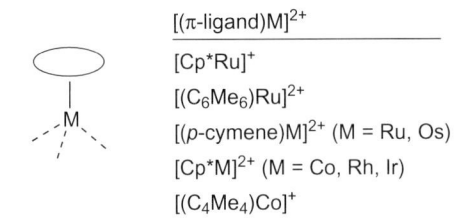

[Cp*Ru]$^+$

[(C$_6$Me$_6$)Ru]$^{2+}$

[(p-cymene)M]$^{2+}$ (M = Ru, Os)

[Cp*M]$^{2+}$ (M = Co, Rh, Ir)

[(C$_4$Me$_4$)Co]$^+$

Figure 7.1 Structures of representative half-sandwich units.

Although the above comments are directed to [Cp*Rh]$^{2+}$ and [Cp*Ir]$^{2+}$, they also apply to the corresponding [(C$_6$R$_6$)Ru]$^{2+}$ derivatives [3]. Additionally, related vertices [Cp*Co]$^{2+}$, [CpCo]$^{2+}$, [Cp*Ru]$^+$ and [(C$_4$Me$_4$)Co]$^+$ have also received intermittent attention (Figure 7.1) [4].

The 12e$^-$ species [(π-ligand)M]$^{2+}$ forms unsaturated products upon binding bidentate ligands XY^{n-}. The resulting 16e$^-$ species, [(π-ligand)M(XY)]^m tend to condense further via the formation of inter-complex donor–acceptor interactions to give [(π-ligand)M(XY)]$_x^{mx-}$. The possibility of adjusting the charge of the ensemble allows one to manipulate the strength of both the intermetallic and the host–guest interactions. The power of this design is especially well illustrated by the work involving organic ligands (see Sections 7.2 and 7.3.2).

The half-sandwich reagents [Cp*MCl(μ-Cl)]$_2$ are commercially available and readily prepared [5]. The chloro-bridged dimers themselves are conveniently reactive because of the easy scission of the chloride bridges by even weak Lewis bases. The solvated salts, which are often generated *in situ*, represent more potent electrophiles. The salts [Cp*Rh(MeCN)$_3$](PF$_6$)$_2$ and [(arene)Ru(MeCN)$_3$](PF$_6$)$_2$ are typical reagents of choice [5]. It is sometimes necessary to generate *in situ* complexes derived from acetone [6] or nitromethane [7] when examining very weakly basic ligands. These salts are prepared by the usual AgCl precipitation methods. Relative to the simple aquo complexes, for example [Rh(H$_2$O)$_6$]$^{3+}$, the solvent ligands on the half-sandwich complexes are more labile [3].

7.2
Macrocycles

Rectangular macrocycles with (π-ligand)M complexes in the corner can be obtained by utilization of linear, difunctional ligands such as diisocyanides [8,9], 4,4′-bipyridine [9,10], cyanamide [11] or cyanide [12]. These macrocycles are generally synthesized in a stepwise fashion from the chloro-bridged dimers [(π-ligand)MCl(μ-Cl)]$_2$ [(π-ligand)M = (arene)Ru, Cp*Rh, Cp*Ir]. Three representative examples are shown in Figure 7.2.

An alternative synthesis of macrocycles entails the combination of (π-ligand)M complexes with trifunctional ligands. In this case, two of the donor atoms form a chelate complex with one metal fragment and the remaining donor atom

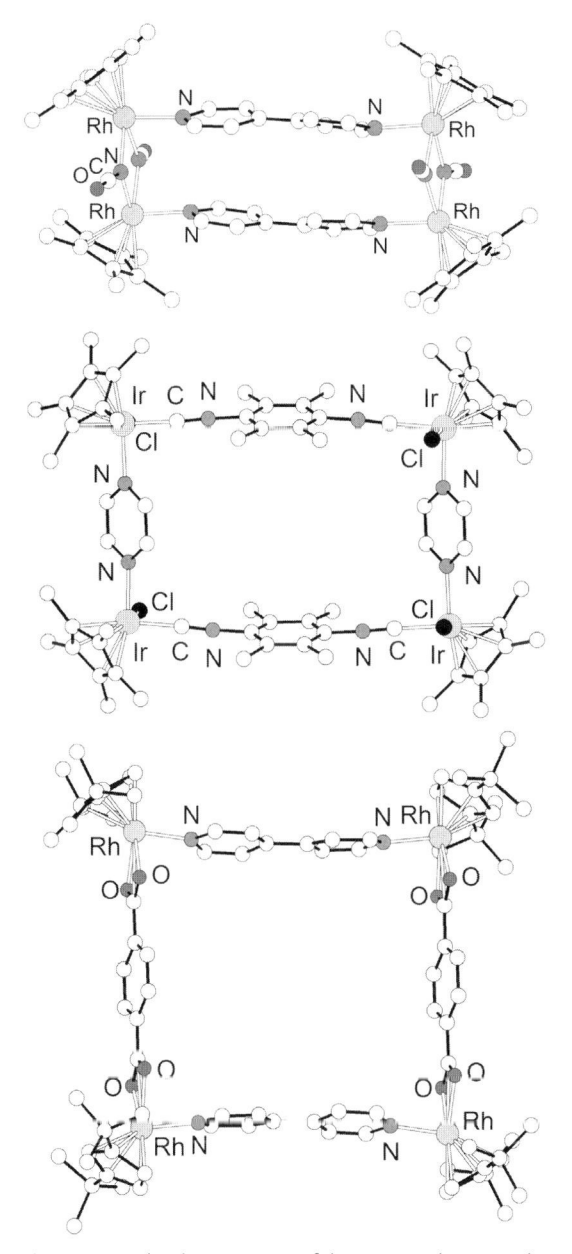

Figure 7.2 Molecular structures of the macrocycles [(Cp*Rh)$_2$ (μ-NCO)$_2$(μ-4,4′-dipyridyl)]$_2$(OTf)$_4$ (top), [(Cp*IrCl)$_2$(μ-C$_4$H$_4$N$_2$) (μ-1,4-(CN)$_2$C$_6$Me$_4$)]$_2$(OTf)$_4$ (middle) and [{(1, 3-^{t}Bu$_2$C$_5$H$_3$) Rh}$_2$(μ-4,4′-dipyridyl)(μ-terephthalate)]$_2$(OTf)$_4$ (bottom) in the crystal. The hydrogen atoms and the triflate anions are not shown for clarity.

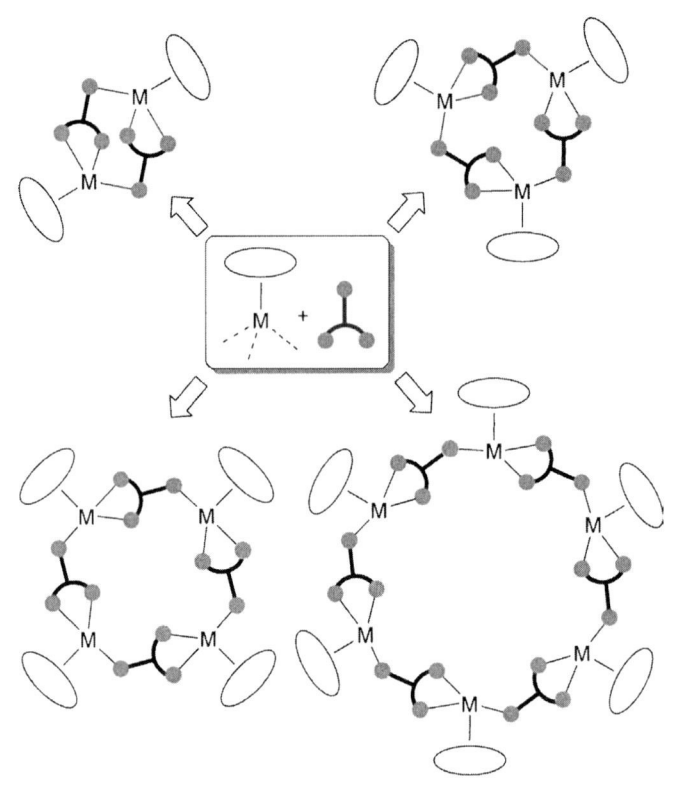

Scheme 7.1 Formation of macrocycles by reaction of organometallic half-sandwich complexes with trifunctional ligands.

coordinates to an adjacent metal fragment. Using this strategy, di-, tri-, tetra- and hexanuclear metallamacrocycles have been obtained (Scheme 7.1).

Cationic complexes comprised of three Cp*Rh fragments connected by three nucleobase derivatives were studied extensively by Fish's group [13]. Structurally related trimers with (arene)Ru^{II}, Cp*Ir^{III} (Figure 7.3) or Cp*Rh^{III} complexes were reported by the groups of Sheldrick [14,15], Yamanari [16] and Vogler [17]. For certain nucleobase ligands, still larger assemblies were observed. Tetranuclear complexes were obtained when adenine [15] or 6-purinethione [18] were used as the bridging ligands and hexanuclear Cp*Rh and Cp*Ir complexes were obtained with the thio derivative 6-purinethione riboside [19].

Amino acidate ligands are also suited to build trimeric organometallic assemblies [20,21]. In this case, the metal fragments are connected via the two carboxylate O atoms and the amino group (Figure 7.3). It is interesting that these trimers can be used as catalysts for the enantioselective hydrogen transfer reactions [21]. The catalytically active species, however, was suggested to be a mononuclear complex.

The macrocycles discussed so far are mostly polycationic species and several of them are soluble in water. In terms of host–guest chemistry, this hydrophilicity

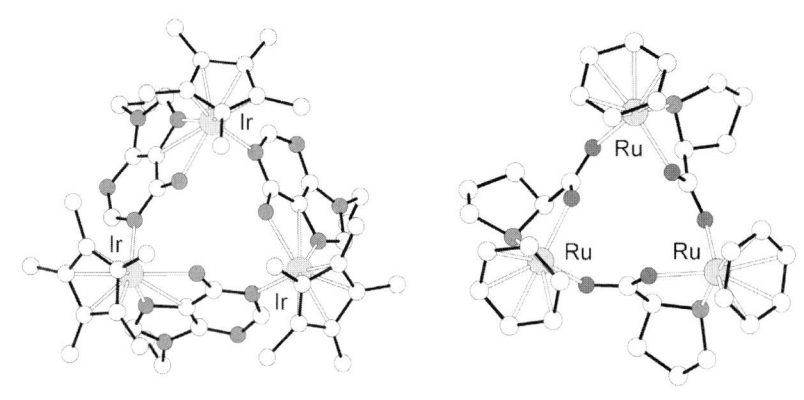

Figure 7.3 Molecular structures of the macrocycles [Cp*Ir(L)]$_3$(OTf)$_3$ (L = deprotonated 9-ethyladenine) (left) and [(p-cymene)Ru(prolinato)]$_3$(BF$_4$)$_3$ (right) in the crystal. The hydrogen atoms, the side-chains of the p-cymene ligands and the counter-anions are not shown for clarity.

may be advantageous if hydrophobic interactions are the main driving force for guest inclusion. Trimeric Cp*Rh complexes, for example, have been shown to act as hosts for aromatic carboxylic acids including amino acids [13e]. Strong complexation-induced chemical shifts were observed in the ^{1}H NMR spectra. The Cp*Rh trimers may therefore be of interest as ^{1}H NMR shift reagents [13c].

Macrocyclic complexes with a net charge of zero were obtained with the following ligands: 2,3-dihydroxypyridine (Figure 7.4a) [22,23], 3-acetamido-2-hydroxypyridine [23], 2,3-dihydroxyquinoline [24], 2,3-dihydroxyquinoxaline [24], 6-methyl-2,3-phenazinediol (Figure 7.4b) [24], 3,4-dihydroxy-2-methylpyridine (Figure 7.4c) [25] and 4-imidazolecarboxylic acid (Figure 7.4d) [24]. Using half-sandwich reagents, both trimeric and tetrameric assemblies were generated.

Both the neutral and the cationic macrocycles based on trifunctional ligands are obtained in a highly diastereoselective fashion. Trimeric aggregates were found to adopt a (pseudo)-C_3 symmetric structures with all three metal centers having the same absolute configuration. Resolution of the macrocycles was achieved by using chiral bridging ligands [16] or the chiral guests Li(Δ-TRISPHAT) [26]. Tetrameric aggregates, on the other hand, display a (pseudo)-S_4 symmetry with the metal centers having alternating configurations [15,18,24].

Trinuclear complexes derived from half-sandwich complexes and 2,3-dihydroxypyridine ligands represent organometallic analogues of 12-crown-3 [27]. It was found that these complexes display a very high affinity for lithium and sodium salts (Scheme 7.2) [22,23]. The alkali metal ion M$^+$ is coordinated to the three adjacent oxygen atoms of the receptor. In the solid state and in apolar organic solvents, the salt MX is bound as an ion pair. Complexation of potassium salts was not observed and this was explained by the steric constraints imposed by the π-ligands.

The affinity of the 12-metallacrown-3 complexes for lithium and sodium salts is remarkably high. Competition experiments with organic ionophores have revealed

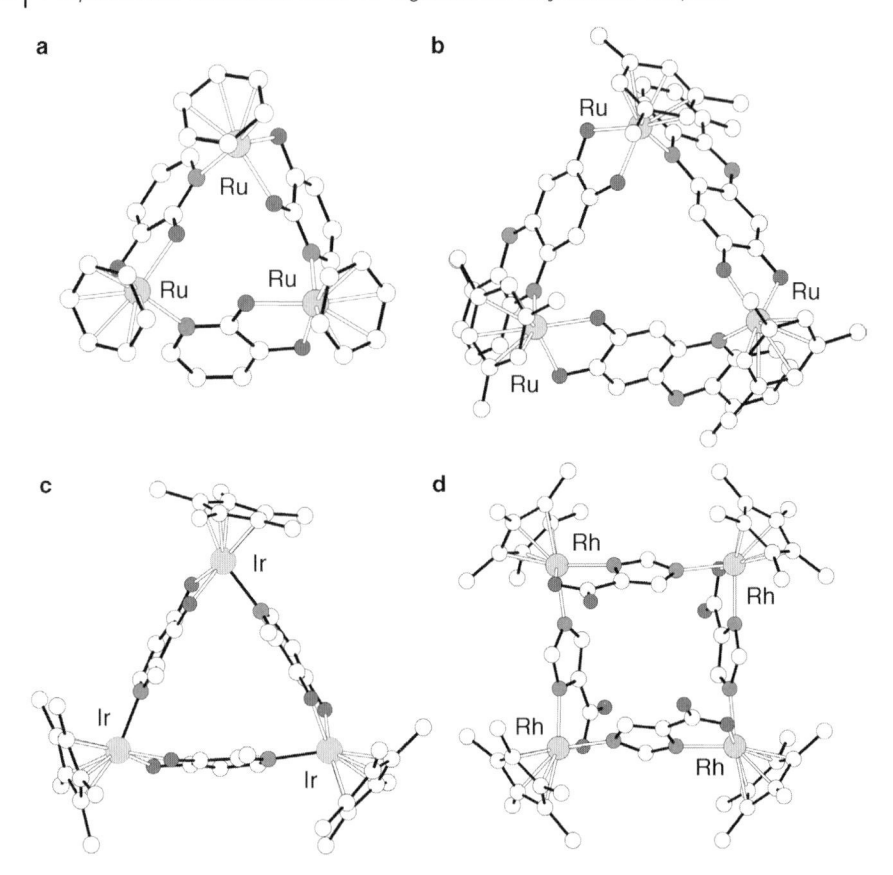

Figure 7.4 Molecular structures of selected organometallic macrocycles in the crystal. They were obtained by combination of (a) (benzene)Ru complexes with 2,3-dihydroxypyridine ligands; (b) (1,3,5-trimethylbenzene)Ru complexes with 6-methyl-2,3-phenazinediol ligands; (c) Cp*Ir complexes with 3,4-dihydroxy-2-methyl-pyridine ligands; and (d) Cp*Rh complexes with 4-imidazolecarboxylic acid ligands. The hydrogen atoms are not shown for clarity.

that in chloroform, the binding affinity for LiCl and NaCl is significantly higher than that of classical crown ethers and comparable to that of cryptands [23]. The high affinity can be attributed to several facts: (a) the receptors are well preorganized to bind lithium or sodium ions; (b) the salts are bound as an ion pair; (c) the energetic costs for the desolvation of the receptors are very low because only one solvent molecule can fit inside the binding cavity; and (d) the oxygen donor atoms display a high partial negative charge [28]. The outstanding affinity of the 12-metallacrown-3 complexes for lithium and sodium salts was utilized to capture molecular LiF [29] and Na_2SiF_6 (Figure 7.5) [30]. It should be noted that the stabilization of these compounds in molecular form represents a challenging task due to the high lattice energy of the salts. 12- Metallacrown-3 complexes have also been investigated for their ability to differentiate between the isotopes $^6Li^+$ and $^7Li^+$ using isotope-selective laser

Scheme 7.2 Metallamacrocycles composed of half-sandwich complexes of 2,3-dihydroxypyridine ligands are analogues of 12-crown-3 [(π-ligand)M = (arene)Ru, Cp*Rh, Cp*Ir]. They are able to bind lithium and sodium salts with very high affinity and selectivity.

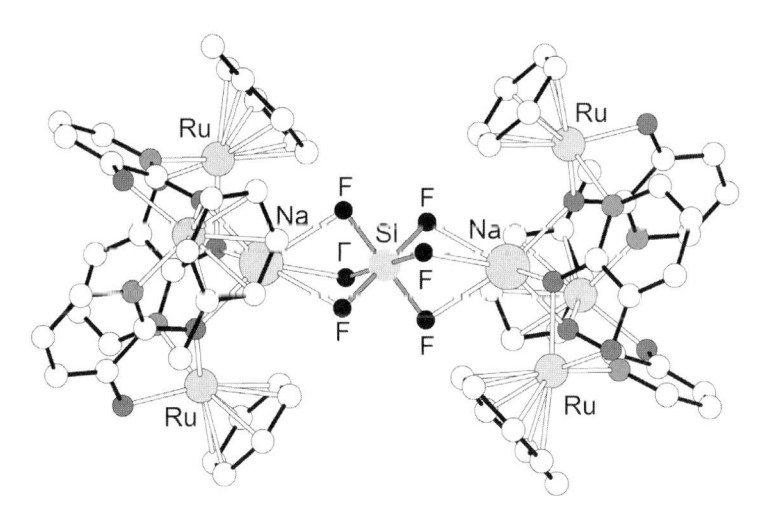

Figure 7.5 Stabilization of molecular Na_2SiF_6 by encapsulation between two [(p-cymene)Ru(3-oxo-2-pyridonate)]$_3$ receptors. Hydrogen atoms and the side-chains of the p-cymene π-ligands are not shown for clarity.

absorption spectroscopy. It was found that the macrocycle [(C_6Me_6)Ru(3-oxo-2-pyridonate)]$_3$ preferentially binds to the smaller isotope $^7Li^+$ (separation factor α 0.945) [31].

The host–guest chemistry of the trimer [($C_6H_5CO_2Et$)Ru(3-oxo-2-pyridonate)]$_3$ proved to be of special interest. Although this receptor is able to bind Na^+ ions, it shows a very pronounced affinity and selectivity for Li^+ salts as demonstrated in an extraction experiment. When an aqueous solution containing LiCl (50 mM) and a large excess of NaCl, KCl, CsCl, $MgCl_2$ and $CaCl_2$ (each 1 M) was shaken with a chloroform solution of this metallacrown complex, the exclusive and quantitative extraction of LiCl was observed after 24 h [32].

The possibility of obtaining metallacrown complexes with a high specificity for Li^+ is of potential interest for analytical applications, because of the pharmacological importance of lithium salts (Li_2CO_3 is a frequently used drug for patients with bipolar disorder) [33]. To construct Li^+-specific sensors, a receptor that could be used directly in water would be advantageous. To render 12-metallacrown-3 complexes water-soluble, dialkylaminomethyl groups were attached to the bridging 2,3-dihydroxypyridine ligands. With the resulting ligands it is possible to generate macrocycles in water at neutral pH simply by dissolving the aminomethyl-substituted ligand with the corresponding [(π-ligand)MCl_2]$_2$ complex in phosphate buffer. The macrocycles are then formed by self-assembly in quantitative yield [34]. The binding constant for the complexation of Li^+ in water depends on the nature of the (π-ligand)M fragment. With (p-cymene)Ru it was possible obtain to a receptor which binds Li^+ with an association constant of $K = 6 \times 10^4 M^{-1}$ (Figure 7.6) [34a]. This value is sufficient to achieve nearly quantitative complexation of Li^+ at the pharmacologically relevant concentration of ∼1 mM. Na^+ ions do not interfere with the complexation because the binding constants are four orders of magnitude lower.

Trimeric macrocycles based on hydroxypyridine ligands were also investigated in the context of dynamic combinatorial chemistry [35]. It was found that macrocycles

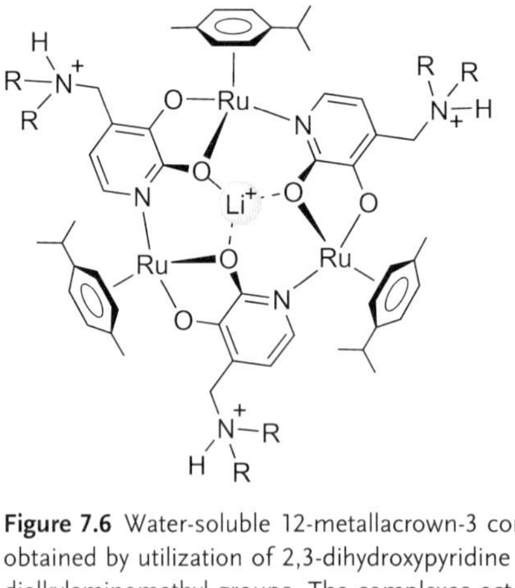

Figure 7.6 Water-soluble 12-metallacrown-3 complexes can be obtained by utilization of 2,3-dihydroxypyridine ligands with dialkylaminomethyl groups. The complexes act as Li^+-specific receptors with binding constants of up to $K = 6 \times 10^4 M^{-1}$.

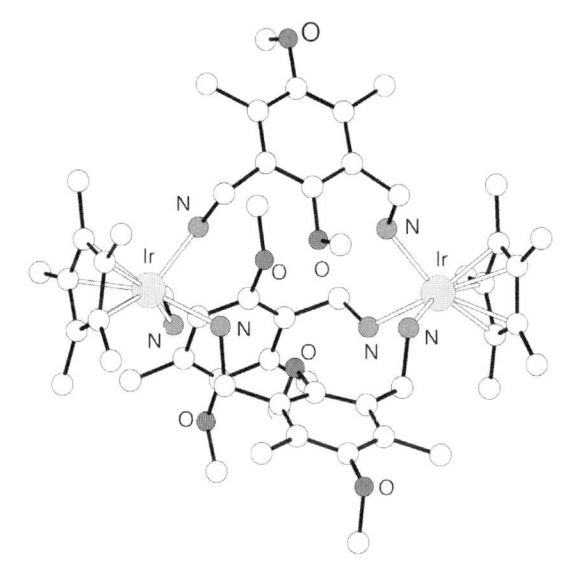

Figure 7.7 Molecular structure of a macrobicyclic Cp*Ir complex with bridging 1,3-bis(aminomethyl)-2,5-dimethoxy-4,6-dimethylbenzene ligands in the crystal. The hydrogen atoms and the counter-anions are not shown for clarity.

with different (π-ligand)M complexes undergo exchange processes to generate mixed-metal macrocycles [36]. Quantitative analysis of these dynamic equilibria revealed that the product distribution is controlled by steric constraints and can be modulated by addition of guests. These results were used to draw conclusions about the adaptive behavior of dynamic combinatorial libraries in general.

An interesting approach to make *macrobicyclic* organometallic complexes was reported by Amouri's group [37]. They reacted [Cp*M(acetone)$_3$](BF$_4$)$_2$ (M = Rh, Ir) with *m*-xylylenediamine or derivatives as the bridging ligands and obtained cryptand-like structures (Figure 7.7). Some of these complexes tightly encapsulate a BF$_4^-$ anion.

7.3
Coordination Cages

7.3.1
Cyanometallate Cages

The utility of the polymeric cyanometallates was recognized for centuries before the underlying supramolecular chemistry was appreciated [38]. The Hofmann clathrates {[Ni(NH$_3$)$_2$][Ni(CN)$_4$](guest)$_2$} have been used for the size-specific separation of aromatic compounds [39] and Prussian Blue [Fe$_7$(CN)$_{18}$(H$_2$O)$_x$] and its analogues are versatile dyes [38]. Related coordination compounds and polymers continue to be topical, because of their potential for the separation of gases [40] and toxic metals [41]. The selectivity of the host–guest behavior of cyanometallates can be attributed in part to the rigidity of the polar framework.

Prussian Blue and its many analogues are derived from the hexacyanometallates, which give rise to three-dimensional polymers [38]. When three of the cyanide ligands are replaced by a π-ligand, the building block retains its ability to form three-dimensional structures but the tendency to polymerize is suppressed. Furthermore, the resulting half-sandwich tricyanides are convenient reagents since, typically as monoanions, they are soluble in a range of nonaqueous solvents. The versatility of the half-sandwich motif is demonstrated by the use of such modules as both the Lewis basic and Lewis acidic components in the synthesis of cyanide-based cages. The prototypical cage complex in this series is $\{[CpCo(CN)_3]_4[Cp^*Rh]_4\}^{4+}$, which forms quantitatively upon combining equimolar amounts of $[CpCo(CN)_3]^-$ and $[Cp^*Rh(MeCN)_3]^{2+}$ (Scheme 7.3) [42]. The cuboidal cage, which has been described as a "molecular box", has idealized T_d symmetry. It is now clear that numerous related boxes can be generated by similar condensations [4]. NMR analysis shows that the condensation proceeds without scission of the Co–CN bonds, thus the building block concept and of course the angles imposed by the box match the coordination preference of the octahedral vertices.

The extensive library of half-sandwich reagents allows one to prepare cyanometallate boxes with a range of charges. The charge on these cages influences the ionophilicity of the resulting cages: anionic and neutral cage are excellent receptors for alkali metal cations. Thus, the condensation of $[CpCo(CN)_3]^-$ and $[Cp^*Ru(NCMe)_3]^+$ and of $[Cp^*Rh(CN)_3]^-$ and $[Mo(CO)_3(NCMe)_3]$ requires the presence of templating cations such as K^+ or Cs^+ [43]. The resulting cages, $\{M \subset [Cp^*Rh(CN)_3]_4[Mo(CO)_3]_4\}^{3-}$ and $\{M \subset [CpCo(CN)_3]_4[Cp^*Ru]_4\}^+$, feature cations inside the cage. The high affinity of these cages for cations arises from the attractive interactions between the π-bonds of the cyanide ligands and the alkali metal cation [44]. Entropy also contributes because encapsulation liberates solvent ligands from

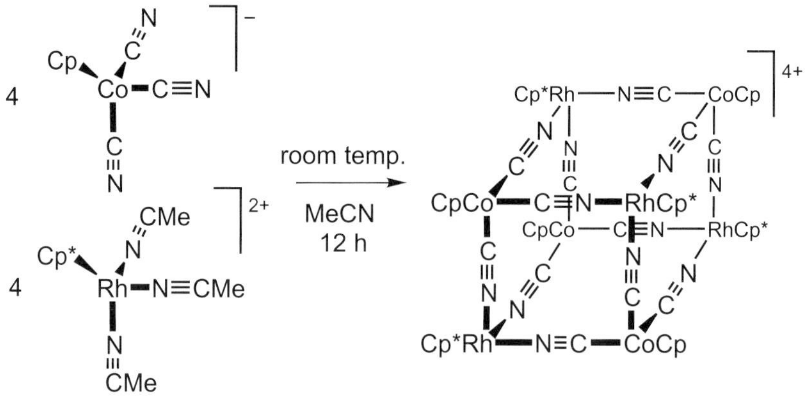

Scheme 7.3 Synthesis of a cyanometallate cage by reaction of $[CpCo(CN)_3]^-$ with $[Cp^*Rh(MeCN)_3]^{2+}$.

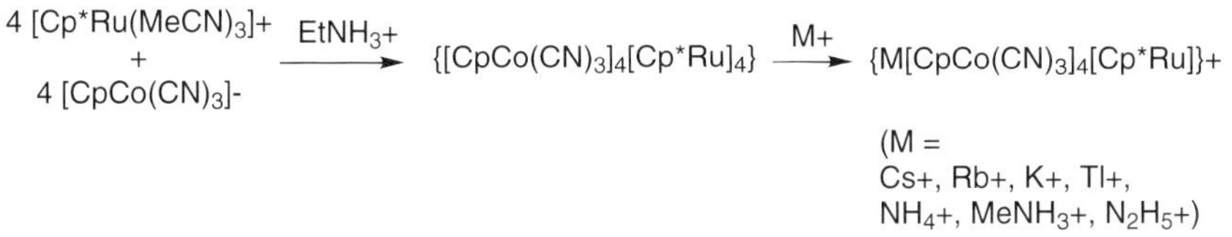

$$4\ [Cp^*Ru(MeCN)_3]^+ + 4\ [CpCo(CN)_3]^- \xrightarrow{EtNH_3^+} \{[CpCo(CN)_3]_4[Cp^*Ru]_4\} \xrightarrow{M^+} \{M[CpCo(CN)_3]_4[Cp^*Ru]\}^+$$

(M = Cs+, Rb+, K+, Tl+, NH$_4$+, MeNH$_3$+, N$_2$H$_5$+)

Scheme 7.4 The charge-neutral box $\{[CpCo(CN)_3]_4[Cp^*Ru]_4\}$ acts as a host for cationic guests.

the alkali metal cation in addition to the solvent that is coordinated to the three cationic half-sandwich reagents. The boxes distort upon complexation, but the individual guest–CN interactions are weak, as indicated by long bonds, typically >3.4 Å. Once inside the cyanometallate, the guests are unable to engage in their characteristic reactions. For example, $\{NH_4 \subset [CpCo(CN)_3]_4[Cp^*Ru]_4\}^+$ is unreactive towards D$_2$O, whereas in solution, of course, NH$_4^+$ and D$_2$O exchange protons at diffusion-controlled rates. Ion exchange between included metal ions and those free in solution can also be very slow.

The ion EtNH$_3^+$ templates the formation of $\{[CpCo(CN)_3]_4[Cp^*Ru]_4\}$, a charge-*neutral* box that lacks guests at its interior. The ability of EtNH$_3^+$ to assist in the condensation reflects the strength of the hydrogen bonding interaction MCN$\cdots$H$^{\delta+}$–NR$_3$. EtNH$_3^+$ is an ideal template because it is highly effective but too large to be contained within the ultimate product. Many ions insert into $\{[CpCo(CN)_3]_4[Cp^*Ru]_4\}$: Cs$^+$, Rb$^+$, K$^+$, Tl$^+$, NH$_4^+$, MeNH$_3^+$ and N$_2$H$_5^+$, to give the corresponding inclusion complexes (Scheme 7.4).

On the basis of kinetic studies, we estimated that the affinity of $\{[CpCo(CN)_3]_4[Cp^*Ru]_4\}$ for Cs$^+$ is $>10^{10}$ M^{-1} [45], which is probably the highest affinity Cs$^+$ binding agent. Furthermore, the selectivity for Cs$^+$ in preference to Cs$^+$ is 10^6.

In no case, however, has this cage been found to bind Na$^+$, Li$^+$ or any di- or trivalent ion. These smaller cations are apparently unable to bind simultaneously to sufficient CoCNRu sites to compensate for their desolvation required by the inclusion process. This selectivity highlights the remarkable rigidity of these container molecules. Smaller tetrahedral or trigonal prismatic cages are effective for these small ions, an example being $\{Na[Mo(CO)_3]_4(CN)_6\}^{5-}$ [46].

7.3.1.1 Electroactive Boxes

The Co(I)-containing building block $[Cb^*Co(NCMe)_3](PF_6)$ (Cb* = η^4 C$_4$Me$_4$) is intriguing because its cage derivatives are susceptible to redox reactions. Condensation of $[Cb^*Co(NCMe)_3]PF_6$ and K[(C$_5$R$_5$)M(CN)$_3$] [(C$_5$R$_5$)M = CpCo, Cp*Rh] affords $\{K \subset [(C_5R_5)M(CN)_3]_4[Cb^*Co]_4\}^+$ [47]. Once again, the success of the condensation depends on the presence of a cation, otherwise one obtains insoluble polymeric solids. The structure of $\{K \subset [CpCo(CN)_3]_4[Cb^*Co]_4\}(PF_6)$ reveals that eight of the Co–CN–Co linkages are bent towards the K$^+$ guest and the remaining four Co–CN–Co edges relieve this distortion by bowing away from the cage interior (Figure 7.8). This all-cobalt cage can be oxidized with ferrocenium to give the tetracationic derivative concomitant with release of the alkali metal guest. In principle, other half-sandwich receptors could be made switchable electrochemically. Also

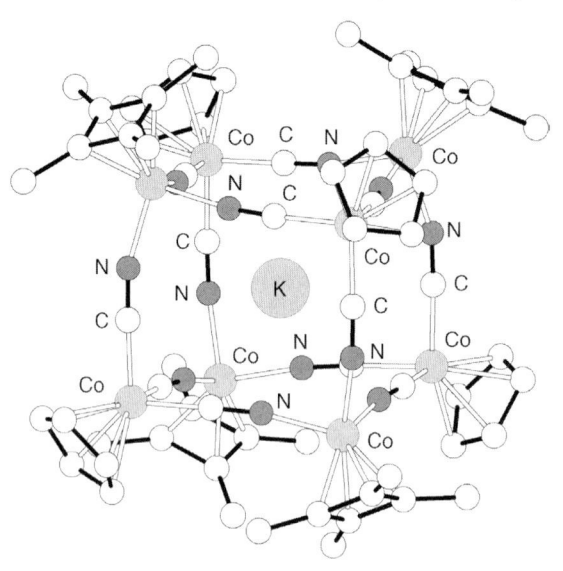

Figure 7.8 Molecular structure of $\{K \subset [CpCo(CN)_3]_4[Cb^*Co]_4\}^+$ in the crystal. The hydrogen atoms are not shown for clarity.

intriguing, the redox potential of the cage complex depends on the nature of the alkali metal, thus indicating that such cages could be used as electrochemical sensors for alkali metal cations.

7.3.1.2 Defect Boxes $\{[(C_5R_5)M(CN)_3]_4[Cp^*M]_3\}^z$

Condensation of $[Cp^*Rh(CN)_3]^-$ and $[Cp^*Rh(NCMe)_3]^{2+}$ affords exclusively the seven-vertex cage $[(Cp^*Rh)_7(CN)_{12}]^{2+}$ [48]. This type of cage, which is better described with the formula $\{[Cp^*Rh(CN)_3]_4[Cp^*Rh]_3\}^{2+}$, is called a "defect box," because it is related to the eight-vertex boxes by removal of one Cp*M vertex. Unlike the $M_8(CN)_{12}$ boxes, however, defect boxes have three terminal cyanides. The stereochemistry of these cyanide ligands is strongly influenced by the presence of a guest at the cage interior. Formation of $\{[Cp^*Rh(CN)_3]_4[Cp^*Rh]_4\}^{4+}$ is prevented by inter-Cp* steric interactions. Even in the absence of strong steric forces, the defect-box motif is kinetically favored relative to the box. Thus, condensation of $[CpCo(CN)_3]^-$ and $[Cp^*Rh(NCMe)_3]^{2+}$ in a 4 : 3 ratio gives $\{[CpCo(CN)_3]_4[Cp^*Rh]_3\}^{2+}$. In these dicationic Rh_7 and Co_4Rh_3 cages, the three terminal cyanide ligands are exocyclic, i.e. they radiate away from the open vertex to give the cage idealized C_{3v} symmetry. In order to convert $\{[CpCo(CN)_3]_4[Cp^*Rh]_3\}^{2+}$ into the box $\{[CpCo(CN)_3]_4[Cp^*Rh]_4\}^{4+}$, the three exocyclic cyanide ligands must reorient, so the final step in box formation is slow.

Condensations of $[CpCo(CN)_3]^-$ with <1 equiv. of the monocationic Lewis acid Cp^*Ru^+ affords defect boxes, but only in the presence of Cs^+, which occupies the interior of the cage (Figure 7.9). In solution, the resulting cage, $\{Cs \subset [CpCo(CN)_3]_4[Cp^*Ru]_3\}$, exists as a pair of interconverting isomers: one has two exo terminal cyanides and the other has one exo terminal cyanide. The "endo" terminal CN groups bind to Cs^+ using their C–N π-bonds [43].

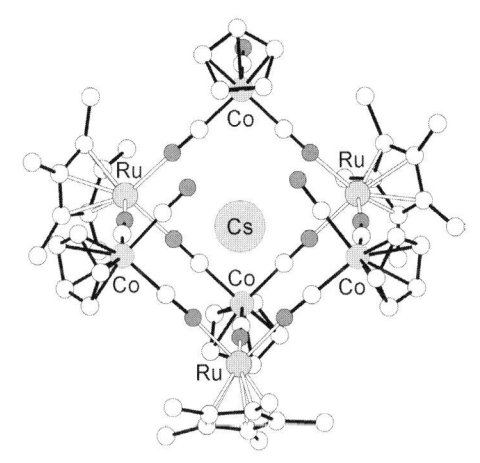

Figure 7.9 Molecular structure of $\{Cs \subset [CpCo(CN)_3]_4[Cp^*Ru]_3\}$ in the crystal. The hydrogen atoms are not shown for clarity.

The species $\{Cs \subset [CpCo(CN)_3]_4[Cp^*Ru]_3\}$ is a versatile triaza ligand, similar in denticity and binding preferences to tris(pyrazolyl)borates and triazacyclononane. It is a rare example of a ligand that contains an alkali metal. In binding the eighth metal, the exo cyanide ligands in $\{Cs \subset [CpCo(CN)_3]_4[Cp^*Ru]_3\}$ reorient to the endo orientation, but this reaction is very fast and obviously intramolecular. Thus, upon treatment with tritopic Lewis acids, $\{Cs \subset [CpCo(CN)_3]_4[Cp^*Ru]_3\}$ efficiently affords heteronuclear boxes $\{Cs \subset [CpCo(CN)_3]_4[Cp^*Ru]_3ML_n\}^z$, where, for example, $ML_n^z =$ RuH(PPh$_3$)$_2$ and [CuPPh$_3$]$^+$. With weakly solvated metal cations, double boxes form with the formula $\{Cs \subset [CpCo(CN)_3]_4[Cp^*Ru]_3\}_2M\}^{n+}$ (M = Na$^+$, Fe^{2+}, Ni^{2+}; Scheme 7.5) [43,49].

Particularly instructive were experiments on the corresponding all-Cp* anion-binding defect box $\{M \subset [Cp^*Rh(CN)_3]_4[Cp^*Ru]_3\}$ (M = Cs$^+$ or NH$_4^+$). Because all vertices carry the Cp* ligand, these cages enjoy enhanced solubility in organic solvents, which in turn has allowed us to monitor cage formation by *in situ* electrospray ionization mass spectrometry (ESI-MS). Such ESI-MS measurements revealed the intermediacy of CsRh$_3$Ru$_2$ and CsRh$_2$Ru$_2^+$ clusters, consistent with a role of the alkali metal cation in guiding the assembly of the cage [50]. Because of their more open architecture, the defect boxes exchange guest ions more rapidly than the related ionophilic boxes [50]. Furthermore, the affinity of $\{NH_4 \subset [Cp^*Rh(CN)_3]_4[Cp^*Ru]_3\}$ for Cs$^+$ dwarfs that of the best organic complexant, calix[4]arene-bis(benzocrown-6) (BC6B):

CsBC6B$^+$ + $\{NH_4 \subset [Cp^*Rh(CN)_3]_4[Cp^*Ru]_3\}$ → $\{Cs \subset [Cp^*Rh(CN)_3]_4[Cp^*Ru]_3\}$ NH$_4$BC6B$^+$

7.3.2
Expanded Organometallic Cyano Cages

The versatility of the CN$^-$ group as a linker ligand has encouraged a search for related building blocks that could stabilize nanoscale architectures. Ishii's group has

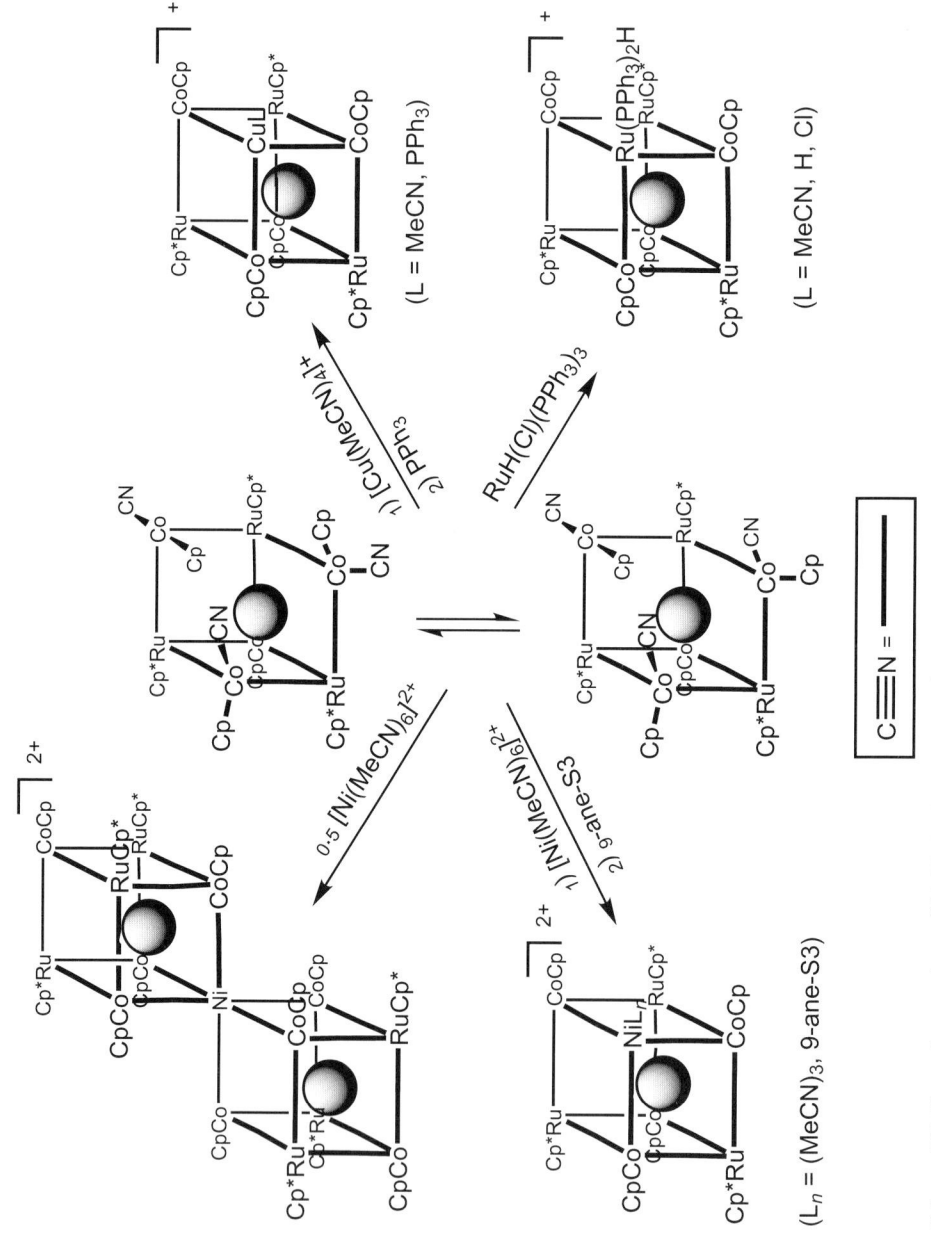

Scheme 7.5 Illustrative reactions of {Cs[CpCo(CN)$_3$]$_4$[Cp*Ru]$_3$}.

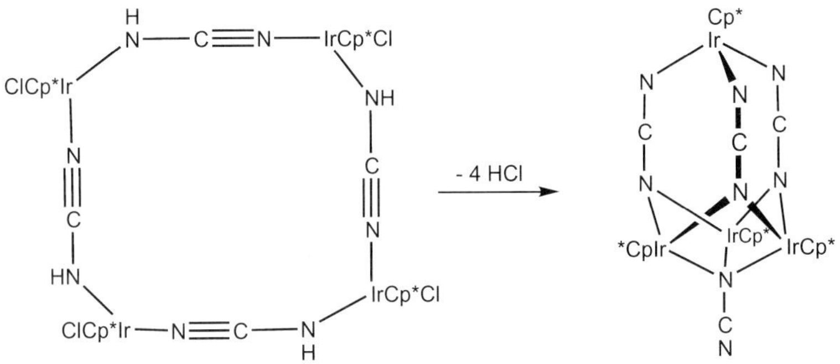

Scheme 7.6 Condensation of an expanded tetrametallic cage via dehydrohalogenation of a cyanamide-linked ring.

shown that cyanamide dianion NCN^{2-} and the related monoanion $NCNH^-$ in combination with half-sandwich building blocks affords a variety of novel cage structures [11]. Thus, treatment of NaNCNH with [Cp*IrCl$_2$]$_2$ affords [Cp*IrCl (μ_2-NCNH-N,N')]$_4$ consisting of 16-membered macrocycle [IrNCN]$_4$ cores. Dehydrohalogenation converts these rings into elongated cubanes (Scheme 7.6).

The $M_8(CN)_{12}$ boxes have internal volumes that are ideal for the selective binding of very small ions. The scope for host–guest behavior would be greater for larger cages and the possibility of such expanded cyanometallate cages is increasingly apparent. Beltran and Long have described "FCC" boxes wherein the eight tritopic metals are connected via [M(CN)$_4$]$^{2-}$ linkers that cap each of the eight faces of the expanded box [51]. Another expanded cyanometallate derives from the formal self-condensation of [Cp*WS$_3$(CuCN)$_3$]$^-$ and [Cp*WS$_3$Cu$_3$]$^{2+}$. Again, obtaining pure samples has proven to be challenging, but the framework of the new box $\{[Cp*WS_3Cu_3]_8(CN)_{12}\}^{4+}$ was characterized crystallographically [52]. The size of this container molecule is indicated by its contents: 4 Li$^+$ and 8 Cl$^-$.

Half-sandwich tricyanides enforce box-like structures by their orthogonal NC–M–CN angles and the inertness of the M–CN bonds, which precludes more drastic structural rearrangements. The organoboron tricyanides with the formula [RB(CN)$_3$]$^-$ are well suited as building blocks because they are basic and available with a range of R groups. Because the NC–B–CN angle is more open, these species give rise to cages that are larger than cubes, such as the hexagonal prismatic $\{[PhB(CN)_3]_6[Cp*Rh]_6\}^{6+}$ (Figure 7.10) [53].

7.3.3
Cages Based on *N*-Heterocyclic Ligands

Trifunctional *N*-heterocyclic ligands can be used to build not only macrocycles but also coordination cages. The basic requirement is that the three donor atoms are not able to form chelate complexes. When the Ru complex [(*p*-cymene)Ru(NO$_2$)$_2$] was mixed with the trifunctional ligand 3,5-pyridinedicarboxylic acid in water, an orange

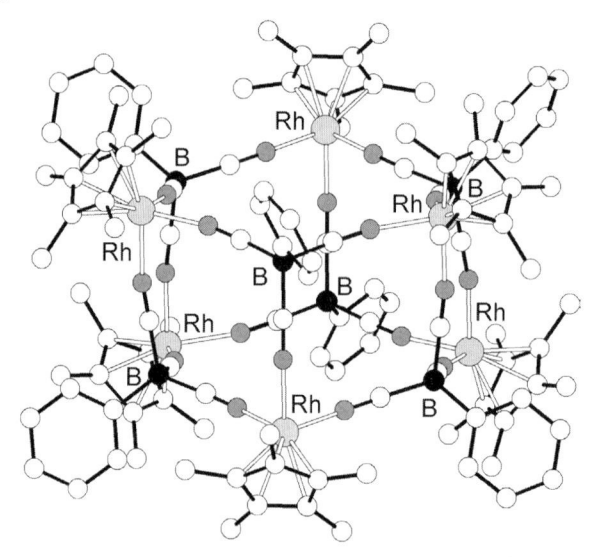

Figure 7.10 Molecular structure of $\{[PhB(CN)_3]_6[Cp^*Rh]_6\}^{6+}$. The hydrogen atoms, the counter-anion and the THF molecule in the cavity are not shown for clarity.

precipitate was formed. This complex turned out to be a hexanuclear cage, in which the (p-cymene)Ru fragments are connected by the 3,5-pyridinedicarboxylate ligands (Scheme 7.7) [54]. A related Cp*Ir complex can be obtained from [Cp*Ir(OAc)$_2$] and 3,5-pyridinecarboxylic acid in methanol [55].

$X^- = NO_3^-$, AcO^-

Scheme 7.7 Synthesis of hexanuclear cage complexes with bridging 3,5-pyridinedicarboxylate ligands.

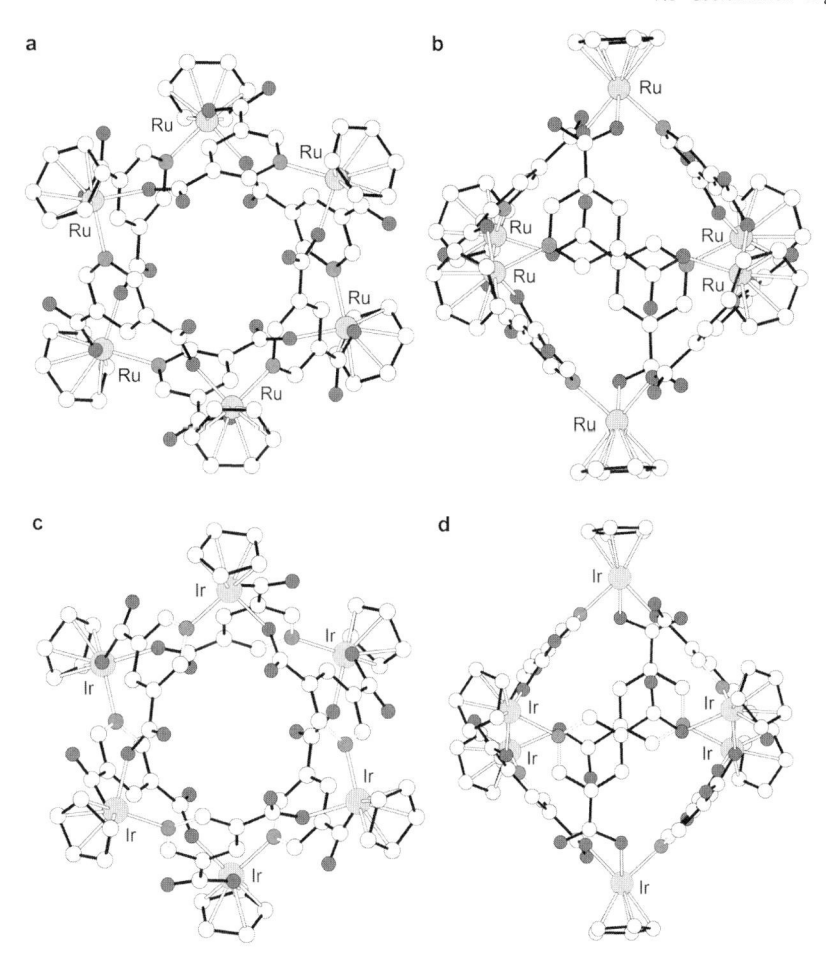

Figure 7.11 Molecular structure of a hexameric (p-cymene)Ru complex [(a) view from the top; (b) view from the side] and a hexameric Cp*Ir complex [(c) view from the top; (d): view from the side] in the crystal. The hydrogen atoms, the co-crystallized solvent molecules and the side chains of the π-ligands are not shown for clarity.

The structures of the hexanuclear cages [(p-cymene)Ru(3,5-pyridinedicarboxylate)]$_6$ and [Cp*Ir(3,5-pyridinedicarboxylate)]$_6$ are closely related (Figure 7.11). Each 3,5-pyridinedicarboxylate ligand is coordinated to three different metal atoms via the carboxylate O atoms and the pyridine N atom. The Ru or Ir atoms are positioned in the corner of an octahedron. Metal atoms in opposite corners are 12 Å apart from each other.

The Ru cage was found to act as an exo-receptor for alkali metal ions such as K^+ and Cs^+ [54]. This transformation was evidenced by NMR titration experiments with KOAc or CsOAc in CD$_3$OD solution. These salts induce a rearrangement of the

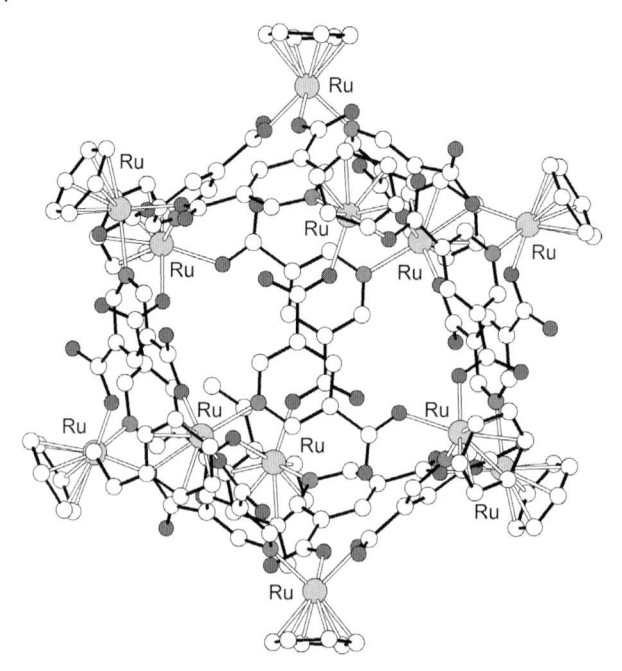

Figure 7.12 Molecular structure of a dodecanuclear (*p*-cymene) Ru complex in the crystal. The hydrogen atoms, the co-crystallized KOAc and solvent molecules and the side-chains of the π-ligands are not shown for clarity.

hexanuclear complex into a dodecanuclear cage (Figure 7.12). The connectivity of the hexa- and dodecanuclear complexes is similar: each (cymene)Ru^{2+} fragment is coordinated to three different heterocyclic ligands via two carboxylates and one pyridine N-atom. The symmetry-related Ru atoms in opposite corners of the cage are separated by 16 Å giving a cavity of approximately 1100 Å^3. The icosahedral geometry adopted by this cage resembles the geometry of natural cage structures such as spherical viruses. Twelve of the 20 faces of the icosahedra are occupied by the bridging 3,5-pyridinecarboxylate ligands. The remaining eight faces are surrounded by three carbonyl groups that constitute a metal binding site. Indeed, in the crystal, all eight sites are occupied by K$^+$ ions (not shown in Figure 7.12). These K$^+$ centers are coordinated to acetate anions, which interconnect the icosahedra in the crystal. The binding sites for K$^+$ and Cs$^+$ are sufficiently better in the dodecanuclear cage to compensate for the entropically disfavored condensation of two hexanuclear cages.

The synthesis of trigonal prismatic cages based on (arene)Ru complexes and bridging 2,4,6-tripyridyl-1,3,5-triazine (tpt) ligands has been investigated by Therrien's group [56,57]. When the dinuclear oxalate complexes [{(*p*-cymene)RuCl}$_2$(C$_2$O$_4$)] or [{(C$_6$Me$_6$)RuCl}$_2$(C$_2$O$_4$)] were reacted with 6 equiv. of AgOTf and 2 equiv. of tpt, cationic metallo-prisms were obtained (Scheme 7.8) [56]. The

Scheme 7.8 Synthesis of trigonal prismatic cage complexes.

hexamethylbenzene complex possesses a helical chirality induced by a twist of the tpt ligands and a concerted tilt of pyridyl moieties. The helicity persists in solution as shown by NMR spectroscopic measurements.

Structurally related complexes with bridging chloro ligands instead of oxalato ligands were obtained by reaction of [(arene)RuCl(μ-Cl)]$_2$ with tpt followed by abstraction of chloride by AgOTf (Figure 7.13) [57]. The metallo-prism are stabilized by π-stacking interactions between the two tpt ligands.

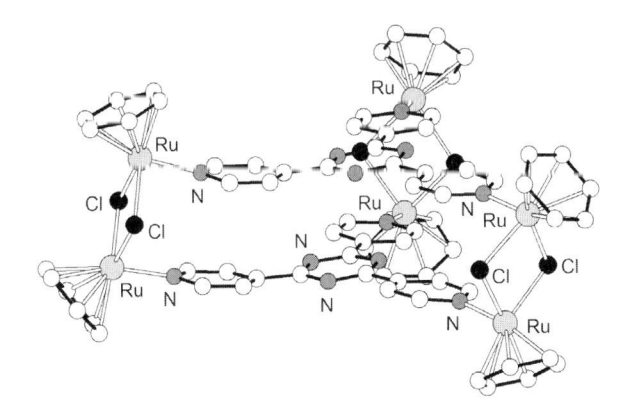

Figure 7.13 Molecular structure of the metallo-prism [{(p-cymene)Ru}$_6$(μ$_3$-tpt)$_2$(μ-Cl)$_6$](OTf)$_6$ in the crystal. The hydrogen atoms, the triflate anions and the side-chains of the π-ligands are not shown for clarity.

7.4
Expanded Helicates

Cylindrical structures with six (π-ligand)M complexes have been obtained by reaction of [(C$_6$H$_5$Me)RuCl(μ-Cl)]$_2$ or [(C$_5$Me$_4$H)RhCl(μ-Cl)]$_2$ with bis(dihydroxypyridine) ligands (Scheme 7.9, Figure 7.14) [58]. These hexanuclear complexes are composed of two 12-metallacrown-3 fragments, which are connected by three flexible spacers. Since the metallacrowns are chiral, the complexes can be regarded as expanded, triple-stranded helicates.

The maximum Rh-to-Rh distance of the (C$_5$Me$_4$H)Rh complex shown in Figure 7.14 is 2.2 nm. The macrocycles formed between two opposite metals have a ring size of 44 atoms containing a total of 18 CH$_2$ groups, only eight of which are part of semi-rigid piperidine units. This complex is therefore a rare example of a discrete, multinuclear complex, which was obtained by metal-based self-assembly with a highly flexible ligand.

More detailed investigations revealed that water-soluble helicates can be obtained upon careful adjustment of the pH [59]. When 5 equiv. of CsOH were added to an aqueous solution of a mixture of [(benzene)RuCl(μ-Cl)]$_2$ and the piperazine-bridged dihydroxypyridine ligand [{(C$_5$H$_2$NO$_2$)CH$_2$}$_2$(C$_4$H$_{10}$N$_2$)](OTf)$_2$ (see Scheme 7.9), a hexanuclear helicate was obtained in over 90% yield. The utilization of LiOH instead of CsOH resulted in the formation of the bis-Li$^+$ adduct, which displayed an even higher solubility in water. Interestingly, the latter complex is able to act as a specific receptor for the phosphate anion (Scheme 7.10). NMR titrations revealed a binding constant of 900 M^{-1} for H$_2$PO$_4^-$ whereas no interaction was detected for halides, nitrate and sulfate [59]. The NMR data suggest that the phosphate anion is bound in the vicinity of the bridging piperidine groups.

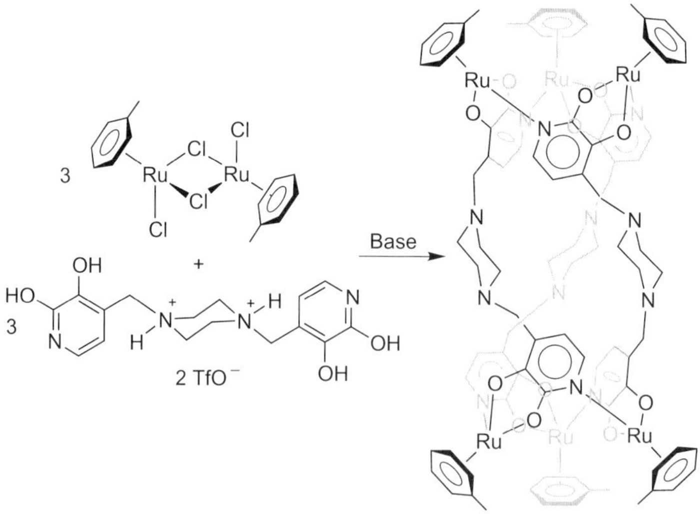

Scheme 7.9 Synthesis of an expanded helicate by base-induced assembly of (toluene)Ru complexes with bridged dihydroxypyridine ligands.

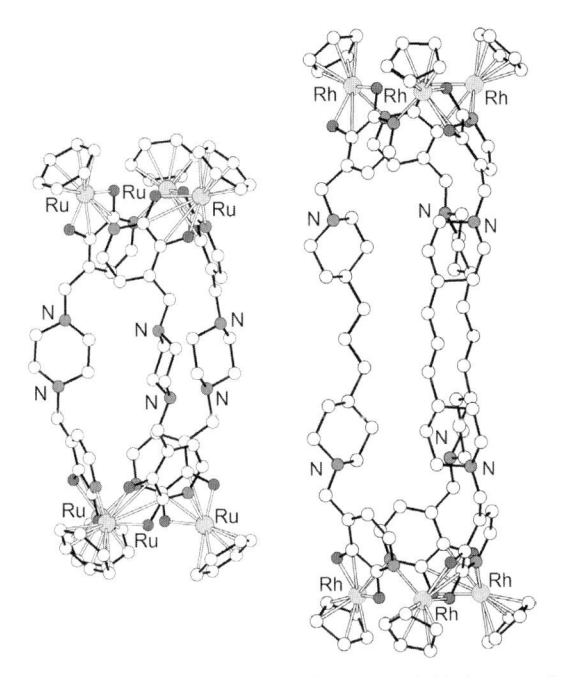

Figure 7.14 Molecular structure of two expanded helicates in the crystal. The hydrogen atoms, the solvent molecules and the side-chains of the π-ligands are not shown for clarity.

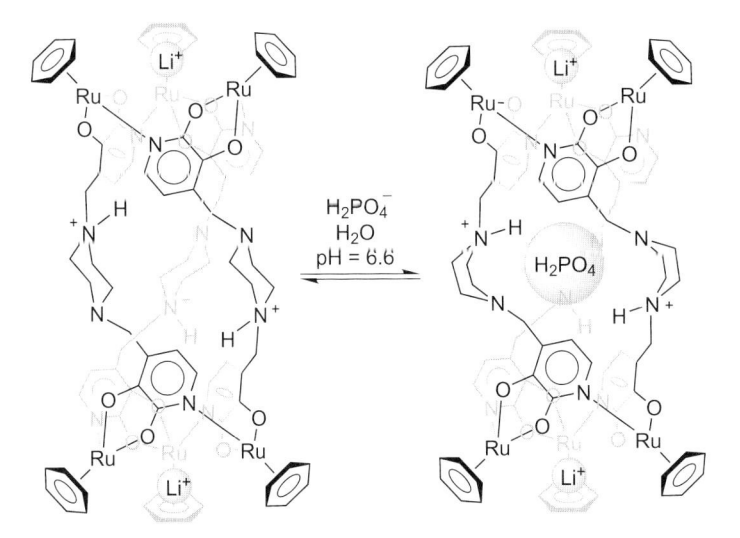

Scheme 7.10 A (benzene)Ru-based helicate is able to act as a specific receptor for the $H_2PO_4^-$ anion in aqueous solution.

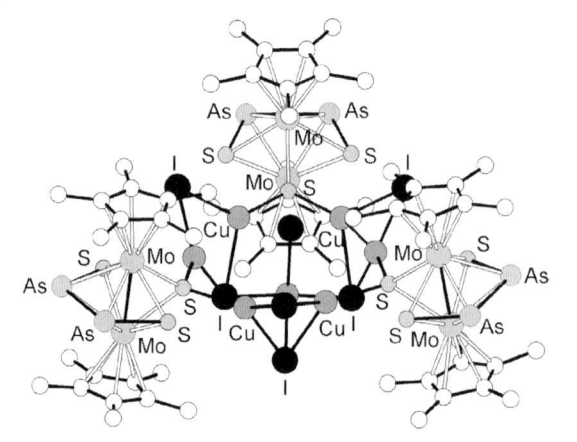

Figure 7.15 Molecular structure of [{(Cp*Mo)$_2$As$_2$S$_3$}$_3$(CuI)$_7$] in the crystal. The hydrogen atoms are not shown for clarity.

7.5
Clusters

The utility of [Cp*Rh]$^{2+}$ in the stabilization of nanostructures was demonstrated 30 years ago for modifying the surfaces of polyoxometallate clusters. The resulting ensembles were proposed as models for oxide-supported metal catalysts. The first species of this type was (Bu$_4$N)$_2$[Cp*Rh(Nb$_2$W$_4$O$_{19}$)], which results from the derivatization of cis-[Nb$_2$W$_4$O$_{19}$]$^{4-}$ with [Cp*Rh]$^{2+}$ [60]. The Rh(III) center binds in a tridentate manner to M$_3$(μ_2-O)$_3$ faces of the oxometallate to give all three possible diastereoisomers. In a complementary way, the species [Cp*Rh(OH)$_2$]$_2$ was employed as both a ligand and as a mineralizer to depolymerize the layered binary V$_2$O$_5$. The resulting 10-vertex cluster is [Cp*Rh]$_4$(V$_6$O$_{19}$), a derivative of the otherwise unknown anion [V$_6$O$_{19}$]$^{8-}$. Four alternating faces have been capped by the rhodium dication [61]. Half-sandwich complexes, although often derived from early metals, are valuable in developing supramolecular structures based on M—S cores. These developments are illustrated by [(Cp*Mo)$_8$Ni$_8$S$_{16}$](PF$_6$)$_4$ [62], Li$_2$(THF)$_2$Cp*$_3$Ta$_3$S$_6$ [63] and [{(Cp*Mo)$_2$As$_2$S$_3$}$_3$(CuI)$_7$] (Figure 7.15) [64].

7.6
Conclusions

Organometallic chemistry and supramolecular coordination chemistry have different roots, but the combination of these two approaches has proven powerful. The enabling advance is the exploitation of half-sandwich reagents, which are truly hemilabile – the π-ligand adheres very tightly to the metal and the other three coordination sites bind diverse ligands, ranging through hydrocarbons, amino acids, small inorganic anions and heterocycles. The half-sandwich motif strongly suppresses polymerization reactions to a remarkable extent and the resulting molecular assemblies engage in specific host–guest interactions. It is clear that opportunities exist for further expansion of this area. Some general themes that can be expected to be particularly fruitful include nanostructures relevant to homogeneous catalysis [65], medicine [66] and sensors [67].

References

1 Shaw, B.L. (1975) *J. Am. Chem. Soc.*, **97**, 3856–3857.
2 Contakes, S.M., Klausmeyer, K.K. and Rauchfuss, T.B. (2004) *Inorg. Synth.*, **34**, 166–171.
3 Severin, K. (2006) *Chem. Commun.*, 3859–3867.
4 Boyer, J.L., Kuhlman, M.L. and Rauchfuss, T.B. (2007) *Acc. Chem. Res.*, **40**, 233–242.
5 White, C., Yates, A. and Maitlis, P.M. (1992) *Inorg. Synth.*, **29**, 228–234.
6 Zhu, B., Ellern, A., Sygula, A., Sygula, R. and Angelici, R.J. (2007) *Organometallics*, **26**, 1721–1728.
7 (a) Vecchi, P.A., Alvarez, C.M., Ellern, A., Angelici, R.J., Sygula, A., Sygula, R. and Rabideau, P.W. (2005) *Organometallics*, **24**, 4543–4552. (b) Zhu, B., Ellern, A., Sygula, A., Sygula, R. and Angelici, R.J. (2007) *Organometallics*, **26**, 1721–1728.
8 (a) Yamamoto, Y., Nakamura, H. and Ma, J.-F. (2001) *J. Organomet. Chem.*, **640**, 10–20. (b) Suzuki, H., Tajima, N., Tatsumi, K. and Yamamoto, Y. (2000) *Chem. Commun.*, 1801–1802.
9 Yamamoto, Y., Suzuki, H., Tajima, N. and Tatsumi, K. (2002) *Chem. Eur. J.*, **8**, 372–379.
10 (a) Zhang, Q.-F., Adams, R.D. and Leung, W.-H. (2006) *Inorg. Chim. Acta*, **359**, 978–983. (b) Wang, J.-Q., Ren, C.-X. and Jin, G.-X. (2006) *Organometallics*, **25**, 74–81. (c) Han, W.S. and Lee, S.W. (2004) *Dalton Trans.*, 1656–1663. (d) Yan, H., Süss-Fink, G., Neels, A. and Stoeckli-Evans, H. (1997) *J. Chem. Soc. Dalton Trans.*, 4345–4350.
11 (a) Takahata, K., Iwadate, N., Kajitani, H., Tanabe, Y. and Ishii, Y. (2007) *J. Organomet. Chem.*, **692**, 208–216. (b) Tanabe, Y., Kuwata, S. and Ishii, Y. (2002) *J. Am. Chem. Soc.*, **124**, 6528–6529.
12 Klausmeyer, K.K., Rauchfuss, T.B. and Wilson, S.R. (1998) *Angew. Chem. Int. Ed.*, **37**, 1694–1696.
13 (a) Fish, R.H. and Jaouen, G. (2003) *Organometallics*, **22**, 2166–2177. (b) Ogo, S., Buriez, O., Kerr, J.B. and Fish, R.H. (1999) *J. Organomet. Chem.*, **589**, 66–74. (c) Ogo, S., Nakamura, S., Chen, H., Isobe, K., Watanabe, Y. and Fish, R.H. (1998) *J. Org. Chem.*, **63**, 7151–7156. (d) Bakhtiar, R., Chen, H., Ogo, S. and Fish, R.H. (1997) *Chem. Commun.*, 2135–2136. (e) Chen, H., Ogo, S. and Fish, R.H. (1996) *J. Am. Chem. Soc.*, **118**, 4993–5001. (f) Chen, H., Olmstead, M.M., Smith, D.P., Maestre, M.F. and Fish, R.H. (1995) *Angew. Chem. Int. Ed. Engl.*, **34**, 1514–1517. (g) Smith, D.P., Baralt, E., Morales, B., Olmstead, M.M., Maestre, M.F. and Fish, R.H. (1992) *J. Am. Chem. Soc.*, **114**, 10647–10649.
14 Korn, S. and Sheldrick, W.S. (1997) *J. Chem. Soc. Dalton Trans.*, 2191–2199.
15 (a) Annen, P., Schildberg, S. and Sheldrick, W.S. (2000) *Inorg. Chim. Acta*, **307**, 115–124. (b) Korn, S. and Sheldrick, W.S. (1997) *Inorg. Chim. Acta*, **254**, 85–91.
16 (a) Yamanari, K., Ito, R., Yamamoto, S., Konno, T., Fuyuhiro, A., Kobayashi, M. and Arakawa, R. (2003) *Dalton Trans.*, 380–386. (b) Yamanari, K., Ito, R., Yamamoto, S. and Fuyuhiro, A. (2001) *Chem. Commun.*, 1414–1415.
17 Kunkely, H. and Vogler, A. (2002) *Inorg. Chim. Acta*, **338**, 265–267.

18 Yamanari, K., Ito, R., Yamamoto, S., Konno, T., Fuyuhiro, A., Fujioka, K. and Arakawa, R. (2002) *Inorg. Chem.*, **41**, 6824–6830.

19 Yamanari, K., Yamamoto, S., Ito, R., Kushi, Y., Fuyuhiro, A., Kubota, N., Fukuo, T. and Arakawa, R. (2001) *Angew. Chem. Int. Ed.*, **40**, 2268–2271.

20 (a) Sünkel, K., Hoffmüller, W. and Beck, W. (1998) *Z. Naturforsch., Teil B*, **53**, 1365–1368. (b) Ogo, S., Chen, H., Olmstead, M.M. and Fish, R.H. (1996) *Organometallics*, **15**, 2009–2013. (c) Krämer, R., Polborn, K., Robl, C. and Beck, W. (1992) *Inorg. Chim. Acta*, **198–200**, 415–420.

21 (a) Carmona, D., Lamata, M.P., Viguri, F., Dobrinovich, I., Lahoz, F.J. and Oro, L.A. (2002) *Adv. Synth. Catal.*, **344**, 499–502. (b) Kathó, Á., Carmona, D., Viguri, F., Remacha, C.D., Kovács, J., Joó, F. and Oro, L.A. (2000) *J. Organomet. Chem.*, **593–594**, 299–306. (c) Carmona, D., Lahoz, F.J., Atencio, R., Oro, L.A., Lamata, M.P., Viguri, F., José, E.S., Vega, C., Reyes, J., Joó, F. and Kathó, Á. (1999) *Chem. Eur. J.*, **5**, 1544–1564.

22 Piotrowski, H., Polborn, K., Hilt, G. and Severin, K. (2001) *J. Am. Chem. Soc.*, **123**, 2699–2700.

23 Piotrowski, H., Hilt, G., Schulz, A., Mayer, P., Polborn, K. and Severin, K. (2001) *Chem. Eur. J.*, **7**, 3196–3208.

24 Lehaire, M.-L., Scopelliti, R., Herdeis, L., Polborn, K., Mayer, P. and Severin, K. (2004) *Inorg. Chem.*, **43**, 1609–1617.

25 Habereder, T., Warchhold, M., Nöth, H. and Severin, K. (1999) *Angew. Chem. Int. Ed.*, **38**, 3225–3228.

26 (a) Missami, L., Cordier, C., Guyard-Duhayon, C., Mann, B.E. and Amori, H. (2007) *Organometallics*, **26**, 860–864. (b) Missami, L., Guyard-Duhayon, C., Rager, M.N. and Amouri, H. (2004) *Inorg. Chem.*, **43**, 6644–6649.

27 For a review on metallacrown complexes, see Mezei, G., Zaleski, C.M. and Pecoraro, V.L. (2007) *Chem. Rev.*, **107**, 4933–5003.

28 Lehaire, M.-L., Schulz, A., Scopelliti, R. and Severin, K. (2003) *Inorg. Chem.*, **42**, 3576–3581.

29 (a) Lehaire, M.-L., Scopelliti, R., Piotrowski, H. and Severin, K. (2002) *Angew. Chem. Int. Ed.*, **41**, 1419–1422. (b) Lehaire, M.-L., Scopelliti, R. and Severin, K. (2002) *Inorg. Chem.*, **41**, 5466–5474.

30 Lehaire, M.-L., Scopelliti, R. and Severin, K. (2002) *Chem Commun.*, 2766–2767.

31 Grote, Z., Wizemann, H.-D., Scopelliti, R. and Severin, K. (2007) *Z. Anorg. Allg. Chem.*, **633**, 858–864.

32 Piotrowski, H. and Severin, K. (2002) *Proc. Natl. Acad. Sci. USA*, **99**, 4997–5000.

33 (a) Pilcher, H.R. (2003) *Nature*, **425**, 118–120. (b) Birch, N.J. (1999) *Chem. Rev.*, **99**, 2659–2682. (c) Bartsch, R.A., Ramesh, V., Bach, R.O., Shono, T. and Kimura, K. (1995) in *Lithium Chemistry*, (eds A.-M. Sapse and P.v.R. Schleyer),Wiley, New York, pp. 393–476.

34 (a) Grote, Z., Scopelliti, R. and Severin, K. (2004) *J. Am. Chem. Soc.*, **126**, 16959–16972. (b) Grote, Z., Lehaire, M.-L., Scopelliti, R. and Severin, K. (2003) *J. Am. Chem. Soc.*, **125**, 13638–13639.

35 Corbett, P.T., Leclaire, J., Vial, L., West, K.R., Wietor, J.-L., Sanders, J.K.M. and Otto, S. (2006) *Chem. Rev.*, **106**, 3652–3711.

36 (a) Grote, Z., Scopelliti, R. and Severin, K. (2007) *Eur. J. Inorg. Chem.*, 694–700. (b) Saur, I., Scopelliti, R. and Severin, K. (2006) *Chem. Eur. J.*, **12**, 1058–1066. (c) Saur, I. and Severin, K. (2005) *Chem. Commun.*, 1471–1473. (d) Grote, Z., Scopelliti, R. and Severin, K. (2003) *Angew. Chem. Int. Ed.*, **42**, 3821–3825.

37 Amouri, H., Rager, M.N., Cagnol, F. and Vaissermann, J. (2001) *Angew. Chem. Int. Ed.*, **40**, 3636–3638.

38 Dunbar, K.R. and Heintz, R.A. (1997) *Prog. Inorg. Chem.*, **45**, 283–391.

39 Evans, R.F., Ormrod, O., Goalby, B.B. and Staveley, L.A.K. (1950) *J. Chem. Soc.*, 3346.

40 (a) Boxhoorn, G., Moolhuysen, J., Coolegem, J.G.F. and van Santen, R.A. (1985) *J. Chem. Soc., Chem. Commun.*, 1305–1306. (b) Chapman, K.W., Southon,

P.D., Weeks, C.L. and Kepert, C.J. (2005) *Chem. Commun.*, 3322–3324. (c) Kaye, S.S. and Long, J.R. (2005) *J. Am. Chem. Soc.*, **127**, 6506–6507. (d) Ramprasad, D., Markley, T.J. and Pez, G.P. (1997) *J. Mol. Catal. A*, **117**, 273–278. (e) Meier, I.K., Pearlstein, R.M., Ramprasad, D. and Pez, G.P. (1997) *Inorg. Chem.*, **36**, 1707–1714. (f) Ramprasad, D., Pez, G.P., Toby, B.H., Markley, T.J. and Pearlstein, R.M. (1995) *J. Am. Chem. Soc.*, **117**, 10694–10701.

41 Prout, W.E., Russell, E.R. and Groh, H.J. (1965) *J. Inorg. Nucl. Chem.*, **27**, 473–479.

42 Klausmeyer, K.K., Rauchfuss, T.B. and Wilson, S.R. (1998) *Angew. Chem. Int. Ed.*, **37**, 1694–1696.

43 Contakes, S.M., Kuhlman, M.L., Ramesh, M., Wilson, S.R. and Rauchfuss, T.B. (2002) *Proc. Natl. Acad. Sci. U.S.A.*, **99**, 4889–4893.

44 Bryan, J.C., Kavallieratos, K. and Sachleben, R.A. (2000) *Inorg. Chem.*, **39**, 1568–1572.

45 Hsu, S.C.N., Ramesh, M., Espenson, J.H. and Rauchfuss, T.B. (2003) *Angew. Chem. Int. Ed.*, **42**, 2663–2666.

46 (a) Contakes, S.M. and Rauchfuss, T.B. (2000) *Angew. Chem. Int. Ed.*, **39**, 1984–1986. (b) Contakes, S.M. and Rauchfuss, T.B. (2001) *Chem. Commun.*, 553–554.

47 Boyer, J.L., Ramesh, M., Yao, H., Rauchfuss, T.B. and Wilson, S.R. (2007) *J. Am. Chem. Soc.*, **129**, 1931–1936.

48 Contakes, S.M., Klausmeyer, K.K., Milberg, R.M., Wilson, S.R. and Rauchfuss, T.B. (1998) *Organometallics*, **17**, 3633–3635.

49 Boyer, J.L., Yao, H., Kuhlman, M.L., Rauchfuss, T.B. and Wilson, S.R. (2007) *Eur. J. Inorg. Chem.*, 2721–2728.

50 Kuhlman, M.L. and Rauchfuss, T.B. (2003) *J. Am. Chem. Soc.*, **125**, 10084–10092.

51 Beltran, L.M.C. and Long, J.R. (2005) *Acc. Chem. Res.*, **38**, 325–334.

52 Lang, J.-P., Xu, Q.-F., Chen, Z.-N. and Abrahams, B.F. (2003) *J. Am. Chem. Soc.*, **125**, 12682–12683.

53 Kuhlman, M.L., Yao, H. and Rauchfuss, T.B. (2004) *Chem. Commun.*, 1370–1371.

54 Brasey, T., Scopelliti, R. and Severin, K. (2006) *Chem. Commun.*, 3308–3310.

55 Mirtschin, S., Krasniqi, E., Scopelliti, R. and Severin, K. in preparation.

56 Govindaswamy, P., Linder, D., Lacour, J., Süss-Fink, G. and Therrien, B. (2006) *Chem. Commun.*, 4691–4693.

57 Govindaswamy, P., Süss-Fink, G. and Therrien, B. (2007) *Organometallics*, **26**, 915–924.

58 Grote, Z., Bonazzi, S., Scopelliti, R. and Severin, K. (2006) *J. Am. Chem. Soc.*, **128**, 10382–10383.

59 Olivier, C., Grote, Z., Solari, E., Scopelliti, R. and Severin, K. (2007) *Chem. Commun.*, 4000–4002.

60 Besecker, C.J., Day, V.W., Klemperer, W.G. and Thompson, M.R. (1984) *J. Am. Chem. Soc.*, **106**, 4125–4136.

61 (a) Chae, H.K., Klemperer, W.G. and Day, V.W. (1989) *Inorg. Chem.*, **28**, 1423–1424. (b) Chae, H.K., Klemperer, W.G., Paez-Loyo, D.E., Day, V.W. and Eberspacher, T.A. (1992) *Inorg. Chem.*, **31**, 3187–3189.

62 Takei, I., Suzuki, K., Enta, Y., Dohki, K., Suzuki, T., Mizobe, Y. and Hidai, M. (2003) *Organometallics*, **22**, 1790–1792.

63 Tatsumi, K., Inoue, Y., Kawaguchi, H., Kohsaka, M., Nakamura, A., Cramer, R.E., VanDoorne, W., Taogoshi, G.J. and Richmann, P.N. (1993) *Organometallics*, **12**, 352–364.

64 Pronold, M., Scheer, M., Wachter, J. and Zabel, M. (2007) *Inorg. Chem.*, **46**, 1396–1400.

65 (a) Fiedler, D., Leung, D.H., Bergman, R.G. and Raymond, K.N. (2005) *Acc. Chem. Res.*, **38**, 349–358. (b) Thomas, C.M. and Ward, T.R. (2005) *Chem. Soc. Rev.*, **34**, 337–346.

66 (a) Allardyce, C.A. and Dyson, P. (2006) *Top. Organomet. Chem.*, **17**, 177–210. (b) G. Jaouen, (ed.) (2005) *Bioorganometallics*, Wiley-VCH, Weinheim.

67 Severin, K. (2006) *Top. Organomet. Chem.*, **17**, 123–142.

8
Endochemistry of Self-assembled Hollow Spherical Cages
Takashi Murase and Makoto Fujita

8.1
Introduction

Nature has developed a variety of macromolecules with high catalytic efficiency and extraordinary versatility in reactions whose rates can be accelerated. Such chemical reactions take place in well-defined and confined environments, which vary from nanometer-sized and relatively simple systems, such as enzymes, to micrometer-sized and extremely complex assemblies, such as cells [1a]. In this context, chemistry inside a cage-like architecture (endohedral chemistry, "endochemistry") prevails everywhere in biological systems (Figure 8.1). The specificity and precision displayed by biological systems are derived from the highly directed mutual recognition process displayed by the components of a structure, so-called "self-assembly" [1b]. Self-assembly of molecules is a ubiquitous strategy to create functional assemblages in nature. Recent developments in noncovalent synthesis employing self-assembly of multiple constituent molecules have made it possible to prepare readily large and hollow cage architectures, which can hardly be obtained otherwise by conventional covalent methods [1]. However, in molecular nanotechnology, the technique of molecular self-assembly is not yet fully exploited for the functionalization of interior surfaces of hollow cage compounds.

To date, a self-assembled spherical cage compound can be roughly classified into two categories: (1) *biomacromolecular cages*, which have various sizes and can be redesigned by altering their chemical composition to attain a desired structure and function, and (2) *polymer micelles*, self-assembled from multi-block copolymers, which consist of hydrophilic and hydrophobic chain segments with pre- or post-attached functionalities [2]. The above spherical compounds are truly prominent and fascinating materials/scaffolds for endohedral functionalization of a restricted nanosized region and have been summarized elsewhere. In this chapter, we will briefly describe the aspects of endohedral functionalization of these existing cage compounds and then focus on the latest findings from our own laboratory.

Organic Nanostructures. Edited by Jerry L. Atwood and Jonathan W. Steed
Copyright © 2008 WILEY-VCH Verlag GmbH & Co. KGaA, Weinheim
ISBN: 978-3-527-31836-0

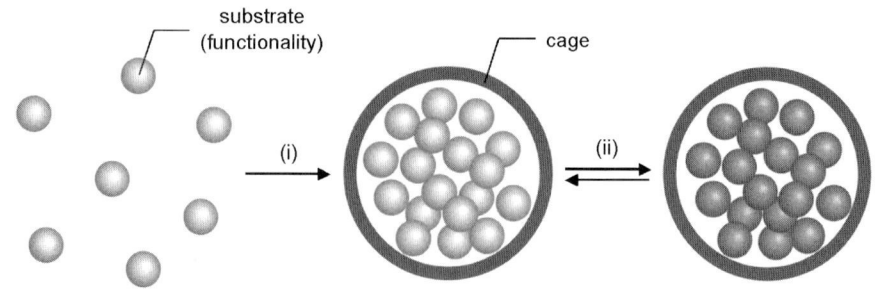

Figure 8.1 Schematic image of endochemistry in a confined region: (i) concentration of substrate/functionality (ii) chemical reaction and dynamic change.

8.2
Biomacromolecular Cages

Biomimetic chemistry offers a new approach to synthesize nanomaterials [3]. Protein cages and viral protein cages (capsides) share the common characteristics of self-assembly from a specific set of subunits into well-defined and highly symmetrical architectures. One example is an iron-storage protein, ferritin [4a]. Iron-free ferritin molecules (apo-ferritin) are composed of 24 polypeptide subunits, which self-assemble into a hollow spherical cage with a molecular mass of 450 kDa. Ferritin has an outer diameter of 12 nm and an inner cavity diameter of 8 nm that stores iron in the form of microcrystalline ferric oxyhydroxide [4b]. Another example is Cowpea Chlorotic Mottle Virus (CCMV) capside, which is composed of 180 identical 20-kDa subunits and has an outer diameter of 28 nm and an inner diameter ranging from 18 to 24 nm [5]. These organized biomacromolecular architectures can serve as not only nanoreactors or nanotemplates for crystallizations and other reactions, but also spatially defined scaffolds for the attachment of new chemical functionalities. Therefore, they possess the potential to be used to express and adjust multivalent presentations [6]. When functional groups are confined in these biomolecular cages by interior modification, the number and position of the functional groups are precisely controlled. So far, interior modification has lagged behind exterior modification, but intriguing and sophisticated studies have already been conducted [7].

However, the cavities of these protein cages, especially capsides, are too large to place functional groups at the cores. The types of functionalities introduced by genetic modification are limited to amino acids that mainly serve as "reactive sites" on the interior surfaces of cages. Chemical modification is mainly conducted to the already self-assembled spherical cages, which demands excess reagents to attach functionalities at all reactive sites, resulting in a poor introduction efficiency. Therefore, the current usage of the cavities of the protein cages remains as nanotemplates for inorganic materials.

8.3
Polymer Micelles

Polymer micelles are formed spontaneously by amphiphilic block copolymers in aqueous solution. By tailoring the relative lengths between the hydrophobic and hydrophilic blocks, compositions and self-assembly conditions, micellar aggregates with various morphologies (including star micelles, crew-cut micelles, rods and vesicles) can be designed and synthesized [8]. The spherical polymer micelle possesses a unique core–shell structure, which can be modified both chemically and structurally by the introduction of functionality at particular positions within the nanoarchitecture. In a typical spherical core–shell diblock polymer micelle, the introduction of functionality at the hydrophilic and hydrophobic chain blocks corresponds to the surface/shell and core functionalization, respectively [9].

The endohedral core functionalization of polymer micelles has attracted much interest, due to their potential applications, such as drug delivery carriers or nanoreactors. However, compared with the surface/shell functionalization, the functionalization of the core domain within polymer micelles has received limited attention, perhaps because of the difficulty in introducing and maintaining functionality within the hydrophobic block [9]. The functional groups in the hydrophobic block are supposed to be closely packed and concentrated at the core domain in the self-assembly process. Therefore, not only the hydrophobic properties but also the molecular sizes of the functional groups seem to affect greatly the formation of polymer micelles. Some notable studies on core functionalization of polymer micelles have been reported [10]. However, as a whole, endohedral functionalization of polymer micelles is regarded as a difficult but challenging task. A significant difference between polymer micelles and biomacromolecules is that in polymer micelles it is impossible to control precisely the number and position of the introduced functional groups. The molecular weight of the polymer micelles is not monodispersive and hydrophobic interaction under the self-assembly process does not have an orientation with a clear direction. Therefore, in the true sense, the polymer micelles do not have well-defined structures and shapes.

8.4
$M_{12}L_{24}$ Spheres

8.4.1
Self-assembly of $M_{12}L_{24}$ Spheres

To control precisely the number and position of functional groups introduced in a confined region, it is necessary to develop a novel nanosized cage compound having a well-defined chemical structure and an inner diameter that is suitable to confine the functional groups. Moreover, the introduction of functionality should be carried out in the subunit level before the self-assembly process, where the desired functionality is attached appropriately and efficiently to each component. However, to date, it has

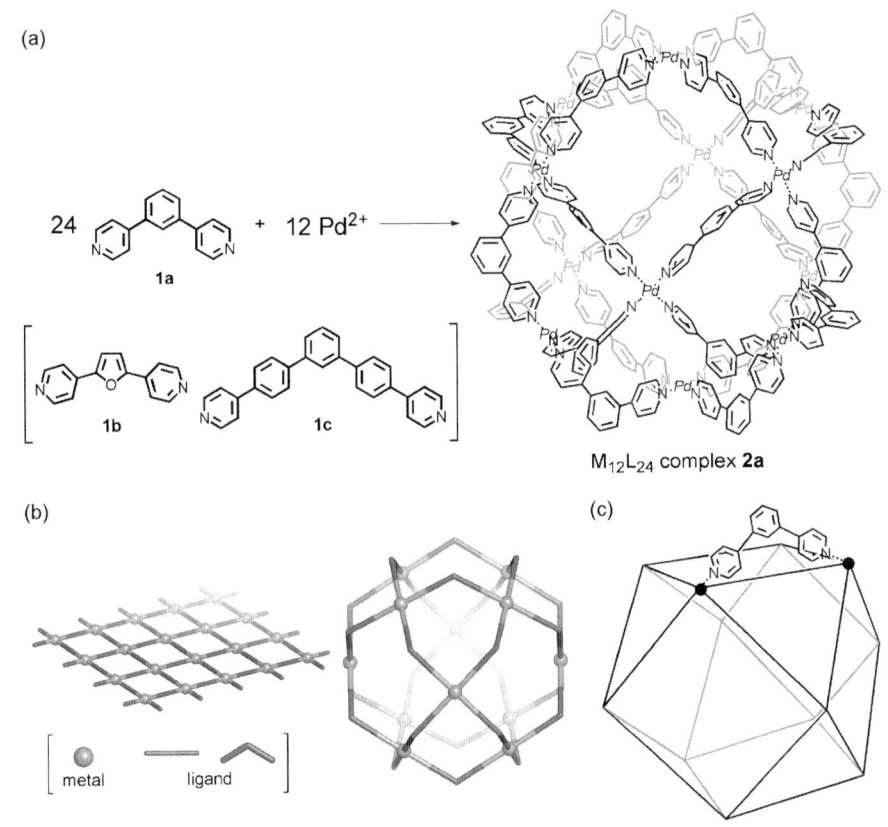

Figure 8.2 (a) Self-assembly of an $M_{12}L_{24}$ spherical complex. (b) Linear ligand (2D grid infinite network) vs. bent ligand (spherical finite complex). (c) Schematic representation of the cubo-octahedral frameworks of **2a** [11].

been difficult to synthesize artificially virus-like huge but well-defined spherical hollow cages.

In 2004, we demonstrated that a spherical coordination network, $M_{12}L_{24}$, can be prepared in a quantitative yield from 36 components, i.e. 12 metal ions (M) and 24 bridging ligands (L) (Figure 8.2a) [11]. The complexation of the bent bidentate ligands with naked square-planar palladium(II) ions affords a discrete spherical complex, in contrast to the 2D infinite network formation from linear ligands (Figure 8.2b). The symmetry of self-assembled $M_{12}L_{24}$ spherical complexes is depicted by a cuboctahedron, which is formed by truncating each of the eight vertices of a cube to produce eight triangular faces (Figure 8.2c). The 12 equivalent vertices and 24 equivalent edges of the cuboctahedron can be superimposed on the 12 palladium(II) centers and 24 bridging ligands, respectively. Such a highly symmetrical and huge structure has been unambiguously confirmed by X-ray crystallographic analysis of spherical

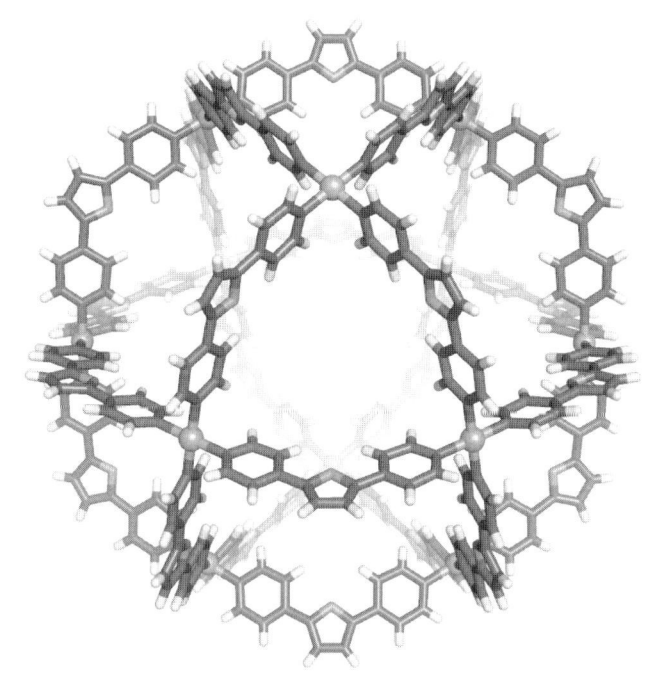

Figure 8.3 X-ray crystal structure of $M_{12}L_{24}$ spherical complex **2b**.

complex **2b**, which is an analogue of **2a** where the bending center of the ligand, m-phenylene, is replaced with 2,5-furanylene (Figure 8.3).

8.4.2
Endohedral Functionalization of $M_{12}L_{24}$ Spheres

The surface functionalization of $M_{12}L_{24}$ spherical complexes is readily conducted by attaching a functional group on the convex side of ligand **1a**, resulting in the equivalent alignment of 24 functional groups at the periphery of the spherical complex. It has been demonstrated that large functional groups, such as fullerene [11], porphyrin [11] and oligosaccharide [12], can be introduced on the surface of spherical complex **2a**.

It was expected that if a functional group was attached to the concave side of the ligand **1a**, the endohedral functionalization of the $M_{12}L_{24}$ spherical complex would be accomplished. However, ligand **1a**, having a methyl group at the concave side (ligand **1d**), did not assemble into the $M_{12}L_{24}$ complex upon complexation with Pd(II) ions [13]. It is considered that the steric repulsion between the pyridyl groups and the attached methyl group causes a nonplanar conformation of the ligand and that such a twisted conformation is not favorable for the complexation (Figure 8.4a). The drawback was removed by inserting an acetylene spacer between the pyridine coordination sites and the core benzene ring to expand the cavity of the complex (ligand **1e**) (Figure 8.4b). Hence the above result indicates that, at each Pd(II) center of

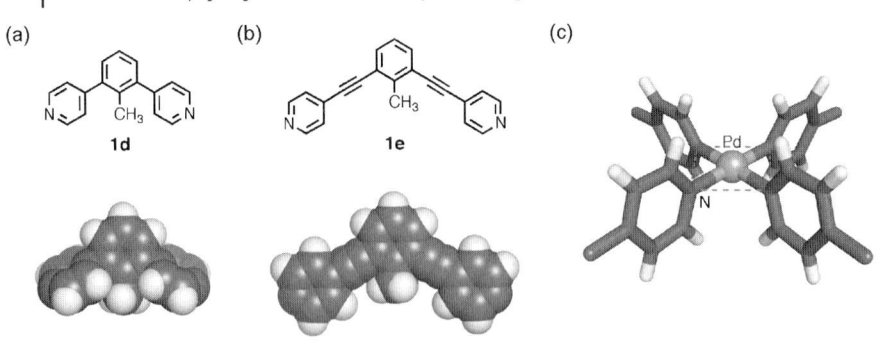

Figure 8.4 Molecular modeling of (a) **1d** and (b) **1e** optimized by a force-field calculation. (c) Partial structure around the Pd coordination center.

the $M_{12}L_{24}$ complex, a perpendicular array of four pyridyl groups with respect to the PdN_4 plane should be essential to the self-assembly of the spherical complex (Figure 8.4c). The acetylene-mediated $M_{12}L_{24}$ spherical complex has an outer diameter of 4.6 nm and an inner diameter of approximately 3.7 nm and serves as a versatile scaffold for endohedral functionalization, as shown schematically in Figure 8.5 [13].

8.4.3
Fluorous Nanodroplets

The term "fluorous" was introduced as an analogy to "aqueous" or "aqueous media" for highly fluorinated alkanes, ethers and tertiary amines [14]. Fluorous solvents do not mix with most common organic solvents at room temperature. Therefore, fluorous molecules have been widely used to separate products and catalysts or products and reagents, purify mixtures and control reactions [14,15]. The creation of a

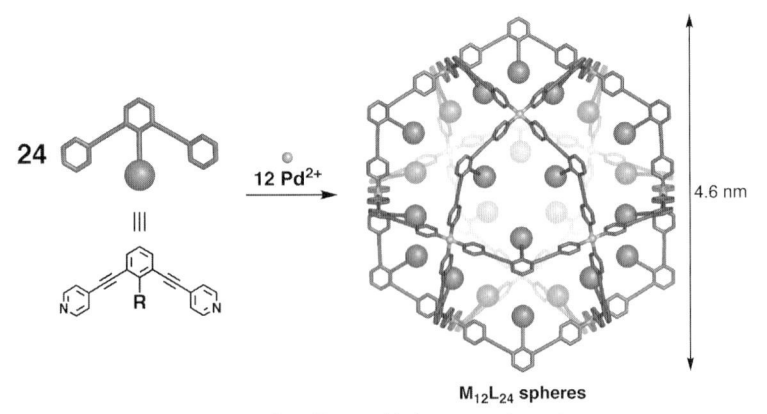

Figure 8.5 Schematic image of a self-assembled $M_{12}L_{24}$ spherical complex with 24 endohedral functional groups [13].

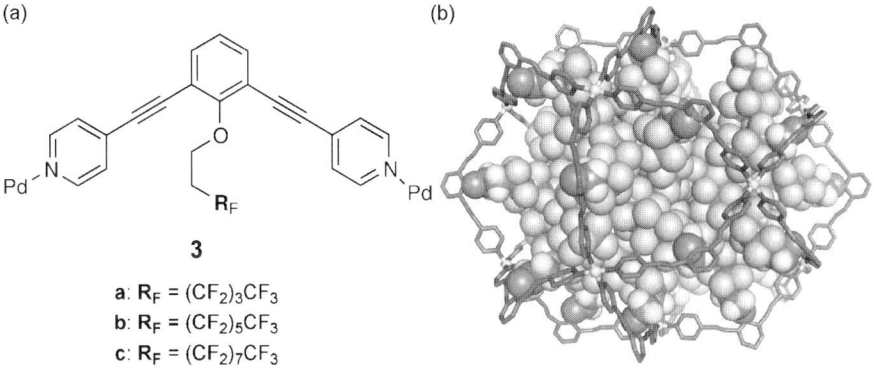

Figure 8.6 (a) $M_{12}L_{24}$ spheres **3a–c** with 24 perfluoroalkyl chains [17]. (b) Molecular drawing prepared by combining the X-ray crystal structure of the shell framework and the optimized $C_6F_{13}(CH_2)_2-$ side-chains. The side-chains are shown as CPK representation.

3

a: $R_F = (CF_2)_3CF_3$
b: $R_F = (CF_2)_5CF_3$
c: $R_F = (CF_2)_7CF_3$

fluorous microenvironment within organic cage compounds, such as vesicles, micelles and dendrimers, has already been demonstrated by some groups [16]. However, the fluorous phases are not distinctly determined physically and structurally.

A distinct endofluorous environment is created in an $M_{12}L_{24}$ spherical cage where 24 perfluoroalkyl chains reside (Figure 8.6a) [17]. $M_{12}L_{24}$ spheres have sufficiently large cavities and the perfluoroalkyl C_6F_{13} chains of complex **3b**, for example, do not completely fill the sphere, leaving a void space at the core. The fluorous core of sphere **3b** can extract an average of 5.8 molecules of perfluorooctane **4** from the suspension in DMSO solution. Detailed analysis of ^{19}F NMR spectra showed that the incorporated guest molecules were accumulated at the core of the spherical hollow complex and surrounded by the terminal portions of the perfluoroalkyl chains. The crystallographic analysis of the sphere **3b** containing the guest **4** revealed that the rigid shell framework and amorphous interior, just like a "raw egg" (Figure 8.6b). This observation indicates that the fluorinated segments furnish a fluid-like or "nanodroplet" environment. The shell framework in the crystalline state was not spherical but oval with dimensions of 4.9 nm by 4.2 nm probably because of the aggregation of the fluorous chains in the shell. Perfluorooctane is miscible in CH_3CN. Therefore, the guest molecules can be re-extracted from the core of sphere **3b** by addition of CH_3CN to the solution. The reversible uptake of fluorous compounds assures that the sphere **3b** can be recycled as a nanosized fluorous medium.

The ability to dissolve perfluoroalkanes in the endofluorous phase largely depends on the lengths of the attached perfluoroalkyl tails (Figure 8.7). Sphere **3c** with longer perfluoroalkyl tails should have a less effective void volume at the core. Therefore, the sphere **3c** can accommodate a smaller amount of perfluorooctane (ca. 2.5 guest molecules per sphere **3c**). In contrast, sphere **3a** with shorter perfluoroalkyl tails has too large a void space to confine perfluorooctane, because of insufficient fluorine

sphere	3a	3b	3c
the number of accommodated guest 4	none	5.8	2.5

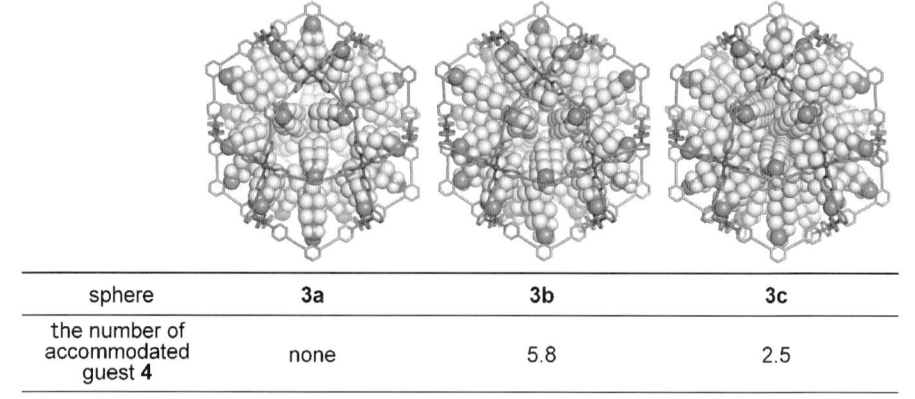

Figure 8.7 The number of accommodated guest 4 in endofluorous spheres 3a–c.

density to define the fluorous atmosphere. The amount of the encapsulated guests in the spheres also depends on the molecular sizes of the guests. For example, a smaller size of a fluorocarbon guest, perfluorohexane, can be extracted in larger amounts by sphere **3b** (ca. 8.6 guest molecules per sphere **3b**). The endofluorous spheres offer a finely tunable environment for fluorous chemistry.

8.4.4
Uptake of Metal Ions into a Cage

Pyridine is a well-known monodentate ligand that can interact with a variety of metal ions. However, pyridine and pyridine-related ligands are not used as functional groups attached to a ligand of an $M_{12}L_{24}$ sphere because such functional groups bind to a Pd(II) ion and inhibit the formation of the sphere. Therefore, to encapsulate metal ions in an $M_{12}L_{24}$ cage, it is necessary to choose a functional group that has the properties to bind strongly to metal ions other than a Pd(II) ion and not to prevent the formation of the $M_{12}L_{24}$ shell structure. This requirement can be satisfied by considering that a Pd(II) ion is a soft acid in Pearson's hard and soft acids and bases (HSAB) theory [18].

When an oligo(ethylene oxide) chain is attached to each ligand, the $M_{12}L_{24}$ sphere **5** is obtained quantitatively (Figure 8.8) [13]. The cavity of complex **5** is filled with a "pseudo-nanoparticle" of poly(ethylene oxide) whose 120 (=5×24) ether oxygen donors (hard bases) in total can bind rare earth and alkaline earth metal ions (hard acids) (Figure 8.9). The ^1H NMR spectrum of the acetonitrile solution of complex **5** and $La(OTf)_3$ revealed that the signals of ethylene oxide chain, especially terminal $-OCH_3$, were shifted downfield, whereas the signals of the aromatic shell of complex **5** remained almost unchanged. These results indicates that La(III) ions are absorbed into the poly(ethylene oxide) core to form La(III)–ethylene oxide complexes inside the complex **5**. The complexation ratio of La(III) ion to each $(OCH_2CH_2)_4OCH_3$ chain

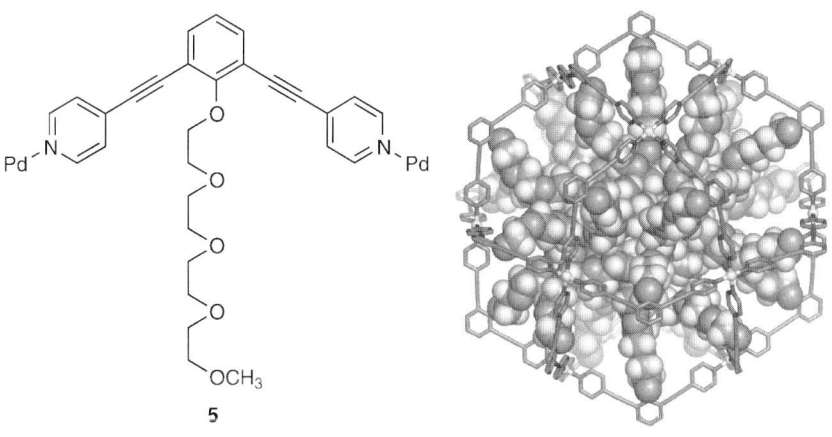

Figure 8.8 Molecular modeling of $M_{12}L_{24}$ sphere **5** with 24 oligo (ethylene oxide) chains. Side-chains are shown as CPK representation [13].

was estimated to be roughly 1:1 from the Job's plot. Thus complex **5** can take up to ca. 20 La(III) ions within the sphere. The incorporated metal ions are completely expelled from the cavity by adding 5 vol.% dimethyl sulfoxide (which strongly solvates many kinds of metal cations). The reversible uptake of metal ions can be accomplished by precisely designing a coordinative functional group and selecting solvents.

8.4.5
Polymerization in a Nutshell

Polymerization in a closed shell gives size- and/or conformation-restricted products. If such a polymerization proceeds in a nanosized region without any volumetric shrinkage, the physical and optical properties of the polymerized region will change dramatically with nanoscale resolution. Therefore, endohedral polymerization in a restricted nanosized region is an important technique to develop the high-density

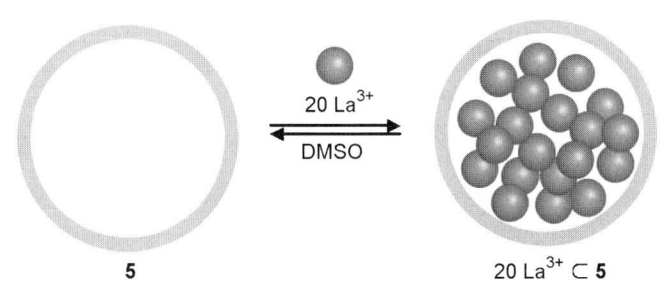

Figure 8.9 Schematic image of the reversible uptake of La(III) ions into sphere **5**.

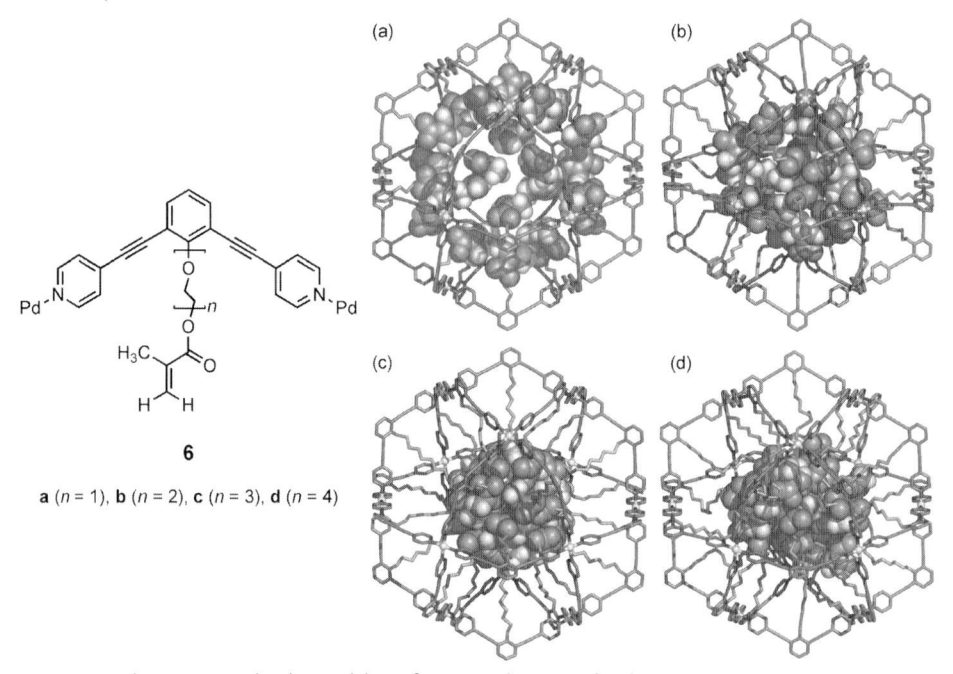

Figure 8.10 Molecular modeling of $M_{12}L_{24}$ spheres **6a–d** with 24 MMA units: (a) **6a**; (b) **6b**; (c) **6c**; (d) **6d**. MMA units are shown as CPK representation [22].

data-storage materials of the next generation [19]. Well-defined nanosized spaces or channels have been exploited to promote polymerization, including zeolites [20], porous organic crystals [20] and porous coordination polymers [21]. However, none of these host materials are discrete and it is impossible to handle them as "molecular-sized" nanoparticles.

When a polymerizable functional group, methyl methacrylate (MMA), is attached to the terminal end of each ligand via an oligo(ethylene oxide) linker, 24 MMA units are confined within molecular spheres **6a–d** (Figure 8.10) [22]. Molecular modeling of spheres **6a–d** illustrates that the relative position of MMA units in the spheres can be precisely controlled by tuning the linker length. Radical polymerization in the spherical shell **6c** proceeds in DMSO solution at 70°C using 2,2′-azobis(isobutyronitrile) (AIBN) as radical initiator. After the polymerization for 17 h, 73% of the MMA units can be converted into PMMA oligomers (Figure 8.11). The MMA monomers tethered to the shell are not completely fluid. A radical center at the end of the growing polymer should be hardly accessible to unreacted MMA monomers far from the radical. Therefore, the value of approximately 70% conversion is the maximum value of the endohedral polymerization in the shell. The aromatic shell structure is maintained during the polymerization. Diffusion-ordered NMR spectroscopy (DOSY) data support the view that the polymerization proceeds only in

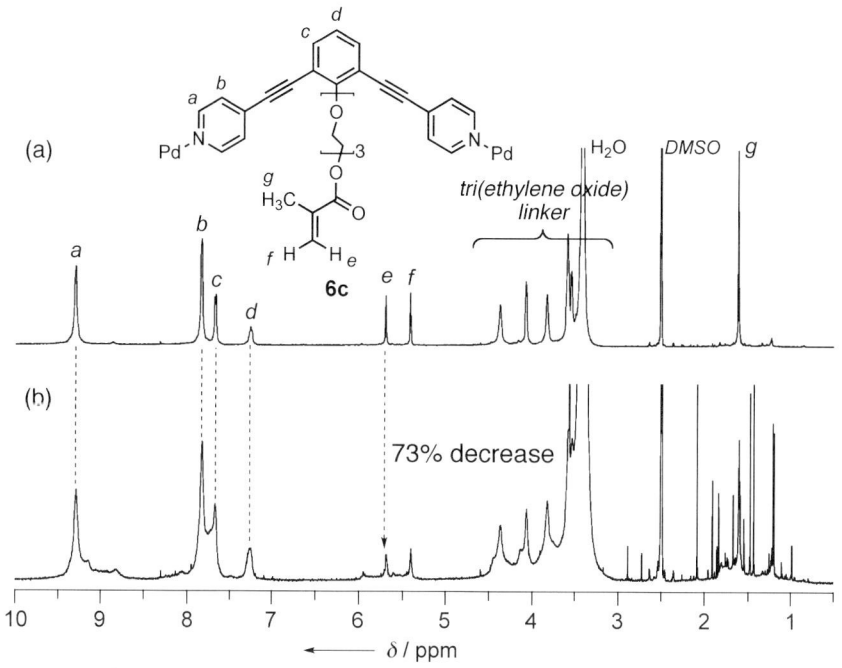

Figure 8.11 ^{1}H NMR spectra of complex **6c** (a) before and (b) after polymerization (500 MHz, DMSO-d_6, 300 K, TMS) [22].

the shell because the diffusion coefficient of the complex **6c** did not change before and after the polymerization. For application to memory storage materials, neither stereoregulation nor molecular weight control is particularly important, but polymerization only in a nanoscopically restricted region is very important. In this context, the polymerizable sphere **6c** is a highly promising material for such an application. The practical concentration of MMA units in the sphere **6c** becomes as high as 1.5 M. Therefore, even at very low concentrations of monomer, the monomers tethered to the shell are concentrated and efficiently polymerized in the sphere.

The length of the oligo(ethylene oxide) linker is extremely important in the endohedral polymerization of spheres. The tri(ethylene oxide) linker affords the best efficiency (73% conversion), because the 24 MMA units are the most closely packed at the core of the sphere **6c**. The mono- and di(ethylene oxide) linkers are too short for polymerization in the sphere, preventing the frequent close approach of MMA units (22 and 29% conversion, respectively). Meanwhile, the tetra(ethylene oxide) linker is slightly too long, resulting in the repulsion of the monomer units at the core (62% conversion). The length of the oligo(ethylene oxide) linker is adjustable to other reactive monomer units with different molecular sizes. Therefore, the present method is widely applicable to the preparation of a variety of endo-reactive molecular spheres.

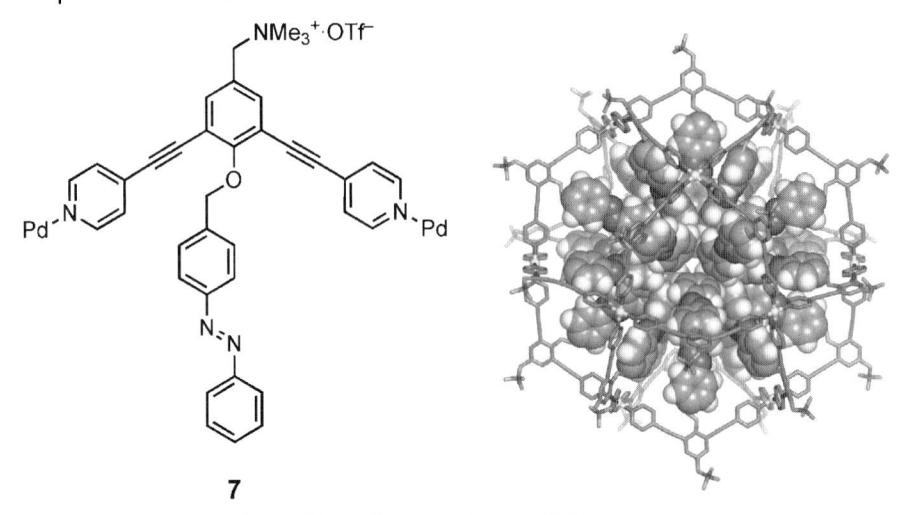

Figure 8.12 Molecular modeling of $M_{12}L_{24}$ sphere **7** with 24 azobenzene chromophores [25]. Azobenzene moieties are shown as CPK representation.

8.4.6
Photoresponsive Molecular Nanoballs

Azobenzene is a well-known chromophore that responds to light and undergoes *cis–trans* isomerization, resulting in large changes in its size and polarity [23]. Azobenzene-containing copolymers and dendrimers are promising photoresponsive materials and the changes of their local structures and properties have attracted a great deal of attention [24]. However, it is synthetically troublesome to confine a restricted number of azobenzene chromophores in their core regions.

Azobenzene-containing $M_{12}L_{24}$ sphere **7** is readily and quantitatively obtained from the corresponding ligand (Figure 8.12) [25]. The quaternary ammonium cations of the surface of the sphere are useful to enhance the water durability. Not all the *trans*-azobenzenes in sphere **7** can be converted to the *cis*-azobenzenes upon UV irradiation. The absorption of UV light by the shell structure is supposed to suppress the *trans–cis* isomerization to be ca. 20% conversion at the photostationary state. Thermally unstable *cis*-azobenzene moieties in sphere **7** are completely returned to the initial *trans* forms by heating at 50 °C.

The sphere **7** can accommodate ca. 16–20 molecules of pyrene in a CH_3CN–H_2O mixed solvent. Neither free ligand nor empty sphere can exert such an effect. Therefore, hydrophobic accumulation of *trans*-azobenzene moieties in the sphere is essential to the guest uptake. Molecular modeling of sphere **7** suggests that the assembled 24 azobenzene moieties construct triangular and square cavities in the sphere, which serve as hydrophobic pockets or portals. The encapsulated pyrenes are magnetically shielded in the sphere and the degree of shielding increases with

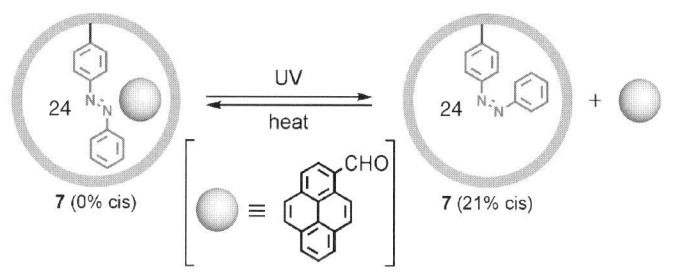

Figure 8.13 Reversible uptake of hydrophobic guests into sphere **7** by alternatively applying UV light and heat.

increase in the ratio of H_2O. These results support the existence of host–guest hydrophobic interaction.

The hydrophobic environment of the interior of sphere **7** is switched by the reversible isomerization of the azobenzene moieties (Figure 8.13). A hydrophobic guest molecule, 1-pyrenecarboxaldehyde, can be incorporated into the sphere in CH_3CN-H_2O (1:1) solvent. Because *cis*-azobenzene is more polar than *trans*-azobenzene, the interior of sphere **7** becomes less hydrophobic. Under UV irradiation, the hydrophobic interaction between the sphere and the guests is weakened and the encapsulated guest is expelled outside the sphere. Almost 80% of the azobenzene moieties remain in the *trans* form in the sphere and therefore the guest release is attributed to the polarity change rather than structural change in the sphere. The hydrophobic environment is completely recovered by heating at 50°C and the sphere again takes up the guest.

8.4.7
Peptide-confined Chiral Cages

Enzyme pockets furnish chiral hollow environments where amino acid residues are precisely arranged and asymmetric molecular recognition and chemical transformations are achieved [26]. There are several approaches on the *de novo* design and control of peptide 3D structures [27]. However, the construction of chiral spatial environments from peptides by synthetic means has never been achieved. It is possible to create a library of spheres that tether various kinds of amino acid residues with different lengths (Figure 8.14). $M_{12}L_{24}$ spheres can accommodate up to 96 ($=4\times24$) amino acid residues within the hollows (sphere **8f**), which might be regarded as an artificial "mini-protein". Synchrotron X-ray studies of the sphere **8a** confining 24 L-alanine (L-Ala) residues clarified the endohedral arrangement of the amino acid residues in the sphere (Figure 8.15). Some L-Ala residues are in mutual proximity via the terminal protective groups in the crystalline state.

Aromatic shells of $M_{12}L_{24}$ spheres are highly symmetric and do not have any asymmetric points in themselves. Therefore, the circular dichroism (CD) spectrum of empty sphere shows no peaks. However, in the L-Ala-confined sphere **8a**, chirality of each amino acid is transferred to the spherical shell, displaying a strong negative Cotton effect in the absorption region of the shell (300–350 nm)

Figure 8.14 Library of peptide-confined $M_{12}L_{24}$ spheres.

(Figure 8.16). The intensity of the induced CD of the sphere is almost 30 times as strong as that of the corresponding ligand. The CD spectra of D/L-Ala-confined spheres have identical shapes and intensities, but opposite signs. The interiors of the spheres where 24 asymmetric centers are accumulated can afford a unique chiral environment. When two kinds of ligands that tether different amino acid residues are used in the formation of the sphere, the internal chiral environment of the obtained sphere should change, depending on the mixing ratio of the two ligands. Although it is difficult to control precisely the ratio and relative positions

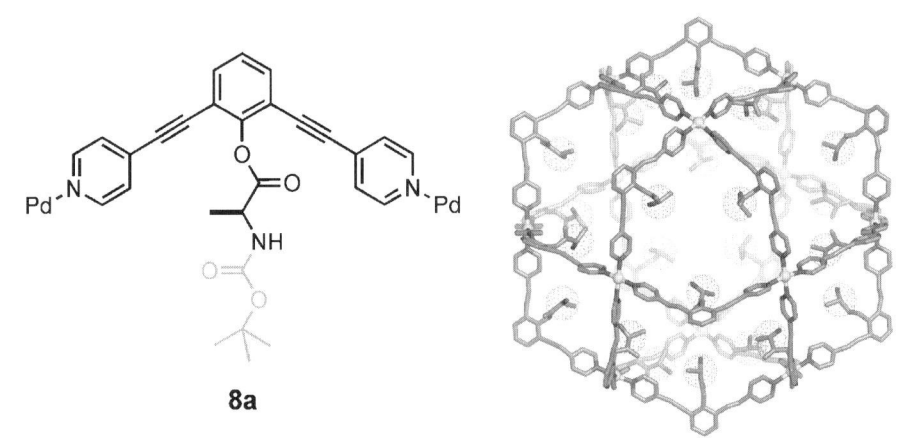

Figure 8.15 X-ray crystal structure of L-Ala-confined sphere **8a**. The terminal Boc groups are disordered and could not be located. Asymmetric centers are shown as dotted spheres.

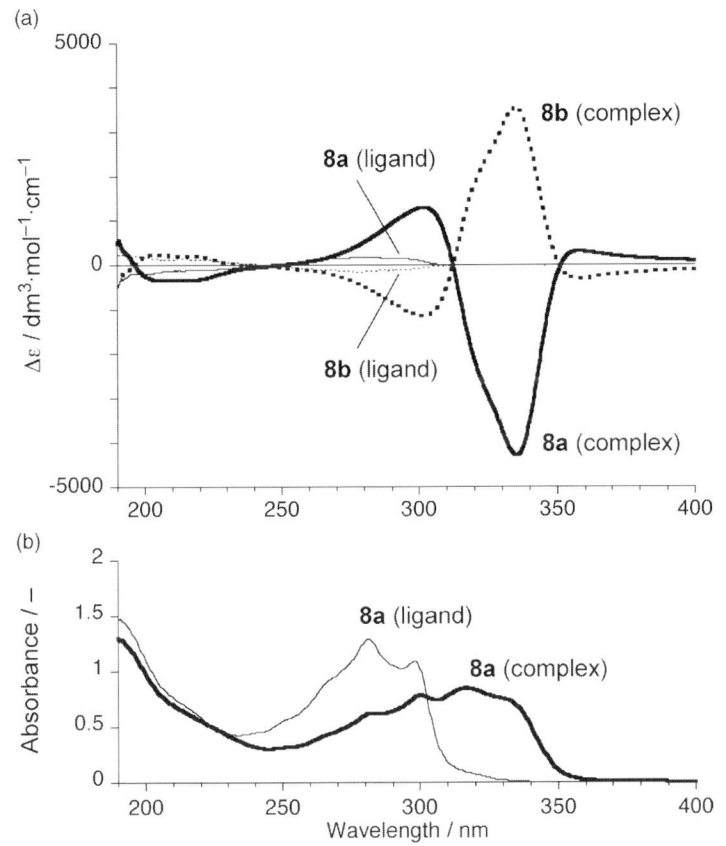

Figure 8.16 (a) CD and (b) UV–Vis spectra of spheres **8a** and **b**. The CD and UV–Vis spectra of the corresponding ligands to spheres **8a** and **b** are also shown as references. Concentration of ligand, 24 μM; concentration of complex, 1 μM.

of the two kinds of ligands introduced, desired peptides can be combinatorially anchored in the sphere. Therefore, the peptide-confined spherical hollows have the potential to be developed as artificial enzyme pockets.

8.5
Conclusions and Outlook

Natural building blocks such as protein-based giant amphiphiles and virus capsides have been redesigned by altering their chemical composition to control precisely the number and position of the introduced functional groups. Molecular self-assembly that is ubiquitously found in biological systems is a powerful method for developing endochemistry of hollow spherical cages. $M_{12}L_{24}$ spherical complexes are versatile nanosized cages whose internal local environments can be chemically modified on demand. At the stage of ligand designing of $M_{12}L_{24}$ spheres, a desired functionality can be attached to each ligand, resulting in the accurate and efficient introduction of 24 functional groups in the spheres. The positions of functional groups in the spheres are changed by the length of linkers that tether the functional groups to the shells of the spheres. This handling flexibility indicates that the functional groups can be placed at any positions, depending on their molecular sizes. The introduced functional groups are concentrated at the cores of spheres. Therefore, it is possible to develop novel properties and functions that are not otherwise accessible by conventional methods.

Further directions of endohedral functionalization of self-assembled hollow spherical cages are contributions to organic synthesis, materials science and biology. The last topic of peptide-based cages tells us that it is possible to encapsulate biomacromolecules, particularly proteins, to direct the control of biofunctions. The elongation of the pyridyl arms of ligands leads to the expansion of the overall cavity sizes of $M_{12}L_{24}$ spheres. Endochemistry with well-defined spherical cages is just getting started.

References

1 (a) Vriezema, D.M., Aragonès, M.C., Elemans, J.A.A.W., Cornelissen, J.J.L.M., Rowan, A.E. and Nolte, R.J.M. (2005) *Chem. Rev.*, **105**, 1445–1490. (b) Philp, D. and Stoddart, J.F. (1996) *Angew. Chem. Int. Ed. Engl.*, **35**, 1154–1196. (c) Caulder, D.L. and Raymond, K.N. (1999) *Acc. Chem. Res.*, **32**, 975–982. (d) MacGillivray, L.R. and Atwood, J.L. (1999) *Angew. Chem. Int. Ed.*, **38**, 1018–1033. (e) Prins, L.J., Reinhoudt, D.N. and Timmerman, P. (2001) *Angew. Chem. Int. Ed.*, **40**, 2382–2426. (f) Hof, F., Craig, S.L., Nuckolls, C. and Rebek, J. Jr. (2002) *Angew. Chem. Int. Ed.*, **41**, 1488–1508. (g) Seidel, S.R. and Stang, P.J. (2002) *Acc. Chem. Res.*, **35**, 972–983. (h) Fujita, M., Tominaga, M., Hori, A. and Therrien, B. (2005) *Acc. Chem. Res.*, **38**, 369–378.

2 Arora, P.S. and Kirshenbaum, K. (2004) *Chem. Biol.*, **11**, 418–420.

3 (a) Douglas, T. and Young, M. (1998) *Nature*, **393**, 152–155. (b) Douglas, T. and Young, M. (1999) *Adv. Mater.*, **11**, 679–681. (c) Niemeyer, C.M. (2001) *Angew. Chem. Int. Ed.*, **40**, 4128–4158. (d) Flynn, C.E., Lee, S.-W., Peelle, B.R. and Belcher, A.M. (2003) *Acta Mater.*, **51**, 5867–5880. (e) Douglas, T. and Young, M. (2006) *Science*, **312**, 873–875. (f) Singh, P., Gonzalez, M.J. and Manchester, M. (2006) *Drug Dev. Res.*, **67**, 23–41.

4 (a) Chasteen, N.D. and Harrison, P.M. (1999) *J. Struct. Biol.*, **126**, 182–194. (b) Lawson, D.M., Artymiuk, P.J., Yewdall, S.J., Smith, J.M.A., Livingstone, J.C., Treffry, A., Luzzago, A., Levi, S., Arosio, P., Cesareni, G., Thomas, C.D., Shaw, W.V. and Harrison, P.M. (1991) *Nature*, **349**, 541–544.

5 (a) Speir, J.A., Munshi, S., Wang, G., Baker, T.S. and Johnson, J.E. (1995) *Structure*, **3**, 63–78. (b) Reddy, V.S., Natarajan, P., Okerberg, B., Li, K., Damodaran, K.V., Morton, R.T., Brooks, C.L. III and Johnson, J.E. (2001) *J. Virol.*, **75**, 11943–11947.

6 Gillitzer, E., Suci, P., Young, M. and Douglas, T. (2006) *Small*, **2**, 962–966.

7 (a) Kramer, R.M., Li, C., Carter, D.C., Stone, M.O. and Naik, R.R. (2004) *J. Am. Chem. Soc.*, **126**, 13282–13286. (b) Flenniken, M.L., Liepold, L.O., Crowley, B.E., Willits, D.A., Young, M.J. and Douglas, T. (2005) *Chem. Commun.*,

447–449. (c) Douglas, T., Strable, E., Willits, D., Aitouchen, A., Libera, M. and Young, M. (2002) *Adv. Mater.*, **14**, 415–418. (d) Hooker, J.M., Kovacs, E.W. and Francis, M.B. (2004) *J. Am. Chem. Soc.*, **126**, 3718–3719.

8 (a) Zhang, L. and Eisenberg, A. (1995) *Science*, **268**, 1728–1731. (b) Stupp, S.I., LeBonheur, V., Walker, K., Li, L.S., Huggins, K.E., Keser, M. and Amstutz, A. (1997) *Science*, **276**, 384–389. (c) Jenekhe, S.A. and Chen, X.L. (1998) *Science*, **279**, 1903–1907. (d) Won, Y.-Y., Davis, H.T. and Bates, F.S. (1999) *Science*, **283**, 960–963. (e) Förster, S. and Plantenberg, T. (2002) *Angew. Chem. Int. Ed.*, **41**, 688–714. (f) Discher, D.E. and Eisenberg, A. (2002) *Science*, **297**, 967–973. (g) Bucknall, D.G. and Anderson, H.L. (2003) *Science*, **302**, 1904–1905. (h) Jean-François, L. (2006) *Polym. Int.*, **55**, 979–993.

9 O'Reilly, R.K., Hawker, C.J. and Wooley, K.L. (2006) *Chem. Soc. Rev.*, **35**, 1068–1083.

10 (a) Bae, Y., Fukushima, S., Harada, A. and Kataoka, K. (2003) *Angew. Chem. Int. Ed.*, **42**, 4640–4643. (b) O'Reilly, R.K., Joralemon, M.J., Wooley, K.L. and Hawker, C.J. (2005) *Chem. Mater.*, **17**, 5976–5988. (c) O'Reilly, R.K., Joralemon, M.J., Hawker, C.J. and Wooley, K.L. (2006) *Chem. Eur. J.*, **12**, 6776–6786.

11 Tominaga, M., Suzuki, K., Kawano, M., Kusukawa, T., Ozeki, T., Sakamoto, S., Yamaguchi, K. and Fujita, M. (2004) *Angew. Chem. Int. Ed.*, **43**, 5621–5625.

12 Kamiya, N., Tominaga, M., Sato, S. and Fujita, M. (2007) *J. Am. Chem. Soc.*, **129**, 3816–3817.

13 Tominaga, M., Suzuki, K., Murase, T. and Fujita, M. (2005) *J. Am. Chem. Soc.*, **127**, 11950–11951.

14 (a) Horváth, I.T. and Rábai, J. (1994) *Science*, **266**, 72–75. (b) Horváth, I.T. (1998) *Acc. Chem. Res.*, **31**, 641–650.

15 (a) Studer, A., Hadida, S., Ferritto, R., Kim, S.-Y., Jeger, P., Wipf, P. and Curran, D.P.(1997) *Science*, **275**, 823–826. (b) Studer, A., Jeger, P., Wipf, P. and Curran, D.P. (1997) *J. Org. Chem.*, **62**, 2917–2924. (c) Richard, H.F. (1999) *Chem. Eur. J.*, **5**, 1677–1680. (d) Yoshida, J. and Itami, K. (2002) *Chem. Rev.*, **102**, 3693–3716. (e) Zhang, W. (2003) *Tetrahedron*, **59**, 4475–4489.(f) Curran, D.P. (2005) in *Handbook of Fluorous Chemistry*, (eds J.A. Gladysz, D.P. Curran and I.T. Horváth), Wiley-VCH, Weinheim, pp. 101–127.

16 (a) Garcia-Bernabé, A., Krämer, M., Olàh, B. and Haag, R. (2004) *Chem. Eur. J.*, **10**, 2822–2830. (b) Percec, V., Imam, M.R., Bera, T.K., Balagurusamy, V.S.K., Peterca, M. and Heiney, P.A. (2005) *Angew. Chem. Int. Ed.*, **44**, 4739–4745.

17 Sato, S., Iida, J., Suzuki, K., Kawano, M., Ozeki, T. and Fujita, M. (2006) *Science*, **313**, 1273–1276.

18 Pearson, R.G. (1963) *J. Am. Chem. Soc.*, **85**, 3533–3539.

19 (a) Irie, M. (2000) *Chem. Rev.*, **100**, 1685–1716. (b) Fullerton, E.E., Margulies, D.T., Moser, A. and Takano, K. (2001) *Solid State Technol.*, **44**, 87–94. (c) Sun, X., Huang, Y. and Nikles, D.E. (2004) *Int. J. Nanotechnol.*, **1**, 328–346. (d) Tang, Q., Shi, S.-Q. and Zhou, L. (2004) *J. Nanosci. Nanotechnol.*, **4**, 948–963. (e) Tian, H. and Yang, S. (2004) *Chem. Soc. Rev.*, **33**, 85–97. (f) Mayes, E.L. and Mann, S. (2005) in *Nanobiotechnology*, (eds C.M. Niemeyer and C.A. Mirkin), Wiley-VCH, Weinheim, 278–287. (g) Naito, K., Hieda, H., Ishino, T., Tanaka, K., Sakurai, M., Kamata, Y., Morita, S., Kikitsu, A. and Asakawa, K. (2005) in *Progress in Nano-Electro-Optics III*, (ed. M. Otsu), Springer, Berlin, 127–144.

20 (a) Farina, M. (1988) in *Encyclopedia of Polymer Science and Engineering*, (ed. J.I. Kroschwitz), Wiley, New York, Vol. 12, 486–504. (b) Miyata, M. (1996) in *Comprehensive Supramolecular Chemistry*, (ed. D. Reinhoudt), Pergamon Press, Oxford, Vol. 10, 557–582. (c) Moller, K. and Bein, T. (1998) *Chem. Mater.*, **10**, 2950–2963. (d) Tajima, K. and Aida, T. (2000) *Chem. Commun.*, 2399–2412.

(e) Cardin, D.J. (2002) *Adv. Mater.*, **14**, 553–563.

21 (a) Uemura, T., Kitagawa, K., Horike, S., Kawamura, T., Kitagawa, S., Mizuno, M. and Endo, K. (2005) *Chem. Commun.*, 5968–5970. (b) Uemura, T., Kitaura, R., Ohta, Y., Nagaoka, M. and Kitagawa, S. (2006) *Angew. Chem. Int. Ed.*, **45**, 4112–4116. (c) Uemura, T., Horike, S. and Kitagawa, S. (2006) *Chem. Asian J.*, **1**, 36–44.

22 Murase, T., Sato, S. and Fujita, M. (2007) *Angew. Chem. Int. Ed.*, **46**, 1083–1085.

23 (a) Kumar, G.S. and Neckers, D.C. (1989) *Chem. Rev.*, **89**, 1915–1925. (b) Natansohn, A. and Rochon, P. (2002) *Chem. Rev.*, **102**, 4139–4175. (c) Dugave, C. and Demange, L. (2003) *Chem. Rev.*, **103**, 2475–2532. (d) Yager, K.G. and Barrett, C.J. (2006) *J. Photochem. Photobiol. A*, **182**, 250–261.

24 (a) Wang, G., Tong, X. and Zhao, Y. (2004) *Macromolecules*, **37**, 8911–8917. (b) Tong, X., Wang, G., Soldera, A. and Zhao, Y. (2005) *J. Phys. Chem. B*, **109**, 20281–20287. (c) Lee, H., Pietrasik, J. and Matyjaszewski, K. (2006) *Macromolecules*, **39**, 3914–3920. (d) Liu, X. and Jiang, M. (2006) *Angew. Chem. Int. Ed.*, **45**, 3846–3850.

25 Murase, T., Sato, S. and Fujita, M. Angew. (2007) *Chem. Int. Ed.*, **46**, 5133–5136.

26 (a) Adams, M.J., Buehner, M., Chandrasekhar, K., Ford, G.C., Hackert, M.L., Liljas, A., Rossmann, M.G., Smiley, I.E., Allison, W.S., Everse, J., Kaplan, N.O. and Taylor, S.S. (1973) *Proc. Natl. Acad. Sci. USA*, **70**, 1968–1972. (b) Auerbach, G., Ostendorp, R., Prade, L., Korndörfer, I., Dams, T., Huber, R. and Jaenicke, R. (1998) *Structure*, **6**, 769–781. (c) Read, J.A., Winter, V.J., Eszes, C.M., Sessions, R.B. and Brady, R.L. (2001) *Proteins: Struct. Funct. Genet.*, **43**, 175–185.

27 (a) Blanco, F.J., Jimenez, M.A., Herranz, J., Rico, M., Santoro, J. and Nieto, J.L. (1993) *J. Am. Chem. Soc.*, **115**, 5887–5888. (b) Haque, T.S. and Gellman, S.H. (1997) *J. Am. Chem. Soc.*, **119**, 2303–2304. (c) Peczuh, M.W., Hamilton, A.D., Sánchez-Quesada, J., Mendoza, J., Haack, T. and Giralt, E. (1997) *J. Am. Chem. Soc.*, **119**, 9327–9328. (d) Kelso, M.J., Hoang, H.N., Oliver, W., Sokolenko, N., March, D.R., Appleton, T.G. and Fairlie, D.P. (2003) *Angew. Chem. Int. Ed.*, **42**, 421–424. (e) Tashiro, S., Tominaga, M., Yamaguchi, Y., Kato, K. and Fujita, M. (2006) *Angew. Chem. Int. Ed.*, **45**, 241–244. (f) Tashiro, S., Kobayashi, M. and Fujita, M. (2006) *J. Am. Chem. Soc.*, **128**, 9280–9281.

9
Polynuclear Coordination Cages
Michael D. Ward

9.1
Introduction

In the last couple of decades, the development of self-assembly methods in transition metal coordination chemistry has led to a large number of beautiful structures of a complexity which would previously have been inconceivable. From the first examples such as simple helicates in the 1980s [1,2] and small molecular grids in the 1990s [3,4], the repertoire of structural types that is accessible from a combination of polydentate bridging ligands and kinetically labile metal ions has expanded to include cages [5–7], rings [8,9], cylinders [10] and interlocked systems such as catenates [11,12] and knots [13]. Many of the first examples were the result of serendipity; however, there are equally many cases where the structures were the result of careful design and an understanding of how the geometric properties of a particular ligand would combine with the stereoelectronic preferences of a specific metal ion to yield the desired result.

One area which has been of particular recent interest is that of the preparation of polyhedral coordination cages. These compounds are attractive for several reasons. Their high-symmetry architectures have been of aesthetic interest and fascination since Plato first described the regular polyhedral solids (tetrahedron, octahedron, cube, dodecahedron and icosahedron), and it is no accident that such structures arise from self-assembly processes in nature in areas as superficially disparate as solid-state chemistry and biology [14,15]. Their high symmetry means that such structures can be particularly susceptible to rational design and there are many beautiful examples of polyhedral cage complexes in which careful matching of the symmetry properties of metal ions and rigid ligands whose geometric properties are fairly inflexible, have resulted in the planned synthesis of new cage architectures. The work of the groups of Raymond [5], Fujita [6] and Stang [7] is particularly notable in this regard. Finally, the fact that such cage complexes have – by definition – large central cavities means that they display intriguing host–guest chemistry, which at its simplest involves incorporation of solvent molecules or counter-ions and at its most sophisticated allows the cages to be used as "microreactors" in which new reactions can be catalyzed and hitherto inaccessible molecules stabilized [5,6].

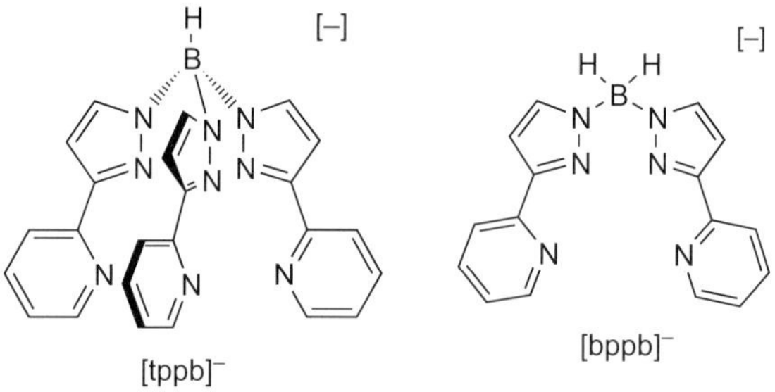

9 Polynuclear Coordination Cages

Scheme 9.1

This chapter presents a personal account of work in the author's research group over the last decade, in which relatively simple ligands based on bidentate pyrazolylpyridine chelating groups have been used as the basis for assembly of polyhedral cages. These ligands were originally developed as second-generation tris(pyrazolyl) borates in which the addition of 2-pyridyl groups to the pyrazolyl rings resulted in a hexadentate binding pocket in the ligand [tppb]$^-$ (Scheme 9.1) [16]. However, it quickly became apparent that this deceptively simple ligand had coordination behavior that was more complicated than we had anticipated and we consequently developed a series of ligands (Scheme 9.2) in which two or three of the bidentate

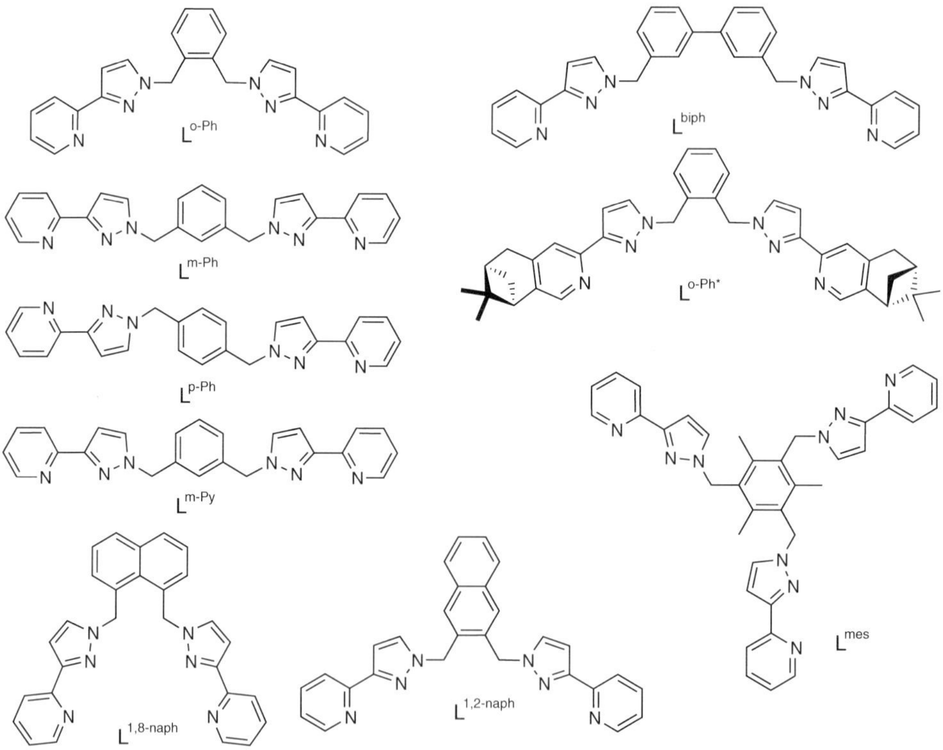

Scheme 9.2

pyrazolylpyridine groups were connected to organic spacers, which are resistant to the hydrolysis that occasionally plagues tris(pyrazolyl)borate chemistry. Serendipity has been a constant ally during this work. The ligands shown in Scheme 9.2 are inherently highly flexible because of the presence of saturated methylene spacers between the pyrazolylpyridine groups and the aromatic spacers, which were introduced for ease of synthesis. This precludes any possibility of control of the relative orientation of the binding sites and consequent deliberate design and synthesis of polyhedral cages, but it has resulted in many surprises, with many examples of unusual high-nuclearity cage structures appearing whose complexity is such that they could never, realistically, have been rationally designed.

9.2
Complexes Based on Poly(pyrazolyl)borate Ligands

Reaction of 3(2-pyridyl)pyrazole with KBH_4 readily afforded the bis- or tris(pyrazolyl) borate ligands [bppb]$^-$ or [tppb]$^-$ (Scheme 9.1) according to the stoichiometry and reaction temperature [16,17]. These ligands were originally planned as tetradentate or hexadentate chelates, respectively, for lanthanide(III) ions and indeed acted in this capacity perfectly well, generating an extensive series of complexes whose photophysical properties we studied [18].

With first-row transition metal ions, however, more complicated behavior emerged. Thus [bppb]$^-$ could act as a tetradentate chelate to lanthanide(III) ions and generate simple mononuclear complexes [18], but it could also act as a bridging ligand spanning two metal ions, which resulted in the unexpected formation of the octanuclear cyclic helicate $[Co_8(\mu\text{-bppb})_{12}(ClO_4)]^{3+}$ (Figure 9.1), whose assembly

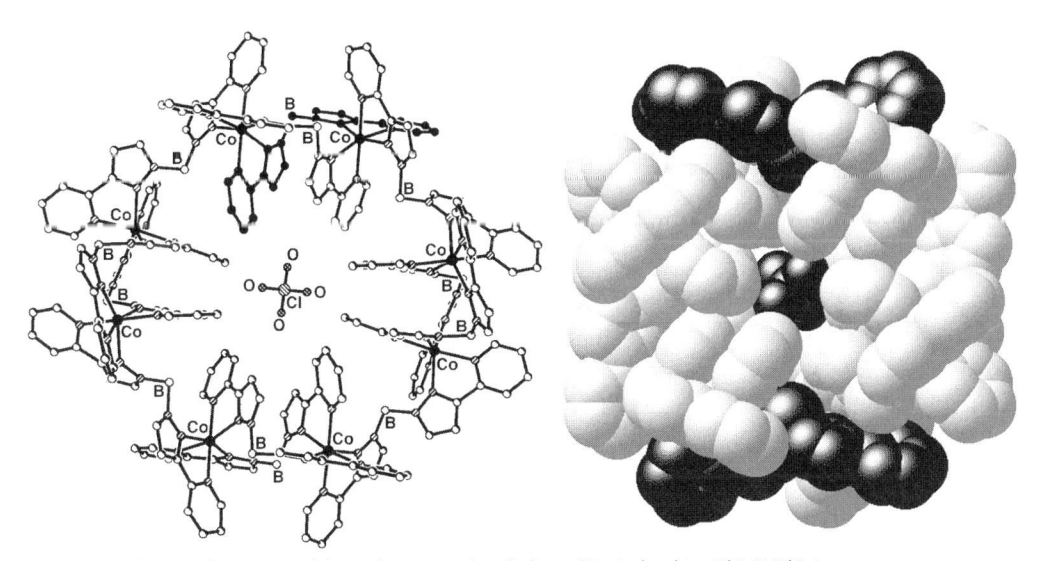

Figure 9.1 The complex cation of the cyclic octanuclear helicate $[Co_8(\mu\text{-bppb})_{12}(ClO_4)](ClO_4)_3$.

appears to be templated by the central perchlorate anion, which is a good fit for the central cavity [19]. The metal:ligand ratio of 2:3 is significant and a point that will be returned to later; it arises, necessarily, from the combination of a metal ion having a preference for octahedral coordination with a ligand which has four donor atoms. Thus, in the absence of coordinating anions or solvent molecules, 1.5 ligands are necessary to satisfy each metal ion, giving a stoichiometry of M_2L_3 or some higher multiple thereof.

The hexadentate ligand [tppb]$^-$ proved equally unpredictable (Figure 9.2). Its Co(II) complex [Co(tppb)]$^+$ is mononuclear with the metal ion in a rather unusual trigonal prismatic coordination geometry which is imposed by the ligand [20]; for this ligand to provide an octahedral coordination environment would require a high degree of twisting such that the N_3 plane provided by the pyridyl donors was staggered with respect to the N_3 plane provided the pyrazolyl donors. Clearly, it is not worth the cost in this complex. However, with metal ions such as Zn(II) and Mn(II), tetrahedral cages

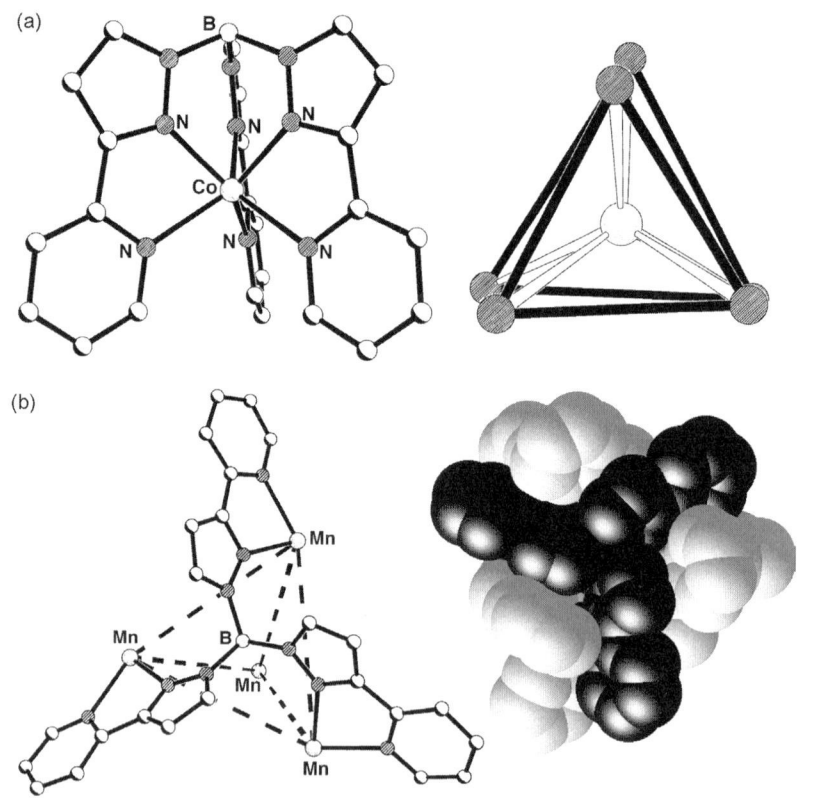

Figure 9.2 (a) The monomeric complex cation [Co(tppb)]$^+$, with trigonal-prismatic coordination of the metal ion; (b) the tetrameric complex cation [Mn$_4$(tppb)$_4$]$^{4+}$, with pseudo-octahedral coordination of the metal ion.

[M₄(tppb)₄]⁴⁺ arise in which the ligand coordinates each of its three arms to a separate metal ion, effectively capping one face of an M₄ tetrahedron [20,21]. This previously unseen coordination mode for a tris(pyrazolyl)borate allows each metal ion, which interacts with three different ligands, to adopt an approximately octahedral coordination geometry that cannot be provided by one ligand alone. This pair of complexes provides a nice demonstration of how different stereoelectronic requirements of metal ions exerts a controlling influence on the course of the metal–ligand self-assembly process. A feature of the tetrahedral structure which may be significant in stabilizing it is the presence of aromatic π-stacking between overlapping aromatic rings of different ligands.

9.3
Complexes Based on Neutral Ligands with Aromatic Spacers

Following the above results, we prepared the simple bridging ligands shown in Scheme 9.2, in which an aromatic spacer is used instead of a borate group to connect the two pyrazolylpyridine chelates. The syntheses of these are straightforward, requiring reaction of 3-(2-pyridyl)pyrazole under basic conditions with a bis(bromomethyl) aromatic compound, such as 1,2-C₆H₄(CH₂Br)₂, to give L$^{o\text{-Ph}}$. The ready availability of bis(bromomethyl) aromatic compounds allowed the preparation of many related examples.

9.3.1
Complexes Based on L$^{o\text{-Ph}}$ and L$^{12\text{-naph}}$

The simplest members of this series, L$^{o\text{-Ph}}$ and L$^{12\text{-naph}}$, are the only ones in which the two bidentate arms are close enough together to chelate to a single metal ion and, as we saw with [bppb]⁻, the coordination mode of the ligand varies. With Cu(II), a range of simple mononuclear complexes form in which L$^{o\text{-Ph}}$ acts as a tetradentate chelate [22]. Cu(I), however, has a preference for pseudo-tetrahedral coordination with two mutually perpendicular bidentate units, which cannot be met by a single L$^{o\text{-Ph}}$ ligand. The result is bridging behavior of the ligand in the dinuclear double helicate [Cu₂(L$^{o\text{-Ph}}$)₂]²⁺ [22]. With six-coordinate metal ions a 2M:3L ratio must arise, as explained earlier, and it is here that unexpected self-assembly behavior arises. There are many ways in which a 2M:3L ratio can be realized in a complex, of which the best known is a dinuclear triple helicate in which three bis-bidentate ligands each span two metal ions. We found instead two alternative types of structure depending on the size of the metal ion and the nature of the counter-ion.

Reaction of L$^{o\text{-Ph}}$ or its structurally similar analogue L$^{12\text{-naph}}$ with either Co(II) or Zn(II) as their fluoroborate or perchlorate salts afforded in each case tetrahedral cages [M₄L₆X]X₇ (M = Co, Zn; L = L$^{o\text{-Ph}}$, L$^{12\text{-naph}}$; X = BF₄, ClO₄) in which a tetrahedral array of metal ions is connected by a bridging ligand along each edge [23–25] (Figure 9.3). Note that the 2M:3L required to complete the coordination around octahedral metal ions is perfectly met by a tetrahedron which has four vertices (metal ions) and six edges (bridging ligands). Each metal ion is therefore approximately

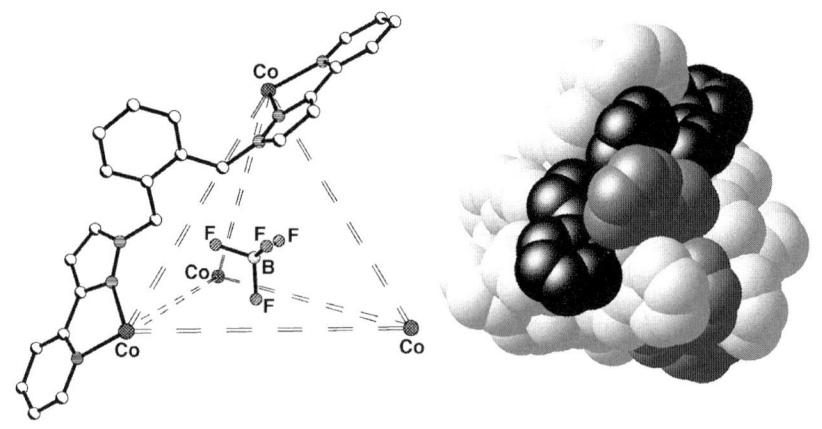

Figure 9.3 The tetranuclear cage complex cation of [Co$_4$(L$^{o\text{-}Ph}$)$_6$(BF$_4$)](BF$_4$)$_7$, showing the tetrahedral superstructure with the encapsulated anion (left) and a space-filling picture emphasizing the close packing of ligands around the periphery (right).

octahedrally coordinated by one bidentate unit from each of three different ligands. In addition, several other features of these complexes are noteworthy:

1. The tetrahedral anion in the central cavity appears to be a perfect fit in terms of size, shape and charge. Each O atom (from perchlorate) or F atom (from tetrafluoroborate) occupies the space at the center of a triangular face of the tetrahedron, such that the tetrahedral anion is inverted with respect to the tetrahedral metal cage. The terminal O/F atoms interact with the cage superstructure via CH···O or CH···F hydrogen bonds with the CH$_2$ groups of the ligands. It is also clear from a space-filling view of the structures that the central anion is completely encapsulated by the metal–ligand cage, with no "windows" in the cage which would allow diffusion of the anions into or out of the cavity.

2. The structure is chiral, with all four metal centers having the same tris-chelate optical configuration; in fact the cage has (non-crystallographic) T symmetry with a C_3 axis through each vertex but no mirror planes. The crystals, however, are racemic with equal numbers of ΔΔΔΔ/ΛΛΛΛ enantiomers in the unit cell.

3. There is extensive π-stacking between ligands around the cage involving overlap of aromatic groups between different ligands. This must provide some stabilization of the structure.

We return to each of these points in turn. The excellent fit of the perchlorate or tetrafluoroborate anion for the cage cavity, the involvement of the anion in hydrogen bonding to the cage superstructure and the complete encapsulation of the anions in every case all imply that the anion has acted as a template around which the cage assembles. Diffusion of an anion into a preformed but empty cage seems unlikely given the absence of "windows" in the cages. ^{19}F and ^{11}B NMR spectra on the fluoroborate-containing cages show the presence of two signals in a 7:1 ratio for the external and encapsulated anions, respectively [23,25], and the spectra do not change significantly on warming up to the limit allowed by the solvent; thus any exchange of internal and external anions is slow on the NMR time-scale. That the central anion does act as a template was demonstrated conclusively by a simple ^{1}H NMR experiment on the Co(II) cages [24]. The paramagnetism of these cages shifts the ^{1}H NMR signals over a wide range between about −50 and 90 ppm and simple spectra were observed consistent with the high symmetry of the cages in solution, with all ligands equivalent and having twofold symmetry. In contrast a mixture of Co(II) acetate and L^{o-Ph} or $L^{12-naph}$ in solution in the correct proportions (2:3) gave only a broad, ill-defined set of signals between 6 and 10 ppm. Addition of NaBF$_4$ or NaClO$_4$ to the NMR sample resulted in the immediate appearance of the characteristic simple, highly shifted set of peaks characteristic of the cage, which therefore only forms after addition of the tetrahedral anion which acts as a template for cage assembly.

The chirality of the cages makes them an appealing target to be resolved into their separate enantiomers. We have not yet been successful at this, but it is clear from NMR studies that diastereoisomers form in solution by ion pairing of the cage cations with the optically pure anion "trisphat" [tris(tetrachlorobenzenediolato)phosphate(V)] [26]. This results in some of the signals associated with the ligands in the ^{1}H spectrum splitting into two components (Figure 9.4) and – more intriguingly – results in the ^{19}F NMR signal for the encapsulated anion also splitting into two peaks with a separation of 2 ppm between the components arising from the two diastereoisomers (Figure 9.4). The chirality of the cage superstructure is therefore manifested through enantiodifferentiation of an achiral guest in the chiral cavity.

Since we could not separate the cage enantiomers by crystallization to get an optically pure sample, we adopted the alternative approach of adding a chiral auxiliary to the ligand (L^{o-Ph^*}, Scheme 9.2). The presence of two equivalent pinene

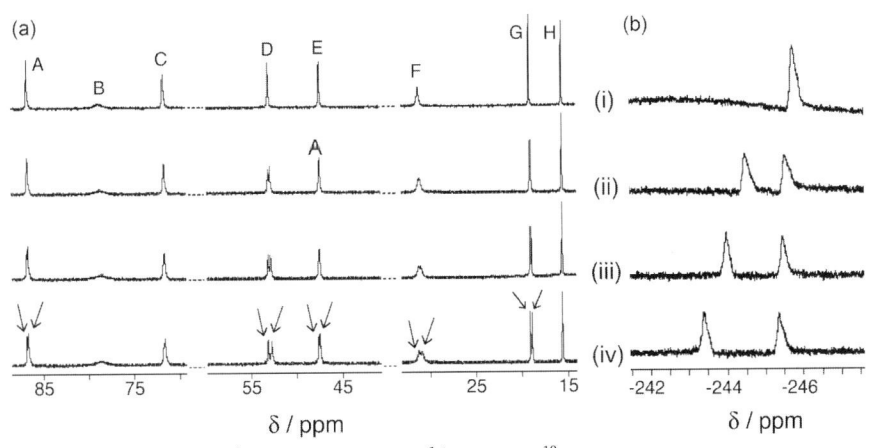

Figure 9.4 (a) 500-MHz ^{1}H NMR spectra and (b) 470-MHz ^{19}F NMR spectra of racemic [Co$_4$(L^{o-Ph})$_6$(BF$_4$)](BF$_4$)$_7$ in the presence of added portions of (i) 0, (ii) 2, (iii) 4 and (iv) 8 equiv. of (Bu$_4$N)(Δ-trisphat). Solvent: 5% CD$_3$NO$_2$ in CDCl$_3$.

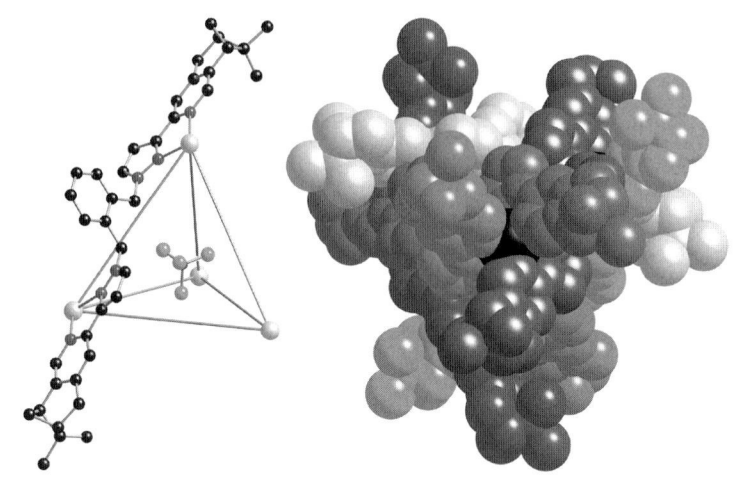

Figure 9.5 Two views of the optically pure cage complex [Zn$_4$(L$^{o\text{-}Ph^*}$)$_6$(ClO$_4$)](ClO$_4$)$_7$: left, a view showing one ligand and the encapsulated anion; right, a space-filling picture viewed down a threefold rotation axis, showing the packing of the ligands.

substituents makes the ligand chiral, such that the two different forms of the cage – based on different tris-chelate configurations of the metal centers – would be diastereoisomers. The resulting cage complex [Zn$_4$(L$^{o\text{-}Ph^*}$)$_6$(ClO$_4$)](ClO$_4$)$_7$ exists as a single diastereoisomer in solution, according to its ^{1}H NMR spectrum, and also crystallizes as a single diastereoisomer in the acentric space group C_2 (Figure 9.5) [27]. The specific molar rotation of this using 589-nm light is 30 times higher than the free ligand (13 400° in contrast to 432°). Given that the cage contains six ligands, it follows that there is an additional fivefold increase in the specific molar rotation arising from the fact that the chirality of the pinene groups on the ligands has dictated the chirality of the cage superstructure. Thus a set of six ligands undergoes a 500% amplification of specific molar rotation when the cage assembles and the ligands each adopt a helical twist; the magnitude of the molar rotation is comparable to those of compounds such as helicenes and a resolved trefoil knot.

The aromatic stacking between ligands around the periphery of the cage has significant consequences for the luminescence behavior of the Zn(II) cage based on L$^{12\text{-}naph}$. This ligand shows the characteristic fluorescence of the naphthyl group with an emission peak in the UV/blue region. In the cage [Zn$_4$(L$^{12\text{-}naph}$)$_6$][BF$_4$]$_8$, however, the participation of the naphthyl group in aromatic stacking interactions with adjacent pyrazolylpyridine groups on either side of it lowers the energy of the emissive state in a manner similar to that seen in excimers (when two units of the same type stack together) to give an excited state that is delocalized over both. This results in red shifted, "excimer-like" luminescence from the naphthyl groups at about 440 nm in the cage compared to the free ligand L$^{12\text{-}naph}$ (Figure 9.6) [28]. The appearance of this red-shifted luminescence can be used as a probe to monitor cage formation in solution by the anion templation affect; titration of NaBF$_4$

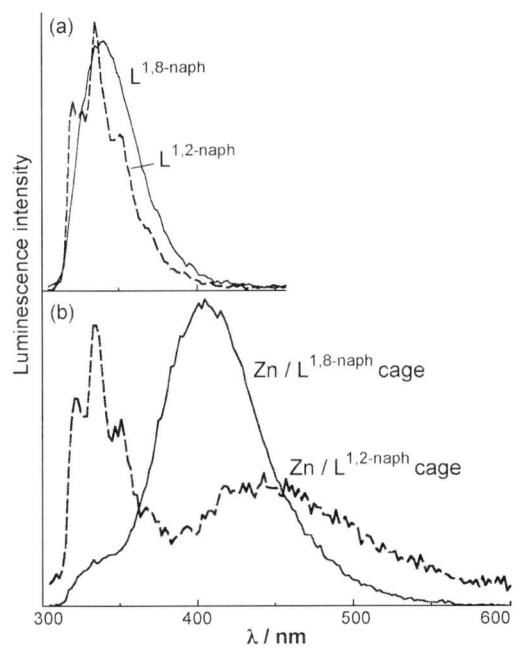

Figure 9.6 Naphthyl-based fluorescence of (a) free ligands L$^{12\text{-naph}}$ and L$^{18\text{-naph}}$ and (b) their tetranuclear and dodecanuclear (respectively) Zn(II) cages in which π-stacking of the naphthyl groups with other aromatic units results in red-shifted fluorescence.

into a mixture of Zn(II) acetate and L$^{12\text{-naph}}$ in solution results in a steady decrease in the intensity of the fluorescence associated with free L$^{12\text{-naph}}$ and the grow-in of the red-shifted fluorescence associated with aromatic stacking in the cage.

The highly intertwined structure of these cages results in remarkable kinetic stability. High-spin Co(II) centers are kinetically labile, as a simple demonstration shows [29]. [Co(bipy)$_3$]$^{2+}$ and [Co(Me$_2$bipy)$_3$]$^{2+}$ have, in their paramagnetically shifted ^{1}H NMR spectra, four and three aromatic signals, respectively, between 10 and 90 ppm. The signals for the aromatic protons that are in common between the two compounds [H(3), H(5), H(6)] have similar chemical shifts. When the two compounds are mixed in a 1:1 ratio, the resulting ^{1}H NMR spectrum – recorded as fast as possible after mixing, i.e. within about 2 min – is not a simple superposition of the two components, but shows that complete equilibration of the ligands between the metals has occurred with a statistical 1:3:3:1 mixture of [CoA$_3$]$^{2+}$, [CoA$_2$B]$^{2+}$, [CoAB$_2$]$^{2+}$ and [CoB$_3$]$^{2+}$. This is clearly shown by the presence of each of the H(3), H(5) and H(6) protons in eight environments with equal likelihood (one environment in each of the homoleptic complexes and three environments in each of the mixed-ligand complexes) (Figure 9.7). Thus in a mixture of kinetically labile [Co(bipy)$_3$]$^{2+}$-type complexes, ligand exchange is complete on a time-scale of a few minutes.

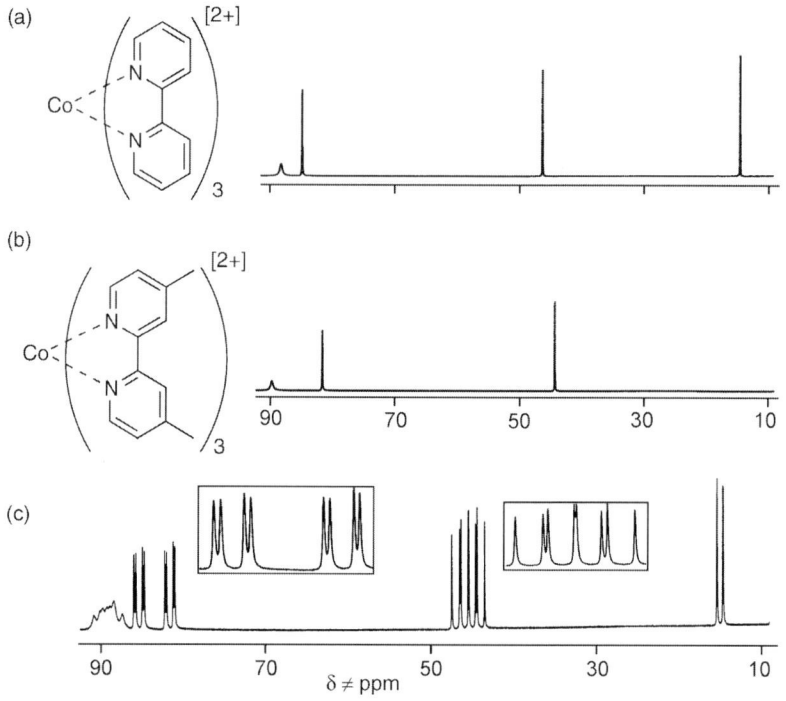

Figure 9.7 ^{1}H NMR spectra between 10 and 90 ppm of (a) [Co(bipy)$_3$]$^{2+}$, (b) [Co(Me$_2$bipy)$_3$]$^{2+}$ and (c) a 1:1 mixture of the above 2 min after mixing, showing a statistical equilibration of ligands between the metal ions.

In contrast, a 1:1 mixture of [Co$_4$(L$^{12\text{-naph}}$)$_6$(BF$_4$)][BF$_4$]$_7$ and [Co$_4$(L$^{o\text{-Ph}}$)$_6$(BF$_4$)][BF$_4$]$_7$, which have essentially identical structures, takes several *months* to reach equilibrium. The spectra of a 1:1 mixture of the complexes in the 10–90 ppm region are shown in Figure 9.8. The presence of two protons for the phenyl spacer versus three for the naphthyl spacer at around 20 ppm results in five signals in this region from simple superposition [spectrum (a)]; the remaining signals, arising from the pyrazolylpyridine groups and the methylene spacers, are almost identical between the two. If we abbreviate these complexes as Co$_4$A$_6$ and Co$_4$B$_6$ then at statistical equilibrium following ligand scrambling there should be seven species present: Co$_4$A$_6$, Co$_4$A$_5$B, Co$_4$A$_4$B$_2$, Co$_4$A$_3$B$_3$, Co$_4$A$_2$B$_4$, Co$_4$AB$_5$ and Co$_4$B$_6$. The simple 1:6:15:20:15:6:1 binomial distribution will be complicated by the fact that there could be two isomers for Co$_4$A$_4$B$_2$ and Co$_4$A$_2$B$_4$, according to whether the pair of ligands of the same type share a vertex or lie along opposed edges of the tetrahedron, and three isomers for Co$_4$A$_3$B$_3$. If all possible isomers exist at equilibrium then the mixture could contain up to 40 different environments for each type of proton.

After mixing, the ^{1}H NMR spectrum is just the sum of the two complexes, with no ligand scrambling evident. After several days, small additional peaks start to appear; after a couple of weeks they are fairly significant; after 3 months the spectrum stopped

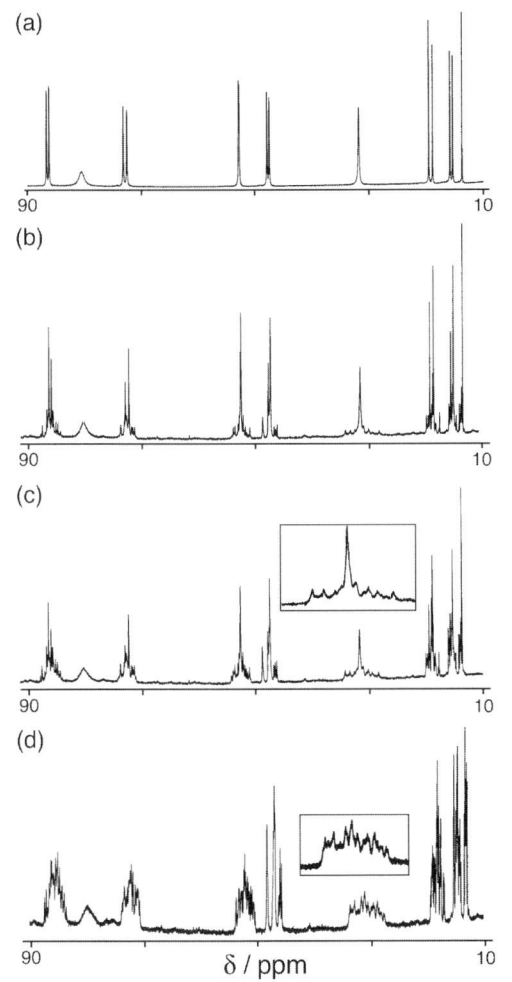

Figure 9.8 ^{1}H NMR spectra in CD$_3$CN between 10 and 90 ppm of a 1:1 mixture of [Co$_4$(L$^{12\text{-naph}}$)$_6$(BF$_4$)][BF$_4$]$_7$ and [Co$_4$(L$^{o\text{-Ph}}$)$_6$(BF$_4$)][BF$_4$]$_7$, at (a) 0, (b) 12, (c) 25 and (d) 79 days after mixing.

changing and it is clear that each type of proton now exists in a large number of different environments (Figure 9.8). Compared with the mononuclear [Co(bipy)$_3$]$^{2+}$ derivatives, it is obvious that formation of the tetrahedral cage assembly results in substantial kinetic stability, which will arise from several factors such as inter-ligand stacking interactions, hydrogen-bonding interactions with the central anion and the fact that the ligands are tetradentate rather than bidentate. All of these factors will inhibit dissociatively activated ligand exchange and the cumulative effect is substantial [29]. Raymond and coworkers have likewise noted remarkable kinetic inertness for some of their cages based on nominally labile metal centers [30].

Finally in this section, it should be pointed out that the formation of these cages between L$^{o\text{-Ph}}$ or L$^{12\text{-naph}}$ and Co(II) or Zn(II) is sensitive to the ionic radius of the metal ion. Both Co(II) and Zn(II) have essentially the same ionic radius in octahedral coordination (89 pm); Ni(II), however, is significantly smaller (ionic radius in octahedral coordination, 83 pm) [31] and this prevents formation of the tightly packed cage complexes. Instead, reaction of L$^{o\text{-Ph}}$ or L$^{12\text{-naph}}$ with Ni(BF$_4$)$_2$ affords [Ni$_2$L$_3$][BF$_4$]$_4$. In these complexes, the necessary 2M:3L ratio is maintained, but in a simpler structure; each Ni(II) ion has one ligand (either L$^{o\text{-Ph}}$ or L$^{12\text{-naph}}$) acting as a tetradentate chelate occupying four of the six sites, with the remaining ligand acting as a bridge, donating one bidentate site to each Ni(II) ion. Even though the tetrafluoroborate anion is present the tetrahedral cages do not form, presumably because they would be too sterically crowded.

9.3.2
Larger Tetrahedral Cages Based on L^{biph}

We next used a biphenyl group as spacer, to make L^{biph}, with the intention of making similar tetrahedral cages but with larger cavities which might accommodate larger guest anions. Reaction of L^{biph} with a range of Co(II) salts afforded [Co$_4$(L^{biph})$_6$X]X$_7$ with a range of anions (X$^-$ = iodide, ClO$_4^-$, BF$_4^-$, PF$_6^-$), all of which were structurally characterized (Figure 9.9) [32]. Although these cages have the same basic topology as the smaller ones described in the previous section, there are important differences.

These cages no longer have *T* symmetry, because one vertex (nominally the apical one) has a *fac* tris-chelate configuration, whereas the three in the basal plane have a *mer* tris-chelate configuration. Hence there is only one (non-crystallographic) C_3 axis, through the apex. Accordingly, one-third of the complex is unique, with two independent ligand environments (apex-to-base and along the edges of the base), such that there are 44 inequivalent protons in the NMR spectrum. Not all of these are resolved but the paramagnetic shift effect of high-spin Co(II) spreads out the signals enough to make it clear that there are about this number of separate signals.

The longer ligands compared with L$^{o\text{-Ph}}$ or L$^{12\text{-naph}}$ result in a larger central cavity which accommodates equally well a range of anions of different sizes, the largest of which we have characterized to date is hexafluorophosphate. None of the anions used is a good match for the central cavity – all are too small to fill it effectively – which implies that a templating effect is unlikely to be operative. In addition, the anions are no longer completely encapsulated as there are windows in the centers of the triangular faces. In consequence, the internal anions are in fast exchange with the external ones at room temperature, with single signals appearing in the ^{19}F NMR spectra for both fluoroborate and hexafluorophosphate complexes. However, cooling results in the exchange becoming frozen out, with separate signals for the internal and external anions becoming apparent (Figure 9.9). From the linewidths of the ^{19}F NMR signals at different temperatures, we could estimate that the ΔG of activation for anion exchange is about 50 kJ mol^{-1} in each case [32]. This value suggests that the exchange mechanism involves diffusion of the anions through the windows of the intact cage; if the mechanism involved dissociation of a bidentate chelating group,

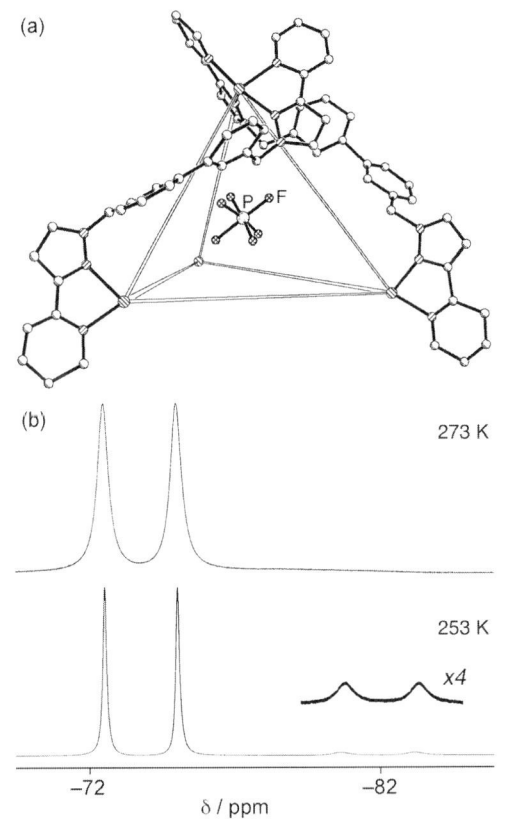

Figure 9.9 (a) Crystal structure of the complex cation of [Co$_4$(L^{biph})$_6$(PF$_6$)](PF$_6$)$_7$; (b) ^{19}F NMR spectra at different temperatures showing the "freezing out" of internal/external anion exchange (the doublets arise from coupling to ^{31}P of the hexafluorophosphate).

opening up the cage, the activation ΔG value would be higher as two Co–N bonds would have to break.

9.3.3
Higher Nuclearity Cages Based on Other Ligands

Additional simple changes to the ligands, by using different aromatic spacers between the pyrazolylpyridine arms, have afforded a series of unexpected high-nuclearity cages, all based on the simple 2:3 metal:ligand ratio, reflected in formation of polyhedra in which there is a 2:3 ratio of vertices (metal ions) to edges (bridging ligands).

Reaction of L$^{m\text{-Ph}}$ or L$^{m\text{-Py}}$ with Co(II) or Zn(II) salts affords molecular cubes [M$_8$L$_{12}$]$^{16+}$, with a metal ion at each vertex and a bridging ligand spanning each edge (Figure 9.10) [33,34]. In both cases the ligand coordinates in a bis-bidentate bridging

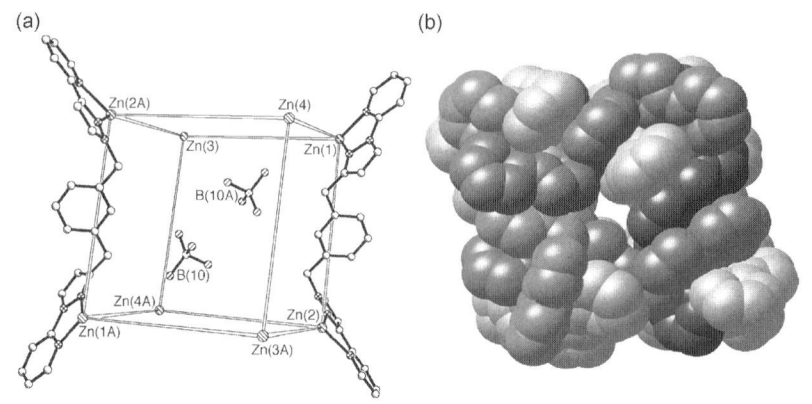

Figure 9.10 Crystal structures of the complex cations of (a) $[Zn_8(L^{m-Ph})_{12}][BF_4]_{16}$ (a view emphasizing the cubic array of metal ions and the encapsulated anions) and (b) $[Zn_8(L^{m-Py})_{12}](ClO_4)_{16}$ (a space-filling view).

manner; the central pyridyl unit of L^{m-Py} does not participate in coordination [33], such that the pyridine-2,6-diyl spacer behaves just like the m-phenylene spacer of L^{m-Ph} [34]. The cubes are slightly slanted, with angles at the corners in the range ca. 76–103°. The central cavity contains either one perchlorate anion, in $[Zn_8(L^{m-Py})_{12}](ClO_4)_{16}$, or two tetrafluoroborate anions, in $[Zn_8(L^{m-Ph})_{12}][BF_4]_{16}$; the windows in the centers of the faces, which are obvious in a space-filling representation of the structure, permit rapid exchange of internal and external anions as shown by ^{19}F NMR spectra of $[Zn_8(L^{m-Ph})_{12}][BF_4]_{16}$, for which a single signal occurs even at low temperatures.

The symmetry of these "cubic" cages is interesting. Unlike the tetrahedral cages, the tris-chelate metal centers in these cubes do not all have the same optical configuration, with a crystallographic inversion center lying at the center of the cube in each case such that the assemblies are achiral. There is a C_3 axis in each case lying along the long diagonal, with the two Zn(II) centers on this axis having a *fac* tris-chelate coordination and the others all having a *mer* geometry. The combination of a C_3 axis and an inversion center means that these cages actually have (non-crystallographic) S_6 symmetry. Extensive aromatic π-stacking between parallel, overlapping sections of ligands around the periphery of the complex is clear.

An additional and unexpected product which was isolated during the preparation of some of these cubic assemblies with L^{m-Ph} is an "open-book" structure, exemplified by $[Co_6(L^{m-Ph})_9][ClO_4]_{12}$, which achieves the necessary 2:3 metal:ligand ratio in a different way (Figure 9.11) [34]. The assembly contains nine ligands associated with six octahedral metal ions; there are two L^{m-Ph} ligands spanning each of the terminal pairs of Co(II) ions (the opposed open edges of the book – red and blue ligands in the figure) in a double helical arrangement, with all remaining Co—Co vectors (from each corner of the book to the spine and along the spine) having one bridging ligand. The two double helical sections are homochiral as they are related by a C_2 rotation

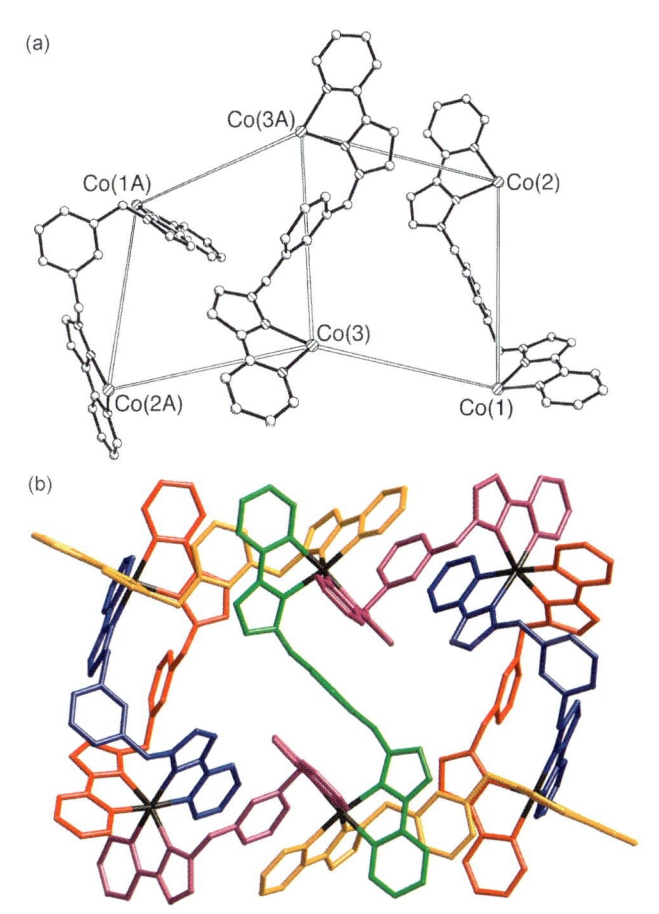

9.3 Complexes Based on Neutral Ligands with Aromatic Spacers

Figure 9.11 Two views of the crystal structure of the complex cation $[Co_6(L^{m\text{-}Ph})_9][ClO_4]_{12}$: (a) a view emphasizing the "open-book" arrangement of metal centers; (b) a view showing all of the ligands.

through the center of the complex. Although this complex is not in itself of great significance from the point of view of our investigations into polyhedral cages, it does illustrate how a single combination of a metal salt and a ligand can follow two different self-assembly pathways to give a mixture of different products which nevertheless obey the same basic stoichiometric principle of having a 2M:3L ratio.

Reaction of $L^{1,8\text{-}naph}$ with a range of M(II) ions [Cd(II), Co(II), Cu(II), Zn(II)] and either BF_4^- or ClO_4^- as counter-ion, affords $[M_{12}(L^{1,8\text{-}naph})_{18}]^{24+}$ cages having the core structure of a truncated tetrahedron (Figure 9.12) [35,36]. Slicing off the vertices of a notional tetrahedron generates four new triangular faces (shaded in the figure); the original triangular faces become hexagons when the vertices are removed. The truncated tetrahedral structure is the simplest of the series of Archimedean solids and conveniently provides 18 edges to go with the 12 vertices, in keeping with the

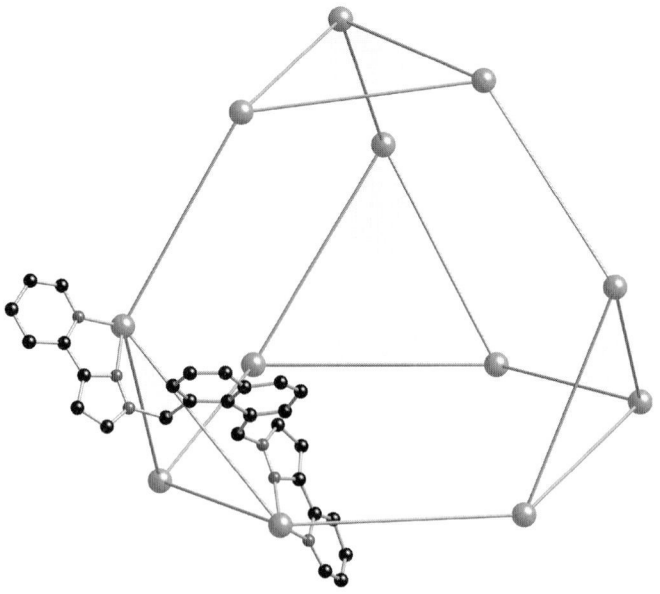

Figure 9.12 A view of the polyhedral metal core of [Cu$_{12}$(L$^{1,8\text{-naph}}$)$_{18}$](ClO$_4$)$_{24}$; each edge of the polyhedron is spanned by a bridging ligand, of which one example is included. The four faces shaded gray are those derived from truncating the apices of a notional tetrahedron.

stoichiometric requirements; there is an M(II) ion at each vertex and a bridging ligand spanning each edge, connecting a pair of metal ions. Around each of the triangular and hexagonal faces the array of bridging ligands forms a cyclic helical structure (Figure 9.13).

It is not obvious why such a complex structure should form when there must be so many simpler alternatives with the same net number of metal–ligand interactions but higher entropy. One possibility is the stabilization afforded by extensive aromatic stacking between ligands around the periphery. Stacks of seven components – an alternating sequence of naphthyl and pyrazolylpyridine units in a seven-layer sandwich – occur, with six such stacks arranged in a roughly cubic array around the outside of the complex. As with the smaller tetranuclear cage based on L$^{1,2\text{-naph}}$, this involvement of the naphthyl groups in extensive aromatic

Figure 9.13 Three views of the structure of [Cd$_{12}$(L^{naph})$_{18}$](BF$_4$)$_{24}$: (a) a view of the polyhedral metal cage and the four encapsulated [BF$_4$]$^-$ anions; (b) a view down one of the triangular faces, emphasizing the cyclic helical array of ligands around the face and the presence of an anion in the center of the face; (c) a view down one of the Cd$_6$ pseudo-hexagonal faces, emphasizing the cyclic helical array of ligands around the face and the presence of an anion in the center of the face.

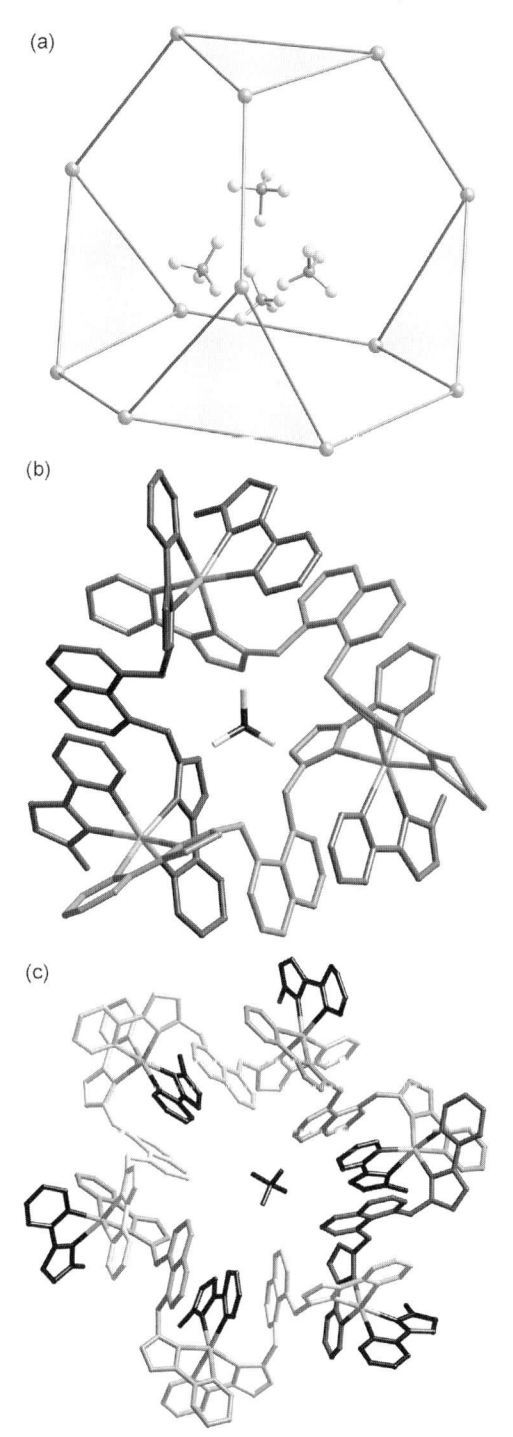

9.3 Complexes Based on Neutral Ligands with Aromatic Spacers | **239**

stacking stabilizes the fluorescent excited state of the ligands and results in strongly red-shifted luminescence from $[Zn_{12}(L^{1,8\text{-naph}})_{18}][BF_4]_{24}$ compared with free $L^{1,8\text{-naph}}$ (Figure 9.6) [28].

Two other features of these cages are noteworthy. First, the central cavity is now large enough to accommodate four BF_4^- or ClO_4^- anions. These are themselves disposed in a roughly tetrahedral array and are fairly close together with their peripheral atoms separated by about the sum of the van der Waals radii. Apparently the unfavorable anion–anion interactions between these four guest anions are more than offset by the electrostatic advantage of accommodating four anions in a cage superstructure with a charge of 24^+. There are also anions associated with the windows in the centers of the triangular and hexagonal faces (Figure 9.13); since there are eight faces, this makes a total of 12 anions closely associated with the 24^+ cage. Second, all 12 metal tris-chelate centers have the same optical configuration, which is necessary for the roughly spherical surface to achieve closure. Altering the configuration at any metal site would result in one of the ligands extending into space away from the core and unable to bridge to a second metal ion. Thus, 72 metal–ligand bonds have to form with the correct optical configuration during the cage assembly. The bulk material of course is racemic. The view on to each type of face (triangular or hexagonal) is that of a cyclic helicate; hence, looking down on a hexagonal face one sees that the six bridging ligands associated with that face have the "over and under" sequence characteristic of a cyclic helicate and the same is true of the set of three ligands associated with each triangular face.

The largest homoleptic cage that we have yet isolated was provided by reaction of $L^{p\text{-Ph}}$ with M(II) salts to give the hexadecanuclear cages $[M_{16}(L^{p\text{-Ph}})_{24}]X_{32}$ (M = Zn, X = BF_4; M = Cd, X = ClO_4) whose polyhedral core of metal ions may be approximately described as a tetra-capped truncated tetrahedron [37]. The structure of $[Cd_{16}(L^{p\text{-Ph}})_{24}](ClO_4)_{32}$ is shown as an example (Figures 9.14 and 9.15). Each apex of a tetrahedron is sliced off to reveal a triangular face; the resulting truncated tetrahedron has 12 vertices, with four triangular faces and four hexagonal faces, as described above. The four triangular faces are then twisted in the same sense, such that the mirror planes through the truncated tetrahedron are removed but the C_3 axes are retained. Finally, a capping atom is added to the center of each of the original four faces. This M_{16} polyhedral array, with (non-crystallographic) T symmetry, has a bridging ligand $L^{p\text{-Ph}}$ along each of the 24 edges, providing the necessary 2:3 M:$L^{p\text{-Ph}}$ ratio. The large central cavity contains eight $[ClO_4]^-$ anions and six MeCN molecules. In contrast to the behavior observed with much smaller cages, this is a very "open" structure with the anions clearly not completely encapsulated by the cage superstructure.

The Cd(II) centers display a mix of facial and meridional tris-chelate geometries; the 12 Cd(II) centers associated with the four triangular faces of the truncated tetrahedron have a meridional arrangement and the four "capping" metal centers are facial. Remarkably, and as we saw with the dodecanuclear truncated-tetrahedral complexes, all 16 metal centers have the same optical configuration, which appears to be essential for the closed cage to form; thus the assembly has occurred with correct

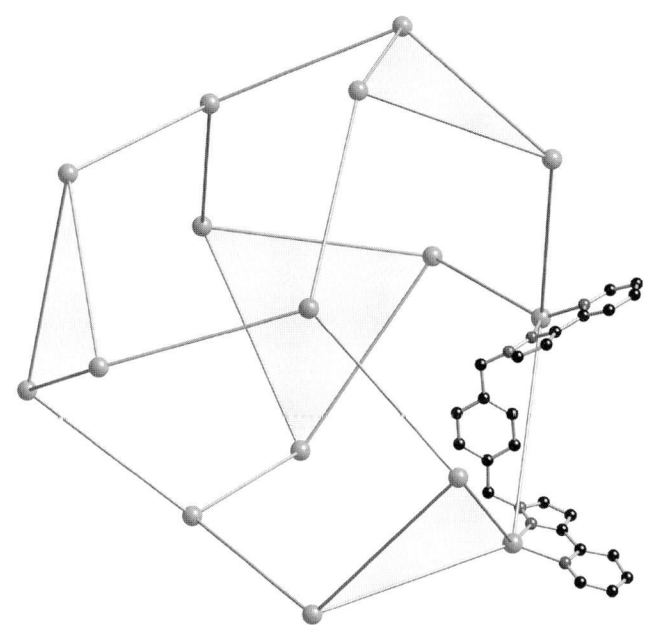

Figure 9.14 A view of the polyhedral metal core of [Cd$_{16}$(μ-L$^{p\text{-}Ph}$)$_{24}$](ClO$_4$)$_{32}$; each edge of the polyhedron is spanned by a bridging ligand, of which one example is included. The four faces shaded gray are those notionally derived from truncating the apices the parent tetrahedron.

control of 96 metal–ligand bonds. The crystal is racemic, containing equal numbers of opposite enantiomers of the cage.

So far, the polyhedral cages that we have observed to date based on these bis-bidentate pyrazolylpyridine ligands (M$_4$L$_6$ tetrahedron, Figures 9.3 and 9.9; M$_8$L$_{12}$ cube, Figure 9.10; M$_{12}$L$_{18}$ truncated tetrahedron, Figure 9.12; M$_{16}$L$_{24}$ tetra-capped truncated tetrahedron, Figure 9.14) all necessarily contain a 2M:3L ratio as a result of combining octahedral metal ions (vertices) with bis-bidentate bridging ligands (edges). Clearly there could be an infinite number of possible larger structures which obey the same stoichiometric principle, although there must come a point at which the entropic cost becomes prohibitive. The prevalence of T-symmetry structures (observed for the M$_4$L$_6$, M$_{12}$L$_{18}$ and M$_{16}$L$_{24}$ cages) is interesting; Cotton et al. have pointed out that T-symmetric species may be derived in a wide variety of ways by "downgrading" assemblies with tetrahedral, octahedral or icosahedral symmetry by removal of mirror planes [38].

It is also important to emphasize the fact that crystals can grow under kinetic control and the structures may not represent thermodynamic minima in solution; there are many well-established examples in the field of self-assembly of transition metal complexes where a mixture of interconverting species in solution (a "dynamic combinatorial library") generates a single (kinetic) product on crystallization. We know for the smaller tetrahedral cages, from a combination

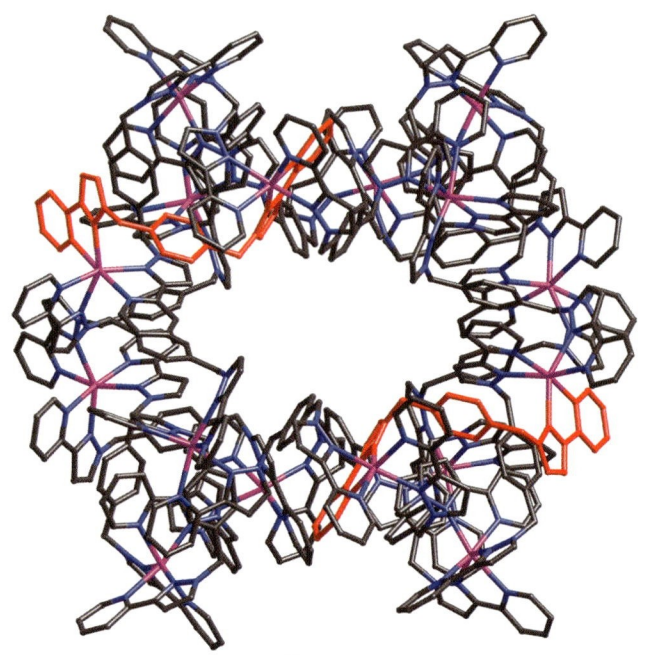

Figure 9.15 A view of $[Cd_{16}(\mu\text{-}L^{p\text{-}Ph})_{24}](ClO_4)_{32}$ showing all atoms in the cage, with two of the bridging ligands colored red (Cd, purple; N, blue).

of NMR and mass spectrometric studies, that the cage structures are retained in solution [23–25]. For the larger cages, it becomes even more important to use solution methods to confirm the existence of the cages in solution and electrospray mass spectrometry is an ideal tool (a ^{1}H NMR spectrum will not distinguish effectively between different metal–ligand assemblies, especially those based on diamagnetic metal cations where all of the signals of interest associated with the ligands come in a narrow range). Examples of an electrospray mass spectrum of one of the 12-nuclear cage complexes is shown in Figure 9.16; such spectra provide clear proof that the cage structures are retained *to some extent* in solution. The presence of smaller and simpler assemblies cannot be ruled out, however: a peak corresponding to the presence of a M_2L_3 assembly, for example, may arise from fragmentation of a larger cage – or it may genuinely correspond to the presence of a simple dinuclear complex as part of an equilibrium mixture. The presence of alternative metal–ligand assemblies has been unequivocally established in one case by isolation and structural characterization of both $[M_8(L^{m\text{-}Ph})_{12}]^{16+}$ and $[M_6(L^{m\text{-}Ph})_9]^{12+}$ complexes from the same metal–ligand combination [34]. The possibility to exploit these polyhedral cages as hosts for, e.g., size- or shape-selective anion binding will require kinetically inert cages which do not dissociate in solution. So far this behavior has only been established unequivocally for the series $[M_4L_6X]X_7$ (M = Co, Zn; L = $L^{o\text{-}Ph}$, $L^{12\text{-}naph}$; X = BF_4, ClO_4), in which

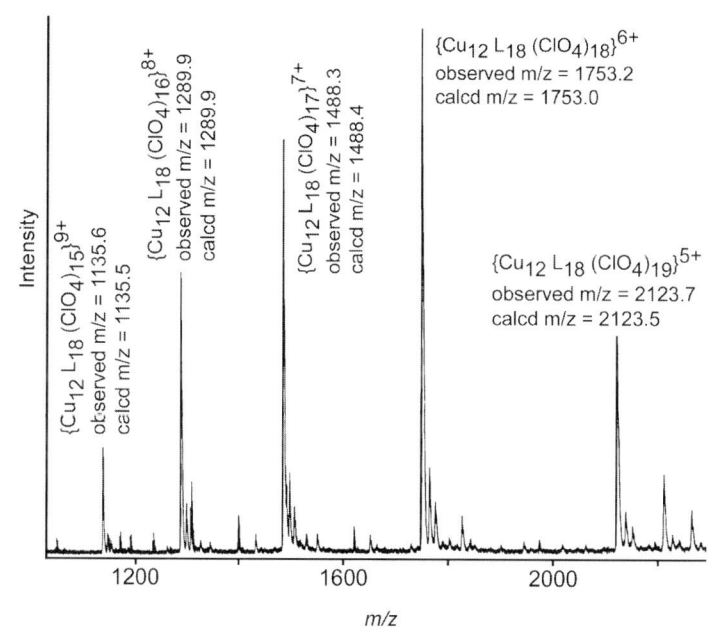

Figure 9.16 Part of the electrospray mass spectrum of $[Cu_{12}(L^{1,8-naph})_{18}](ClO_4)_{24}$ showing the intact cage with loss of 5, 6, 7, 8 and 9 perchlorate anions.

Peak labels:
- $\{Cu_{12}L_{18}(ClO_4)_{15}\}^{9+}$ observed m/z = 1135.6, calcd m/z = 1135.5
- $\{Cu_{12}L_{18}(ClO_4)_{16}\}^{8+}$ observed m/z = 1289.9, calcd m/z = 1289.9
- $\{Cu_{12}L_{18}(ClO_4)_{17}\}^{7+}$ observed m/z = 1488.3, calcd m/z = 1488.4
- $\{Cu_{12}L_{18}(ClO_4)_{18}\}^{6+}$ observed m/z = 1753.2, calcd m/z = 1753.0
- $\{Cu_{12}L_{18}(ClO_4)_{19}\}^{5+}$ observed m/z = 2123.7, calcd m/z = 2123.5

the anion exerts a templating effect and remains tightly bound in the central cavity even at elevated temperatures according to NMR spectra [23–25].

9.4
Mixed-ligand Complexes: Opportunities for New Structural Types

All of the above examples – and, indeed, most polynuclear coordination complexes formed by self-assembly methods – contain a single type of ligand. Using a mixture of ligands L^A and L^B in a reaction with a labile metal ion introduces a much higher degree of complexity to the problem. In addition to the possibilities that could arise from reaction of a metal ion with one ligand on its own, which (as we have seen above) are extensive and unpredictable, there is the additional possibility of mixed-ligand complexes occurring if there are any favorable interactions between L^A and L^B which make it likely that they will occur together in the same assembly. Notable examples of reactions in which a metal ion reacts with a specific combination of different ligands to generate a single mixed-ligand product are provided by (i) Lehn's cylindrical stacks which combine both linear bridging ligands along the edges and triangular tritopic bridging ligands in the core [39] and (ii) mixed-ligand rectangular grid complexes which form in preference to the homoleptic grids [40].

A simple illustration of the issues is provided by the bis-bidentate bridging ligands L^1 and L^2 which contain two N,O-chelating pyrazolylphenolate units (Scheme 9.3).

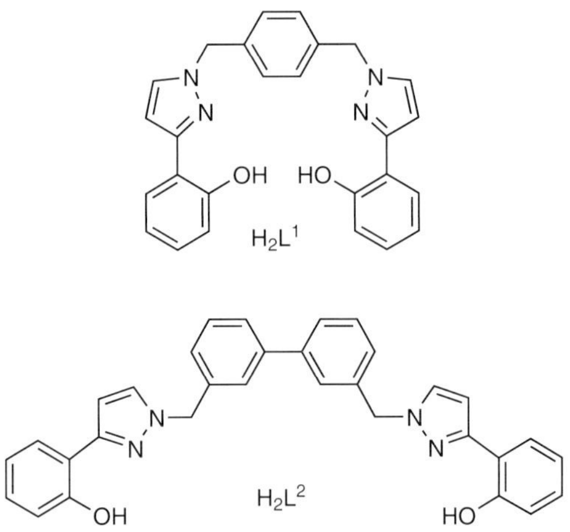

Scheme 9.3

These ligands both form simple dinuclear double helicates M_2L_2 with M = Cu and Zn in which each metal ion is four coordinate [41]. When $M_2(L^1)_2$ and $M_2(L^2)_2$ are mixed in a 1:1 ratio in solution there are three possible outcomes. First, the homoleptic complexes could remain as they are with no significant amounts of mixed-ligand complexes forming; this might happen if there were any unfavorable interactions between L^1 and L^2. If there were no significant interactions between L^1 and L^2 then we might expect a statistical mixture of $M_2(L^1)_2$, $M_2(L^1)(L^2)$ and $M_2(L^2)_2$ in a 1:2:1 ratio. If there were an attractive interaction between L^1 and L^2 then the mixed-ligand system would dominate and there would be relatively little of the homoleptic complexes. In these cases, with both Cu(II) and Zn(II), the mixed-ligand complexes $M_2(L^1)(L^2)$ dominate according to mass spectrometry and could be preferentially crystallized (Figure 9.17) [42]. The Cu(II) complex is a conventional double helicate, but contains in the solid state inter-ligand aromatic stacking interactions with the phenyl spacer of one ligand sandwiched between the two pyrazolylphenolate termini of the other. Since no such inter-ligand interactions were present in either of the homoleptic complexes $Cu_2(L^1)_2$ and $Cu_2(L^2)_2$ and there is no obvious electronic reason why two equivalents of $Cu_2(L^1)(L^2)$ should be preferred over a mixture of the homoleptic complexes, we suggest that this stacking plays a significant role in stabilizing the mixed-ligand helicate. In contrast the mixed-ligand Zn(II) complex $Zn_2(L^1)(L^2)$ is a "mesocate" with the two ligands in a side-by-side arrangement and not helically intertwined, unlike the parent helicates $Zn_2(L^1)_2$ and $Zn_2(L^2)_2$. The reason for preferential formation of the mixed-ligand Zn(II) complex compared with the homoleptic complexes is not obvious.

Two examples of mixed-ligand complexes which combine tetradentate and hexadentate bridging ligands show how new structural types, which are not accessible from either ligand on its own, can be isolated. The bis-bidentate bridging ligands form polyhedral arrays with metal cations in which the ligands span the *edges* of the polyhedron; in contrast the tris-bidentate ligands coordinate to three

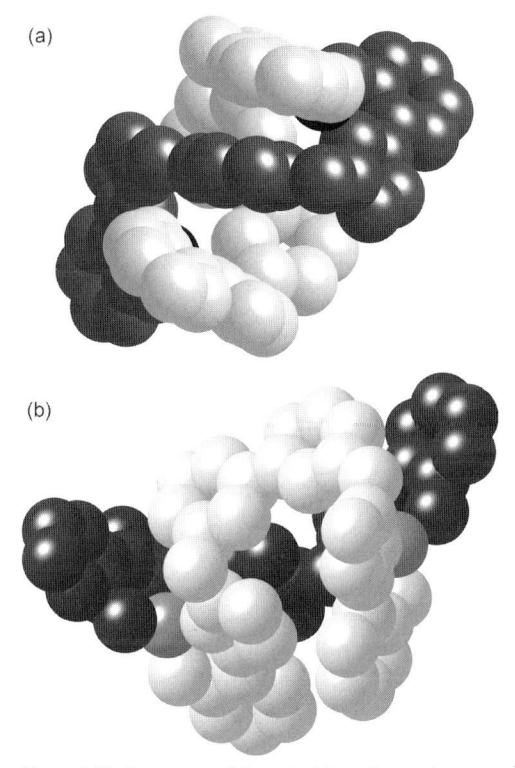

Figure 9.17 Structures of the mixed-ligand complexes M(L^1)(L^2): (a) a double helicate with M = Cu; (b) a mesocate with M = Zn with the ligands in a "side-by-side" arrangement rather than helically twisted around one another.

metal ions at a time, capping the *faces* of the polyhedron, as shown in [M$_4$(tppb)$_4$]$^{4+}$ [20,21]. A mixture of the bis- and tris-pyrazolylborate ligands [bppb]$^-$ and [tppb]$^-$ reacts with Mn(II) to form, in addition to homoleptic complexes, the mixed-ligand trinuclear complex [Mn$_3$(tppb)(bppb)$_3$]$^{2+}$, a triangular complex with one face-capping [tppb]$^-$ ligand and three edge-bridging [bppb]$^-$ ligands (Figure 9.18) [43]. Clearly, such a structure could not arise in a homoleptic complex based on these components.

A more spectacular example based on the same principles is provided by reaction of M(BF$_4$)$_2$ (M = Cu, Zn, Cd) with a mixture of L^{mes} (a three-armed triangular ligand with a mesityl core) and L$^{p\text{-Ph}}$. {Recall that use of L$^{p\text{-Ph}}$ alone afforded the hexadecanuclear cages [M$_{16}$(L$^{p\text{-Ph}}$)$_{24}$]$^{32+}$ with a 2:3 M:L ratio; we could not (yet) isolate a homoleptic complex with L^{mes} but it must have a 1:1 M:L ratio with an octahedral metal cation}. The crystalline products from these reactions are the mixed-ligand complexes [M$_{12}$(μ-L$^{p\text{-Ph}}$)$_{12}$(μ^3-L^{mes})$_4$](BF$_4$)$_{24}$ (Figures 9.19 and 9.20) which has a cuboctahedral metal framework containing eight triangular and six square faces [37]. The complex lies on a C_2 axis. Four of the eight

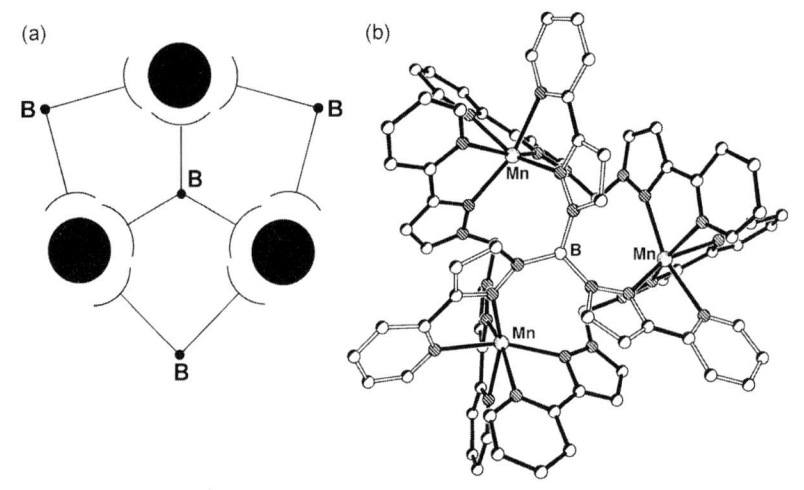

Figure 9.18 (a) A sketch showing how one face-capping (tris-bidentate) and three edge-bridging (bis-bidentate) ligands can afford a trinuclear valence-satisfied complex based on six-coordinate metal ions; (b) crystal structure of the complex cation [Mn$_3$(tppb)(bppb)$_3$]$^{2+}$ which conforms to this principle.

triangular faces are capped by a triply bridging ligand L^{mes} and the remaining vacant edges are spanned by twelve doubly bridging ligands L$^{p\text{-}Ph}$. Numerous counter-ions and solvent molecules occupy the open space in the center of the complex. It is interesting that the 1:3 ratio of face capping:edge bridging ligands in this complex is the same as was observed in the simpler complex [Mn$_3$(tppb)(bppb)$_3$]$^{2+}$.

All 12 tris-chelate metal centers have meridional geometry and again all have the same chirality, indicating that the same chiral configuration at each metal center is necessary for the closed cage to form. The crystal is racemic. In this case electrospray mass spectra on solutions of redissolved crystals show a clear sequence of peaks corresponding to the intact mixed-ligand cage with loss of increasing numbers of anions, but *no* peaks for the homoleptic cages which might also be expected. In other words, the mixed-ligand complex is the only product, even in solution, implying a cooperative interaction between the two types of ligand that renders formation of the mixed-ligand cage more favorable than formation of distinct homoleptic complexes with L$^{p\text{-}Ph}$ and L^{mes}.

With hindsight, one can see how the formation of this complex conforms to the simple requirement that the number of binding sites available from the ligand set must exactly match those required by the set of metal ions, and also that the spatial arrangement of ligand binding sites must match the stereoelectronic properties of the metal ions. This principle has been termed "avoidance of valence frustration" by Nitschke and coworkers [44]. Apart from that factor, which is important in all of the complexes described, the prediction that such an elaborate mixed-ligand assembly would form is clearly beyond our capabilities at the moment, particularly

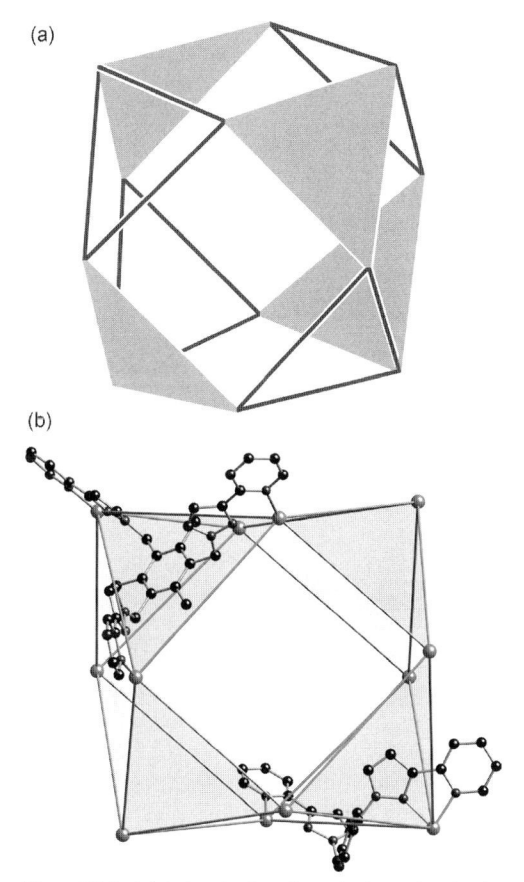

Figure 9.19 (a) A sketch of a cuboctahedron, showing how it can be delineated using four face capping ligands (shaded faces) and 12 edge-bridging ligands; (b) a view of the cuboctahedral metal core of $[Cu_{12}(\mu\text{-}L^{p\text{-}Ph})_{12}(\mu^3\text{-}L^{mes})_4](BF_4)_{24}$, which conforms to this principle (only one of each type of ligand is shown for clarity).

given the geometric flexibility of these ligands arising from the methylene spacers that were employed for synthetic convenience. Isolation of such beautiful cages must therefore – for the moment – rely to a large extent on serendipity. Having isolated them, however, the opportunities that they offer for studying properties such as host–guest chemistry and photophysical and chiroptical properties are extensive.

Acknowledgments

I am indebted to the many talented and patient research students and post-docs who have carried out this work; their names appear in the reference list. Financial support

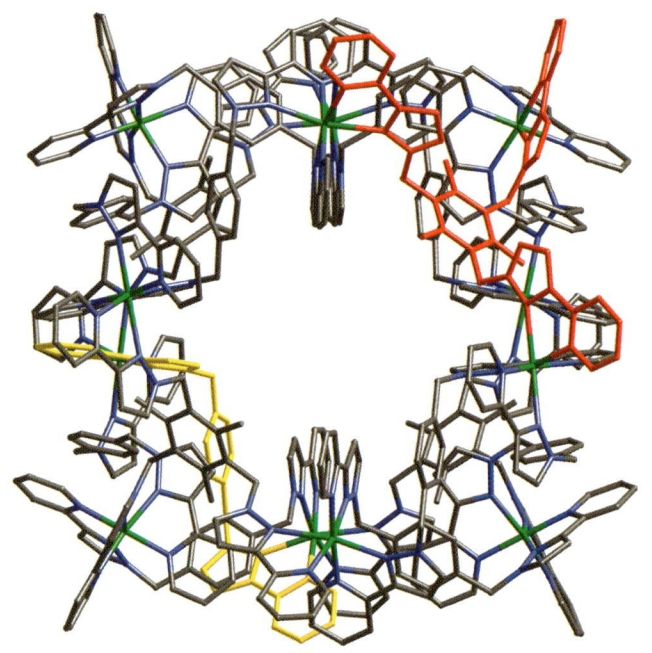

Figure 9.20 A view of $[Cu_{12}(\mu\text{-}L^{p\text{-}Ph})_{12}(\mu^3\text{-}L^{mes})_4](BF_4)_{24}$ showing all atoms in the cage, with one of the face-capping ligands of the bridging ligands colored red and one of the edge-bridging ligands colored yellow (Cu, green; N, blue).

from the EPSRC, the Leverhulme Trust and the Universities of Sheffield and Bristol is also gratefully acknowledged.

References

1 Lehn, J.-M., Rigault, A., Siegel, J., Harrowfield, J., Chevrier, B. and Moras, D. (1987) *Proc. Natl. Acad. Sci. USA*, **84**, 2565–2569.

2 Barley, M., Constable, E.C., Corr, S.A., McQueen, R.C.S., Nutkins, J.C., Ward, M.D. and Drew, M.G.B. (1988) *J. Chem. Soc., Dalton Trans.*, 2655–2662.

3 Youinou, M.-T., Rahmouni, N., Fischer, J. and Osborn, J.A. (1992) *Angew. Chem. Int. Ed. Engl.*, **31**, 733–735.

4 Baxter, P.N.W., Lehn, J.-M., Fischer, J. and Youinou, M.-T. (1994) *Angew. Chem. Int. Ed. Engl.*, **33**, 2284–2287.

5 Fiedler, D., Leung, D.H., Bergman, R.G. and Raymond, K.N. (2005) *Acc. Chem. Res.*, **38**, 349–358.

6 Fujita, M., Tominaga, M., Hori, A. and Therrien, B. (2005) *Acc. Chem. Res.*, **38**, 369–378.

7 Seidel, S.R. and Stang, P.J. (2002) *Acc. Chem. Res.*, **35**, 972–983.

8 Hasenknopf, B., Lehn, J.-M., Kneisel, B.O., Baum, G. and Fenske, D. (1996) *Angew. Chem. Int. Ed. Engl.*, **35**, 1838–1840.

9 Mamula, O., von Zelewsky, A. and Bernardinelli, G. (1996) *Angew. Chem. Int. Ed. Engl.*, **37**, 290–293.

10. Baxter, P.N.W., Lehn, J.-M., Baum, G. and Fenske, D. (1999) *Chem. Eur. J.*, **5**, 102–112.
11. Sauvage, J.-P. (1998) *Acc. Chem. Res.*, **31**, 611–619.
12. Fuller, A.M.L., Leigh, D.A., Lusby, P.J., Slawin, A.M.Z. and Walker, D.B. (2005) *J. Am. Chem. Soc.*, **127**, 12612–12619.
13. Dietrich-Buchecker, C., Colasson, B.X. and Sauvage, J.-P. (2005) *Top. Curr. Chem.*, **249**, 261–283.
14. Alvarez, S. (2005) *Dalton Trans.*, 2209–2233.
15. Rossmann, M.G., Arisaka, F., Battisti, A.J., Bowman, V.D., Chipman, P.R., Fokine, A., Hafenstein, S., Kanamaru, S., Kostyuchenko, V.A., Mesyanzhinov, V.V., Shneider, M.M., Morais, M.C., Leiman, P.G., Palermo, L.M., Parrish, C.R. and Xiao, C. (2007) *Acta Crystallogr., Sect. D*, **63**, 9–16.
16. Amoroso, A.J., Cargill Thompson, A.M.W., Jeffery, J.C., Jones, P.L., McCleverty, J.A. and Ward, M.D. (1994) *J. Chem. Soc., Chem. Commun.*, 2751–2752.
17. Bardwell, D.A., Jeffery, J.C., Jones, P.L., McCleverty, J.A., Psillakis, E., Reeves, Z. and Ward, M.D. (1997) *J. Chem. Soc., Dalton Trans.*, 2079–2086.
18. Armaroli, N., Accorsi, G., Barigelletti, F., Couchman, S.M., Fleming, J.S., Harden, N.C., Jeffery, J.C., Mann, K.L.V., McCleverty, J.A., Rees, L.H., Starling, S.R. and Ward, M.D. (1999) *Inorg. Chem.*, **38**, 5769–5776.
19. Jones, P.L., Byrom, K.J., Jeffery, J.C., McCleverty, J.A. and Ward, M.D. (1997) *Chem. Commun.*, 1361–1362.
20. Paul, R.L., Amoroso, A.J., Jones, P.L., Couchman, S.M., Reeves, Z.R., Rees, L.H., Jeffery, J.C., McCleverty, J.A. and Ward, M.D. (1999) *J. Chem. Soc., Dalton Trans.*, 1563–1568.
21. Amoroso, A.J., Jeffery, J.C., Jones, P.L., McCleverty, J.A., Thornton, P. and Ward, M.D. (1995) *Angew. Chem. Int. Ed. Engl.*, **34**, 1443–1446.
22. Fleming, J.S., Mann, K.L.V., Couchman, S.M., Jeffery, J.C., McCleverty, J.A. and Ward, M.D. (1998) *J. Chem. Soc., Dalton Trans.*, 2047–2052.
23. Fleming, J.S., Mann, K.L.V., Carraz, C.-A., Psillakis, E., Jeffery, J.C., McCleverty, J.A. and Ward, M.D. (1998) *Angew. Chem. Int. Ed.*, **37**, 1279–1281.
24. Paul, R.L., Bell, Z.R., Jeffery, J.C., McCleverty, J.A. and Ward, M.D. (2002) *Proc. Natl. Acad. Sci. USA*, **99**, 4883–4888.
25. Paul, R.L., Bell, Z.R., Jeffery, J.C., Harding, L.P., McCleverty, J.A. and Ward, M.D. (2003) *Polyhedron*, **22**, 781–787.
26. Frantz, R., Grange, C.S., Al-Rasbi, N.K., Ward, M.D. and Lacour, J. (2007) *Chem. Commun.*, 1459.
27. Argent, S.P., Riis-Johannessen, T., Jeffery, J.C., Harding, L.P. and Ward, M.D. (2005) *Chem. Commun.*, 4647–4649.
28. Al-Rasbi, N.K., Sabatini, C., Barigelletti, F. and Ward, M.D. (2006) *Dalton Trans.*, 4769–4772.
29. Bell, Z.R. and Ward, M.D. unpublished results.
30. Terpin, A.J., Ziegler, M., Johnson, D.W. and Raymond, K.N. (2001) *Angew. Chem. Int. Ed.*, **40**, 157.
31. http://www.webelements.com.
32. Paul, R.L., Argent, S.P., Jeffery, J.C., Harding, L.P., Lynam, J.M. and Ward, M.D. (2004) *Dalton Trans.*, 3453–3458.
33. Bell, Z.R., Harding, L.P. and Ward, M.D. (2003) *Chem. Commun.*, 2432–2433.
34. Argent, S.P., Adams, H., Harding, L.P. and Ward, M.D. (2006) *Dalton Trans.*, 542–544.
35. Bell, Z.R., Jeffery, J.C., McCleverty, J.A. and Ward, M.D. (2002) *Angew. Chem. Int. Ed.*, **41**, 2515–2518.
36. Argent, S.P., Adams, H., Riis-Johannessen, T., Jeffery, J.C., Harding, L.P., Mamula, O. and Ward, M.D. (2006) *Inorg. Chem.*, **45**, 3905–3919.
37. Argent, S.P., Adams, H., Riis-Johannessen, T., Jeffery, J.C., Harding, L.P. and Ward, M.D. (2006) *J. Am. Chem. Soc.*, **128**, 72–73.
38. Cotton, F.A., Murillo, C.A. and Yu, R. (2005) *Dalton Trans.*, 3161–3165.

39 Baxter, P.N.W., Lehn, J.-M., Baum, G. and Fenske, D. (1999) *Chem. Eur. J.*, **5**, 102–112.

40 Baxter, P.N.W., Lehn, J.-M., Kneisel, B.O. and Fenske, D. (1997) *Angew. Chem. Int. Ed. Engl.*, **36**, 1978–1981.

41 Ronson, T.K., Adams, H. and Ward, M.D. (2005) *Inorg. Chim. Acta*, **358**, 1943–1954.

42 Ronson, T.K., Adams, H., Riis-Johannessen, T., Jeffery, J.C. and Ward, M.D. (2006) *New J. Chem.*, **30**, 26–28.

43 Ward, M.D., McCleverty, J.A. and Jeffery, J.C. (2001) *Coord. Chem. Rev.*, **222**, 252–272.

44 Hutin, M., Bernardinelli, G. and Nitschke, J.R. (2006) *Proc. Natl. Acad. Sci. USA*, **103**, 17655–17660.

10
Periodic Nanostructures Based on Metal–Organic Frameworks (MOFs): En Route to Zeolite-like Metal–Organic Frameworks (ZMOFs)

Mohamed Eddaoudi and Jarrod F. Eubank

10.1
Introduction

The quest for solid-state materials possessing desired functionalities for specific applications is ever increasing. The discovery of novel functional materials or new functions for existing materials has been fundamental in their development for application processes, in addition to the advances in synthetic strategies towards such materials. However, the design of targeted functional solid-state materials for desired applications remains a scientific challenge [1]. In recent years, the successful introduction of new synthetic strategies, particularly the molecular building block (MBB) approach, has offered great potential for the eventual design of solid-state materials with targeted functions [2].

Porous or open-framework materials, such as zeolites and metal–organic assemblies (MOAs), are one of the most prominent groups of functional solid-state materials. They exhibit a broad range of properties that are ideal for critical applications, such as ion exchange, gas storage, separations or even catalysis [3,4], which depend greatly on the framework charge and composition, and also the shape and dimensions of the pores. The potential scope of natural and synthetic zeolites is well established; nevertheless, difficulties, such as expanding the pore size above the 1 nm prison and/or decorating the pores with organic functionalities, has restricted their application [5,6]. On the other hand, MOAs have burgeoned in recent years due, primarily, to their facile tunability (ease of modification), mild synthesis, multifunctional capacity and inorganic–organic hybrid nature [7], which allow the simple construction of functional materials, including zeolite-like metal–organic frameworks (ZMOFs), from myriad metal ion and organic linker combinations, to offer practically limitless properties for equally boundless applications.

Organic Nanostructures. Edited by Jerry L. Atwood and Jonathan W. Steed
Copyright © 2008 WILEY-VCH Verlag GmbH & Co. KGaA, Weinheim
ISBN: 978-3-527-31836-0

10.2
Historical Perspective

10.2.1
Metal–Cyanide Compounds

Metal–organic assemblies, themselves, are not new to the solid-state scientific community. As early as 1897, a nickel ammine cyanide compound was discovered by Hofmann and found to crystallize around benzene as $Ni(CN)_2 \cdot NH_3 \cdot C_6H_6$ [8,9]. Its ability to trap benzene led Hofmann to study the inclusion properties of this material, which indicated its selectivity for other small aromatic molecules depending on their size and shape [9,10]. Although a pivotal discovery, it was several years later before the crystal structure was solved for Hofmann's clathrate by Powell and Rayner [9,11]. They determined that the structure, $Ni(NH_3)_2Ni(CN)_4 \cdot 2C_6H_6$, was extended and consisted of alternating square-planar and octahedral Ni(II) cations bridged by cyanide anions, NiC_4 and NiN_6 (two axial ammonia molecules), resulting in layers of square grid sheets (Figure 10.1). The axial ammonia molecules are pointed into the interlayer spacing between sheets, where the benzene molecules are trapped perpendicular to the sheets.

Around the same time that Hofmann was developing his clathrates, interest in another type of metal–cyanide compound, Prussian blue (PB) and its analogues, also began to intensify [12]. Since its fortuitous discovery in 1704, the dye-related properties have resulted in numerous practical applications, which, combined with the discovery that the framework could be dehydrated while maintaining its structural integrity and could reversibly sorb small molecules, has attracted scientific interest in this class of metal–cyanide materials. As a result, there were many attempts to elucidate the chemical formula and metal oxidation state(s), including efforts to synthesize crystals for solid-state structure studies. It was not until 1970 that Prussian blue was determined to be a mixed-valence iron(III) hexacyanoferrate(II) compound and the first crystal structure was reported two years later, revealing a

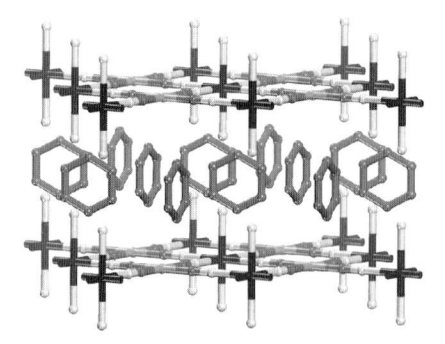

Figure 10.1 Crystal structure of Hofmann's clathrate. Hydrogen atoms have been omitted for clarity. Octahedral Ni = black, square-planar Ni = gray, C = dark gray and N = light gray.

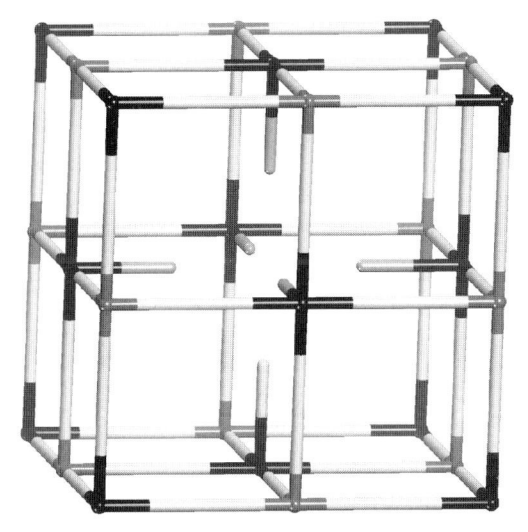

Figure 10.2 Structure of Prussian blue. Hydrogen atoms have been omitted for clarity. Fe(III) = black, Fe(II) = dark gray, CN light gray and O = gray.

cubic structure with the general formula $Fe_4[Fe(CN)_6]_3 \cdot xH_2O$ ($x = 14$–16), where defects in the structure are occupied by water molecules, resulting in $Fe^{II}C_6$ and $Fe^{III}N_{4.5}O_{1.5}$ coordination (Figure 10.2).

The unique host–guest properties and application potential of Hofmann's clathrates and Prussian blue have led to numerous attempts to target and construct analogous materials of each type. As with the MOAs of today, it was seen early on that these materials exhibited the potential for facile tunability based on substitution of the metal ions and/or organic ligands. Such substitutions would, no doubt, introduce novel properties to the materials while maintaining similar structural characteristics associated with each material type.

Specifically, the inclusion capability and selectivity of Hofmann's clathrate, along with the then newly known crystal structure, sparked interest in the development of similar materials. As early as the 1960s, researchers were attempting to substitute the Ni(II) ions with other metals [13], which eventually resulted in the development of a entire series of analogous compounds, aptly named Hofmann-type clathrates after their originator [14]. These Hofmann-type clathrates were then given the general formula $M(NH_3)_2M'(CN)_4 \cdot 2G$, where M = Cd, Co, Cu, Fe, Mn, Ni or Zn, M' = Ni, Pd, Pt and G = C_4H_5N, C_4H_4S, C_6H_6 and $C_6H_5NH_2$ [14]. Further studies revealed that the square-planar M' could be substituted with a tetrahedral metal ion, such as Cd or Hg [15], to give new three-dimensional metal–cyanide structures that were still capable of guest enclathration. Likewise, metal-substituted Prussian blue analogues were synthesized and have since been assigned the general formula $M_x[M'(CN)_6]_y$, including numerous transition metal, mixed transition metal and rare earth transition metal analogues with a range of properties, especially related to magnetism [12].

It has also been shown that other amine organic ligands could be utilized to modify the Hofmann-type structures by replacing the terminal ammonia ligands [9]. The most common of these alterations involves the use of diaminoalkanes, such as ethylenediamine ($NH_2CH_2CH_2NH_2$) and trimethylenediamine ($NH_2CH_2CH_2CH_2NH_2$) [12], which are ditopic ligands intended to replace two terminal ammonia ligands from neighboring metal–cyanide sheets, in effect pillaring the sheets and creating a 3D structure. In most cases the Hofmann-type metal–cyanide layers are maintained, merely the interlayer spacing and/or guest inclusion are altered. Longer diamines, $NH_2(CH_2)_nNH_2$ ($n = 4$–9), have been utilized to make modified Hofmann-type clathrates that can trap molecules larger than the typical small aromatic guests [12]. Although the metal–cyanide inclusion compounds can readily trap molecules, many are not stable upon loss of the guest molecules or heating [9,12]. There have been many efforts to synthesize more stable derivatives with unique properties or larger cavities, with limited success. The utilization of diaminoalkanes to pillar the Hofmann-type clathrate metal–cyanide sheets does add to their stability in some cases, but the ligands remain relatively flexible compared with the rigid cyanide bridges and can result in chelation of the metal instead of pillaring, and also collapse of the structure [9,12]. However, metal–cyanide chemistry has continued to progress with efforts to use rigid pillars, linear metal spacers [16] and even angular ditopic [17] or polytopic [18] amines to synthesize similar, and also unique [19], architectures. In addition, there have been numerous efforts to assess their potential for novel applications, including hydrogen storage [20].

10.2.2
Werner Complexes

Another series of solid-state organic–inorganic hybrid compounds that should not be overlooked are the Werner complexes (Figure 10.3), represented by the general formula MX_2A_4, where M = a divalent cation (Cd, Co, Cr, Cu, Fe, Hg, Mn, Ni, Zn) with octahedral coordination, X = an anionic ligand (Cl^-, Br^-, I^-, CN^-, NCO^-, NCS^-, NO_2^-, NO_3^-) and A = neutral pyridine-based ligands or α-arylalkylamines [21]. Like the Hofmann-type clathrates and Prussian blue-type materials, Werner complexes are inclusion compounds, having the ability to trap guest molecules due to the inefficient packing of the complexes in the solid state. The diversity of compositions allows for the inclusion of a variety of guest molecules, ranging from simple noble gases to complex aromatic molecules. The generality of the Werner complex formula, the scope of sorbed species and the occurrence of β-phases that exhibit porosity upon loss of guest molecules are indicative of the versatility of coordination chemistry and the utility of ligand substitution on the overall properties of metal–organic materials.

Although these Werner complexes and the previously mentioned metal–cyanide compounds have primarily focused on metals with octahedral coordination (and square planar in Hofmann-type clathrates), with few exceptions, it is evident that a variety of crystalline (periodic) inorganic–organic hybrid materials can be constructed. In addition, the dual composition of this class of materials offers the potential to alter

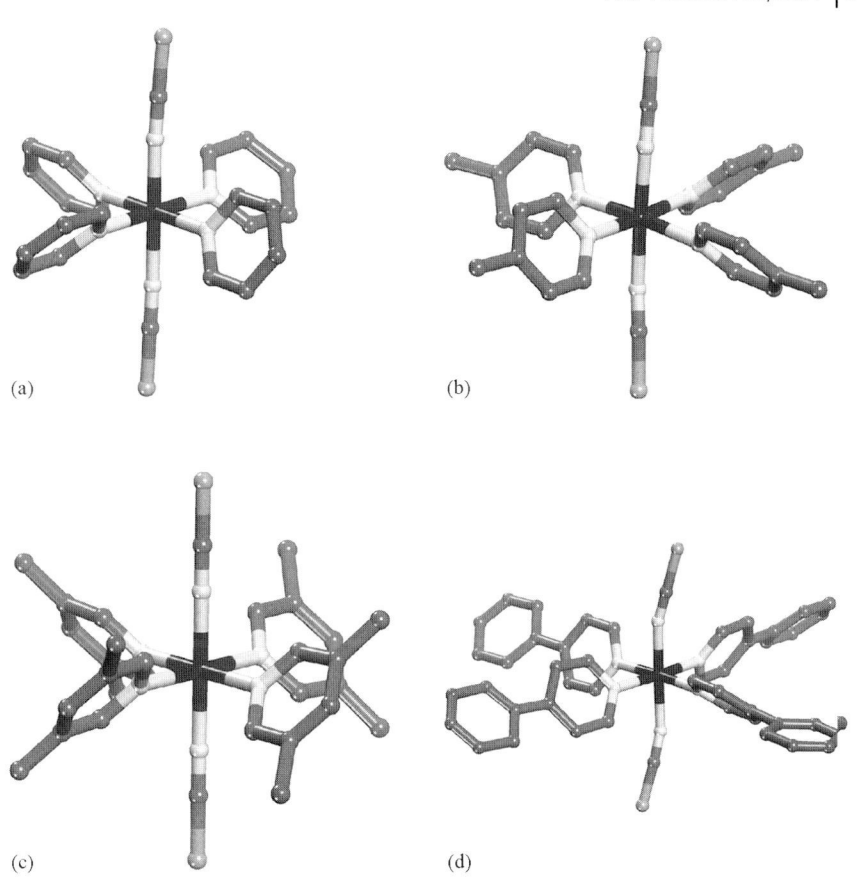

Figure 10.3 Typical MX_2A_4 Werner complexes: (a) $X = NCS^-$, $A =$ pyridine; (b) $X = NCS^-$, $A =$ 4-picoline; (c) $X = NCS^-$, $A =$ 3,5-lutidine; and (d) $X = NCS^-$, $A =$ 4-phenylpyridine.

the organic and inorganic components. Thus, the ability to introduce new coordination modes and geometries of metals, in combination with predesigned ligands, can only expand the scope and function of periodic hybrid materials.

10.2.3
Expanded Nitrogen-donor Ligands

However, it was not until 1989 that the potential for design was visualized and recognized for these inorganic–organic hybrid materials, when Hoskins and Robson first reported the deliberate design and synthesis of a cubic diamond-like MOA

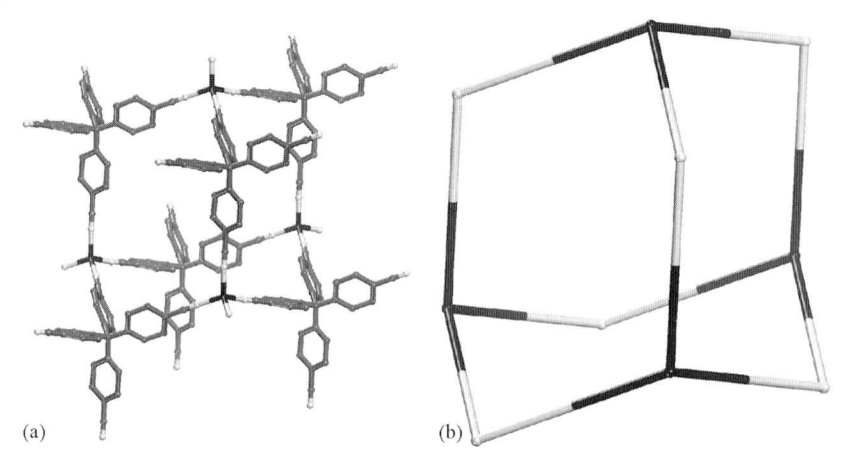

Figure 10.4 (a) Hoskins' and Robson's Cu-tetracyanotetraphenylmethane MOA (hydrogen atoms have been omitted for clarity; Cu = black, C = dark gray, N = light gray) with (b) cubic diamond topology.

constructed *in situ* from tetrahedral Cu(I) ions and 4,4′,4″,4‴-tetracyanotetraphenylmethane molecules (Figure 10.4) [22]. They proposed that if tetrahedral centers could be linked by molecular rods, an extended network could be constructed and expected to have an enlarged open structure related to cubic diamond or hexagonal Lonsdaleite that may be tailor-made (via variations in and modifications to the tetrahedral centers and molecular rods) to exhibit interesting properties. In a later publication, they elaborated on the subject, suggesting other possible molecular rods and coordination modes that might also be utilized to target particular frameworks, in addition to the potential for the resultant open materials to complement zeolites for related applications [23].

The hypotheses and results of Hoskins and Robson mark the beginnings of the molecular building block (MBB) approach, where the organic ligand and the metal ion each serve as building blocks *in situ*. These building blocks can then combine in numerous fashions to give a variety of architectures, depending on the size and shape of the organic linker and the coordination environment and geometry of the metal ion(s) [24]. These concepts opened the door to a new and interesting class of solid-state materials.

A variety of molecular rods could be utilized to target analogous and unique structures, but probably the most prominent example is 4,4′-bipyridine (bipy). Like the cyanide anion of the metal–cyanides mentioned previously, bipy can act as a linear ditopic ligand, possessing monodentate nitrogen donors at opposite ends of the molecule. The increased length of bipy, compared with cyanide, should result in larger cavities and thus more open structures.

The similarity of neutral bipy to the anionic cyanide linker suggests that structures with network topologies analogous to Hofmann-type clathrates could be targeted when coordinated to square-planar or octahedral metal ions in an MN_4 or MN_4L_2

fashion, respectively, where L = axial terminal ligands; octahedral MN_6 building blocks could lead to Prussian blue-type analogues. Indeed, the synthesis of materials consisting of layered square grid networks, like that of Hofmann-type clathrates, has been achieved by the combination of transition metal salts and bipy in various solvents, although anionic terminal ligands or guests are necessary to balance the charge of the cationic metals (i.e. neutral or cationic metal–bipy coordination polymers or frameworks) [25]. Likewise, cationic Prussion Blue-related octahedral assemblies have been synthesized from bipy and octahedral metal ions [26] and, as predicted, coordination to a tetrahedral metal center (MN_4) results in the formation of materials with frameworks related to the cubic diamond net [27].

In addition to structures composed of square grid sheets and linearly connected octahedra, a variety of networks have been synthesized from bipy and single metal ions depending on several factors, including the metal:ligand ratio, counter ions, guests and the coordination environment and geometry of the metal. Since bipy is a simple linear linker, the determination of each structure's network topology is directly governed by the coordination environment and geometry of the metal ion. A diversity of 3D structures can be targeted, including those analogous to the Si net in α-$ThSi_2$ [28], and also additional 2D [29,30], 1D [27,30,31] and discrete [32] topologies.

Other linear ditopic nitrogen-donor organic ligands (Figure 10.5), such as pyrazine, 1,2-bis(2-pyridyl)ethylene, 1,2-bis(4-pyridyl)ethane and many others, have

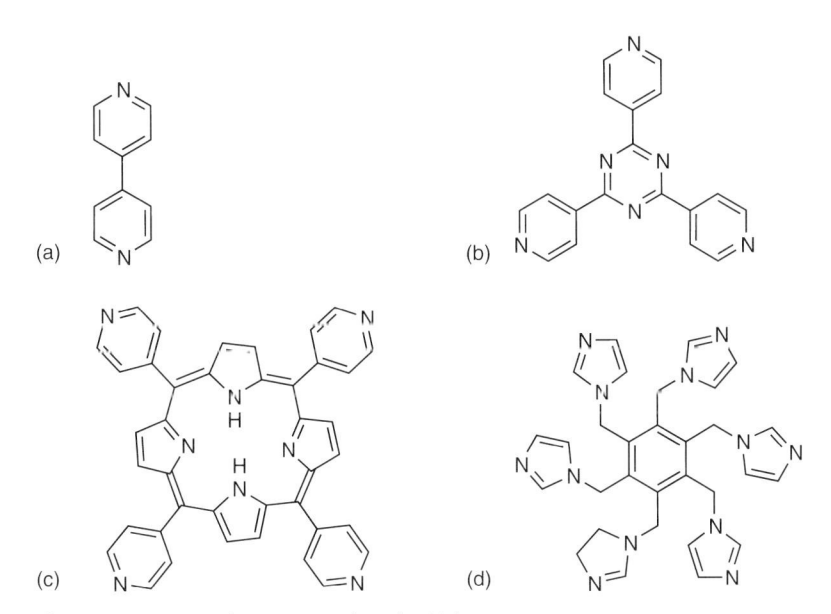

Figure 10.5 Nitrogen-donor organic ligands: (a) linear 4,4'-bipyridine; (b) 3-connector 2,4,6-tris(4-pyridyl)-1,3,5-triazine; (c) 4-connector 5,10,15,20-tetrakis(4-pyridyl)porphyrin; and (d) 6-connector 1,2,3,4,5,6-hexakis(imidazol-1-ylmethyl)benzene.

also been utilized to target compounds with analogous and novel topologies [33]. The coordination of these ligands to metal ions results in structures that have topologies ranging from discrete molecular squares [34] to 3D nets [35,36].

The success of the development of open metal–organic structures based on bipy-like ligands has led to the utilization of polytopic nitrogen-donor ligands for similar purposes. Although ditopic bipy merely serves as a linear spacer between metal ion nodes in the networks, other ditopic nitrogen donor ligands have been targeted with specific built-in angular nature to orient the metal ions at specific angles and target novel structures [37–39], including metal–organic frameworks with a sodalite-like topology [40]. In addition, nitrogen-donor ligands have been utilized with the potential to act as 3-, 4- or 6-connected linkers [41–43] (Figure 10.5). These non-linear ligands can then serve as additional nodes when coordinated to the metal ions, allowing for the synthesis of analogous and novel topologies when compared with bipy, depending on the size and shape of the ligand (limited only by the synthesis capabilities of the organic chemist) and the coordination environment and geometry of the metal ion [38].

Although recent developments have led to porous bipy-based MOAs [44], these materials typically remain unstable and collapse upon exchange or removal of guests and have been plagued by interpenetration [24]. Many of the bipy-like ligands are fairly rigid; however, lack of permanent porosity has traditionally been correlated with the flexible nature of the M–N coordination angles, which usually results in more flexible frameworks and limits their utility as porous materials [45].

10.2.4
Carboxylate-based Ligands

The quest for hybrid materials with large cavities and permanent porosity has led to the exploration for new coordination modes via carboxylate-based ligands to construct MOAs. Unlike the neutral bipy-like ligands, these anions can counter the cationic charge of the metal ions [45]. In addition, organic synthesis offers a vast repertoire of carboxylate-based ligands, many analogous in size and shape to the previously utilized nitrogen-donor molecules (Figure 10.6).

Due to the flexible nature of the monodentate coordination of nitrogen-donor ligands and, subsequently, the decreased predictability of the final structure, carboxylate ligands began to be explored as potentially multidentate moieties [24]. The ability of the carboxylate functionalities to bind metal ions in a bis-monodentate fashion (Figure 10.7) and form metal–carboxylate clusters has long been known [46], but only recently have researchers begun to target these clusters in the formation of extended MOAs from multi-carboxylate ligands [45].

Since they possess multiple metal–oxygen coordination bonds that result in the generation of rigid nodes with fixed geometry, these metal–carboxylate clusters are ideal molecular building blocks well suited for the construction of MOAs [45]. One of the most commonly used metal–carboxylate MBBs is the so-called paddlewheel cluster $[M_2(CO_2)_4L_2$, where L = ancillary ligands] [47]. This dinuclear MBB is

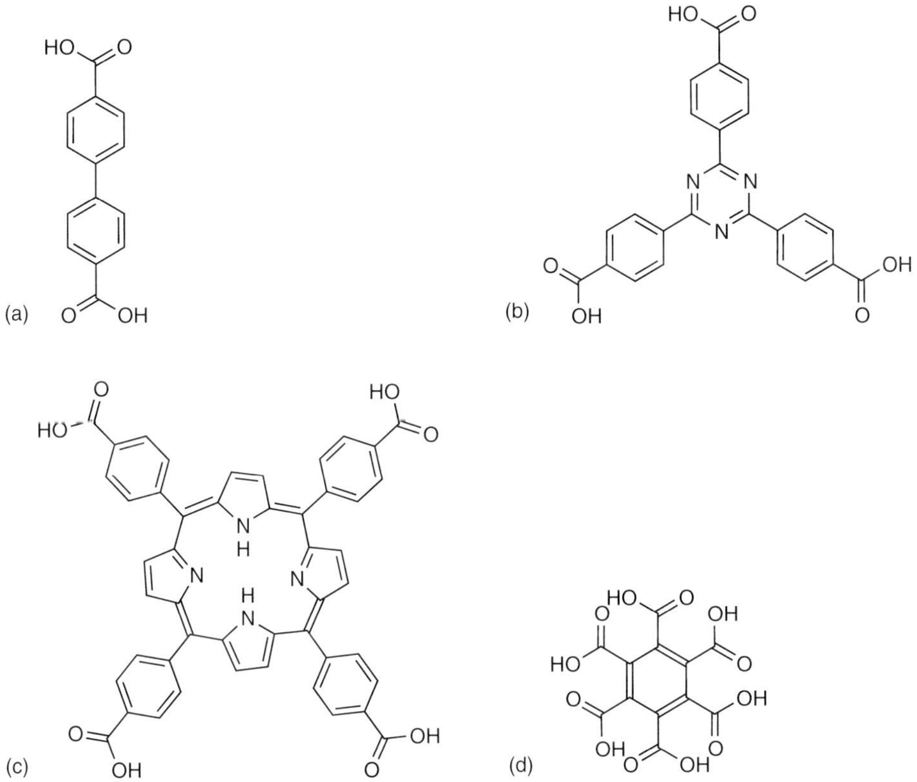

Figure 10.6 Carboxylate-based organic ligands: (a) linear 4,4′-biphenyldicarboxylic acid; (b) 3-connector 2,4,6-tris(4-carboxyphenyl)triazine; (c) 4-connector 5,10,15,20-tetrakis(4-carboxyphenyl)porphyrin; and (d) 6-connector mellitic acid.

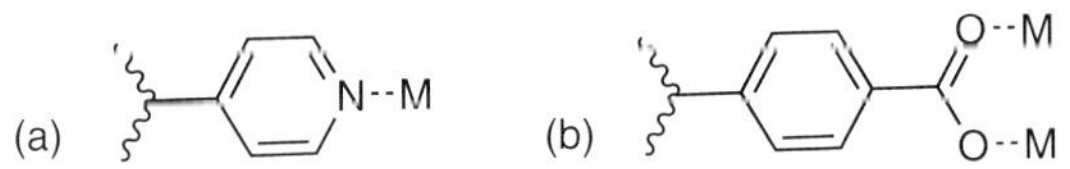

Figure 10.7 Common coordination modes in (a) bipy-like ligands and (b) carboxylate-based ligands.

most often targeted as a molecular square for building block purposes [7a,48,49] (Figure 10.8), but can be utilized as a linear spacer via the axial positions when equatorial carboxylates are terminal [48,49] or as an octahedral MBB if all the metal coordination sites are saturated with organic linkers [49]. However, there are a variety of other metal–carboxylate clusters that can be targeted as MBBs to construct stabile materials, including other dinuclear [50], trinuclear [$M_3O(CO_2)_6L_3$, basic

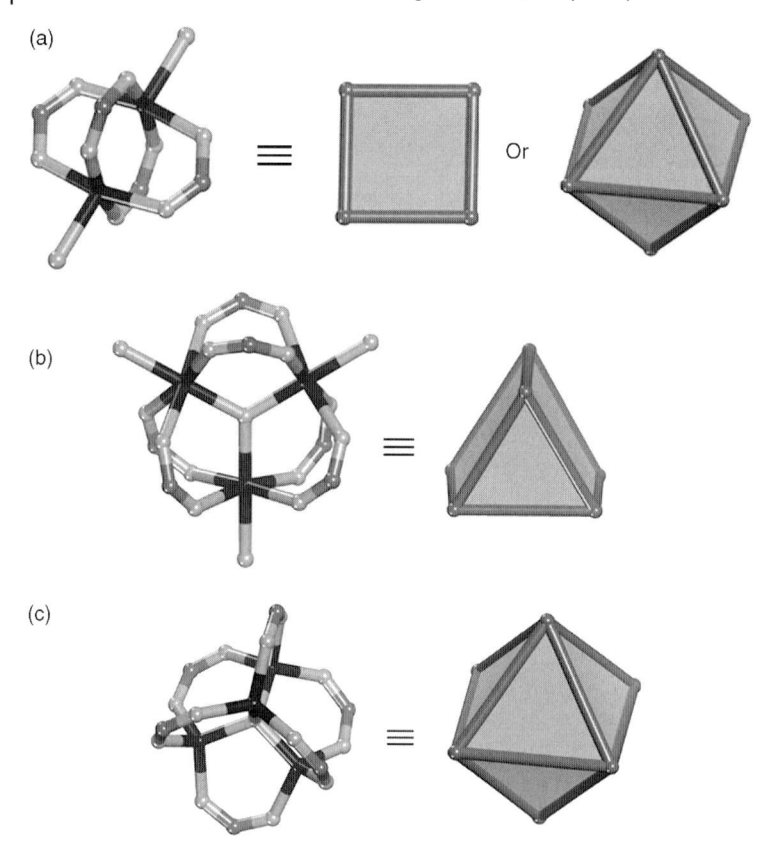

Figure 10.8 Common metal–carboxylate clusters and the resulting molecular building block(s): (a) paddlewheel can act as a linear (not shown), square or octahedral building block; (b) basic chromium acetate forms a trigonal prism building block; and (c) basic zinc acetate serves an octahedral building block.

chromium acetate] [51] and tetranuclear [$M_4O(CO_2)_6$, basic zinc acetate] clusters [52], through either ligand exchange or cluster generation *in situ*.

The ability to target, i.e. consistently generate *in situ*, these MBBs has attracted much attention since it provides the material designer with a prospective method to construct systematically functionalized porous materials. In general, if a given MBB can be targeted, the utilization of expanded or functionalized ligands can be used in conjunction with that MBB to construct analogous frameworks with varied pore size, shape and functionality [7c]. The viability and versatility of this metal–carboxylate approach has been proven many times over and has led to the burgeoning of this field when permanent porosity was finally achieved and proven for this class of materials in 1998 [53].

The metal–carboxylate cluster approach is exemplified through the basic zinc acetate MBB and its subsequent use in the construction of MOF-5 and numerous other extended CaB_6 analogues with unprecedented lower densities than encountered in any crystalline material [7b]. In addition, metal–carboxylate cluster MBBs have been utilized to synthesize homochiral MOFs for enantioselective separation and catalysis [51c]. Porous materials with some of the highest capacities for storage of carbon dioxide [54] and hydrogen [55] and the highest observed surface areas on any porous materials, up to about five times higher than the most open inorganic zeolite [51h,56]. In addition, rational synthesis of numerous MOAs [57] with various other predicted architectures [58,59] are on the rise. Although clear progress has been achieved in the design of MOAs using metal–carboxylates, the attainment of frameworks with default topologies is still predominant. Hence, new strategies are necessary to complement those already established in order to permit access to more complex structures.

10.3
Single-metal Ion-based Molecular Building Blocks

Combined with the desirable properties (inorganic–organic hybrid, mild synthesis, facile tunability and multifunctional capacity) and aesthetic architectures of the metal–organic materials themselves, the success of the MBB approach has promoted a surge not only in the synthesis of metal–organic assemblies, but also in the development of strategies toward the eventual design of desired functional porous materials for specific applications. As mentioned earlier, the utilization of metal–carboxylate clusters is one route to generate stability and target non-default structures, but methods that generate rigid MBBs from single-metal ions have not been realized. In the past, the use of single-metal ions and simple organic linkers as building blocks has consistently led to the construction of default structures depending on the shape(s) of the building blocks [57], which is supported by the synthesis of 2D square grid Hofmann-type clathrates and square grid and diamond nets from bipy-based ligands. These default structures are typically obtained due to the flexibility and lability of the (single-metal)–ligand coordination. Our group has recently developed and established a design strategy that involves a single-metal ion-based MBB approach that promotes the rational construction of atypical MOAs by forcing rigidity and directionality through control of the metal coordination sphere and judicious selection of suitable organic ligands [60].

Our group, among others, has utilized hetero-coordination of multifunctional organic molecules that combine both nitrogen donors and carboxylate groups with great success in the design and synthesis of novel functional materials [61], and our approach involves prior judicious selection of these would-be ligands based on the quantity and relative position of the functional groups on the molecule depending on the desired angle and target structure [60]. An aromatic ring must contain the nitrogen atoms and at least one nitrogen atom must have a carboxylic acid located

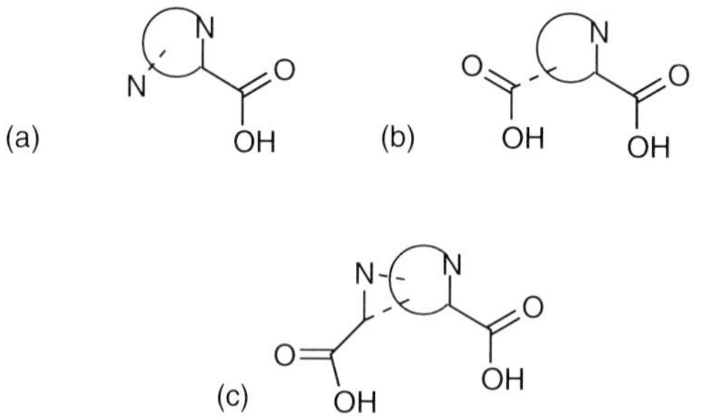

Figure 10.9 Possible variations of the $MN_{(x+y)}(CO_2)_{(x+z)}$ coordination, where M is any metal with coordination of 6–8, $x =$ number of N,O-hetero-chelating moieties, $y =$ the number of ancillary nitrogen-donors and $z =$ the number of ancillary carboxylic acids: (a) $x = 1, y = 1, z = 0$; (b) $x = 1, y = 0, z = 1$; (c) $x = 2, y = z = 0$. The ancillary group position depends on the desired angle and target structure.

one carbon away. This α-position allows for N,O-hetero-chelation to the metal ion. As part of the aromatic ring, the nitrogen atoms direct the framework topology, while the carboxylate groups secure the geometry of the metal by locking it into its position through the formation of rigid five-membered rings. In addition to having one hetero-chelating moiety, the organic molecule must also be polytopic, possessing at least one ancillary coordinating moiety, i.e. an additional carboxylate, nitrogen atom or hetero-chelate (Figure 10.9). The ability of a ligand to saturate the coordination sphere of the selected metal ion is also ideal, as this precludes coordination of any solvent, template or other guest molecules and allows directionality to be induced entirely dependent on the organic linker.

The use of potentially 6–8-coordinate metal ions allows the targeting of numerous structures, depending on the ligand shape and multiplicity of functionalities; this gives $MN_{(x+y)}(CO_2)_{(x+z)}$, where M is any metal with coordination of 6–8, $x =$ the number of N,O-hetero-chelating moieties, $y =$ the number of ancillary nitrogen-donors and $z =$ the number of ancillary carboxylic acids. Each hetero-coordinated single-metal ion can be rendered rigid and directional while simultaneously its coordination sphere is saturated via the coordination of the hetero-chelate and/or secondary functionalities of the ligands. The relative position of the ancillary functional group on the organic molecule dictates the directionality and plays a vital role in providing the angles necessary to target particular structures.

10.3.1
Discrete, 2D and 3D Metal–Organic Assemblies

Our design strategy has proven effective in synthesizing metal–organic polyhedra. The first example was a metal–organic cube using $Ni(NO_3)_2 \cdot 6H_2O$ and 4,5-imidazoledicarboxylic acid (H_3ImDC) [62]. H_3ImDC possesses concurrently the

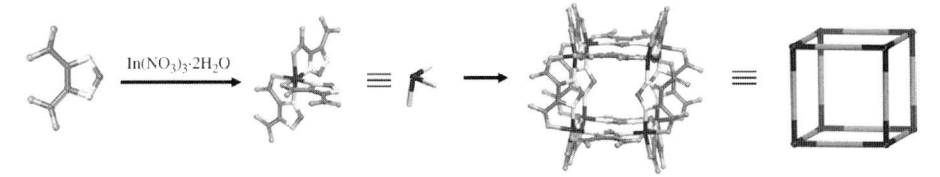

Figure 10.10 The metal–organic cube is constructed from 12 ImDC ligands and eight octahedral single-indium ion MBBs.

initial *N,O*-hetero-chelating moiety and an additional *N,O*-hetero-chelate. The H$_3$ImDC molecules are doubly deprotonated *in situ* and three HImDC^{2-} anions coordinate each Ni^{2+} cation in a facial (*fac*) manner to complete the Ni(II) octahedral coordination sphere (Figure 10.10). Each ligand chelates two metal ions at a large angle, which allows the ligand to function as the edges of the cube while the *fac*-NiN$_3$(CO$_2$)$_3$ MBB occupies the vertices. As the metal–organic cube consists of 12 doubly deprotonated ligands and only eight divalent nickel ions, cationic amines used in the solvothermal synthesis conditions fill the inter-cube space and neutralize the anionic polyhedra, [Ni$_8$(HImDC)$_{12}$]$^{8-}$·[(H$_2$TMDP)$_4$]$^{8+}$·(DMF)$_4$·(EtOH)$_4$·(H$_2$O)$_6$.

The versatility of this approach was realized when we were able to synthesize supramolecular isomers using 2,5-pyridinedicarboxylic acid (H$_2$PDC) [63]. H$_2$PDC possesses concurrently the initial *N,O*-hetero-chelating moiety and an additional carboxylic acid in the 5-position, which can provide a 120° angle upon coordination. The angularity of this ligand can then be used to target discrete metal–organic octahedra or 2D metal–organic frameworks with Kagomé lattice topology, which are based on 4-connected vertices (Figure 10.11). Indeed, the H$_2$PDC molecules can be doubly deprotonated *in situ* and four PDC^{2-} anions can coordinate each metal cation [MN$_2$(CO$_2$)$_4$ MBB] to generate the 4-connected nodes necessary [MN$_2$(CO$_2$)$_2$

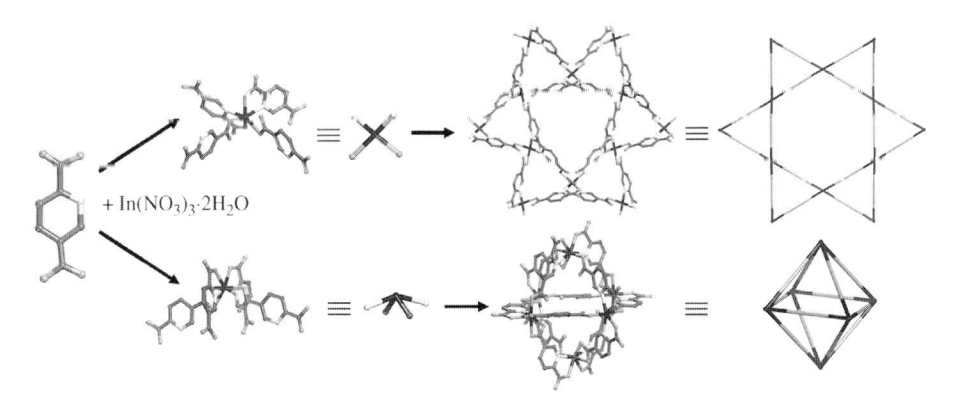

Figure 10.11 H$_2$PDC and In(NO$_3$)$_3$·2H$_2$O were reacted in the presence of different SDAs to give a MOF with Kagomé lattice or metal–organic octahedra.

vertex]. As the coordination of four PDC^{2-} ligands to a single divalent or trivalent metal ion {[M^{II}(PDC)$_4$]$^{2-}$, [M^{III}(PDC)$_4$]$^{1-}$} will yield an anionic framework, cationic structure directing agents (SDAs) can be utilized to direct the formation of the material.

In fact, the solvothermal reaction of H$_2$PDC and In(NO$_3$)$_3$·2H$_2$O in the presence of the SDA 4,4′-trimethylenedipiperidine (TMDP) yields a MOF with the Kagomé lattice topology, [In(PDC)$_2$]$_n^-$·(HTMDP)$^+$·EtOH·H$_2$O [63]. In this case, the single trivalent indium ions are 6-coordinate, each chelated by two ligands and coordinated through the ancillary carboxylates in a monodentate fashion by two additional ligands to give an InN$_2$(CO$_2$)$_4$ MBB, which can be simplified as a quasi-planar *cis*-InN$_2$(CO$_2$)$_2$ building unit. The anionic framework is neutralized by the protonation of the organic amine SDA, which occupies the hexagonal channels and interlayer space between the 2D Kagomé sheets along with guest solvent molecules.

The reaction of the same reactants in the presence of a different SDA, 1,2-diaminocyclohexane (DACH), results in the synthesis of an MOA consisting of discrete M$_6$L$_{12}$ metal–organic octahedra and guests of the formula [In$_6$(PDC)$_{12}$]$^{6-}$·[(H$_2$DACH)$_2$]$^{4+}$·[(H$_3$O)$_2$]$^{2+}$·(DMF)$_5$·(EtOH)$_5$ [63]. Here, the single trivalent indium ions are actually 7-coordinate, each chelated by two ligands and one ancillary carboxylate in a monodentate fashion in the same manner as in the Kagomé lattice, but the ancillary carboxylate of one ligand binds in a bidentate fashion to the In(III) center. The coordination still gives an InN$_2$(CO$_2$)$_4$ MBB, as the points of extension are considered to be the carboxylate carbon atoms rather than the coordinated oxygen atoms, but the building unit is now quasi-pyramidal, *trans*- InN$_2$(CO$_2$)$_2$. As with the metal–organic cube, these polyhedra are anionic and the amine SDA molecules serve to neutralize the negative charge and actually link the octahedra together through N–H···O hydrogen bonds to form a 3D network.

Thus, this method has proven effective in targeting and synthesizing discrete metal–organic polyhedra, metal–organic frameworks with Kagomé lattice topology and potentially 3D MOAs. One can envision using the same route to target other unique metal–organic assemblies from 6-, 7- or 8-coordinate single metal ions, including those with discrete, 2D and 3D architectures. Of great interest would be MOFs with zeolite-like topologies, due to the vast application potential of such materials associated with their porous and ionic nature.

10.3.2
Zeolite-like Metal–Organic Frameworks (ZMOFs)

Our single-metal ion-based MBB approach can also be applied to the decoration and/or expansion of tetrahedral-based zeolite-like nets [64] by lengthening the edges of the net with a longer functional organic linker [5,6], similar to the previously mentioned enlargement of metal–cyanide compounds by using an extended organic linker such as bipy. In effect, the organic ligand will serve as a functionalized multi-atom substitute for the single oxygen atoms that bridge the tetrahedral (T) silicon and/or aluminum atoms in traditional inorganic zeolites. However, the organic ligand (L) must be judiciously selected to contain two functionalities (at least one N, O-hetero-chelate) that can provide an M–L–M angle comparable to the average T–O–T angle (~145°) found in typical zeolites [65]. In addition, higher coordination (6–8) metal ions are utilized instead of tetrahedral ions, so the ligands must coordinate to the metal in a manner that gives an MBB that can act as a tetrahedral building unit (TBU). Thus, for 6-coordinate metals there must be an $MN_4(CO_2)_2$ or $MN_2(CO_2)_4$ coordination environment and $MN_4(CO_2)_4$ or $MN_2(CO_2)_6$ MBBs for 8-coordinate metals that all result in MN_4 or $MN_2(CO_2)_2$ TBUs, respectively.

According to these criteria, the aforementioned organic molecule, H_3ImDC, is well suited to target metal–organic frameworks with zeolite-like topologies. This ligand concurrently possesses two *N,O*-hetero-chelating moieties with a potential M–L–M angle of 145°. When doubly deprotonated *in situ*, the $HImDC^{2-}$ anions can coordinate two divalent or trivalent metal cations at the appropriate angle (directed by the M–N coordination) and four ligands saturate each metal ion in the proposed manner to complete the respective coordination sphere and give an anionic zeolite-like metal–organic framework (ZMOF). This anionic nature allows for the utilization of different cationic SDAs and exploration of applications akin to inorganic zeolites.

As expected, solvothermal reaction of $In(NO_3)_3 \cdot 2H_2O$ and H_3ImDC in the presence of different SDAs yields metal–organic frameworks with different zeolite-like topologies, specifically *sod* (sodalite) and *rho* [59,65]. Predictably, the expansion of the oxygen atom bridge in existing zeolite topologies to the larger $HImDC^{2-}$ linker results in the two ZMOFs, *sod*-ZMOF and *rho*-ZMOF, that are up to eight times larger than their inorganic analogues. These new ZMOF materials offer great potential for numerous applications, combining the unique properties of both MOFs and zeolites, in addition to novel properties.

10.3.2.1 *sod*-ZMOF

The solvothermal reaction of H_2PDC and $In(NO_3)_3 \cdot 2H_2O$ in the presence of the imidazole (HIm) yields a MOF with a zeolite SOD-like topology, *sod*-ZMOF (Figure 10.12). In the crystal structure of *sod*-ZMOF, each single indium ion is chelated by two $HImDC^{2-}$ ligands and coordinated by the ancillary nitrogen donor of two more $HImDC^{2-}$ ligands, resulting in 6-coordinate In(III) ions that form $InN_4(CO_2)_2$ MBBs. Since the nitrogen atoms direct the topology, this MBB can be regarded as an InN_4 TBU. When the structure is reduced to the simplified InN_4 TBUs, it can be viewed a 3D periodic array of large truncated octahedral cages, analogous to β-cages found in numerous inorganic zeolites. These β-cages connect to neighboring cages through common four- and six-membered rings, which results in an overall structure based on the edge expansion of the inorganic sodalite net.

The presence of four $HImDC^{2-}$ ligands around each trivalent indium ion results in an anionic framework that is neutralized by the SDA molecules, [In(HImDC)$_2$]$_n^-$·[(H'HIm)]$^+$·(DMF)$_4$·(CH$_3$CN).(H$_2$O)$_4$. Therefore, *sod*-ZMOF represents the first example of an MOF with an anionic framework based on the *sod* topology, although a few examples of neutral or cationic sodalite-like MOFs have been synthesized previously [57].

(a)

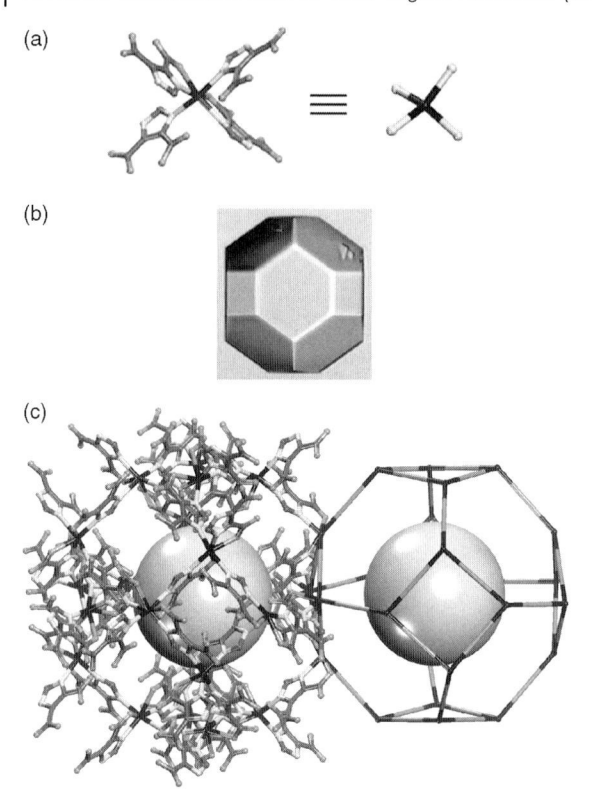

(b)

(c)

Figure 10.12 (a) sod-ZMOF is composed of 6-coordinated MBBs, which also can be viewed as 4-connected TBUs. (b) Optical image obtained on an Olympus MIC-D optical microscope of as-synthesized sod-ZMOF crystals. (c) A fragment of the sod-ZMOF single-crystal structure, where the large light-gray spheres represent the largest sphere that would fit in the β-cavities without touching the van der Waals atoms of the framework and the β-cages of the sodalite network connect through common four- (shown) and six-membered rings. Hydrogen atoms have been omitted for clarity; In = black, C = dark gray, O = gray, N = light gray.

10.3.2.2 *rho*-ZMOF

The same starting materials can be solvothermally reacted in the presence of a different SDA, 1,3,4,6,7,8-hexahydro-2*H*-pyrimido[1,2-*a*]pyrimidine (HPP), to yield a MOF with *rho* topology (Figure 10.13). In this case, each single indium ion is 8-coordinate, saturated by the chelation of four $HImDC^{2-}$ ligands to give $InN_4(CO_2)_4$ MBBs. As with sod-ZMOF, the nitrogen atoms of the ligand direct the topology and, although two additional α-carboxylates coordinate the metal ion, result in a 4-connected InN_4 TBU. When the structure is simplified as connected InN_4 TBUs, it is evident that four-, six- and eight-membered rings fuse together to form α-cages that are linked via double eight-membered rings (D8Rs) in a 3D periodic array to create first material ever to contain an organic component and have a zeolite *rho*-like topology [59]. The purity of the as-synthesized material

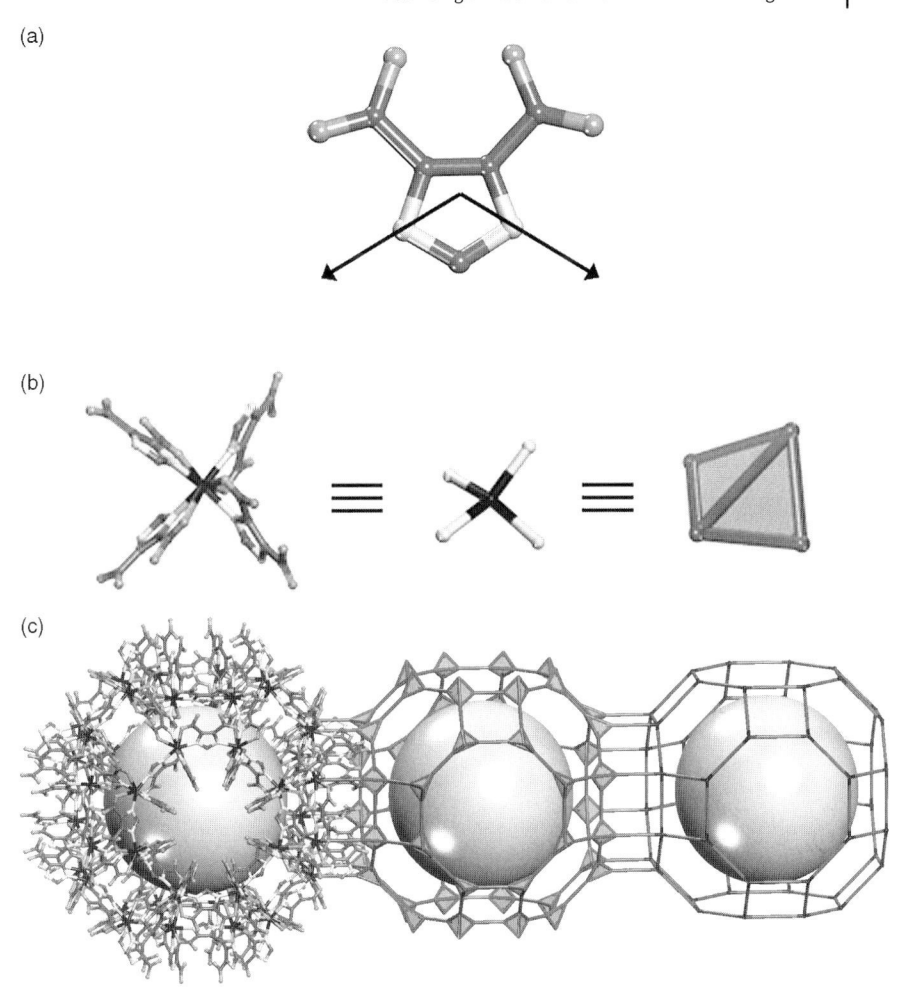

Figure 10.13 (a) The angle in the ImDC ligand. (b) *rho*-ZMOF is composed of 8-coordinated MBBs, which can be viewed as 4-connected TBUs. (c) A fragment of *rho*-ZMOF, where the large light-gray spheres represent the largest sphere that would fit in the α-cavities without touching the van der Waals atoms of the framework and the α-cages of the *rho* network connect through double eight-membered rings. Hydrogen atoms have been omitted for clarity; In = black, C = dark gray, O = gray, N = light gray.

was confirmed by similarities between the experimental and simulated XRPDs (Figure 10.14).

Again, the presence of four HImDC^{2-} ligands around each trivalent indium ion results in the formation of an anionic framework that is neutralized by the SDA molecules, in this case 24 doubly protonated HPP molecules per unit cell, $[In_{48}(HImDC)_{96}]_n^{48-} \cdot [(H'_2HPP)_{24}]^{48+} \cdot (DMF)_{36} \cdot (H_2O)_{192}$. The H'$_2$HPP

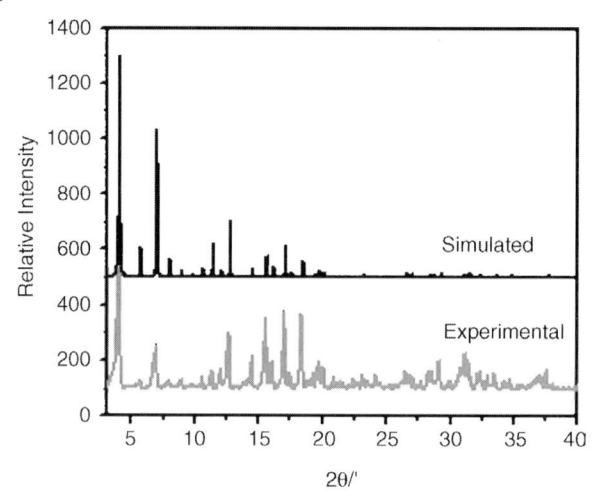

Figure 10.14 Experimental and simulated powder X-ray diffraction patterns for *rho*-ZMOF.

molecules occupy part of the open space in the α-cage cavities and the smaller D8R cages, in addition to the guest solvent molecules. The space occupied by guest molecules represents 56% of the cell volume or 16 718 Å^3 per unit cell and the as-synthesized compound is insoluble in water and common organic solvents.

The D8R cages of *rho*-ZMOF actually represent the largest apertures (~8 Å) of the framework for potential diffusion of small molecules. These pores serve as openings to the extra-large α-cage cavities, which have an internal diameter of 18.2 Å.

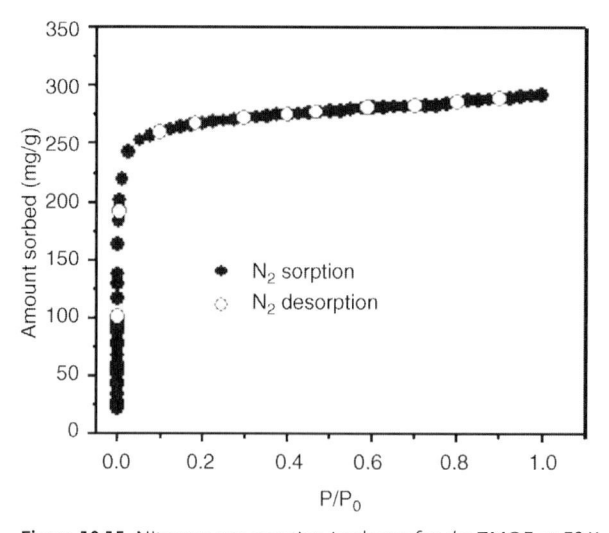

Figure 10.15 Nitrogen gas sorption isotherm for *rho*-ZMOF at 78 K.

The extra-large cavities and the anionic nature of *rho*-ZMOF make it ideal for applications involving cationic exchange, interesting considering that there are very few examples of anionic MOFs and the encapsulation of large molecules, an area where previous anionic porous materials, such as inorganic zeolites, have been limited [3].

Cation Exchange As stated previously, *rho*-ZMOF is an anionic framework; each α-cage is neutralized by 24 doubly protonated HPP molecules. In addition, the large apertures and robust nature (stable up to ∼260 °C) of this material, suggest the potential to perform cationic exchange readily. Na^+ exchange can be achieved at

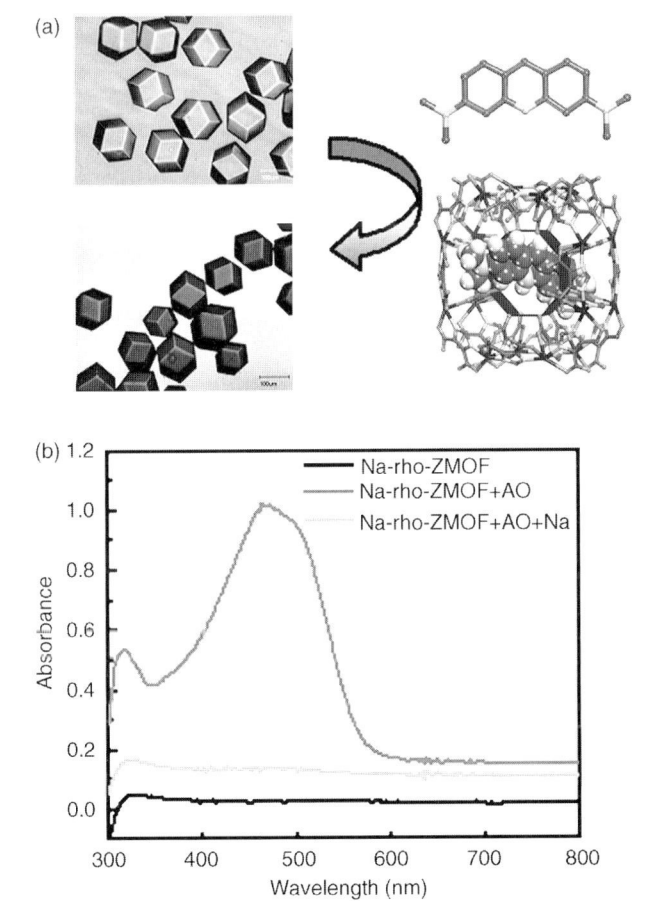

Figure 10.16 Optical images obtained on an Olympus MIC-D optical microscope of as-synthesized *rho*-ZMOF crystals (light gray = colorless) and *rho*-ZMOF crystals after AO exchange (dark gray = red) through the D8R windows. UV/Vis spectra of *rho*-ZMOF exchange derivatives.

room temperature, when 48 sodium ions replace the H′$_2$HPP molecules, which is confirmed by elemental microanalysis and atomic absorption studies. The crystallinity and structural integrity of *rho*-ZMOF are retained after Na$^+$ exchange, as indicated by similarities in the XRPD patterns. Cation exchange had never been successfully achieved in MOFs prior to this example, especially not in an aqueous environment, without dissociation of the framework [59].

The sodium-exchanged *rho*-ZMOF, Na-*rho*-ZMOF, can be fully evacuated at 100 °C, including guest water molecules. Sorption studies were conducted on the as-synthesized material and the sodium-exchanged derivative, revealing that both have type I nitrogen sorption isotherms (Figure 10.15). The isotherms are fully reversible, suggesting that the material contains homogeneous micropores; the apparent Langmuir surface area was estimated as 1067 m^2 g^{-1}.

(Host–Guest)–Guest Sensing Just as Na$^+$ ions exchange cationic H′$_2$HPP molecules out of the *rho*-ZMOF cavities, other cationic organic molecules can also be utilized. There are numerous examples of cationic fluorophores, such as acridine orange (AO), which can be used to probe the physical properties of the *rho*-ZMOF interior. AO is also an excellent choice since it is smaller than the D8R aperture of *rho*-ZMOF and can diffuse freely into the α-cages (Figure 10.16).

The extra-large α-cage cavities allow not only the encapsulation of cationic molecules, but also additional neutral guest molecules. As AO is sensitive to its local environment, it can be utilized to sense a variety of neutral molecules, such as methylxanthines or DNA nucleoside bases. These results demonstrate the ability of anionic ZMOFs to serve as a (host–guest)–guest sensor with the ZMOF providing a periodic porous substrate for fluorescent cations which can then serve as the sensor unit.

10.4
Conclusion

Although serendipity will continue to aid in the advancement of science, the growing need for materials with specific functions will, no doubt, lead to an increased thrust for design strategies and methods. History has provided fundamental discoveries and principles that persist in modern solid-state materials development, such as the molecular building block approach. Our contribution has focused primarily on the utilization of single-metal ions as viable MBBs, depending upon the judicious selection of the organic ligands.

We have already proven the ability of this method to synthesize unique discrete, 2D and 3D metal–organic assemblies with targeted architectures, including the first anionic MOFs with zeolite-like topologies and properties, from higher coordination metals and ligands with specific angles and functionalities. One can envision how the single-metal ion-based MBB approach offers a rational route towards design and will, no doubt, aid the quest for novel functional materials.

References

1. (a) Robson, R., Abrahams, B.F., Batten, S.R., Gable, R.W., Hoskins, B.F. and Liu, J. (1992) *ACS Symp. Ser.*, **499**, 256–273. (b) Batten, S.R. and Robson, R. (1998) *Angew. Chem. Int. Ed.*, **37**, 1461–1494. (c) Champness, N.R. and Schroder, M. (1998) *Curr. Opin. Solid State Mater. Sci.*, **3**, 419–424. (d) Moulton, B. and Zaworotko, M.J. (2001) *Chem. Rev.*, **101**, 1629–1658. (e) Desiraju, G.R. (2003) *J. Molec. Struct.*, **656**, 5–15. (e) Rowsell, J.L.C. and Yaghi, O.M. (2004) *Microporous Mesoporous Mater.*, **73**, 3–14.

2. (a) Stein, A., Keller, S.W. and Mallouk, T.E. (1993) *Science*, **259**, 1558–1564. (b) Yaghi, O.M., Li, G. and Li, H. (1995) *Nature*, **378**, 703–706. (c) Pecoraro, V.L., Bodwin, J.J. and Cutland, A.D. (2000) *J. Solid State Chem.*, **152**, 68–77. (d) Holliday, B.J. and Mirkin, C.A. (2001) *Angew. Chem. Int. Ed.*, **40**, 2022–2043. (e) Pilkington, M. and Decurtins, S. (2003) *Perspect. Supramol. Chem.*, **7**, 275–323. (f) Finn, R.C., Burkholder, E. and Zubieta, J. (2003) *Perspect. Supramol. Chem.*, **7**, 241–274. (g) Brammer, L. (2004) *Chem. Soc. Rev.*, **33**, 476–489.

3. (a) Davis, M.E. (2002) *Nature*, **417**, 813–821. (b) Corma, A. and Davis, M.E. (2004) *Chem. Phys. Chem.*, **5**, 305–313.

4. (a) Kosal, M.E., Chou, J., Wilson, S.R. and Suslick, K.S. (2002) *Nat. Mater.*, **1**, 118–121. (b) Rowsell, J.L.C. and Yaghi, O.M. (2005) *Angew. Chem. Int. Ed.*, **44**, 4670–4679. Kaskel, S. (2005) *Nachr. Chem.*, **53**, 394–399. (c) Lin, W. (2005) *J. Solid State Chem.*, **178**, 2486–2490. (d) Mueller, U., Schubert, M., Teich, F., Puetter, H., Schierle-Arndt, K. and Pastre, J. (2006) *J. Mater. Chem.*, **16**, 626–636.

5. (a) Davis, M.E. (1997) *Chem. Eur. J.*, **3**, 1745–1750. (b) Jones, C.W., Katsuyuki, T. and Davis, M.E. (1998) *Nature*, **393**, 52–54. (c) Yamamoto, K., Sakata, Y., Nohara, Y., Takahashi, Y. and Tatsumi, T. (2003) *Science*, **300**, 470–472.

6. Cheetham, A.K., Férey, G. and Loiseau, T. (1999) *Angew. Chem. Int. Ed.*, **38**, 3268–3292.

7. (a) Eddaoudi, M., Moler, D.B., Li, H., Chen, B., Reineke, T.M., O'Keeffe, M. and Yaghi, O.M. (2001) *Acc. Chem. Res.*, **34**, 319–330. (b) Eddaoudi, M., Kim, J., Rosi, N., Vodak, D., Wachter, J., O'Keeffe, M. and Yaghi, O.M. (2002) *Science*, **295**, 469–472. (c) Yaghi, O.M., O'Keeffe, M., Ockwig, N.W., Chae, H.K., Eddaoudi, M. and Kim, J. (2003) *Nature*, **423**, 705–714.

8. Hofmann, K.A. and Küspert, F.A. (1897) *Z. Anorg. Chem.*, **15**, 204–207.

9. Iwamoto, T. (1984) in *Inclusion Compounds: Structural Aspects of Inclusion Compounds Formed by Inorganic and Organometallic Host Lattices*, Vol. 1, (eds J.L. Atwood, J.E.D. Davies and D.D. MacNicol), Academic Press, London, pp. 29–57.

10. (a) Hofmann, K.A. and Hochtlen, F. (1903) *Ber. Dtsch. Chem. Ges.*, **36**, 1149–1151. (b) Hofmann, K.A. and Arnoldi, H. (1906) *Ber. Dtsch. Chem. Ges.*, **39**, 339–344.

11. (a) Powell, H.M. and Rayner, J.H. (1949) *Nature*, **163**, 566–567. (b) Rayner, J.H. and Powell, H.M. (1952) *J. Chem. Soc.*, 319–328.

12. Dunbar, K.R. and Heintz, R.A. (1997) *Prog. Inorg. Chem.*, **45**, 283–391, and references therein.

13. Baur, R. and Schwarzenbach, G. (1960) *Helv. Chim. Acta*, **43**, 842–847.

14. Iwamoto, T., Nakano, T., Morita, M., Miyoshi, T., Miyamoto, T. and Sasaki, Y. (1968) *Inorg. Chim. Acta*, **2**, 313–316.

15. Iwamoto, T., Kiyoki, M., Ohtsu, Y. and Takeshige-Kato, Y. (1978) *Bull. Chem. Soc. Jpn.*, **51**, 188–191.

16. (a) Abrahams, B.F., Hardie, M.J., Hoskins, B.F., Robson, R. and Sutherland, E.E. (1994) *J. Chem. Soc. Chem. Commun.*, 1049–1050. (b) Soma, T., Yuge, H. and Iwamoto, T. (1994) *Angew. Chem. Int. Ed.*, **106**, 1746–1748. (c) Leznoff, D.B., Xue, B., Stevens, C.L., Storr, A., Thompson, R.C. and Patrick, B.O. (2001) *Polyhedron*, **20**, 1247–1254. (d) Niel, V., Munoz, M.C., Gaspar, A.B., Galet, A., Levchenko, G. and

Real, J.A. (2002) *Chem. Eur. J.*, **8**, 2446–2453.

17 (a) Niel, V., Galet, A., Gaspar, A.B., Munoz, M.C. and Real, J.A. (2003) *Chem. Commun.*, 1248–1249. (b) Niel, V., Thompson, A.L., Munoz, M.C., Galet, A., Goeta, A.E. and Real, J.A. (2003) *Angew. Chem. Int. Ed.*, **42**, 3760–3763.

18 (a) Abrahams, B.F., Hoskins, B.F., Liu, J. and Robson, R. (1991) *J. Am. Chem. Soc.*, **113**, 3045–3051. (b) Pickardt, J. and Gong, G.T. (1994) *Z. Anorg. Allg. Chem.*, **620**, 183–186.

19 (a) Abrahams, B.F., Hoskins, B.F. and Robson, R. (1990) *J. Chem. Soc. Chem. Commun.*, 60–61. (b) Gable, R.W., Hoskins, B.F. and Robson, R. (1990) *J. Chem. Soc. Chem. Commun.*, 762–763. (c) Batten, S.R., Hoskins, B.F. and Robson, R. (1991) *J. Chem. Soc. Chem. Commun.*, 445–447. (d) Hoskins, B.F., Robson, R. and Scarlett, N.V.Y. (1994) *J. Chem. Soc. Chem. Commun.*, 2025–2026. (e) Hoskins, B.F., Robson, R. and Scarlett, N.V.Y. (1995) *Angew. Chem. Int. Ed. Engl.*, **34**, 1203–1204.

20 Kaye, S.S. and Long, J.R. (2007) *Catal. Today*, **120**, 311–316, and references therein.

21 Lipowski, J. (1984) in *Inclusion Compounds: Structural Aspects of Inclusion Compounds Formed by Inorganic and Organometallic Host Lattices*, **Vol.1**, (eds J.L. Atwood, J.E.D. Davies and D.D. MacNicol), Academic Press, London, pp. 59–103, and references therein.

22 Hoskins, B.F. and Robson, R. (1989) *J. Am. Chem. Soc.*, **111**, 5962–5964.

23 Hoskins, B.F. and Robson, R. (1990) *J. Am. Chem. Soc.*, **112**, 1546–1554.

24 Yaghi, O.M., Li, H., Davis, C., Richardson, D. and Groy, T.L. (1998) *Acc. Chem. Res.*, **31**, 474–484.

25 Gable, R.W., Hoskins, B.F. and Robson, R. (1990) *J. Chem. Soc. Chem. Commun.*, 1677–1678. (b) Fujita, M., Kwon, Y.J., Washizu, S. and Ogura, K. (1994) *J. Am. Chem. Soc.*, **116**, 1151–1152. (c) Losier, P. and Zaworotko, M.J. (1997) *Angew. Chem. Int. Ed. Engl.*, **35**, 2779–2782. (d) Lu, J., Paliwala, T., Lim, S.C., Yu, C., Niu, T. and Jacobson, A.J. (1997) *Inorg. Chem.*, **36**, 923–929.

26 Subramanian, S. and Zaworotko, M.J. (1995) *Angew. Chem. Int. Ed. Engl.*, **34**, 2127–2129.

27 (a) MacGillivray, L.R., Subramanian, S. and Zaworotko, M.J. (1994) *J. Chem. Soc. Chem. Commun.*, 1325–1326. (b) Carlucci, L., Ciani, G., Prosperio, D.M. and Sironi, A. (1994) *J. Chem. Soc. Chem. Commun.*, 2755–2766. (c) Yaghi, O.M., Richardson, D.A., Li, G., Davis, C.E. and Groy, T.L. (1995) *Mater. Res. Soc. Symp. Proc.*, **371**, 15–19.

28 (a) Yaghi, O.M. and Li, H. (1995) *J. Am. Chem. Soc.*, **117**, 10401–10402. (b) Yaghi, O.M. and Li, H. (1996) *J. Am. Chem. Soc.*, **118**, 295–296.

29 Yaghi, O.M. and Li, H. (1995) *Angew. Chem. Int. Ed. Engl.*, **34**, 207–209.

30 Woodward, J.D., Backov, R.V., Abboud, K.A. and Talham, D.R. (2006) *Polyhedron*, **25**, 2605–2615.

31 Yaghi, O.M., Li, H. and Groy, T.L. (1997) *Inorg. Chem.*, **36**, 4292–4293.

32 Fujita, M., Yazaki, J. and Ogura, K. (1990) *J. Am. Chem. Soc.*, **112**, 5645–5647.

33 Kitagawa, S. and Kondo, M. (1998) *Bull. Chem. Soc. Jpn.*, **71**, 1739–1753.

34 Fujita, M. (1999) *Acc. Chem. Res.*, **32**, 53–61.

35 Power, K.N., Hennigar, T.L. and Zaworotko, M.J. (1998) *Chem. Commun.*, 595–596.

36 Carlucci, L., Ciani, G., Proserpio, D.M. and Sironi, A. (1995) *Angew. Chem. Int. Ed. Engl.*, **34**, 1895–1898.

37 Kaes, C., Katz, A. and Hosseini, M.W. (2000) *Chem. Rev.*, **100**, 3553–3590.

38 Leininger, S., Olenyuk, B. and Stang, P.J. (2000) *Chem. Rev.*, **100**, 853–908.

39 Navarro, J.A.R. and Lippert, B. (2001) *Coord. Chem. Rev.*, **222**, 219–250.

40 Tabares, L.C., Navarro, J.A.R. and Salas, J.M. (2001) *J. Am. Chem. Soc.*, **123**, 383–387.

41 3-Connected N-donor ligand: Fujita, M., Oguro, D., Miyazawa, M., Oka, H.,

Yamaguchi, K. and Ogura, K.((1995) *Nature*, **378**, 469–471. (b) Batten, S.R., Hoskins, B.F. and Robson, R. (1995) *Angew. Chem. Int. Ed. Engl.*, **34**, 820–822. (c) Batten, S.R., Hoskins, B.F. and Robson, R. (1995) *J. Am. Chem. Soc.*, **117**, 5385–5386. (d) Abrahams, B.F., Batten, S.R., Hamit, H., Hoskins, B.F. and Robson, R. (1996) *Chem. Commun.*, 1313–1314. (e) Abrahams, B.F., Batten, S.R., Hamit, H., Hoskins, B.F. and Robson, R. (1996) *Angew. Chem. Int. Ed. Engl.*, **35**, 1690–1692. (f) Ibukuro, F., Kusukawa, T. and Fujita, M. (1998) *J. Am. Chem. Soc.*, **120**, 8561–8562. (g) Abrahams, B.F., Batten, S.R., Grannas, M.J., Hamit, H., Hoskins, B.F. and Robson, R. (1999) *Angew. Chem. Int. Ed.*, **38**, 1475–1477

42 4-Connected N-donor ligand: Abrahams, B.F., Hoskins, B.F. and Robson, R. (1991) *J. Am. Chem. Soc.*, **113**, 3606–3607. (d) Abrahams, B.F., Hoskins, B.F., Michail, D.M. and Robson, R. (1994) *Nature*, **369**, 727–729. (c) Sharma, C.V.K., Broker, G.A., Huddleston, J.G., Baldwin, J.W., Metzger, R.M. and Rogers, R.D. (1999) *J. Am. Chem. Soc.*, **121**, 1137–1144. (d) Hagrman, D., Hagrman, P.J. and Zubieta, J. (1999) *Angew. Chem. Int. Ed.*, **38**, 3165–3168. (e) Kondo, M., Kimura, Y., Wada, K., Mizutani, T., Ito, Y. and Kitagawa, S. (2000) *Chem. Lett.*, 818–819.

43 6-Connected N-donor ligand:Hoskins, B.F., Robson, R. and Slizys, D.A. (1998) *Angew. Chem. Int. Ed.*, **36**, 2752–2755.

44 (a) Kondo, M., Yoshitomi, T., Seki, K., Matsuzaka, H. and Kitagawa, S. (1997) *Angew. Chem. Int. Ed. Engl.*, **36**, 1725–1727. (b) Ohmura, T., Usuki, A., Fukumori, K., Ohta, T., Ito, M. and Tatsumi, K. (2006) *Inorg. Chem.*, **45**, 7988–7990.

45 Eddaoudi, M., Li, H., Reineke, T.M., Fehr, M., Kelley, D.G., Groy, T.L. and Yaghi, O.M. (1999) *Top. Catal.*, **9**, 105–111.

46 (a) Watanabe, S. (1949) *Nature*, **163**, 225–226. (b) van Niekerk, J.N. and Schoening, F.R.L. (1953) *Nature*, **171**, 36–37. (c) van Niekerk, J.N. and Schoening, F.R.L. (1953) *Acta Crystallogr.*, **6**, 227–232. (d) van Niekerk, J.N., Schoening, F.R.L. and de Wet, J.F. (1953) *Acta Crystallogr.*, **6**, 501–504. (e) Koyama, H. and Saito, Y. (1954) *Bull. Chem. Soc. Jpn.*, **27**, 112–114.

47 Paddlewheel cluster; there are nearly 1800 occurrences in the Cambridge Structural Database (CSD) (January 2007): Allen, F.H. (2002) *Acta Crystallogr.*, **B58**, 380–388.

48 (a) Moulton, B., Lu, J., Hajndl, R., Hariharan, S. and Zaworotko, M.J. (2002) *Angew. Chem. Int. Ed.*, **41**, 2821–2824. (b) Eddaoudi, M., Kim, J., Vodak, D., Sudik, A., Wachter, J., O'Keeffe, M. and Yaghi, O.M. (2002) *Proc. Natl. Acad. Sci. USA*, **99**, 4900–4904. (c) Abourahma, H., Bodwell, G.J., Lu, J., Moulton, B., Pottie, I.R., Walsh, R.B. and Zaworotko, M.J. (2003) *Cryst. Growth Des.*, **3**, 513–519.

49 Cotton, F.A., Lin, C. and Murillo, C.A. (2001) *Acc. Chem. Res.*, **34**, 759–771.

50 Yaghi, O.M., Davis, C.E., Li, G. and Li, H. (1997) *J. Am. Chem. Soc.*, **119**, 2861–2868.

51 Basic chromium acetate trinuclear cluster; 385 occurrences in CSD [42]: Yaghi, O.M., Jernigan, R., Li, H., Davis, C.E. and Groy, T.L. (1997) *J. Chem. Soc. Dalton Trans.*, 2383–2384. (b) Li, H., Davis, C.E., Groy, T.L., Kelley, D.G. and Yaghi, O.M. (1998) *J. Am. Chem. Soc.*, **120**, 2186–2187. (c) Seo, J.S., Whang, D., Lee, H., Jun, S.I., Oh, J., Jin, Y. and Kim, K. (2000) *Nature*, **404**, 982–986. (d) Barthelet, K., Riou, D. and Férey, G. (2002) *Chem. Commun.*, 1492–1493. (e) Férey, G., Serre, C., Mellot-Draznieks, C., Millange, F., Surble, S., Dutour, J. and Margiolaki, I. (2004) *Angew. Chem. Int. Ed.*, **43**, 6296–6301. (f) Serre, C., Millange, F., Surble, S. and Férey, G. (2004) *Angew. Chem. Int. Ed.*, **43**, 6286–6289. (g) Sudik, A.C., Cote, A.P. and Yaghi, O.M. (2005) *Inorg. Chem.*, **44**, 2998–3000. (h) Férey, G., Mellot-Draznieks, C., Serre, C., Millange, F., Dutour, J., Surble, S. and Margiolaki, I. (2005) *Science*, **309**, 2040–2042. (i) Volkringer, C. and Loiseau, T. (2006) *Mater. Res. Bull.*, **41**, 948–954. (j) Ma, S., Wang, X., Manis, E.S., Collier, C.D. and Zhou, H. (2007) *Inorg. Chem.*, **46**,

3432–3434. (k) Liu, Y., Eubank, J.F., Cairns, A.J., Eckert, J., Kravtsov, V.Ch., Luebke, R. and Eddaoudi, M. (2007) *Angew. Chem. Int. Ed.*, **46**, 3278–3283.

52 Basic zinc acetate tetranuclear cluster; 18 occurrences in the CSD [42]: Li, H., Eddaoudi, M., O'Keeffe, M. and Yaghi, O.M. (1999) *Nature*, **402**, 276–279. (b) Kesanli, B., Cui, Y., Smith, M.R., Bittner, E.W., Bockrath, B.C. and Lin, W. (2005) *Angew. Chem. Int. Ed.*, **44**, 72–75. (c) Sun, D., Collins, D.J., Ke, Y., Zuo, J. and Zhou, H. (2006) *Chem. Eur. J.*, **12**, 3768–3776.

53 Li, H., Eddaoudi, M., Groy, T.L. and Yaghi, O.M. (1998) *J. Am. Chem. Soc.*, **120**, 8571–8572.

54 Millward, A.R. and Yaghi, O.M. (2005) *J. Am. Chem. Soc.*, **127**, 17998–17999.

55 Wong-Foy, A.G., Matzger, A.J. and Yaghi, O.M. (2006) *J. Am. Chem. Soc.*, **128**, 3494–3495.

56 Chae, H.K., Siberio-Perez, D.Y., Kim, J., Go, Y., Eddaoudi, M., Matzger, A.J., O'Keeffe, M. and Yaghi, O.M. (2004) *Nature*, **427**, 523–527.

57 Ockwig, N.W., Delgado-Friedrichs, O., O'Keeffe, M. and Yaghi, O.M. (2005) *Acc. Chem. Res.*, **38**, 176–182.

58 (a) Wells, A.F. (1977) *Three-Dimensional Nets and Polyhedra*, Wiley, New York. (b) Wells, A.F. (1979) *Further Studies of Three-Dimensional Nets*, American Crystallographic Association Monograph 8, American Crystallographic Association, Buffalo, NY. (c) O'Keeffe, M. and Hyde, B.G. (1996) *Crystal Structures I. Patterns and Symmetry*, Mineralogical Society of America, Washington, DC.

59 O'Keeffe, M. *Reticular Chemistry Structure Resource*, http://rcsr.anu.edu.au/.

60 Brant, J.A., Liu, Y., Sava, D.F., Beauchamp, D. and Eddaoudi, M. (2006) *J. Mol. Struct.*, **796**, 160–164.

61 (a) Eubank, J.F., Walsh, R.D. and Eddaoudi, M. (2005) *Chem. Commun.*, 2095–2097. (b) Eubank, J.F., Walsh, R.D., Poddar, P., Srikanth, H., Larsen, R.W. and Eddaoudi, M. (2006) *Cryst. Growth Des.*, **6**, 1453–1457.

62 Liu, Y., Kravtsov, V.Ch., Walsh, R.D., Poddar, P., Hariharan, S. and Eddaoudi, M. (2004) *Chem. Commun.*, 2828–2829.

63 Liu, Y., Kravtsov, V.Ch., Beauchamp, D.A., Eubank, J.F. and Eddaoudi, M. (2005) *J. Am. Chem. Soc.*, **127**, 7266–7267.

64 Liu, Y., Kravtsov, V.Ch., Larsen, R. and Eddaoudi, M. (2006) *Chem. Commun.*, 1488–1490.

65 Baerlocher, Ch. and McCusker, L.B. *Database of Zeolite Structures* http://iza-structure.org/databases/.

11
Polyoxometalate Nanocapsules: from Structure to Function
Charalampos Moiras and Leroy Cronin

11.1
Introduction

The assembly of nanoscale capsules or cages using metal coordination represents one of the most interesting and challenging areas of chemical nanoscience today. This is because these capsules are composed of many discrete metal- and ligand-based building blocks which have the ability to self-assemble into a single gigantic [1], even protein-sized [2], species, often rapidly and in high yield; most of these have very high symmetry [3], rendering them aesthetically extremely appealing. However, the interest in these systems goes far beyond aesthetics since understanding of the principles and mechanism for the assembly process which leads to their formation will allow systems to be designed from first principles. In this chapter we discuss how polyoxometalate-based molecular capsules and cages are defining a new class of capsules and nanospaces [4]. Nanoscale polyoxometalate clusters (POMs) provide an arguably unrivalled structural diversity of molecules displaying a wide range of important physical properties and nuclearities; these cover the range from 6 to 368 metal ions in a single molecule and are assembled under "one-pot" reaction conditions. At the extreme, these cluster molecules are truly macromolecular, rivaling the size of proteins, and are thought to be formed by a self-assembly process (Figure 11.1) [5]. The clusters are based on metal–oxide building blocks with the general formula MO_x (where M is Mo, W, V and sometimes Nb and x can be 4, 5, 6 or 7). POM-based materials have many interesting physical properties which result from their versatile structures, the ability to delocalize electrons over the surface of the clusters, the ability to incorporate heteroanions, electrophiles and ligands and to encapsulate guest molecules within a metal–oxide-based cage. POM clusters have been shown to exhibit superacidity and catalytic activity [6], photochemical activity [7], ionic conductivity [7] and reversible redox behavior [8]. Therefore, there are many potential applications for this class of molecules [4–8].

Organic Nanostructures. Edited by Jerry L. Atwood and Jonathan W. Steed
Copyright © 2008 WILEY-VCH Verlag GmbH & Co. KGaA, Weinheim
ISBN: 978-3-527-31836-0

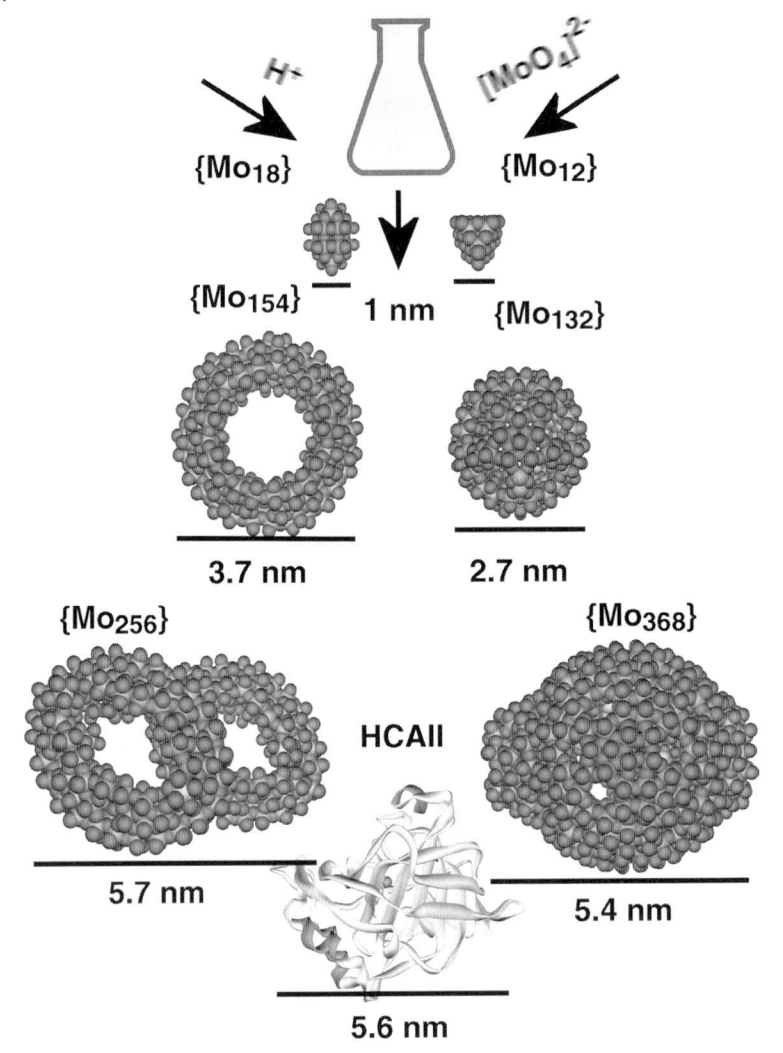

Figure 11.1 Representations of the structures of some POM clusters, all synthesized under "one-pot–one-step" reaction conditions (space filling: Mo large gray spheres, O smaller gray spheres) from the well-known and studied $\{M_{12}\}/\{M_{18}\}$ Keggin–Dawson ions to the $\{Mo_{154}\}/\{Mo_{132}\}$ and $\{Mo_{256}\}/\{Mo_{368}\}$ clusters. These clusters are compared (to scale) with the protein human carbonic anhydrase II (a medium-sized protein with 260 residues, MW 29.6 kDa) to demonstrate their macro dimensions.

Importantly, the assembly of coordination cages involving metal ligand coordination to mediate the self-assembly process depends critically on the building blocks chosen and their respective reactivities. One very important route to the formation of metallosupramolecular architectures involves the selection of building blocks that (1) are preorganized, (2) are kinetically stable, (3) incorporate reactive coordination sites and (4) are complementary. For instance, the exhaustive study of the molybdenum, tungsten and vanadium oxide systems, i.e. polyoxometalate clusters, has emerged as one of very important and versatile building block source in molecular self-assembly [4,5].

11.2
Background and Classes of Polyoxometalates

The large number of structural types in polyoxometalate chemistry [4,5] can be broadly split into three classes:

1. Heteropolyanions are metal–oxide clusters that include heteroanions such as SO_4^{2-} and PO_4^{3-}. These represent by far the most explored subset of POM clusters, with over 5000 papers being reported on these compounds during the last 4 years alone. There is great emphasis on catalysis in this literature, of which the Keggin $[XM_{12}O_{40}]$ and the Wells–Dawson $[X_2M_{18}O_{54}]$ (where M=W or Mo) anions are fundamental examples. Their popularity, reflected in an enormous volume of literature over several decades, can be attributed to a large extent to the ease of their synthesis or commercial availability, but most importantly to the stability of these clusters. In particular W-based POMs are very stable and this has been exploited to develop W-based Keggin ions with vacancies that can be systematically linked using electrophiles to larger aggregates [4,5].

2. Isopolyanions are composed of a metal–oxide framework, but without the internal heteroatom/heteroanion. As a result, they are often much more unstable than their heteropolyanion counterparts. However, they also have interesting physical properties, such as high charges and strongly basic oxygen surfaces, which means that they are attractive units for use as building blocks [9].

3. Mo-blue and Mo-brown reduced Mo-based POM clusters are related to molybdenum blue-type species, which was first reported by Scheele in 1783 [10]. Their composition was largely unknown until Müller et al. reported, in 1995, the synthesis and structural characterization of a very high nuclearity cluster $\{Mo_{154}\}$ crystallized from a solution of Mo-blue, which has a ring topology [11]. The interest generated by this result is partly due to its high nuclearity and partly because of the size of this cluster; with an outer diameter of ca. 34 Å, an inner diameter of 25 Å and a thickness of 14 Å, it is a truly nanoscopic molecule. Using reaction conditions of pH ≈ 1, with a concentration of Na_2MoO_4 of ca. 0.5 M and a degree of reduction of between 1 and 20%, the solution yields the "giant-wheel" $[Mo_{154}O_{462}H_{14}(H_2O)_{70}]^{14-}$ in over 80% yield in 24 h [12]. The building up principle does not stop there: a series of mixed-valence Mo-blue (containing delocalized Mo^V–Mo^{VI}) clusters (e.g. $[Mo_{256}Eu_8O_{776}H_{20}(H_2O)_{162}]^{20-} \equiv \{Mo_{256}\}$) [13] have been reported and also a class of spherical Mo-brown (containing localized Mo^V–Mo^V units) clusters (e.g. $[Mo^{VI}_{72}Mo^V_{60}O_{372}(MeCO_2)_{30}(H_2O)_{72}]^{42-} \equiv \{Mo_{132}\}$) [14] and the highest nuclearity cluster so far found, a "lemon" cluster ($[H_xMo_{368}O_{1032}(H_2O)_{240}(SO_4)_{48}]^{48-} \equiv \{Mo_{368}\}$) [2] (Figure 11.2).

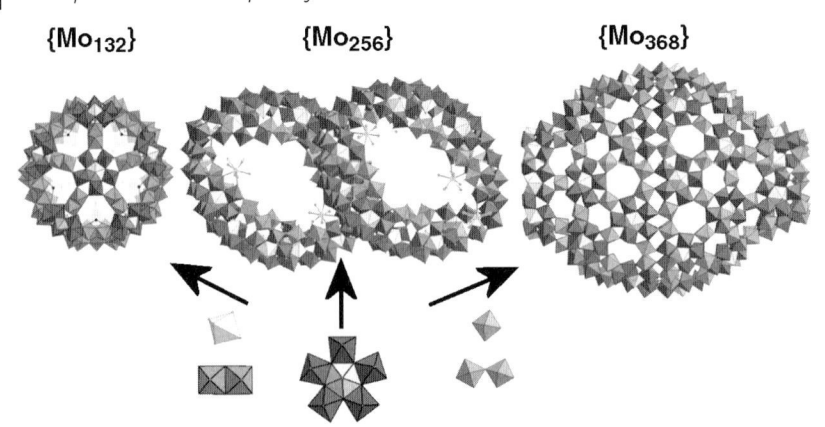

Figure 11.2 Structures of the $\{Mo_{132}\} \equiv [Mo^{VI}_{72}Mo^{V}_{60}O_{372}(MeCO_2)_{30}(H_2O)_{72}]^{42-}$, $\{Mo_{256}\} \equiv [Mo_{256}Eu_8O_{776}H_{20}(H_2O)_{162}]^{20-}$ and $\{Mo_{368}\} \equiv [H_xMo_{368}O_{1032}(H_2O)_{240}(SO_4)_{48}]^{48-}$ clusters shown with polyhedral plots. The transferable building blocks found in these clusters are shown below also as polyhedral plots whereby the metal is at the center of the polyhedron and the oxygen ligands form the vertexes.

11.3
Wells–Dawson $\{M_{18}O_{54}\}$ Capsules

The Wells–Dawson cluster type can be considered to be an $\{M_{18}O_{54}\}^{m-}$ cluster that often incorporates either one or two templating anionic units e.g. PO_4^{3-} and SO_4^{2-} [5], so the cluster can be formulated as $([M_{18}O_{54}(XO_n)_2]^{m-}$; M=Mo or W, X = a main group element, $n = 3$ or 4) (Figure 11.3). The $\{M_{18}O_{54}\}^{m-}$ can be considered to have a hydrophilic cavity and can incorporate many types of different anions; traditionally these have been either PO_4^{3-} or SO_4^{2-} [4,5]. By incorporating electronically interesting anions such as sulfite, it is possible to engineer several types of cluster cage

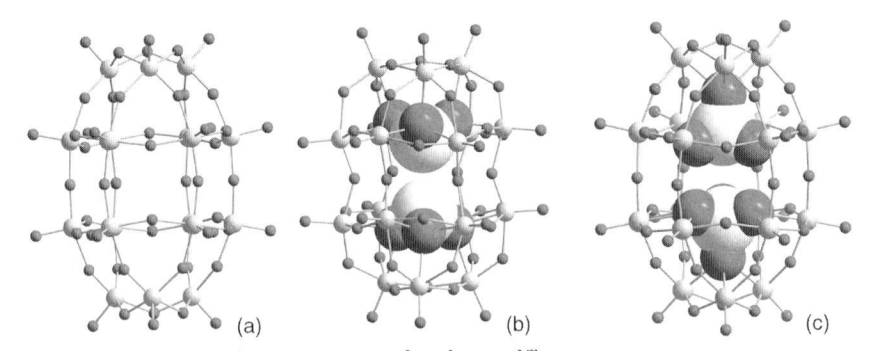

Figure 11.3 Ball and stick representations of (a) $\{M_{18}O_{54}\}^{m-}$, (b) $[Mo_{18}O_{54}(SO_3)_2]^{4-}$ and (c) $[Mo_{18}O_{54}(SO_4)_2]^{4-}$. The M atoms are shown as small gray spheres and the O atoms as dark gray spheres. The S and O atoms of the SO_3 and SO_4 groups are shown in space filling mode.

whereby the anions can themselves interact with each other or undergo reactions inside the cluster cage. This sulfite-containing cluster can be formulated [15] as $[Mo_{18}O_{54}(SO_3)_2]^{4-}$ and the sulfites as positioned within the cluster cage such that they are located only 3.2 Å apart from each other inside the cluster shell (this is 0.4 Å less than the 3.6 Å expected from non-bonded S···S interactions but still longer than the 2.2 Å found in $S_2O_6^{2-}$) and could indicate a strong interaction between the two sulfur atoms; this was also suggested by DFT calculations [15]. In the limit it was thought that engineering such capsules may allow the oxidation of sulfite to yield dithionate $S_2O_6^{2-}$ anions (note: sulfur is the only main group element to form $X_2O_6^{n-}$ analogues with X—X single bonds), e.g.

$$2SO_3^{2-} \rightarrow S_2O_6^{2-} + 2e^-$$

Thus, engineering an intramolecular S···S interaction within the Dawson $\{Mo^{VI}_{18}\}$ matrix is interesting since the formation of a dithionate anion would release two electrons to reduce the surrounding polyoxomolybdate shell to the mixed-valence reduction state $\{Mo^{VI}_{16}Mo^{V}_2\}$, with its characteristic blue color. Although switching has not yet been achieved, the sulfite-based clusters are thermochromic between 77 and 500 K [15].

By altering the synthetic conditions, it is also possible to obtain a sulfite-based polyoxotungstate, α-$[W_{18}O_{54}(SO_3)_2]^{4-}$ [16], which is isostructural with α-$[Mo_{18}O_{54}(SO_3)_2]^{4-}$ [14], and $[W^{VI}_{18}O_{56}(SO_3)_2(H_2O)_2]^{8-}$ [16]. The latter is described as a "Trojan horse" in which a structural rearrangement allows the two embedded pyramidal sulfite (SO_3^{2-}) anions to release up to four electrons (analogous to the "soldiers" hidden inside the "Trojan horse") to the surface of the cluster generating the sulfate-based, deep-blue, mixed-valence cluster $[W_{18}O_{54}(SO_4)_2]^{8-}$ upon heating (Figure 11.4). The sulfite anions adopt a radically different orientation in $[W^{VI}_{18}O_{56}(SO_3)_2(H_2O)_2]^{8-}$,

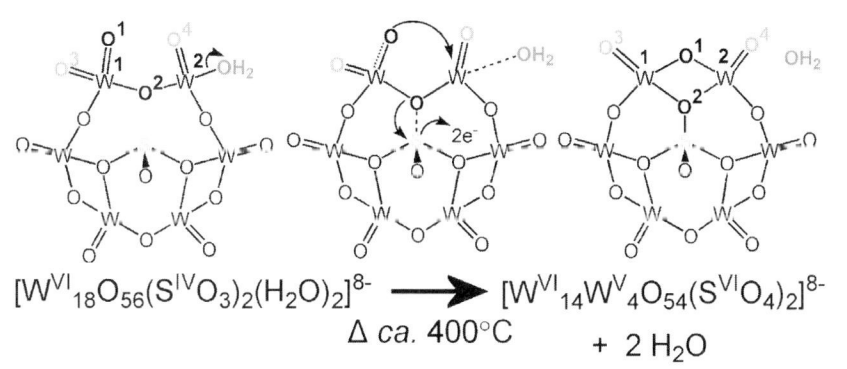

$[W^{VI}_{18}O_{56}(S^{IV}O_3)_2(H_2O)_2]^{8-} \xrightarrow{\Delta \text{ ca. } 400°C} [W^{VI}_{14}W^{V}_{4}O_{54}(S^{VI}O_4)_2]^{8-} + 2H_2O$

Figure 11.4 Scheme showing the change in the metal oxo framework on one half of the cluster upon oxidation of the internal SO_3^{2-} ligand to SO_4^{2-} (shown by the movement of each of the number oxygen atoms on the left to the end position on the right), which is commensurate with the reduction of the cluster shell by four electrons, giving rise to the deep-blue material from the colorless crystals.

whereby they each only ligate to seven metal centers: three from the cap and four (out of six) from the "belt" of the cluster framework.

The orientation for the sulfite anions within the cluster type is somewhat like the coordination mode for the tetrahedral templates (XO_4^{y-}) in conventional Dawson $[M_{18}O_{54}(XO_4)_2]^{2y-}$, i.e. one of the oxo ligand bridges three capping W centers, and the remaining oxo ligands each bridge two of the "belt" W centers. Nevertheless, this leaves two "belt" W atoms uncoordinated to the template SO_3 moiety as SO_3 has one oxo ligand less than XO_4. Hence it can be seen that the sulfite ions are grafted on to the bottom side of the cluster, which resembles a "basket" with four "uncoordinated" "belt" metal centers on the top part and now has a lower C_{2v} symmetry compared with the cluster α-$[Mo_{18}O_{54}(SO_3)_2]^{4-}$, which has a D_{3h} symmetry. To compensate for the coordination, these unique "uncoordinated" "belt" W centers (four for the whole cluster) each have two terminal ligands, rather than one as found for the remaining metal centers in the cluster. These are in addition to the four other μ_2 bridging oxo (O^{2-}) ligands between metal centers and complete a slightly distorted octahedral coordination geometry for each of the four "uncoordinated" "belt" metal centers concerned. Single-crystal structure analysis revealed that two of the four unique metal centers each have two W=O terminals (W−O ≈ 1.7 Å) and the other two each have one W=O terminal and one W−OH_2 terminal (W−O ≈ 1.7 and 2.2 Å, respectively). Furthermore, it is interesting that the unique "belt" μ_2 bridging oxo ligands between the pair "uncoordinated" "belt" W atoms now bends in towards the cluster, rather than outwards as normal, and is located ca. 2.9 Å distant from the sulfur center of the SO_3 moiety, whereas the two sulfur centers are positioned 3.6 Å apart at opposite sides of the cluster shell. In this respect, the mechanism for the reduction of the cluster shell requires an interaction between the sulfur atom and the special belt oxo ligand, which then react to form two sulfate anions located within the $\{W_{18}\}$ cluster shell.

In summary, compounds $[Mo_{18}O_{54}(SO_3)_2]^{4-}$ and $[W_{18}O_{56}(SO_3)_2(H_2O)_2]^{8-}$ demonstrate unprecedented electronic properties, thermochromism and a unique electron-transfer reaction, in which, when heated, a structural rearrangement allows the two embedded pyramidal sulfite ($S^{IV}O_3^{2-}$) anions to release up to four electrons to the surface of the cluster, generating the sulfate-based, deep-blue, mixed-valence cluster $[W_{18}O_{54}(SO_4)_2]^{8-}$. Although electron-transfer reactions and structural rearrangements are known for HPOMs, these electron transfer reactions are "confined" within a molecular nanocage whereby the electrons are "released" from the core of the cluster. Moreover, this property is really intriguing, from the point of view of nanoscience, and $[W_{18}O_{56}(SO_3)_2(H_2O)_2]^{8-}$ can be considered to be a prototype nanodevice which responds according to a stimulus.

11.4
Isopolyoxometalate Nanoclusters

Going one step further, someone could realize the fact that porosity, separation and ion transportation effects start being a noteworthy feature of the POM capsules.

11.4 Isopolyoxometalate Nanoclusters

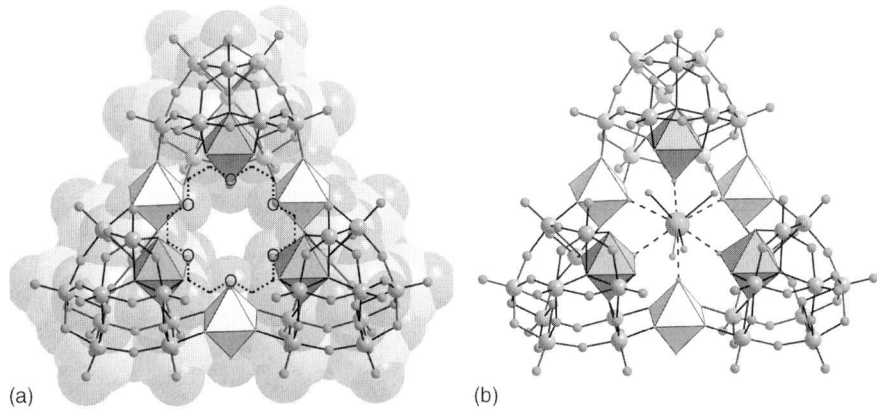

Figure 11.5 (a) Comparative illustration of the {W_{36}} cluster framework $[H_{12}W_{36}O_{120}]^{12-}$ and the 18-crown-6 structure (show to scale). W and O atoms shown as light and dark spheres, respectively, with the crown ether superimposed on O atoms forming the "cavity" of the cluster. The ball and stick structure is also superimposed on the space-filling CPK representation, along with a polyhedral representation of the six W units that provide the "crown" coordinating O atoms. (b) The same ball and stick/polyhedral structure but this time showing the complexed potassium ion and the ligated water molecules within the cavity representing the overall $\{(H_2O)_4K \subset [H_{12}W_{36}O_{120}]\}^{11-}$ complex.

The {W_{36}}-based cluster with the formula $\{(H_2O)_4K_{[H_{12}}W_{36}O_{120}]\}^{11-}$ includes the threefold symmetric cluster anion $[H_{12}W_{36}O_{120}]^{12}$ (Figure 11.5) [17]. The cluster anion complexes a potassium ion at the center of the {W_{36}} cluster in an O_6 coordination environment. The {W_{36}} structure consists of three {W_{11}} subunits; these subunits contain a ring of six basal W positions, an additional W position in the center of this ring and four apical W positions in a butterfly configuration. Every W position around the cluster center has a distorted WO_6 octahedral coordination geometry with one terminal W=O moiety $[d(W=O) \approx 1.70 \text{ Å}]$ extending towards the cluster center where the K ion is located; this arrangement maps extremely well on to the structure of the crown ether [18]crown-6.

The implications for the development of this system in a similar fashion to the crown ethers is interesting, especially the possibilities for discrimination and sensing of metal ions using this cluster framework [18]. The crown ether-like properties for the {W_{36}} cluster can be realized in the presence of other cations such as K^+, Rb^+, Cs^+, NH_4^+ and Sr^{2+}, and it was determined that these ions can bind to the cluster and their structures crystallographically determined (Table 11.1).

As can be seen in the table, the found average oxygen–metal distances in the {W_{36}} host–guest complexes **1–5** are very close to the distances shown in the corresponding 18-crown-6 complexes. The metal cations in the synthesized compounds show different distances to the equatorial plane of the cluster framework which reflect the different sizes of their ionic radii. Although there are some great similarities between the {W_{36}} system and 18-crown-6, there are some features that are not common to both systems. First, the six-coordinated oxygen atoms on {W_{36}} are not

Table 11.1 Comparison of acquired metal–oxygen distances and displacements of metal ions to the cavity center in the {W$_{36}$} cluster and the corresponding figures for the 18-crown-6 ether (e.s.d.s are all with 0.02 Å).

Cation	Ionic radius (Å)	Average metal–oxygen distance (Å)		Distance from cavity center (Å)	
		{W$_{36}$} cluster	18-Crown-6	{W$_{36}$} cluster	18-Crown-6
K$^+$	1.38	2.80	2.80	0.70	0
Rb$^+$	1.52	2.87	2.95	0.84	0.93
Cs$^+$	1.67	3.16	3.18	1.61	1.47
Sr^{2+}	1.18	2.70	2.73	0.53	0
Ba^{2+}	1.35	2.82	2.82	0.73	0

planar, whereas those on 18-crown-6 can adopt a planar conformation. Further, the {W=O}$_6$ donor groups of the {W$_{36}$} crown are rigid, unlike in 18-crown-6, which is much more flexible. This means that 18-crown-6 can deform to form metal complexes with small ions such as Na$^+$, Ca^{2+}, lanthanide ions and d-transition metal ions. This is because 18-crown-6, along with other similar crown ethers, is able to distort and wrap itself around these smaller metal cations in an attempt to maximize the electrostatic interactions. This increases the strain of the ligand, which makes these complexes less stable than those with metal cations of optimal spatial fit. However, the {W$_{36}$} framework, due to its high rigidity, simply cannot change conformation in such a way as to bind these metal cations. This is partly confirmed by the observation that the diameter of the central cavity present in the family of {W$_{36}$} clusters presented here is very well defined and rigid.

11.5
Keplerate Clusters

Reduced polyoxomolybdate clusters represent one of the most interesting cluster classes for nanoscience as these clusters adopt ring and spherical shapes comprising pentagonal {(Mo)Mo$_5$} building blocks [5]. In particular, the spherical icosahedral {Mo$_{132}$} Keplerate cluster as been described as an inorganic superfullerene due to its symmetry and large nanoscale size [19]. The discovery of such a cluster, particularly as it is spherical, water soluble and has such a large cavity, probably represents one of the most significant findings in recent years [20], especially since this discovery has gone on to reveal a whole family of related Keplerates. Within this family [20–33], all the spherical and approximately icosahedral clusters have the form [{(pent)$_{12}$(link)$_{30}$}], e.g. like [{(Mo)(Mo$_5$O$_{21}$(H$_2$O)$_6$}$_{12}${Mo$_2$O$_4$(ligand)}$_{30}$]$^{n-}$ with binuclear linkers where the 12 central pentagonal units span an icosahedron and the linkers form a distorted truncated icosahedron; the highly charged capsule with sulfate ligands and $n = 72$ was used very successfully. For instance, the truly nanoscale capsules (inner cavity diameter ca. 2.5 nm) allow different types of encapsulations, e.g. of well-structured

Figure 11.6 A polyhedral representation of the {Mo$_{132}$} Keplerate cluster. The pentagonal {Mo$_6$} building blocks and the {Mo$_2$} building blocks can be seen. The bridgehead atom of the XO$_2$ bridge of the {Mo$_2$} groups can be seen as black spheres.

large water assemblies (up to 100 molecules) with an "onion"-like layer structure enforced by the outer shell (Figure 11.6) [21]. Most importantly, the capsules have 20 well-defined pores and the internal shell functionalities can be tuned precisely since the nature of the bidentate ligands can be varied. In the special case of binuclear Mo$^V_2$O$_4^{2+}$ linkers the pores are {Mo$_9$O$_9$} rings with a crown ether function (diameters 0.6–0.8 nm) which can be reversibly closed, e.g. by guanidinium cations interacting noncovalently with the rings via formation of hydrogen bonds [22]. In a related smaller capsule with mononuclear linkers, the {Mo$_6$O$_6$} pores can become closed/complexed correspondingly by smaller potassium ions [23].

The most intriguing and exciting property of the highly negatively charged capsules is that they can mediate cation transfer from the solution to the inner nanocavity. Indeed, reaction of the above-mentioned highly charged capsule with different substrates/cations such as Na$^+$, Cs$^+$, Ce^{3+}, C(NH$_2$)$_3^+$ and OC(NH$_2$)NH$_3^+$ in aqueous solution leads to formations/assemblies which exhibit well-defined cation separations *at*, *above* or *below* the capsules channel-landscapes ("nano-ion chromatograph" behavior) [24]. Taking this one step further, a temperature-dependent equilibrium process that involves the uptake/release of Li$^+$ ions through the capsule pores has been observed: the porous capsule behaves as a semipermeable inorganic membrane open for H$_2$O and small cations [25,26]. Furthermore, the 20 pores of the same capsule "shut" by protonated urea as "stoppers" can be opened in solution, thus allowing calcium(II) ion uptake while later closing occurs again (Figure 11.7) [27]. Remarkably,

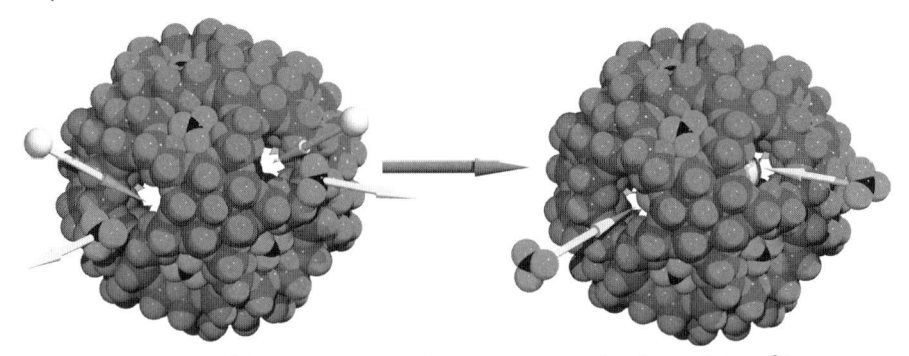

Figure 11.7 Space-filling representation demonstrating a simplified view of the Ca^{2+} ion uptake based on the capsule $[\{(Mo)(Mo_5O_{21}(H_2O)_6\}_{12}\{Mo_2O_4(SO_4)\}_{30}]^{72-}$. Initially the pores are closed, but treating a solution of the capsule with Ca^{2+} ions leads to cation uptake (left) while in the final product the pores again are closed (right; Mo, dark gray, O, medium gray, C, black, N/O(urea), light gray, Ca^{2+}, light gray spheres).

"pore gating" – just modeling biological ion transport – can illustratively be demonstrated: after initial cation uptake, subsequent cations are found hydrated above the pores due to a decrease of negative capsule charge [28].

This type of nanocapsules proved their efficiency in cation separation processes also in the presence of heavier metal atoms, such as Pr^{3+} [29]. The interesting phenomenon of metal cations such as Pr^{3+} entering into channels and the inner capsule, thereby constituting a novel situation with a metal center in two different environments corresponding to a coordination chemistry under confined conditions or, in other words, cation transfer in a controlled and specific fashion, has been reported. The reported findings open up perspectives for a special type of en capsulation chemistry, i.e. coordination chemistry under confined conditions. This can be extended to (1) a variety of ligands like the present one, (2) different capsule charges, (3) different types of porosity, (4) different solvent molecules like water and (5) different metal centers. In addition, the presence of cations such as Pr^{3+} in two different coordination geometries gives a unique opportunity to study the ligand influence on the electronic structure of rare earth compounds.

Furthermore, nanocapsules have given us also the opportunity to study the structures of the simplest chemical reagent we have in our hands, namely pure H_2O. Even though it is the simplest chemically known compound, it gives rise to a plethora of structural motifs which are difficult to study [30]. The use of nanocapsules as "crystallization flasks" has given the advantage of isolating different kinds of water structures, while the same system can also be studied as a probe to investigate "complex system" behavior in general, and further, a "new state of inorganic ions" involved in the formation of a novel type of aggregates [31], even allowing the control of the aggregate size by the change in the cluster charge. It is even possible to incorporate large cluster guests, e.g. $\{PMo_{12}O_{40}\}^{3-}$, within the Fe-substituted Keplerate cluster $\{Mo_{72}Fe_{30}\}$, where the Keplerate host acts rather like a prison, totally encompassing the guest or, more precisely, a hostage molecule to give

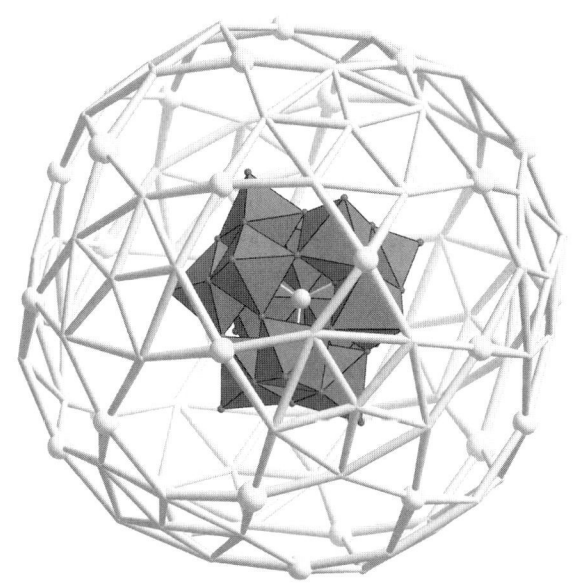

Figure 11.8 (a) Structure of $\{Mo_{12}\}\subset[\{Mo_{132}\}]$. The hostage is shown in polyhedral form and the $\{Mo_{72}Fe_{30}\}$ is shown as the framework with the {Fe} positions shown as white spheres linking the 12 $\{(Mo^{VI})Mo^{VI}{}_5\}$ pentagons, and the Keggin nucleus in polyhedral representation.

the compound $[PMo_{12}O_{40\{(Mo^{VI})}Mo^{VI}{}_5\}_{12}Fe^{III}O_{252}(H_2O)_{102}(CH_3COO)_{15}].\sim 120H_2O$ (Figure 11.8) [32,33].

11.6
Surface-Encapsulated Clusters (SECs): Organic Nanostructures with Inorganic Cores

In an effort to combine inorganic polyoxometalates and organic materials chemistry, cationic surfactants have been applied to improve the surface properties of POMs. The resulting surfactant encapsulated complexes (SECs) are compatible with organic matrixes, improve the stability of the encapsulated cluster against fragmentation, enhance the solubility in nonpolar, aprotic organic solvents and neutralize their charge, thus leading to discrete, electrostatically neutral assemblies, while altering the surface chemical properties in a predictable manner. Importantly, the basic physical and chemical properties of the polyoxometalates are retained [34] while the coexistence of hydrophobic alkyl chains and hydrophilic clusters in SECs gives them an amphiphilic character. Recently, it has been reported that this amphiphilic character is also exhibited in a solution environment and an unusual vesicular assembly of $(DODA)_4H\ [Eu(H_2O)_2SiW_{11}O_{39}]$ (DODA = dimethyldioctadecylammonium) has been observed [35]. An even larger aggregation

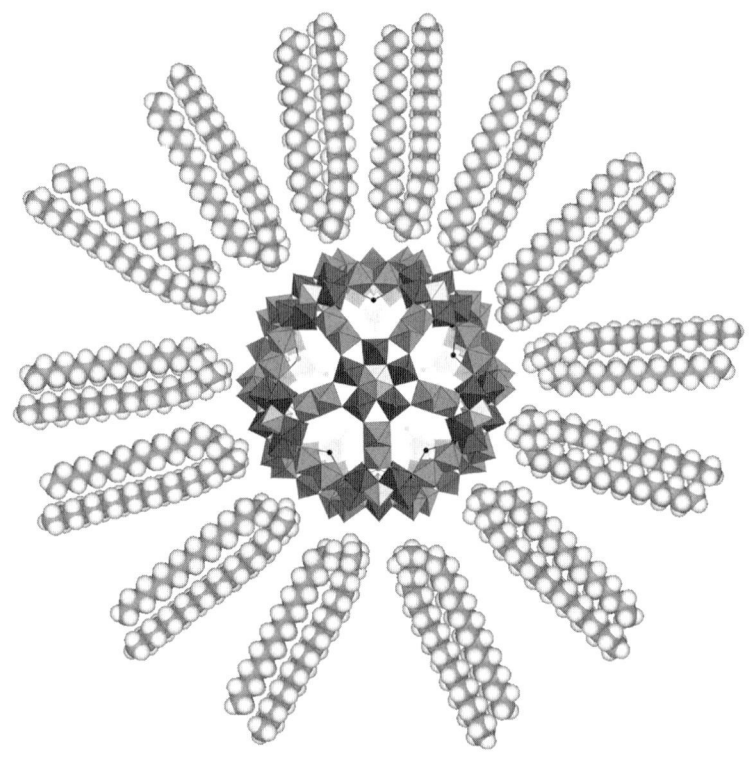

Figure 11.9 {Mo_{132}} SEC cluster $(DODA)_{40}(NH_4)_2$ [$(H_2O)_n Mo_{132}O_{372}(CH_3CO_2)_{30}$ $(H_2O)_{72}$] showing the inorganic core and the "soft" organic outer shell.

has been reported recently, the "onion"-like structure $(DODA)_4 SiW_{12}O_{40}$ [36] and $(DODA)_{40}(NH_4)_2$ [$(H_2O)_n Mo_{132}O_{372}(CH_3CO_2)_{30}$ $(H_2O)_{72}$] [37,38] (Figure 11.9) and also $(DODA)_{20}(NH_4)$ [$H_3Mo_{57}V_6(NO)_6O_{183}(H_2O)_{18}$] [39], where the spherical species can provide a similar microenvironment to vesicles, which makes these assemblies suitable carriers to perform the catalytic and pharmacological functions of polyoxometalates. Moreover, the ability of polyoxometalates to undergo multiple reduction/oxidation steps may be exploited in information storage devices or optical switches. The possibility of fabricating well-defined two-dimensional arrays is an important step towards this goal.

The above-mentioned findings and suggestions reveal the feasibility of utilizing SECs as polyoxometalate-containing amphiphiles to construct regular assemblies from solution, help us to comprehend the catalyzing reactions of polyoxometalates in organic media, promote the effective interaction with organic molecules of biological interest and enhance, potentially, the interaction with biological media. However, the origin of the assembly process and whether it is a general behavior of SECs in solution are both still the subject of much debate and study.

11.7
Perspectives

It is clear that polyoxometalate-based building blocks can provide routes to the designed assembly of nanoscale capsules using coordinative interactions. This in turn is leading to and defining new areas of chemistry, i.e. under confined conditions, with parallels to processes/situations in biological cells regarding cell response and ion transport and even the possibility of engineering systems which exhibit complex and maybe adaptive behavior allowing chemical emergence. One tantalizing objective is the observation of evolving and dissipative inorganic systems. Generally, matter can be studied under confined conditions while the discovery of new systems resulting from encapsulation reveals new phenomena not observable in the bulk. This includes "confined water" with and without electrolytes and also spectacular chemical reactions. The presence of well-defined cavities/nanospaces and gated pores allows specific interactions of the capsules with their environments and subsequent uptake and selective binding of guests, since the architectures of the inner cluster walls can be "programmed/redesigned". Therefore, the design and synthetic approaches to polyoxometalates and the fact that these clusters can be constructed over multiple length scales, along with their almost unmatched range of physical properties, mean that they are serious candidates to be used as the functional part of any nano-device. The challenge now is to design individual POM cluster molecules that can interact both with each other and with the macroscale, in a desired fashion in response to inputs and environmental effects, so a functioning molecular system is really constructed. These capsule systems are truly fascinating and the future is exciting, since great leaps in understanding the principles that underpin the assembly mechanisms of such capsules allow the designed construction of extremely complex and specifically interacting systems – and this in the broadest sense is "bringing inorganic chemistry to life".

References

1 Sato, S., Iida, J., Suzuko, K., Kawano, M., Ozeki, T. and Fujita, M. (2006) *Science*, **313**, 1273.
2 Müller, A., Beckmann, E., Bögge, H., Schmidtmann, M. and Dress, A. (2002) *Angew. Chem. Int. Ed.*, **41**, 1162.
3 Pope M.T. and Muller A. (eds) (2001) *Polyoxometalate Chemistry: from Topology via Self-Assembly to Applications*, Kluwer, Dordrecht. Wassermann, K., Dickman, M.H. and Pope, M.T. (1997) *Angew. Chem. Int. Ed. Engl.*, **36**, 1445.
4 Long, D.-L., Burkholder, E. and Cronin, L. (2007) *Chem. Soc. Rev.*, **36**, 105.
5 Cronin, L. (2004) in *Comprehensive Coordination Chemistry II*, (eds J.A. McCleverty and T.J. Meyer), Vol. 7 Elsevier, Amsterdam, pp. 1–56.
6 Neumann, R. and Dahan, M. (1997) *Nature*, **388**, 353; Mizuno, N. and Misono, M. (1998) *Chem. Rev.*, **98**, 199.
7 Katsoulis, D.E. (1998) *Chem. Rev.*, **98**, 359; Yamase, T. (1998) *Chem. Rev.*, **98**, 307.

8 Rüther, T., Hultgren, V.M., Timko, B.P., Bond, A.M., Jackson, W.R. and Wedd, A.G. (2003) *J. Am. Chem. Soc.*, **125**, 10133.

9 Long, D.L., Kogerler, P., Parenty, A.D.C., Fielden, J. and Cronin, L. (2006) *Angew. Chem. Int. Ed.*, **45**, 4798.

10 Scheele, C.W. (1971) in *Sämtliche Physische und Chemische Werke*, (ed. D.S.F. Hermbstädt), M. Sändig, Niederwalluf/Wiesbaden, Vol. 1 pp. 185–200 (reprint; original published 1783).

11 Müller, A., Krickemeyer, E., Meyer, J., Bögge, H., Peters, F., Plass, W., Diemann, E., Dillinger, S., Nonnenbruch, F., Randerath, M. and Menke, C. (1995) *Angew. Chem. Int. Ed.*, **34**, 2122.

12 Cronin, L., Diemann, E. and Müller, A. (2003) in: *Inorganic Experiments*, (ed. Woollins J.D.), Wiley-VCH, Weinheim, pp. 340–346.

13 Cronin, L., Beugholt, C., Krickemeyer, E., Schmidtmann, M., Bögge, H., Kögerler, P., Luong, T.K.K. and Müller, A. (2002) *Angew. Chem. Int. Ed.*, **41**, 2805.

14 Müller, A., Das, S.K., Talismanov, S., Roy, S., Beckmann, E., Bögge, H., Schmidtmann, M., Merca, A., Berkle, A., Allouche, L., Zhou, Y.S. and Zhang, L.J. (2003) *Angew. Chem. Int. Ed.*, **42**, 5039.

15 Long, D., Abbas, H., Kogerler, P. and Cronin, L. (2004) *Angew. Chem. Int. Ed.*, **43**, 1817.

16 Long, D., Abbas, H., Kogerler, P. and Cronin, L. (2005) *Angew. Chem. Int. Ed.*, **44**, 3415.

17 Long, D., Abbas, H., Kögerler, P. and Cronin, L. (2004) *J. Am. Chem. Soc.*, **126**, 13880.

18 Long, D.L., Brucher, O., Streb, C. and Cronin, L. (2006) *Dalton Trans.*, 2852.

19 Müller, A., Kögerler, P. and Bögge, H. (2000) *Struct. Bonding*, **96**, 203.

20 Müller, A., Krickemeyer, E., Bögge, H., Schmidtmann, M. and Peters, F. (1998) *Angew. Chem. Int. Ed.*, **37**, 3359.

21 Müller, A., Krickemeyer, E., Bögge, H., Schmidtmann, M., Botar, B. and Talismanova, M.O. (2003) *Angew. Chem. Int. Ed.*, **42**, 2085.

22 Müller, A., Krickemeyer, E., Bögge, H., Schmidtmann, M., Roy, S. and Berkle, A. (2002) *Angew. Chem. Int. Ed.*, **41**, 3604.

23 Müller, A., Botar, B., Bögge, H., Kögerler, P. and Berkle, A. (2002) *Chem. Commun.*, 2944.

24 Müller, A., Das, S.K., Talismanov, S., Roy, S., Beckmann, E., Bögge, H., Schmidtmann, M., Merca, A., Berkle, A., Allouche, L., Zhou, Y.S. and Zhang, L.J. (2003) *Angew. Chem. Int. Ed.*, **42**, 5039.

25 Müller, A., Rehder, D., Haupt, E.T.K., Merca, A., Bögge, H., Schmidtmann, M. and Heinze-Brückner, G. (2004) *Angew. Chem. Int. Ed.*, **43**, 4466.

26 Haupt, E.T.K., Wontorra, C., Rehder, D., Müller, A. (2005) *Chem. Commun.*, 3912. Müller, A., Botar, B., Bögge, H., Kögerler, P. and Berkle, A. (2002) *Chem. Commun.*, 2944.

27 Müller, A., Toma, L., Bögge, H., Schäffer, C. and Stammler, A. (2005) *Angew. Chem. Int. Ed.*, **44**, 7757.

28 Müller, A., Zhou, Y., Bögge, H., Schmidtmann, M., Mitra, T., Haupt, E.T.K. and Berkle, A. (2006) *Angew. Chem. Int. Ed.*, **45**, 460.

29 Müller, A., Zhoua, Y., Zhang, L., Bogge, H., Schmidtmann, M., Dresselb, M. and van Slageren, J. (2004) *Chem. Commun.*, 2038.

30 Müller, A., Krickemeyer, E., Bogge, H., Schmidtmann, M., Botar, B. and Talismanova, M.O. (2003) *Angew. Chem. Int. Ed.*, **42**, 2085.

31 Müller, A., Diemann, E., Kuhlmann, C., Eimer, W., Serain, C., Tak, T., Knoechel, A. and Pranzas, P.K. (2001) *Chem. Commun.*, 1928.

32 Müller, A., Das, S.K., Kögerler, P., Bögge, H., Schmidtmann, M., Trautwein, A.X., Schünemann, V., Krickemeyer, E. and Preetz, W. (2000) *Angew. Chem. Int. Ed.*, **112**, 3556.

33 Müller, A., Todea, A.M., Bogge, H., van Slageren, J., Dressel, M., Stammlera, A. and Rusuc, M. (2006) *Chem. Commun.*, 3066.

34 Tao, Y., Yu, Q. and Bu, X.H. (2007) *Chem. Commun.*, **15**, 1527.

35 Li, J.R., Bu, W., Li, H., Sun, H., Yin, S. and Wu, L. (2005) *J. Am. Chem. Soc.*, **127**, 8016.
36 Li, H., Sun, H., Qi, W., Xu, M. and Wu, L. (2007) *Angew. Chem. Int. Ed.*, **46**, 1300.
37 Kurth, D.G., Lehmann, P., Volkmer, D., Müller, A. and Schwahn, D. (2000) *J. Chem. Soc., Dalton Trans.*, 3989.
38 Kurth, D.G., Lehmann, P., Volkmer, D., Cölfen, H., Koop, M.J., Müller, A. and Chesne, A.D. (2000) *Chem. Eur. J.*, **6**, 385.
39 Volkmer, D., Chesne, A.D., Kurth, D.G., Schnablegger, H., Lehmann, P., Koop, M.J. and Müller, A. (1995) *J. Am. Chem. Soc.*, **46**, 122.

12
Nano-capsules Assembled by the Hydrophobic Effect
Bruce C. Gibb

12.1
Introduction

Compartmentalization is a key feature of living systems, helping to control in a temporal manner both where a particular molecule can be found, its concentration and what it can react with. This is in contrast to contemporary solution-based chemistry, where solvent, temperature and reactant concentrations offer some degree of control, but by and large random collision reigns. Moving from solution-based chemistry to systems that are compartmentalized in some manner therefore offers many exciting possibilities. Greater reaction control is one obvious example [1–5], but the complexity of biosystems undoubtedly harbors many as yet unknown "systems chemistry" phenomena that will only be pinpointed with the aid of model systems less complex than biochemical networks.

Compartments, of course, come in many shapes and sizes. They can be composed of a single shell-like molecule [6–9], but more often are engendered by the self-assembly of molecular subunits [10]. The nature of the subunits themselves determines what drives their assembly, whether the inner domain is similar or completely different from the outer domain, how thick the partition is (a few atoms, a bilayer, multiple bilayers?) and how permeable the shell or membrane is. Our focus here is a compartmentalization driven by the hydrophobic effect [11,12]. However, in contrast to "fluid" assemblies such as liposomes and micelles composed of many hundreds or thousands of copies of the subunits, the assemblies in question are dimerizations; dimers of bowl-shaped molecules called cavitands that are extremely well defined structurally (Figure 12.1). These assemblies also differ from micelles and liposomes in as much as a templating guest (or guests) is required to trigger assembly of the nano-capsule. Many analogous self-assembling systems have been reported previously, but the driving forces for their formation have been highly directional non-covalent forces such as hydrogen bonding [13–17] or metal

---

Organic Nanostructures. Edited by Jerry L. Atwood and Jonathan W. Steed
Copyright © 2008 WILEY-VCH Verlag GmbH & Co. KGaA, Weinheim
ISBN: 978-3-527-31836-0

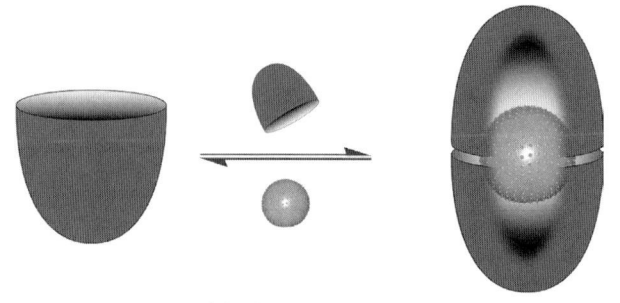

Figure 12.1 Cartoon of the dimerization of two cavitand molecules around a guest.

coordination [4,18–20]. Hence, the system in question offers a unique perspective on how the hydrophobic effect drives assembly and may shed light on biologically relevant assemblies such as protein quaternary structure.

12.2
Synthesis of a Water-soluble, Deep-cavity Cavitand

The structure of the water-soluble cavitand used to form the dimeric nano-capsules (**1**) is shown in Scheme 12.1. The synthesis is accomplished in seven steps, only two of which require purification by chromatography [21]. It begins with formation of the resorcinarene **2**, a reaction that can be performed easily on the hundreds of grams scale [22]. This resorcinarene is then converted to cavitand **3** by bridging using 3,5-dibromobenzal bromide; a compound available in near quantitative yields by treating the commercially available 3,5-dibromobenzaldehyde with BBr_3. Cavitand **3** is awkward to purify so without purification it is protected as its tetrabenzyl ether **4**. The 40% yield for the two steps understates the efficiency of the initial stereoselective bridging that forms four stereogenic centers and eight covalent bonds. Thus, assuming quantitative protection, a 40% yield corresponds to an average 89% yield for each bond formed and an average diastereoselectivity of 80% for each stereogenic center. Subsequently, octabromide **4** is treated with commercially available 3,5-dihydroxybenzyl alcohol to yield, after an eight-fold Ullmann ether reaction, cavitand **5**. Again, the 40% yield for this reaction understates the efficiency of this process; each aryl ether bond is formed in an average 89%. Finally, removal of the benzyl groups and oxidation give octa-acid cavitand **1** in quantitative yield.

12.2.1
Structure of the Cavitand (What It Is and What It Is Not)

A space-filling model of cavitand **1** is shown in Figure 12.2a. The molecule is in essence a curved amphiphile. It has a hydrophobic concave surface and a hydrophilic convex surface. The former defines a pseudo-conical cavity approximately 1 nm in diameter and 1 nm in depth, whereas the latter is decorated with eight carboxylic acid

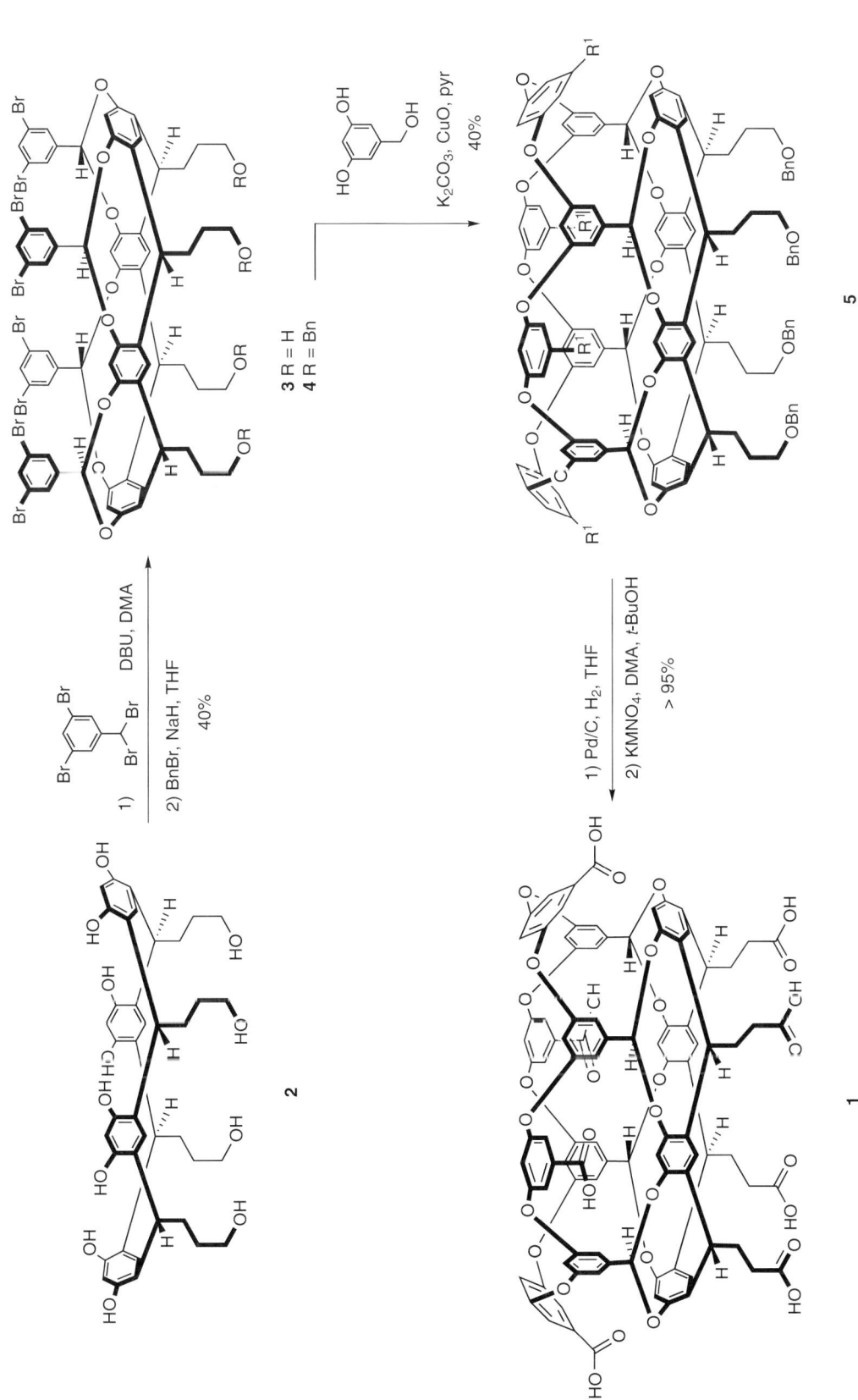

Scheme 12.1 Synthesis of octa-acid cavitand **1**.

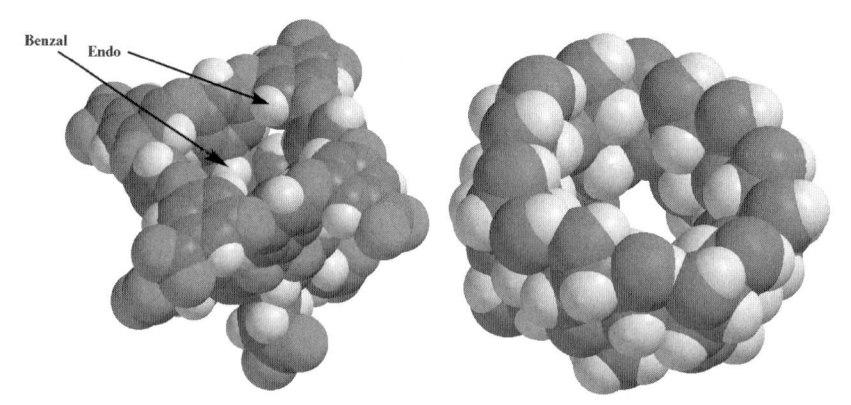

Figure 12.2 Space-filling models of (a) the octa-carboxylate of cavitand **1** and (b) β-cyclodextrin.

groups. This functionality bestows cavitand **1** with sparing water solubility at pH 7, but considerable solubility in basic solutions. Three other structural features should be mentioned. First, what is parochially termed the third row of aromatic rings – those introduced during the Ullmann ether reaction – define a wide hydrophobic rim. It is this rim that acts as an interface between the subunits of the nanocapsule. Second, the cavitand has only one significant portal. There is a small hole at the base of the cavity, but models suggest that it can only act as a portal for small diatoms traveling along the C_4 axis of the host. Essentially the cavity is closed at its base. Third, there are two kinds of H atoms that point into the cavity: the benzal (or acetal) hydrogens near the base of the cavity and what are termed the endo hydrogens (Figure 12.2a). Both are sensitive to guest binding, especially if the guest is halogenated (the guest will form hydrogen bonds with the benzal hydrogens) or a capsule is formed (the endo atoms become part of the interface).

Compare these structural features with β-cyclodextrin (Figure 12.2b). Like cavitand **1**, β-cyclodextrin has a cavity about 1 nm wide. At this point, however, the similarities end. The truncated conical cavities of cyclodextrins are so foreshortened that they are essentially tori; they are open at both ends, giving them a bangle-like structure that automatically diminishes the significance of dissociative mechanisms of guest exchange. Furthermore, both rims of the cyclodextrins are decorated with hydroxy groups. Hence, cyclodextrins are predisposed [23] to form 1:1 complexes in aqueous solution [24], and persuading them to do otherwise requires either a guest of specific shape and rigidity or modification to the cyclodextrins themselves. In summary, although both are water-soluble hosts with cavities about 1 nm in diameter, there are few structural features common to cavitand **1** and β-cyclodextrin.

12.2.2
Assembly Properties of the Cavitand

At relatively low concentrations (∼1 mM), NMR spectroscopy confirms that cavitand **1** is monomeric [21]. Thus, the ^{1}H NMR spectrum of the host in sodium tetraborate

12.2 Synthesis of a Water-soluble, Deep-cavity Cavitand

buffer shows sharp signals for each H atom, while diffusion rate measurements [25] with pulse gradient spin echo (PGSE) NMR reveals a molecule about 8 nm^3 in volume. In contrast, at higher concentrations (~20 mM), the ^{1}H NMR spectrum of the host shows exceptionally broad signals indicative of aggregation. This result was, to a degree, expected. One of the central theses of this program is to determine the degree of predisposition [23] required for a molecule to assemble via the hydrophobic effect and we had assumed that, without highly complementary and preorganized guests (templates), assembly to a well-defined dimer would not occur. Our line of thinking was as follows: if an interface between subunits involves highly directional noncovalent forces such as hydrogen bonds or metal coordination, the precise form of the assembled product is intimately tied to its enthalpy of formation. But what if the interfacial forces are weak and non-directional, such as the presupposed π–π stacking and C–H···π interactions in the dimer of **1** (Figure 12.3)? [26]. In other words, what if the assembly is instead driven by entropy? Can the hydrophobic effect lead to discrete assemblies of molecules possessing relatively small and topologically mundane interfaces? The analogy from Nature is, of course, protein quaternary structure [27], with the most exquisite examples perhaps being viral capsids [28]. By and large, however, the protein–protein interfaces (or hot-spots) are of the order of 600 Å^2 and highly complementary in terms of both their shape and distribution of functionality.[29] In contrast, the rim of cavitand **1** amounts to between 100 and 200 Å^2 of rather flat, uninspiring interface. As it transpires, our concerns were unfounded: cavitand **1** is highly predisposed – verging on

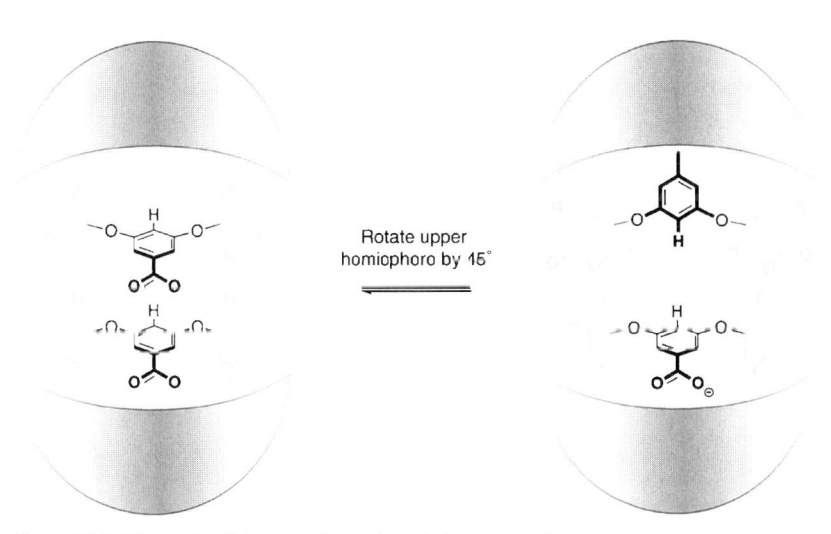

Figure 12.3 Schematic of the capsule interface. It is presumed that the two hemispheres can readily rotate around their common C_4 axis. If the two hemispheres are in register, face-to-face π–π-stacking dominates (left), whereas when out of register by 45° two kinds of C–H···π interactions are possible.

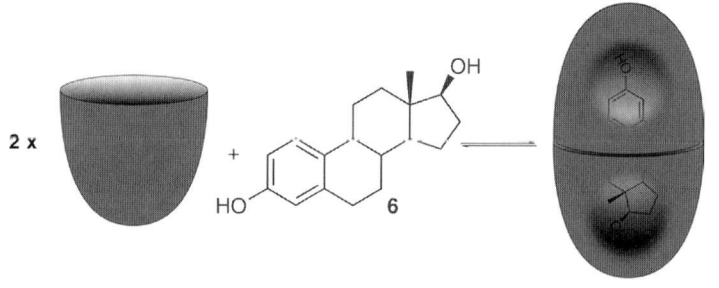

Figure 12.4 Self-assembly of **1** and encapsulation of estradiol (**6**).

"trigger happy" – to undergo assembly. It surpassed our initial expectations with consummate ease.

Our first encapsulation targets were steroids, primarily because molecules such as estradiol (**6**, Figure 12.4) appeared to be complementary in shape to the capsule, in addition to being hydrophobic. Additionally, however, their low symmetry (C_1) would help in confirming assembly [21]. As expected then, when estradiol was sonicated with a solution of cavitand **1**, the steroid was quickly taken up into solution.

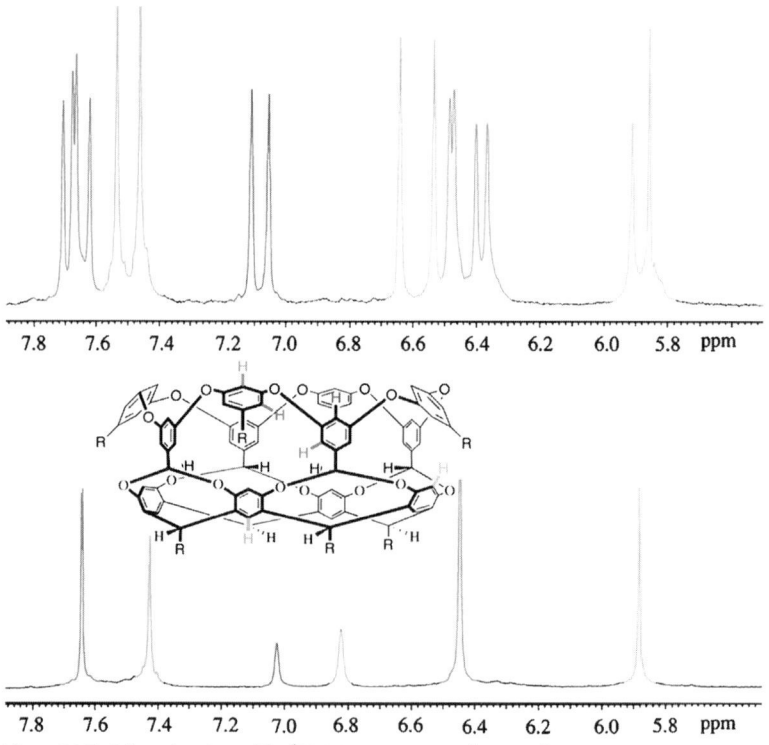

Figure 12.5 Selected region of the ^{1}H NMR spectrum of cavitand **1** (lower trace) and its 2:1 complex with estradiol **6** (upper trace).

NMR spectroscopy clearly showed capsule formation (Figure 12.5). Because the A and D rings of the steroid are aromatic and aliphatic, respectively, encapsulation resulted in two magnetically dissimilar "hemispheres". Consequently, as the exchange between free and bound guest was slow on the NMR time-scale, this led to a doubling of all signals from H atoms proximal to the guest, including the endo hydrogens (orange, Figure 12.5). In addition, though, the H atoms that were enantiotopic in the achiral free host (e.g. blue in Figure 12.5) became diastereotopic in the chiral complex. Consequently, these signals were doubled again. Encapsulation was also evident in the position of the bound guest signals. As expected, these were shifted upfield relative to their free position (in DMSO) because they experience magnetic shielding from the shell of the host. In particular, those guest atoms that were necessarily positioned towards the "poles" of each hemisphere underwent the largest shifts. Finally, capsule formation was also evident from NOESY NMR experiments. These revealed not only NOE interactions between the interface hydrogens of the two hemispheres, but also interactions between each end of the guest and the inward pointing benzal hydrogens of the host.

Estradiol was not actually the best guest for the capsule. Of the eight steroids examined, dehydroisoandrosterone (7) fitted best. In competition experiments it displaced estradiol, presumably because its more voluminous A-ring and C-18 methyl group filled the cavity of 1 better than an aromatic ring. One of the poorest guests examined was cholesterol (8), which although solubilized by the cavitand, did not yield a kinetically stable (500 MHz NMR time-scale) capsule. Models indicate that with its long C-17 chain, this guest was slightly too large for the capsule. As a result, the two hemispheres could not clamp down on one another to form a tight, desolvated interface.

Having demonstrated successful capsule formation, our next task was to determine how preorganized the guest needed to be to form a well-defined capsule. We opted to dive in at the deep-end and study straight-chain alkanes [30]. How less preorganized can you get! The availability of a large number of alkanes in this homologous series offered a detailed analysis of the assembly properties of the cavitand and the carrying capacity of the capsule; the macro-scale equivalent to seeing how many methylene groups can be fitted into a cavity is the use by some car manufacturers of cubes of a defined size to estimate the volume of a trunk (boot). We examined the potential guests pentane through octadecane. Only the latter appeared not to form a stable complex with 1. For the other molecules, the ^{1}H NMR spectra indicated that the smaller guests formed complexes with a host to guest ratio of 1 : 1,

Table 12.1 Summary of the complexes formed by **1** with straight-chain alkanes.

Guest	Ternary complex?	Quaternary complex?	Molecular formula of contents	Percentage occupancy[a]
Pentane	✓	✗	$C_{10}H_{24}$	43
Hexane	✓	✗	$C_{12}H_{28}$	50
Heptane	✓	✗	$C_{14}H_{32}$	57
Octane	✓	✓	$C_8H_{18}/C_{16}H_{36}$	33/65
Nonane	✗	✓	C_9H_{20}	36
Decane	✗	✓	$C_{10}H_{22}$	40
Undecane	✗	✓	$C_{11}H_{24}$	43
Dodecane	✗	✓	$C_{12}H_{26}$	47
Tridecane	✗	✓	$C_{13}H_{28}$	50
Tetradecane	✗	✓	$C_{14}H_{30}$	54
Pentadecane	✗	✓	$C_{15}H_{32}$	58
Hexadecane	✗	✓	$C_{16}H_{34}$	62
Heptadecane	✗	✓	$C_{17}H_{36}$	65

[a] These values assume a capsule volume of ~500 Å^3. However, defining what is termed usable space is highly subjective. Given free reign, workers in the laboratory have estimated cavity volumes from 400 to 700 Å^3. 500 Å^3 therefore is therefore on the conservative side of the average value obtained (600 Å^3).

whereas larger guests formed 2:1 entities. However, PGSE experiments revealed that the former had a stoichiometry of 2:2 rather than 1:1. In other words, small guests such as pentane formed quaternary capsular complexes with two entrapped guests. Table 12.1 summarizes the complexes formed and the percentage occupancy in each complex. From an admittedly biased standpoint, the fact that guests as small as pentane template capsule formation is remarkable; two small molecules, occupying less than half the volume of the capsule, are capable of templating its formation. Even though only weak C–H···π and/or π–π stacking forces exist within the interface (Figure 12.3) and between hosts and guests, a kinetically stable complex is formed.

Octane is a unique guest; on the cusp between ternary and quaternary complexes it forms both. Although it was not possible to "visualize" these two complexes directly with NMR, it was possible to infer this by examining shift data for the endo hydrogens of the host as a function of total guest volume. With the exception of octane, all the host–guest complexes formed two distinct trends for the ternary and quaternary entities. If it was assumed that octane formed a ternary complex, its shift data suggest a much larger guest, whereas if a quaternary complex was assumed, the shift corresponded to a much smaller guest. This observation is best explained by simultaneously invoking both ternary and quaternary complexes. As a result of this duality, octane covers the gamut of binding in terms of the total volume of the guest(s) or the percentage occupancy.

CPK models demonstrated that guests larger than roughly decane cannot reside in the capsule in a fully extended conformation, a point previously noted by Rebek and coworkers for analogous hydrogen bonded capsules [31,32] where long alkanes preferentially adopt helical conformations. We therefore used NOESY NMR to look

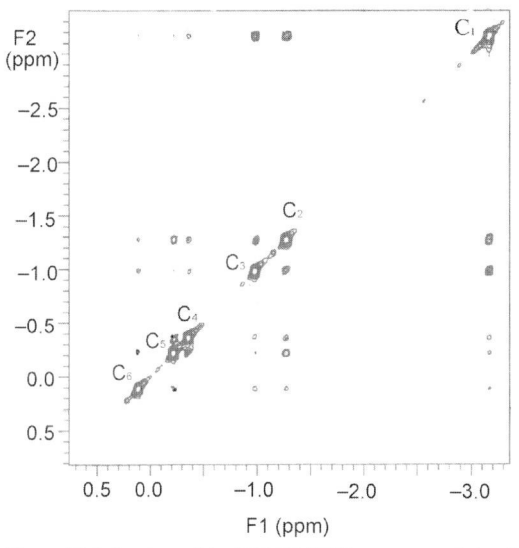

Figure 12.6 Section of the NOESY NMR spectrum of the dodecane complex with cavitand **1**.

for tell-tail signs that the guests inside the capsule formed by **1** also adopted preferred conformations. These studies revealed that only for guests larger than dodecane were 1,3 and 1,4 interactions indicative of helical structure observed (Figure 12.6), but the latter were rather weak, suggesting that a helical conformation was at best only marginally preferred. What is the difference between the capsules here and the related capsules of Rebek and coworkers? We suspect that the capsule 1_2 is slightly wider and therefore is not an ideal template for promoting helix formation.

12.2.3
Photophysics and Photochemistry Within Nano-capsules

We will return to the properties of these capsules and how the hydrophobic effect can promote unusual binding phenomena and separations. In the interim, it is also worth highlighting our collaboration with Ramamurthy's and Turro's groups, where we examine how photochemical or photophysical processes carried out inside the capsule differ from their counterparts in free solution. These experiments not only reveal how photochemical processes can be made more selective and suggest new ways in which organic chemistry can be brought into aqueous solution, but also shine light on the relatively unexplored phenomenon of external or concave templation. This type of templation, where the template is the host rather than the more typical guest, lags behind its more familiar counterpart because of the size requirement of the template; they need to be large enough to encapsulate interesting guests. That said, they should not be so large that guests never encounter the walls of the reaction vessel, for in an extreme case of a capsule filled with many guests, the center of these will resemble the liquid or solution state.

Scheme 12.2 Oxidation of 1-methylcyclohexene with singlet oxygen.

We have published on how the capsule formed by **1** can (1) redirect the Norrish Type I reactivity of dibenzyl ketones to give rearrangement products not observed in solution [33], (2) template the formation of the notoriously difficult to observe anthracene excimer [34] and (3) constrain the photo-Fries reactivity of naphthyl esters to one product instead of the more normal nine [35].

Most recently, we have reported on the reaction between singlet oxygen ($^1O_2$) and encapsulated alkenes [36]. In solution, the regioselectivity of this reaction is poor for substrates with multiple allylic hydrogens (Scheme 12.2) and so we wished to examine whether encapsulation could enhance selectivity. In addition, however, this particular reaction also raises the bar in terms of (encapsulated) reaction complexity. Is a water-soluble sensitizer required or can one be bound within a capsule? Also, can oxygen enter the capsule(s) containing the substrate (and sensitizer)? As it transpires, oxygen migration is not an issue, whether the sensitizer is water-soluble Rose Bengal or encapsulated dimethylbenzil (DMB) (Scheme 12.3). In either case, substrate encapsulation leads to highly regioselective reaction, although the reaction kinetics are slower if the sensitizer is encapsulated.

Whereas only one copy of DMB fits within the capsule formed by **1**, two copies of substrates such as 1-methylcyclohexene (MCH) occupy the host. NMR demonstrated that all of the methylcycloalkenes examined adopted one principal orientation in which the methyl group fills the tapering end of a cavity (Scheme 12.3). NMR also showed that mixing solutions of the two kinds of capsules did not result in guest exchange and the formation of capsules containing one DMB molecule and one substrate.

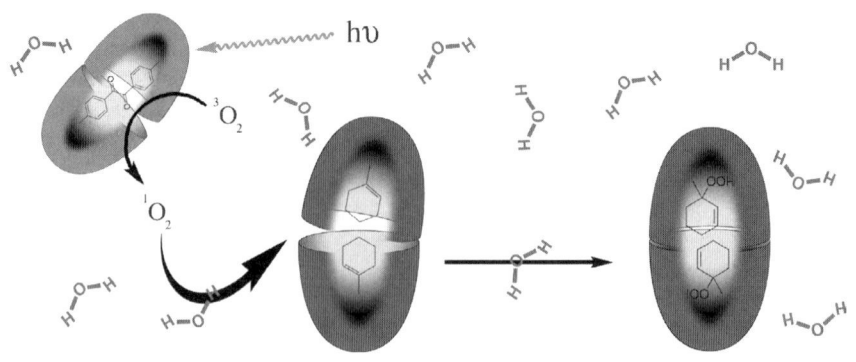

Scheme 12.3 Capsular control of the oxidation of 1-methylcyclohexene.

With this information at hand, we examined the outcome of irradiating (at 310 nm) the mixed solutions. That upon excitation encapsulated DMB generated $^1O_2$ was evident from the characteristic phosphorescence emission of $^1O_2$ at 1270 nm. An analysis of the quenching of the DMB as a function of $[O_2]$ gave a rate for this "bimolecular" energy transfer one order of magnitude less than for diffusion, but four orders of magnitude greater than the rate of energy transfer to O_2 from biacetyl triplet encapsulated in a covalent capsule [37]. Furthermore, kinetic traces of the phosphorescence of DMB and $^1O_2$ monitored at 560 and 1270 nm, respectively, revealed a 30-μs lag (rise time) between excitation of DMB and maximal $^1O_2$ phosphorescence, indicating that oxygen is not automatically present in the DMB capsule, but has to work its way in to undergo excitation. Once excited, however, the $^1O_2$ is free to enter other capsules and if one of these contains substrate, only one of the three allylic positions (the 3-position) undergoes hydrogen abstraction. Consequently, excellent yields of the corresponding tertiary hydroperoxide are obtained. We do not know the full details of this mechanism of attack and indeed there are still many things that we do not understand about the overall system. However, it is evident that capsule 1_2 can successfully bring about fairly complex reactions in which a reagent can move between two different capsules. We are therefore interested in examining further this approach to organic chemistry in water.

12.2.4
Hydrocarbon Gas Separation Using Nano-capsules

We finish this short review by returning to hydrocarbon binding within 1_2 with a twist, a twist that highlights the remarkable ability of **1** to self-assemble into discrete capsules and illustrates a unique way to separate hydrocarbon gases [38].

Our first attempt to bind hydrocarbon gases simply involved the bubbling of butane gas through a buffered, aqueous solution of **1**. Peak shifts for the endo and benzal hydrogen signals that point into the cavity indicated complexation, but there was only a broad signal in the high-field region of the NMR spectrum that could be attributed to bound guest (Figure 12.7a and b). However, these bound guest signals sharpened when the excess butane was allowed to escape from solution (Figure 12.7c). Integration of this kinetically stable complex confirmed a 1:1 ratio of host to guest, but it again took a PGSE experiment to reveal that the stoichiometry was 2:2 and that we were dealing with a stable capsular complex.

If binding is fairly strong, then it should be possible to extract a gas directly from the gas phase. This was shown in a subsequent experiment in which without agitation the quaternary butane complex formed spontaneously. Likewise, propane was also shown to template the formation of the capsule via direct extraction from the gas phase. In contrast to butane and propane, PSGE experiments reveal that ethane only formed a 1:1 complex. Apparently, the templation limit for the capsule is propane, at least in solutions devoid of salts known to increase the hydrophobic effect. In solutions of 14 mM NaCl, butane binding has been observed to be nearly two orders of magnitude greater than in NaCl-free solutions [39]. Hence ethane may ultimately be able to template the formation of the capsule. That point aside, that propane can

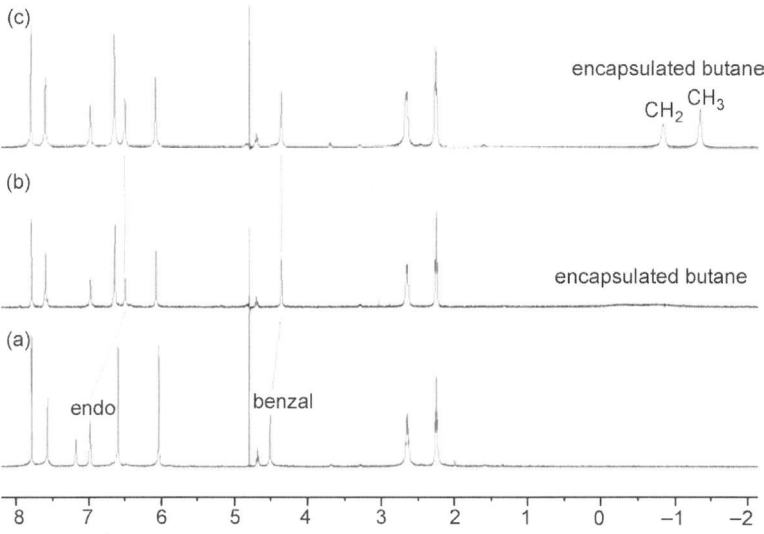

Figure 12.7 ^{1}H NMR spectra of (a) the free host, highlighting the signals from the endo and benzal hydrogens, (b) the 2:2 complex formed between **1** and excess butane and (c) the 2:2 complex of **1** and butane formed by extraction of the hydrocarbon gas directly from the gas phase.

template capsule formation is surprising. When we originally set out to synthesize cavitand **1**, we did not assume that its assembly would be templated by something as small as propane, a guest that results in complexes with an occupancy of 28%. There could be a host deformation that tempers this low value somewhat, but if the pocket is reduced in volume then the pressure of the contents increases; which is problematic because assuming a maximal cavity, (näve) calculations treating the guest as ideal give an internal pressure of 100 atm!

In these experiments, it was noted that propane binding was weaker than that of butane. Consequently, we decided to determine whether an aqueous solution of **1** could be used to separate these gases. We formed a mixture of the two hydrocarbons and exposed a large excess (to give relatively rapid uptake in the absence of agitation) to an aqueous solution of the host. As expected, NMR revealed that only the butane complex was formed, giving a gas phase mixture enriched in propane. As hydrocarbon gases cannot yet be separated by membrane technologies, capsule 1_2 suggests some interesting alternative strategies.

12.3
Conclusions

We have been surprised by the predisposition of cavitand **1** to self-assemble into a dimeric capsule. Two anthracene molecules can template capsule formation packing the cavity to about an 80% occupancy ratio; two propane molecules (~28% occupancy) can do likewise. The tenacity of the hydrophobic effect is sufficient to allow the capsule to act as an external template and bring about unusual reactions and prevent the capsule from "blowing its own top" when filled with hydrocarbon gas. Nevertheless, these capsular assemblies are dynamic, reversibly formed entities. An interjection is required at this point. What do we mean by the hydrophobic effect? There are, after all, several flavors. Our early thermodynamic analyses suggest a non-classical hydrophobic effect ($\Delta H°$ negative, $\Delta S°$ small) driving the formation of 1:1 complexes and a classic hydrophobic ($\Delta H°$ small, $\Delta S°$ positive) effect driving capsule capping. Furthermore, it seems that generally $K_1 < K_2$). More studies are needed to confirm these points, but with an assembly primarily driven by entropy we should repeat the question posed earlier: how can the hydrophobic effect lead to discrete assemblies with molecules with relatively small and topologically mundane interfaces? Towards answering these and other questions, we are continuing to investigate the subtleties behind these assemblies and the properties of the resulting complexes.

Acknowledgement

The author gratefully acknowledges the National Institutes of Health for financial support (GM074031).

References

1 Leung, D.H., Bergman, R.G. and Raymond, K.N. (2007) *J. Am. Chem. Soc.*, **129**, 2746–2747.
2 Kang, J. and Rebek, J. Jr. (1997) *Nature*, **385**, 50–52.
3 Heinz, T., Rudkevich, D.M. and Rebek, J.J. (1998) *Nature*, **394**, 764–766.
4 Fujita, M., Tominaga, M., Hori, A. and Therrien, B. (2005) *Acc. Chem. Res.*, **38**, 371–380.
5 Yoshizawa, M., Tamura, M. and Fujita, M. (2006) *Science*, **312**, 251–254.
6 Cram, D.J. and Cram, J.M. (1994) *Container Molecules and Their Guests*, 1st edn., Royal Society of Chemistry, Cambridge.
7 Liu, X., Liu, Y., Li, G. and Warmuth, R. (2006) *Angew. Chem. Int. Ed.*, **45**, 901–904.
8 Warmuth, R. and Yoon, J. (2001) *Acc. Chem. Res.*, **34**, 95–105.
9 Warmuth, R. (2001) *Eur. J. Org. Chem.*, 423–437.
10 MacGillivray, L.R. and Atwood, J.L. (1999) *Angew. Chem. Int. Ed.*, **38**, 1018–1033.
11 Chandler, D. (2002) *Nature*, **417**, 491.
12 Tanford, C. (1980) *The Hydrophobic Effect. Formation of Micelles and Biological Membranes*, 2nd edn., Wiley, New York.
13 Hof, F. and Rebek, J. Jr. (2002) *Proc. Natl. Acad. Sci. USA*, **99**, 4775–4777.
14 Rebek, J. Jr. (2005) *Angew. Chem. Int. Ed.*, **44**, 2068–2078.
15 MacGillivray, L.R. and Atwood, J.L. (1997) *Nature*, **389**, 469–472.
16 Atwood, J.L., Barbour, L.J. and Jerga, A. (2002) *Proc. Natl. Acad. Sci. USA*, **99**, 4837–4841.
17 McKinlay, R.M., Thallapally, P.K. and Atwood, J.L. (2006) *Chem. Commun.*, 2956–2958.
18 Fujita, M., Umemoto, K., Yoshizawa, M., Fujita, N., Kusukawa, T. and Biradha, K. (2001) *Chem. Commun.*, 509–518.

19 Caulder, D.L. and Raymond, K.N. (1999) *J. Chem. Soc., Dalton Trans.*, 1185–1200.
20 Davis, A.V., Yeh, R.M. and Raymond, K.N. (2002) *Proc. Natl. Acad. Sci. USA*, **99**, 1793–1796.
21 (a) Gibb, C.L.D. and Gibb, B.C. (2004) *J. Am. Chem. Soc.*, **126**, 11408–11409. (b) for a recent review on water-soluble cavitand, see Biros, S. and Rebek, J. Jr. (2007) *Chem. Soc. Rev.*, **36**, 93–104.
22 Gibb, B.C., Chapman, R.G., Sherman, J.C. (1996) *J. Org. Chem.*, **61**, 1505–1509.
23 Rowan, S.J., Hamilton, D.G., Brady, P.A. and Sanders, J.K.M. (1997) *J. Am. Chem. Soc.*, **119**, 2578–2579.
24 Rekharsky, M.V. and Inoue, Y. (1998) *Chem. Rev.*, **98**, 1875–1917.
25 Cohen, Y., Avram, L. and Frish, L. (2005) *Angew. Chem. Int. Ed.*, **44**, 520–554.
26 Meyer, E.A., Castellano, R.K. and Diederich, F. (2003) *Angew. Chem. Int. Ed.*, **42**, (11), 1210–1250.
27 Petsko, G.A. and Ringe, D. (2004) *Protein Structure and Function*, New Science Press, London.
28 Casper, D.L.D. and Klug, A. (1962) *Symp. Quant. Biol.*, **27**, 1–24.
29 Peczuh, M.W. and Hamilton, A.D. (2000) *Chem. Rev.*, **100**, 2479–2494.
30 Gibb, C.L.D. and Gibb, B.C. (2007) *Chem. Commun.*, 1635–1637.
31 Scarso, A., Trembleau, L. and Rebek, J. Jr. (2004) *J. Am. Chem. Soc.*, **126**, 13512–13518.
32 Scarso, A., Trembleau, L. and Rebek, J. Jr. (2003) *Angew. Chem. Int. Ed.*, **42**, 5499–5502.
33 Kaanumalle, L.S., Gibb, C.L.D., Gibb, B.C. and Ramamurthy, V. (2004) *J. Am. Chem. Soc.*, **126**, 14366–14367.
34 Kaanumalle, L.S., Gibb, C.L.D., Gibb, B.C. and Ramamurthy, V. (2005) *J. Am. Chem. Soc.*, **127**, 3674–3675.
35 Kaanumalle, L.S., Gibb, C.L.D., Gibb, B.C. and Ramamurthy, V. (2007) *Org. Biomol. Chem.*, **5**, 236–238.
36 Natarajan, A., Kaanumalle, L.S., Jockusch, S., Gibb, C.L.D., Gibb, B.C., Turro, N.J. and Ramamurthy, V. (2007) *J. Am. Chem. Soc.*, **129**, 4132–4133.
37 (a) Balzani, V., Pina, F.A., Parola, J., Ferreira, E., Maestri, M., Armaroli, N. and Ballardini, R. (1995) *J. Phys. Chem.*, **99**, 12701–12703. (b) Farran, A. and Deshayes, K. (1996) *J. Phys. Chem.*, **100**, 3305–3307. (c) Place, I., Farran, A., Deshayes, K. and Piotrowiak, P. (1998) *J. Am. Chem. Soc.*, **120**, 12626–12633.
38 Gibb, C.L.D. and Gibb, B.C. (2006) *J. Am. Chem. Soc.*, **128**, 16498–16499.
39 Gibb, C.L.D. and Gibb, B.C., unpublished work.

13
Opportunities in Nanotechnology via Organic Solid-state Reactivity: Nanostructured Co-crystals and Molecular Capsules

Dejan-Krešimir Bučar, Tamara D. Hamilton, and Leonard R. MacGillivray

13.1
Introduction

Organic reactions are generally carried out in solution (i.e. liquid phase) [1]. However, a variety of organic reactions (e.g. oxidation, elimination, photoreactions) occur in the solid state, providing regio- and enantioselective access to molecular products in high yields [2]. Photoinduced solid-state reactions [3] were discovered at the beginning of the last century. In the early 1900s, Riiber discovered that cinnamylidineacetic acid and cinnamylidinemalonic acid photodimerize to give cyclobutanes upon UV irradiation [4]. Two decades later, Stobbe's and de Jong's groups discovered that different polymorphs of cinnamic acid yield two different photoproducts upon UV irradiation [5]. This surprising observation was attributed to crystal-packing effects of cinnamic acid molecules in the different polymorphs. The mechanistic details of such solid-state reactions were, at that point, not well understood, but their intriguing nature provoked further studies to elucidate mechanisms.

In the 1960s, Schmidt accomplished crystallographic and photochemical studies of a wide range of cinnamic acids. Schmidt recognized the structural requirements for [2 + 2] photodimerizations to proceed in the solid state. Schmidt proposed that two C=C bonds should be aligned parallel and separated by <4.2 Å to react [6]. These requirements are known as the topochemical postulates. Although the topochemical postulates are extremely valuable for predicting whether a photodimerization will occur, frustrating effects of close packing [7] have largely thwarted efforts to synthesize molecules in solids using the carbon–carbon (C–C) bond-forming reaction with synthetic freedoms (e.g. control of product size) experienced in solution.

13.2
Template-controlled [2 + 2] Photodimerization in the Solid State

In recent years, we have introduced a method to control [2 + 2] photodimerizations in the solid state using molecular templates [8]. We have employed this method as a

Organic Nanostructures. Edited by Jerry L. Atwood and Jonathan W. Steed
Copyright © 2008 WILEY-VCH Verlag GmbH & Co. KGaA, Weinheim
ISBN: 978-3-527-31836-0

Scheme 13.1

means to circumvent the effects of close packing [7]. We have shown that ditopic molecules (e.g. resorcinol, 1,8-naphthalenedicarboxylic acid), in the form of linear templates, can preorganize olefins in positions suitable for the photoreaction. The templates organize the olefins via hydrogen bonds into positions largely independent of long-range packing (Scheme 13.1). Thus, we have shown that co-crystallization of resorcinol (res) with *trans*-1,2-bis(4-pyridyl)ethylene (4,4′-bpe) produces a discrete, four-component hydrogen-bonded assembly, 2(res)·2(4,4′-bpe), in which two C=C bonds are arranged for a [2 + 2] photodimerization (Figure 13.1a). UV irradiation of the solid produced *rctt*-1,2,3,4-tetrakis(4-pyridyl)cyclobutane (4,4′-tpcb) (where *rctt* reference,*cis*,*trans*,*trans*) stereospecifically and in 100% yield (Figure 13.1b) [8]. The photoproduct was isolated via basic extraction. Later, we showed that that this supramolecular approach to control solid-state reactivity is tolerant to molecular size by enabling the construction of complex molecular targets (e.g. ladderanes) [9].

Having developed a means to achieve control of the [2 + 2] photodimerization in the solid state, we now discuss, in this chapter, how the field of solid-state reactivity provides opportunities for emerging studies in the area of nanotechnology. In

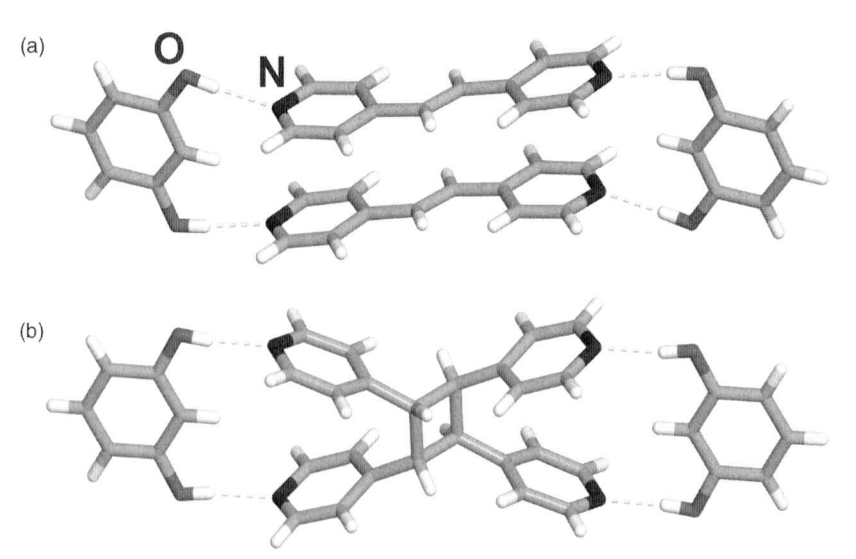

Figure 13.1 X-ray crystal structures of (a) 2(res)·2(4,4′-bpe) and (b) 2(res)·2(4,4′-tpcb).

particular, we will first describe how the template approach has led us to a method that facilitates the construction of nanostructured organic co-crystals using sonochemistry [10]. We will show how sonocrystallization produces nanostructured co-crystals that exhibit a rare phenomenon known as single crystal-to-single crystal (SCSC) reactivity. Second, we will demonstrate how the photoproducts of the template method can be used, *following the solid-state syntheses*, as organic building units of self-assembled molecular capsules [11,12].

13.3
Nanostructured Co-crystals

Upon UV irradiation of crystalline (res)·2(4,4′-bpe), each C-atom of the C=C bond of the olefin undergoes a change in position to form the cyclobutane ring. Such movement is invariably accompanied by the accumulation of strain and stress in the solid [13]. The accumulation of strain and stress can, and most often does, result in a collapse of the crystal lattice. The collapse will cause the single crystals to turn opaque and crack. We determined that macrosized single crystals of (res)·2(4,4′-bpe) crack upon photoreaction (Figure 13.2) [10]. Examples of solid-state reactions in which the integrity of single crystals remain virtually unchanged are rare. In such an SCSC reaction [13], the product forms homogeneously within the solid phase of the reactant.

SCSC reactions are intriguing to study for two general reasons [13]. First, SCSC reactions provide an ability to study reaction mechanisms and pathways by monitoring a reaction via single-crystal X-ray diffraction. Second, such solids have potential to be incorporated into solid-state devices (e.g. ultra-high-density data storage). The rareness of SCSC reactivity combined with the promise in materials science has created a need to generate materials that undergo SCSC reactions by design. Approaches to achieve SCSC reactivity of the [2 + 2] photodimerization have involved either adjusting the UV source for tail-end absorption of the reactant or, as will be discussed in more detail here, the use of crystals of nanometer-scale dimensions [13a].

Figure 13.2 SEM micrographs of macro-sized co-crystals of (res)·2(4,4′-bpe) (a) before and (b) after photoreaction. Adapted from Ref. [10] with permission from the American Chemical Society.

In our laboratory, we have used the tail-end UV absorption method to achieve SCSC reactivity within our templated solids. We have shown, for example, that SCSC reactivity can be achieved with 2-(1,8-naphthalenedicarboxylic acid)·2(3,4′-bpe) [14]. Studies on similar template-based co-crystals, however, have demonstrated that the tail-end method does not ensure SCSC reactivity [10]. To circumvent this problem, we aimed to achieve SCSC reactivity using co-crystals of nanometer-scale dimensions.

13.3.1
Organic Nanocrystals and Single Crystal-to-single Crystal Reactivity

Organic nanocrystals have attracted much attention owing to potential applications in electronics, biotechnology and catalysis [15–20]. The increasing interest in the properties of organic nanocrystals has led to numerous methods to synthesize such materials (e.g. sonication, microemulsion, vapor condensation). In this context, Nakanishi and coworkers have described a method to fabricate organic nanocrystals under relatively mild conditions. The method is based on reprecipitating an organic molecule during a solvent-exchange process [21–25]. They also demonstrated that the [2 + 2] photodimerization can proceed in an SCSC fashion by reducing the crystal size to nanometer-scale dimensions [26]. In particular, pure diolefin crystals of nano- and micrometer dimensions were fabricated via the reprecipitation method. The single crystals were shown to generate a polycyclobutane through an SCSC transformation. Macroscopic single crystals of the diolefin failed to undergo SCSC reaction and collapsed during the photoreaction. From these studies, we anticipated that an SCSC [2 + 2] photoreaction involving our photoactive co-crystals could be achieved by decreasing the crystal size to nanometer-scale dimensions.

Our first experiment to prepare nanometer-sized co-crystals of 2(res)·2(4,4′-bpe) using the reprecipitation method did not succeed [10]. In particular, an ethanolic solution of res and 4,4′-bpe was injected into water and vigorously stirred. As expected, the ethanolic solution created a cloudy suspension owing to precipitation of 2(res)·2(4,4′-bpe), which is poorly soluble in water. SEM images revealed particles of primarily micrometer dimensions (i.e. >5 µm). The majority of the particles were non-uniform in shape and exhibited uneven edges and irregular, flake-like morphologies (Figure 13.3a). The formation of the particles was attributed to an inherent mismatch in the solubilities of the molecular components of the co-crystal. UV irradiation resulted in cracking and destruction of the crystalline solids (Figure 13.3b).

To overcome the limitations of the reprecipitation method, we turned to sonocrystallization [27]. In particular, we combined low-intensity ultrasonic radiation using a water-bath with reprecipitation to synthesize nano- and micrometer-sized co-crystals of 2(res)·2(4,4′-bpe) [10]. In a typical experiment, ultrasonic radiation was immediately applied to a cloudy low-temperature (approximately 10 °C) suspension of res, 4,4′-bpe, ethanol and water obtained after concomitant addition of ethanol solutions of res and 4,4′-bpe into water using a microsyringe. SEM images revealed the formation of well-defined crystals of uniform shape and a size distribution [28] of 500 nm–8 µm (Figure 13.4a). Upon UV irradiation, single crystals of 2(res)·2(4,4′tpcb)

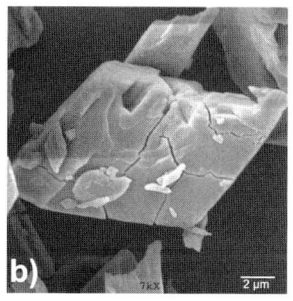

Figure 13.3 SEM images of co-crystals of (res)·2(4,4′-bpe) via the reprecipitation method (a) before and (b) after photoreaction. Adapted from Ref. [10] with permission from the American Chemical Society.

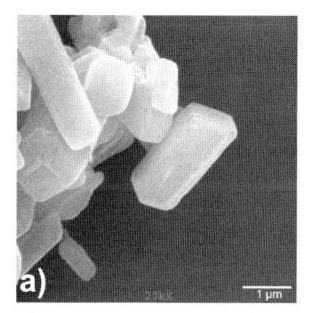

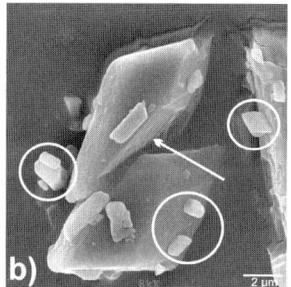

Figure 13.4 SEM images of nanostructured (res)·2(4,4′-bpe) (a) before and (b) after photoreaction. Arrows show cracks in large macrocrystals and circles show intact co-crystals. Adapted from Ref. [10] with permission from the American Chemical Society.

of less than 2 μm underwent an SCSC transformation, whereas the larger macrosized crystals cracked (Figure 13.4b). We attributed the formation of the nanostructured co-crystals of 2(res)·2(4,4′-bpe) to effects of cavitation. The process is associated with high temperatures (i.e. up to 5000 °C) and pressures (i.e. 1000 atm), which result in the formation, growth and rapid collapse of bubbles in a liquid environment [29]. Moreover, we attributed the generation of the nanostructured materials to cavitation being able to rapidly solubilize the components of 2(res)·2(4,4′-bpe) [30] and, at the same time, provide a mechanism for fast precipitation and formation of the solids.

13.4
Self-assembled Capsules Based on Ligands from the Solid State

In recent years, widespread attention has been devoted to self-assembled metal–organic frameworks with structures that conform to polygons and polyhedra [31]. The internal cavities of such discrete frameworks are studied for applications in areas such as logic gates, nano-sized reaction vessels and delivery systems [31]. We have shown that the photoproducts of template-controlled reactions in the solid state can be used to generate metal–organic polygons and polyhedra [11,12]. We determined that rctt-1,2-bis(2-pyridyl)-3,4-bis(4-pyridyl)cyclobutane (2,4'-tpcb) can be an organic connector of both a polygon [11a] and polyhedron [11b]. We have also shown that rctt-1,2-bis(2-pyridyl)-3,4-bis(3-pyridyl)cyclobutane (2,3'-tpcb) can act as a connector of a polyhedron [11c]. It is instructive to note that Nature employs a conceptually similar approach to construct polyhedral hosts (i.e. viruses). Specifically, the linear structure of deoxyribonucleic acid (DNA) directs the formation of organic subunits (i.e. proteins) that, in a second step, form a functional, self-assembled structure [32]. For 2,4'-tpcb and 2,3'-tpcb, each cyclobutane possessed a chelating unit based on the 2-pyridyl group and two monodentate units based on either the 4- or 3-pyridyl group (Figure 13.5). Both cyclobutanes were synthesized in the solid state using a resorcinol stereospecifically, in quantitative yield and gram amounts [11].

Our first report described the ability of 2,4'-tpcb to self-assemble with Cu(II) ions to form a hexanuclear polyhedron in form a trigonal antiprism in $[Cu_6(2,4\text{-tpcb})_6(H_2O)_6][ClO_4]_{12}$ (Figure 13.6). The hexanuclear assembly, $[Cu_6(2,4\text{-tpcb})_6(H_2O)_6]^{12+}$, formed upon reaction of 2,4'-tpcb and $Cu(ClO_4)_2 \cdot 6H_2O$ [11a]. The six Cu(II) ions occupied the vertices of the antiprism. Each Cu(II) ion was coordinated by a 2-pyridyl chelating unit, two 4-pyridyl monodentate units and a water molecule to form a nearly square-prism coordination environment. Consequently, the 4-pyridyl groups corresponded to the edges of the antiprism and the 2-pyridyl groups corresponded to the corners. As a result of the assembly process, a cylindrical cavity that encapsulated two ClO_4^- anions formed.

We also determined that 2,4'-tpcb serves as a building unit of a metal-organic polygon. Reaction of copper(II) hexafluoroacetylacetonate (hfac) with 2,4'-tpcb

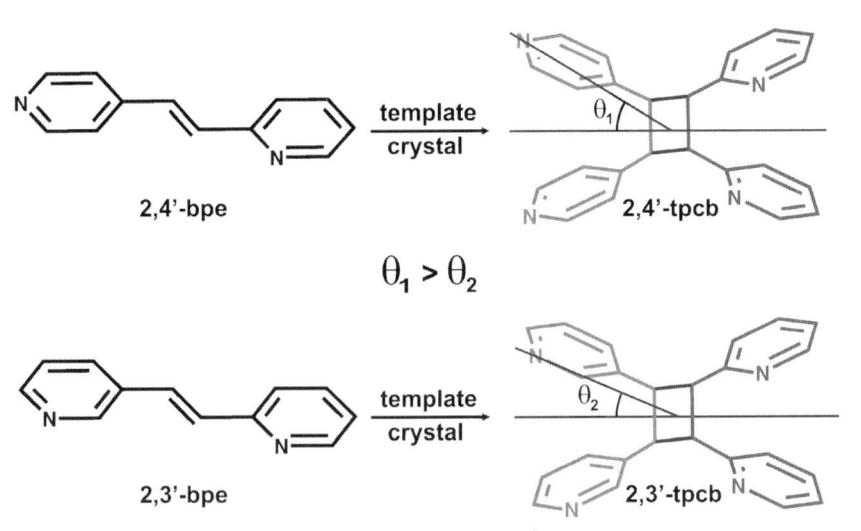

Figure 13.5 Polydentate ligands derived from the solid state that support metal–organic polygons and polyhedra: (a) 2,3'-tpcb and (b) 2,4'-tpcb.

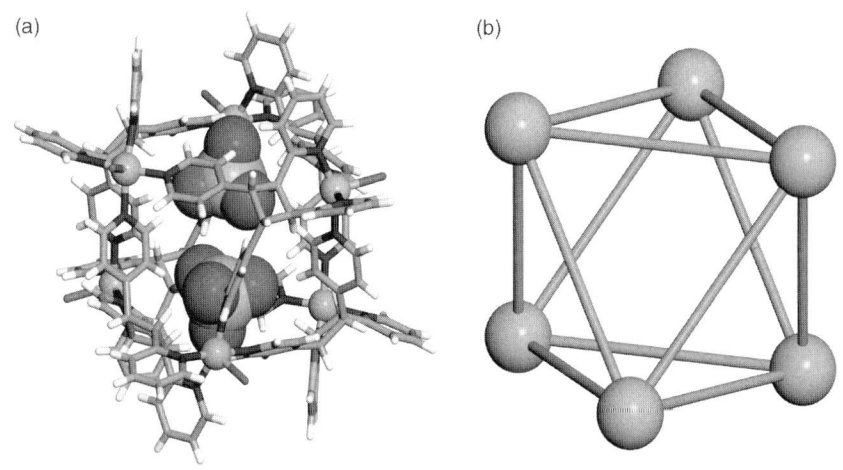

13.4 Self-assembled Capsules Based on Ligands from the Solid State | 311

Figure 13.6 X-ray structure of $[Cu_6(2,4\text{-tpcb})_6(H_2O)_6]^{12+}$: (a) encapsulated ClO_4^- ions and (b) topology of the trigonal antiprism based on the Cu(II) ions (Cu(II) ions are displayed using the ball-and-stick model and the ClO_4^- ions using the space-fill model).

produced a tetranuclear metal–organic polygon $[Cu_4(2,4'\text{-tpcb})_2(hfac)_8]$, with a geometry that conformed to a rhombus (Figure 13.7). In contrast to the polyhedron, each metal atom adopted an octahedral coordination environment. The 4-pyridyl groups provided the edges of the polygon. The cavity of the polygon was too small to accommodate an organic molecule as a guest. This was the first example in which an organic connector unit supported the structures of both and polyhedron and polygon [32].

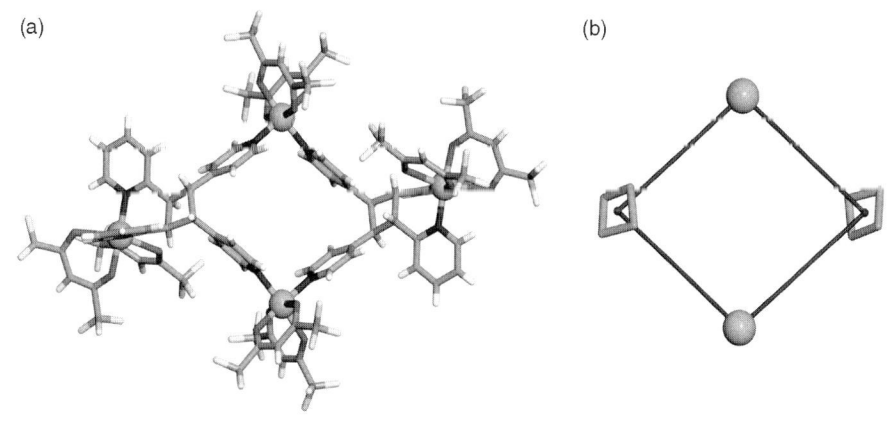

Figure 13.7 X-ray structure of $[Cu_4(2,4\text{-tpcb})_2(hfac)_8]$: (a) stick view and (b) topology of the polygon based on the Cu(II) ions and cyclobutane rings [Cu(II) ions are displayed using the ball-and-stick model].

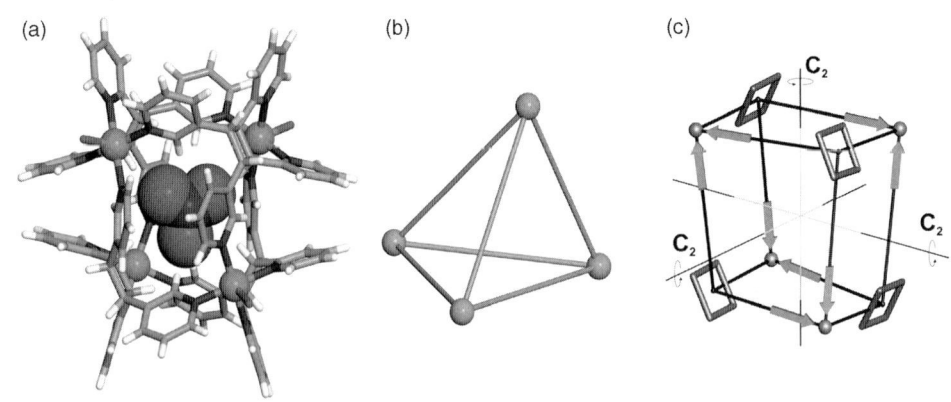

Figure 13.8 X-ray crystal structure of $[Cu_4(2,3'-tpcb)_4(H_2O)_4]^{8+}$ showing: (a) encapsulated NO_3^- ion, (b) topology of the tetrahedron based on the Cu(II) ions and (c) placement of the Cu(II) ions and cyclobutane rings showing how the directionality of the 3-pyridyl groups precludes mirror symmetry [Cu(II) ions are displayed using the ball-and-stick model and the NO_3^- ions using the space-fill model].

To build on our observation involving the polyhedron $[Cu_6(2,4\text{-tpcb})_6(H_2O)_6]^{12+}$, we anticipated that a second polyhedron could form using the related isomer 2,3'-tpcb. We expected that the exchange of the 4-pyridyl group with a 3-pyridyl group would produce a smaller polyhedron owing to a decrease in the angle of the coordination-vector (Figure 13.5). Reaction of 2,3'-tpcb with Cu(II) ions afforded the tetranuclear polyhedron $[Cu_4(2,3'\text{-tpcb})_4(H_2O)_4]^{8+}$ with a structure that conformed to a tetrahedron (Figure 13.8) [11c]. As expected, the polyhedron based on 2,3'-tpcb was smaller than that based on 2,4'-tpcb, with the inner cavity hosting only a single NO_3^- ion as a guest. To our surprise, however, the tetrahedron exhibited a chiral topology [12]. The polyhedron displayed approximate D_2 symmetry. The chirality was a result of the geometric fit of the metal and organic components along the surface of the capsule. As with the trigonal antiprism, the Cu(II) ions occupied the vertices of the tetrahedron. Each Cu(II) ion adopted an approximately square pyramidal coordination environment.

13.5
Summary and Outlook

In this chapter, we have described a method to direct reactivity in the organic solid state using molecular templates. We have shown how nanostructured co-crystals of the template-based materials can be synthesized using sonochemistry. We have also shown how the nanostructured solids support SCSC reactivity involving the [2 + 2] photodimerization. The use of sonochemistry overcomes difficulties of controlled nucleation and crystal growth, presumably caused by mismatching of the solubilities of the components of a co-crystal. We have also demonstrated how the molecules synthesized using templates in the solid state can be used as organic building blocks of self-assembled capsules. The method has provided a route to molecular capsules akin to the approach employed by Nature. Collectively, the studies involving the nanostructured co-crystals and the molecular capsules provide opportunities for studies in nanotechnology via templated reactions directed in organic solids.

References

1 Smith, M.B. and March, J. (2007) in *March's Advanced Organic Chemistry: Reactions, Mechanisms and Structure*, Wiley, Hoboken, NJ.

2 (a) Kaupp, G. (2005) Organic Solid-State. Reactions with 100% Yield. *Top. Curr. Chem.*, **254**, 95–183. (b) Tanaka, K. and Toda, F. (2000) Solvent-Free Organic Syntheses. *Chem. Rev.*, **100**, 1025–1074.

3 Ramamurthy, V. and Venkatesan, K. (1987) Photochemical Reactions of Organic Crystals. *Chem. Rev.*, **87**, 433–481.

4 (a) Riiber, C.N. (1902) Die Synthese der α-Truxillsäure. *Chem. Ber.*, **35**, 2411–2415. (b) Riiber, C.N. (1913) Die Licht-Polymerisation der Cinnamyliden-essigsäure. *Chem. Ber.*, **46**, 335–338.

5 (a) Stobbe, H. and Steinberger, F.K. (1922) Lichtreaktionen der *trans*-und *cis*-Zimtsäuren. *Chem. Ber.*, **55B**, 2225–2245. (b) Stobbe, H. and Lehfeldt, A. (1925) Polymerisationen und Depolymerisationen durch Licht verschiedener Wellenlänge, II.: α- und β-*trans*-Zimtsaure, allo-Zimtsaure und ihre Dimeren. *Chem. Ber.*, **58B**, 2415–2427. (c) de Jong, A.W.K. (1923) Über die Konstitution der Truxill- und Truxinsäuren und über die Einwirkung des Sonnenlichtes auf die Zimtsäuren und Zimtsäure-Salze. *Chem. Ber.*, **56B**, 818–832. (d) de Jong, A.W.K. (1922) Über die Einwirkung des Lichtes auf die Zimtsäuren und über die Konstitution der Truxillsäuren. *Chem. Ber.*, **55B**, 463–474.

6 Schmidt, G.M.J. (1971) Photodimerization in the Solid State. *Pure Appl. Chem.*, **27**, 647–678.

7 Desiraju, G.R. (1995) Supramolecular Synthons in Crystal Engineering – A New Organic Synthesis. *Angew. Chem. Int. Ed.*, **34**, 2311–2327.

8 MacGillivray, L.R., Reid, J.L. and Ripmeester, J.A. (2000) Supramolecular Control of Reactivity in the Solid State Using Linear Molecular Templates. *J. Am. Chem. Soc.*, **122**, 7817–7818.

9 Gao, X., Friscic, T. and MacGillivray, L.R. (2004) Supramolecular Construction of Molecular Ladders in the Solid State. *Angew. Chem. Int. Ed.*, **43**, 232–236.

10 Bučar, D.-K. and MacGillivray, L.R. (2007) Preparation and Reactivity of Nanocrystalline Co-Crystals Formed via Sonocrystallization. *J. Am. Chem. Soc.*, **129**, 32–33.

11 (a) Hamilton, T.D., Papaefstathiou, G.S. and MacGillivray, L.R. (2002) A Polyhedral Host Constructed Using a Linear Template. *J. Am. Chem. Soc.*, **124**, 11606–11607. (b) Papaefstathiou, G.S., Hamilton, T.D. and MacGillivray, L.R. (2004) Self-Assembled Metal–Organic Squares Derived from Linear Templates as Exemplified by a Polydentate Ligand that Provides Access to Both a Polygon and Polyhedron. *Chem. Commun.*, 270–271. (c) Hamilton, T.D., Bučar, D.-K. and MacGillivray, L.R. (2007) Coding a Coordination-Driven Self-Assembly via a Hydrogen-Bond-Directed Solid-State Synthesis: an Unexpected Chiral Tetrahedral Capsule. *Chem. Commun.*, 1603–1604.

12 Hamilton, T.D. and MacGillivray, L.R. (2004) Enclosed Chiral Environments from Self-Assembled Metal–Organic Polyhedra. *Cryst. Growth Des.*, **4**, 419–430.

13 (a) Friščić, T. and MacGillivray, L.R. (2005) Single-Crystal-to-Single-Crystal Transformations Based on the [2+2] Photodimerization: from Discovery to Design. *Z. Kristallogr.*, **220**, 351–363. (b) Halder, G.J. and Kepert, C.J. (2006) Single Crystal to Single Crystal Structural Transformations in Molecular Framework Materials. *Aust. J. Chem.*, **59**, 597–604.

14 Varshney, D.B., Papaefstathiou, G.S. and MacGillivray, L.R. (2002) Site-Directed Regiocontrolled Synthesis of a "Head-to-Head" Photodimer via a Single-Crystal-to-Single-Crystal Transformation Involving a Linear Template. *Chem. Commun.*, 1964–1965.

15 Zhao, Y.S., Yang, W. and Yao, J. (2006) Organic Nanocrystals with Tunable Morphologies and Optical Properties Prepared Through a Sonication Technique. *Phys. Chem. Chem. Phys.*, **8**, 3300–3303.

16 Jang, J. and Oh, J.H. (2003) Facile Fabrication of Photochromic Dye-Conducting Polymer Core-Shell Nanomaterials and Their Photoluminescence. *Adv. Mater.*, **15**, 977–980.

17 Kwon, E., Oikawa, H., Kasai, H. and Nakanishi, H. (2007) A Fabrication Method of Organic Nanocrystals Using Stabilizer-Free Emulsion. *Cryst. Growth Des.*, **7**, 600–602.

18 Zhao, Y.S., Yang, W., Xiao, D., Sheng, X., Yang, X., Shuai, Z., Luo, Y. and Yao, J. (2005) Single Crystalline Submicrotubes from Small Organic Molecules. *Chem. Mater*, **17**, 6430–6435.

19 Chiu, J.J., Kei, C.C., Perng, T.P. and Wang, W.S. (2003) Organic Semiconductor Nanowires for Field Emission. *Adv. Mater.*, **15**, 1361–1364.

20 Masuhara, H., Nakanishi, H. and Sasaki, K. (eds)(2003) *Single Organic Nanoparticles*, Springer, Berlin.

21 Kasai, H., Nalwa, H.S., Oikawa, H., Okada, S., Matsuda, H., Minami, N., Kakuta, A., Ono, K., Mukoh, A. and Nakanishi, H. (1992) A Novel Preparation Method of Organic Microcrystals. *Jpn. J. Appl. Phys.*, **31**, L1132–L1134.

22 Kasai, H., Oikawa, H., Okada, S. and Nakanishi, H. (1998) Crystal Growth of Perylene Microcrystals in the Reprecipitation Method. *Bull. Chem. Soc. Jpn.*, **71**, 2597–2601.

23 Baba, K., Kasai, H., Okada, S., Oikawa, H. and Nakanishi, H. (2000) Novel Fabrication Process of Organic Microcrystals using Microwave Irradiation. *Jpn. J. Appl. Phys.*, **39**, L1256–L1258.

24 Nakanishi, H. and Kasai, H. (1997) Polydiacetylene Microcrystals for Third-Order Nonlinear Optics. *ACS Symp. Ser.*, **672**, 183–198.

25 Chung, H.-R., Kwon, E., Oikawa, H., Kasai, H. and Nakanishi, H. (2006) Effects of Solvent on Organic Nanocrystal Growth Using the Reprecipitation Method. *J. Cryst. Growth*, **294**, 459–463.

26 Takahashi, S., Miura, H., Kasai, H., Okada, S., Oikawa, H. and Nakanishi, H. (2002) *J. Am. Chem. Soc.*, **124**, 10944–10945.

27 (a) Bang, J.H. and Suslick, K.S. (2007) Sonochemical Synthesis of Nanosized Hollow Hematite. *J. Am. Chem. Soc.*, **129**, 2242–2243. (b) Dhas, N.A. and Suslick, K.S. (2005) Sonochemical Preparation of Hollow Nanospheres and Hollow Nanocrystals. *J. Am. Chem. Soc.*, **127**, 2368–2369.

28 Veerman, M., Resendiz, M.J.E. and Garcia-Garibay, M. (2006) Large-Scale Photochemical Reactions of Nanocrystalline Suspensions: a Promising Green Chemistry Method. *Org. Lett.*, **8**, 2615–2617.

29 (a) Suslick, K.S. (1990) Sonochemistry. *Science*, **247**, 1439–1445. (b) Suslick, K.S. and Price, G.J. (1999) Applications of Ultrasound to Materials Chemistry. *Annu. Rev. Mater. Sci.*, **29**, 295–326.

30 Ruecroft, G., Hipkiss, D., Ly, T., Maxted, N. and Cains, P.W. (2005) Sonocrystallization: the Use of Ultrasound for Improved Industrial Crystallization. *Org. Process Res. Dev.*, **9**, 923–932.

31 (a) Seidel, S.R. and Stang, P.J. (2002) High-Symmetry Coordination Cages via Self-Assembly. *Acc. Chem. Res.*, **35**, 972–983. (b) Gianneschi, N.C., Masar, M.S., III. and Mirkin, C.A. (2005) Development of a Coordination Chemistry-Based Approach for Functional Supramolecular Structures. *Acc. Chem. Res.*, **38**, 825–837. (c) Caulder, D.L. and Raymond, K.N. (1999) Supermolecules by Design. *Acc. Chem. Res.*, **32**, 975–982.

32 Hamilton, T.D., Papaefstathiou, G.S. and MacGillivray, L.R. (2005) Template-Controlled Reactivity: Following Nature's Way to Design and Construct Metal–Organic Polygons and Polyhedra. *J. Solid State Chem.*, **178**, 2409–2413.

14
Organic Nanocapsules
Scott J. Dalgarno, Nicholas P. Power, and Jerry L. Atwood

14.1
Introduction

By exploiting the principles of supramolecular chemistry, nature has mastered the ability to form remarkably complex molecular capsules on a widely varied scale. This chapter is focused on reviewing the principles upon which molecular capsules are constructed at all length scales. Research in our group has involved: (1) the synthesis of new building blocks for capsules; (2) the engineering of functionality into such capsules; (3) the extension of length scales from the nano- to the micro-level, providing robust capsules; and (4) the understanding of the structural relationship of spheres (capsules) to tubes. Success in these endeavors will afford nanocpasules for applications in drug delivery and catalysis. The increased understanding of the principles governing the assembly of larger capsules will provide insight into aspects of living systems.

Dimeric capsules held together by covalent bonds were prepared in the 1980s by Collet [1] and Cram et al. [2], and self-assembled dimeric capsules were reported by Conn and Rebek [3] and others [4–8] in the 1990s. Dimeric capsules generally possess an internal volume in the range 100–300 $Å^3$. Monomeric capsules have also been characterized [9]. In this chapter, we will summarize the development of larger capsules, capsules which may be measured on the nano-length scale, capsules possessing an internal volume exceeding 800 $Å^3$.

14.2
First Generation Nanocapsules

In 1997, we discovered a spherical assembly consisting of [(C-methylresorcin[4]arene)$_6$(H$_2$O)$_8$], **1**. This assembly (Figure 14.1), with an enclosed volume of 1375 $Å^3$, was characterized by a single-crystal X-ray diffraction study and was found to be stable in non-polar solvents [10]. The evidence that **1** maintains the capsule structure in

Organic Nanostructures. Edited by Jerry L. Atwood and Jonathan W. Steed
Copyright © 2008 WILEY-VCH Verlag GmbH & Co. KGaA, Weinheim
ISBN: 978-3-527-31836-0

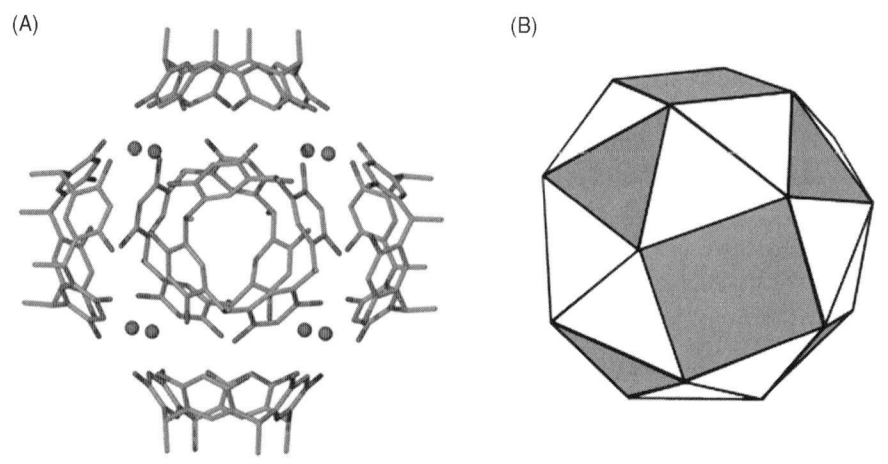

Figure 14.1 (A) The near-spherical nanocapsule assembly, **1**, based on C-methylresorcin[4]arene and water, formed from wet nitrobenzene [10]. The capsule is composed of six resorcin[4]arenes and eight structural water molecules. (B) The hydrogen bonding pattern representing the Archimedean solid, the *snub cube*.

solution was obtained from both one- and two-dimensional ^{1}H NMR studies. The spectrum of **1** in benzene-d_6 at increasing concentrations shows resonances attributed to the chiral and achiral calixarenes as well as the eight water molecules. These observations, coupled with the molecular mass determination in benzene (7066 g mol^{-1}), [11], provide convincing evidence supporting the nanocapsule structure of **1** in solution. Many subsequent studies have supported this initial conclusion of solution stability for the nanocapsule [12,13].

We were also able to link the geometry of the nanocapsule **1** to the Archimedean solid known as the *snub cube* (Figure 14.1B). In a review, we have set forth structural classifications and general principles for the design of nanocapsules based, in part, on the solid geometry ideas of Plato and Archimedes [14]. Indeed, we have used the well-known solid geometry principles embodied in Platonic and Archimedean solids to design new, large spherical container assemblies. A recent success involved the construction and characterization of an *icosahedron* made up of *p*-sufonatocalix[4]arenes, pyridine *N*-oxide, metal ions and water [15].

The discovery of the link between the solid geometry principles of Plato and Archimedes and the chemical assembly of small building blocks into large supramolecular structures was important. Specifically, the discovery that members of the resorcin[4]arene family self-assemble to form the capsule shown as **1** in Figure 14.1 prompted our research group to examine the topologies of related spherical hosts with a view to understanding their structures on the basis of symmetry. In addition to providing a basis for classification, it was anticipated that such an approach would allow one to identify similarities at the structural level, which, at the chemical level, may not seem obvious and may be used to design large, spherical host assemblies similar to **1**.

Our group has now described the results of this analysis which we regard as the development of a general strategy for the construction of spherical molecular hosts. In these reports we began by presenting the idea of self-assembly in the context of spherical hosts and then, after summarizing the Platonic and Archimedean solids, we provided examples of cubic symmetry-based hosts, from both the laboratory and nature, with structures that conform to these polyhedra.

To construct a spherical host from two subunits ($n = 2$), each unit must cover half of the sufrace of the sphere. This can only be achieved if the subunits exhibit cruvature and they are placed such that their centroids lie at a maximum distance from each other. These criteria place two points along the surface of a sphere separated by a distance equal to the diameter of the shell. As a consequence of this arrangement, there exist two structure types: one with two identical subunits attached at the equator and one belonging to the point group D_{nd} which is topologically equivalent to a tennis ball.

To construct a spherical host from three subunits ($n = 3$), each must cover one-third of the surface of the sphere. Following the design conditions described previously, placing three identical subunits along the surface of a sphere results in an arrangement in which their centroids constitute the vertices of an equilateral triangle. As a result, there is only one structure type, that belonging to D_{3h}. Each of the subunits must exhibit curvature.

For $n = 4$, positioning four points along the surface of a sphere such that they lie a maximum distance from each other places the points at the vertices of a tetrahedron. This is the first case in which joining the points via line segments gives rise to a closed surface container. The container, a tetrahedron, is comprised of four identical subunits, in the form of equilateral triangles where surface curvature is supplied by edge-sharing of regular polygons rather than by the subunits themselves.

The Platonic solids comprise a family of five convex uniform polyhedra which possess cubic symmetry and are made of the same regular polygons (equilateral triangle, square, pentagon) arranged in space such that the vertices, edges and three coordinate directions of each solid are equivalent. That there is a finite number of such polyhedra is due to the fact that there exists a limited nuber of ways in which identical regular polygons may be adjoined to construct a convex corner. There are thus only five such isometric polyhedra, all of which are achiral.

In addition to the Platonic solids, there exists a family of 13 convex uniform polyhedra known as the Archimedean solids. Each member of this family is made up of at least two different regular polygons and may be derived from at least one Platonic solid through either truncation or twisting of faces. In the case of the latter, two chiral members, the *snub cube* and the *snub dodecahedron*, are realized. The remaining Archimedean solids are achiral.

It is important to realize the limitations of the Platonic and Archimedean solid models for supramolecular assemblies. For the *snub dodecahedron*, a total of 60 triangles are called for, but triangles can be simply the result of hydrogen bonds from adjacent triangles or pentagons. For example, the *snub cube* in Figure 14.1 is composed of 32 triangles and six squares, with the triangles being represented by water molecules and the squares, resorcin[4]arenes. However, in nanocapsule **1** only eight of the 32 triangles contain water molecules, as one can observe from the shading. Therefore, in the search for the *snub dodecahedron*, the ratio of water to calix[5]arene pentacarboxylic acid cannot be higher than 60:12 and it is likely to be lower.

14.3
Second Generation Nanocapsules

In the theme of control of the guest species by external forces, a discovery was made which relates to the *p*-sulfonatocalix[4]arene capsules first reported in 1999 [15]. In the solid state, truncated cone-shaped *p*-sulfonatocalix[4]arene, **2**, favors the formation of infinite bilayer structures in which neighboring calixarenes are orientated in an up–down fashion relative to one another (Figure 14.2) [16]. However, it is possible to circumvent the formation of such bilayer structures under controlled conditions for a ternary system containing pyridine *N*-oxide (PNO), a lanthanide(III) nitrate salt and the pentasodium salt of *p*-sulfonatocalix[4]arene, Na_5**2** (Figure 14.3) [15].

Parallel (up–up) packing of neighboring calixarenes imparts significant curvature to the overall assembly, resulting in the formation of either spheroidal or tubular arrays. The spheroidal array consists of 12 calixarenes arranged at the vertexes of an icosahedron (Figure 14.3). Indeed, this is the most symmetrical way in which to arrange 12 like objects efficiently about a Platonic/Archimedean solid. The calixarenes enclose a central core comprised of two sodium ions and 30 water molecules that form an extensively hydrogen-bonded regime within the capsule [15]. Since icosahedra are unable to pack closely in three dimensions by sharing vertexes, edges or faces, the near spheroids do not form a close-packed arrangement in the extended solid. The spheroidal structure is based on a C-shaped dimer, consisting of two molecules of **2** linked to a common trivalent lanthanide ion by way of metal–sulfonate coordination. Furthermore, each of the calixarene cavities contains a PNO molecule, which is also coordinated to the bridging lanthanide ion. Each component of the dimer forms part of a separate spheroid and neighboring superstructures can therefore be considered to be multiply bridged by means of first and second sphere coordination of lanthanide ions.

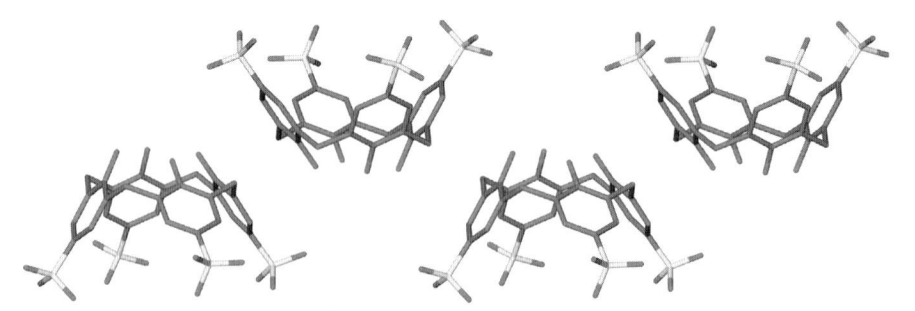

Figure 14.2 Schematic of *p*-sulfonatocalix[4]arene, **2**, and an example of a favorable bilayer anti-parallel arrangement found in the solid state [16].

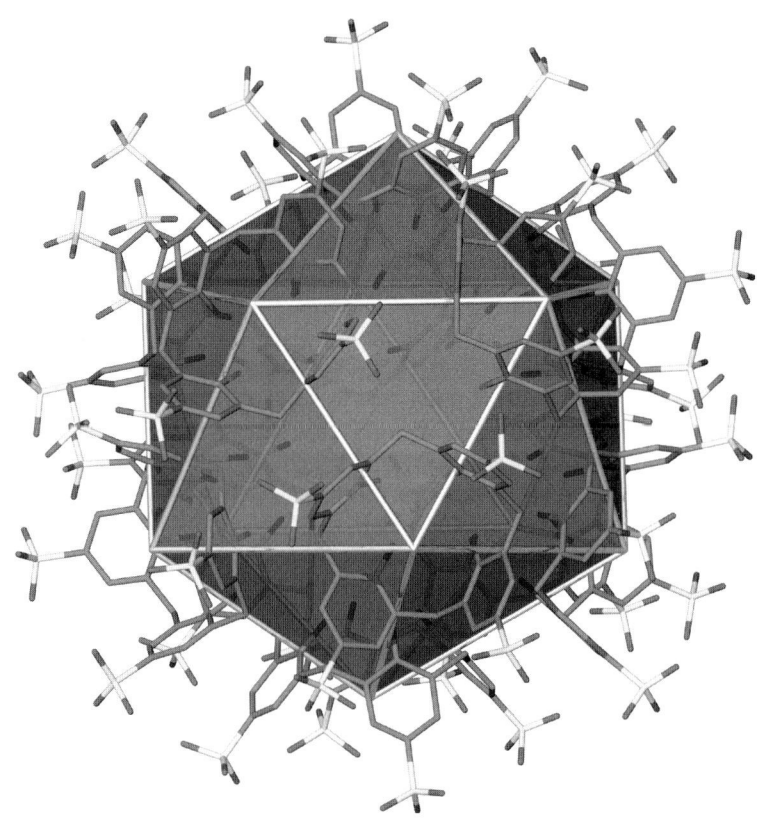

Figure 14.3 Arrangement of 12 p-sulfonatocalix[4]arene molecules at the vertexes of an icosahedron to form a compact spherical assembly [15].

We have now shown that replacement of PNO by 18-crown-6 in the above-mentioned ternary system results in a remarkably dissimilar spheroidal array consisting of 12 calixarenes arranged at the vertexes of a cuboctahedron (Figure 14.4) [17,18]. We attribute the formation of this new type of spheroid to the constraint placed upon facing calixarenes of neighboring spheres by their shared 18-crown-6 guests. The relative rigidity of the shared crown ether guest forces facing calixarenes of the {2–crown ether–2} dimer to be eclipsed relative to one another, whereas the {2–PNO–lanthanide–PNO–2} dimer is bridged by multisphere coordination (as the C-shaped dimer). We believe that a discrete spheroidal entity comprising 12 molecules of 2 would favor placement of the components at the vertexes of an icosahedron, since this arrangement is the most compact of the 12-vertex Platonic and Archimedean solids. However, when the spheres are arranged to form a three-dimensional solid, either cubic or hexagonal close-packed arrangements of the multicomponent entities should be favored. Therefore, we understand that the extended structure has arranged itself such that the spheres are situated along vectors representing the vertexes of either a cuboctahedron (Archimedean solid,

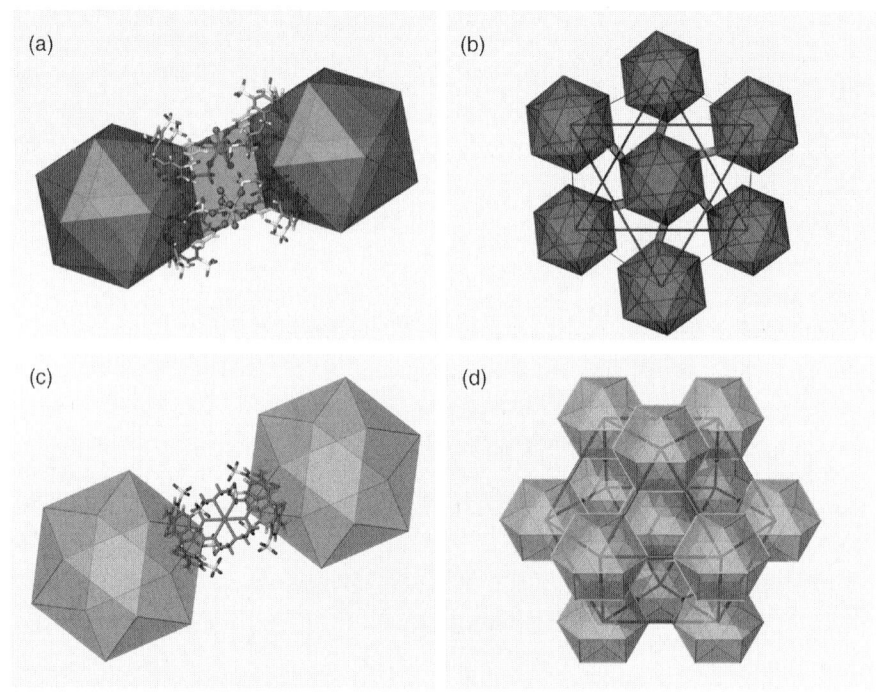

Figure 14.4 Guests external to the 12-calixarene assemblies based on **2** determine the packing and symmetry of the capsule [17,18]. (a) Icosahedral nanocapsules linked by C-shaped dimers. (b) Packing of icosahedra in an anti-prismatic fashion. (c) Linking of cuboctahedral arrays by {**2**–crown ether–**2**} linkages. (d) Cubic close packing of neighboring cuboctahedra.

cubic close-packed) or a triangular orthobicupola (Johnson solid capable of hexagonal close packing by vertex–vertex alignment).

The internal volume of the icosahedral arrangement is calculated to be 975 Å^3, whereas the cuboctahdral geometry (Figure 14.4) contains a volume of 1260 Å^3. This represents an increase of about 30% in enclosed volume upon reorganization from icosahedral to cuboctaheral packing. This disparate packing of calixarenes has a dramatic effect on both the chemical composition and nature of the outer shell. The calixarenes are arranged such that the shell of each sphere contains six pores in which disordered water molecules are situated. These pores have a van der Waals diameter of 4.2 Å and are arranged at the vertexes of an octahedron centered at the core of the capsule. This arrangement can be thought of as a channel for the communication of molecular material between the hydrophilic interior of the capsule, through the hydrophobic shell, to the hydrophilic exterior [17,18]. The amount of 'chemical space' enclosed by either of the two capsules is about 1000 Å^3. As discussed above, this space (in the icosahedral case) houses 30 water molecules and two sodium ions. However, the van der Waals volume of these capsules is about 11 000 Å^3 (this calculation is based on the approximate nanocapsule diameter). The second generation capsules are differentiated from **1**, the first generation capsules, by several factors. First, the supramolecular forces used to hold together capsule **1** together are hydrogen bonds, while a combination of van der Waals forces, π-stacking interactions and metal ion coordinate covalent bonds is employed for the *p*-sulfonatocalix[4]arenes of the second generation. Second, the surface which encloses the chemical space is essentially one atom thick for **1**, while it is the thickness of the *p*-sulfonatocalix[4]arene building block in the second generation nanocapsules (hence the 11 000 Å^3 volume of the assembly with only 1000 Å^3 of space within). Third, the contents of the capsule are rather completely ordered for the second generation capsules (by the hydrogen bonds from the enclosed water to the phenolic oxygen atom hydrogen bond acceptors at the base of the *p*-sulfonatocalix[4]arene), but the contents are completely disordered for **1** (because of the lack of any directional bonding force connecting the skeleton of the assembly to the contents therein).

An important outgrowth of the work described above was the discovery of a method of control of molecular architecture such that in one example a spherical assembly (an *icosahedron*, a Platonic solid) was converted into a tubular structure [15]. This will be discussed along with other similar examples at the end of the chapter. It is important to emphasize that the second generation nanocapsule can be formed from a wide range of metal ions, both 3+ (lanthanides) and 2+ (cadmium), and the stoichiometry of the reaction mixture controls the architecture of the final product, sphere or tubule.

14.4
Third Generation Nanocapsules

In 1999 there was a report [19] of a large supramolecular assembly related to our resorcin[4]arene work. The synthesis of *C*-isobutylpyrogallol[4]arene (Figure 14.5A), **3**, was accomplished under mild conditions [19]. However, the authors reported that

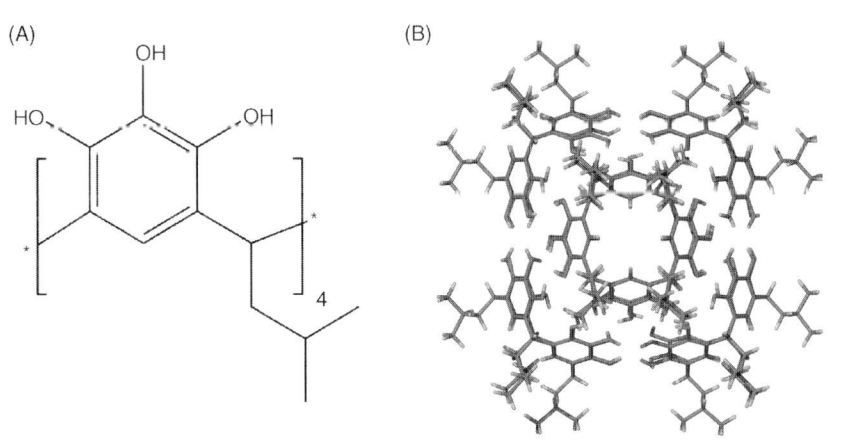

Figure 14.5 (A) Schematic of *C*-isobutylpyrogallol[4]arene. (B) Self-assembled capsule of *C*-isobutylpyrogallol[4]arene, as crystallized from acetonitrile [19].

they obtained the self-assembled hexamer of **3** only one time out of many attempts (Figure 14.5B). Further, this lack of reproducibility was used as evidence that the hexamer is unstable compared with our [(C-methylresorcin[4]arene)$_6$(H$_2$O)$_8$], **1** [10]. This did not seem credible, since the hexamer appears to be held together by 72 hydrogen bonds, 48 of which are intermolecular. This means that the assembly is bound together by eight intermolecular hydrogen bonds per bonded entity. Capsule **1** is bound together by 36 intermolecular hydrogen bonds or 2.6 per bonded molecule. For comparison, the tennis ball of Rebek's group [20] is bound together by four hydrogen bonds per bonded molecule. It seemed to us that the pyrogallol[4]arene hexamer should be more, not less, stable than our [(C-methylresorcin-[4]arene)$_6$(H$_2$O)$_8$] capsule in the same solvents.

We therefore synthesized a variety of pyrogallol[4]arenes by the acid-catalyzed condensation of pyrogallol with appropriate aldehydes [21–23]. The yields are high, approaching quantitative, and although pyrogallol[4]arenes can often be crystallized in bilayer motifs [24], we obtained X-ray structural data for the hexamer **3** with R = ethyl, propyl, isobutyl, butyl, hexyl and others. The hexamer for R = butyl is shown in Figure 14.6 and the formation of this nanocapsule is fairly reproducible [24]. We have obtained various hexamer from a variety of solvents, including Et$_2$O

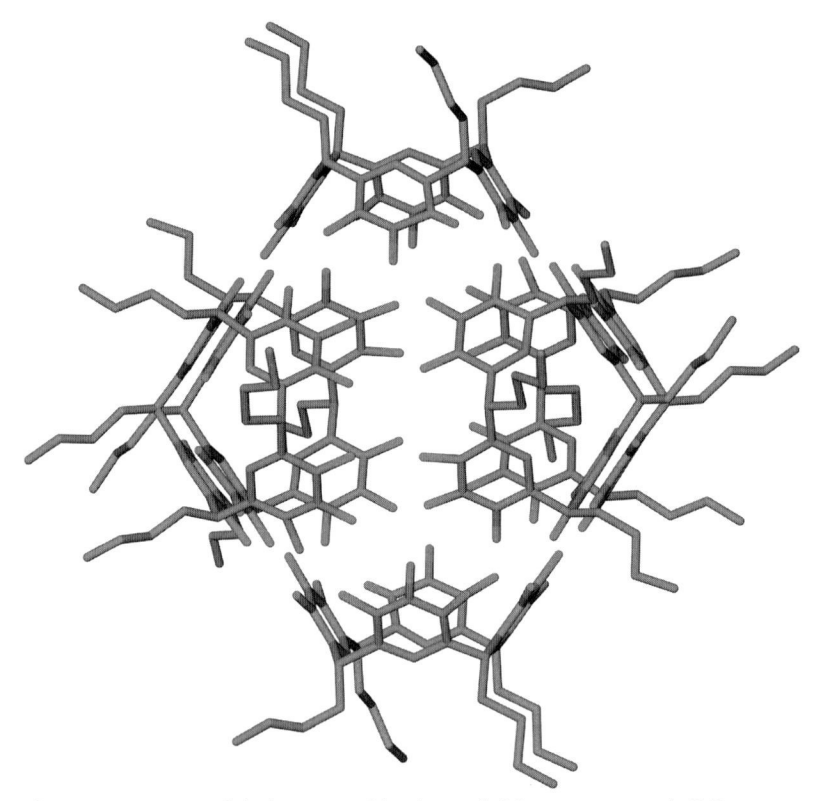

Figure 14.6 Structure of the hexameric C-butylpyrogallol[4]arene nanocapsule [24].

with nitrobenzene and, surprisingly, methanol. We have also studied these pyrogallol[4]arenes by NMR techniques in a variety of solvents in order to assess their stability and their binding properties (as have others) [25–27]. The volume available for guests is ~1500 Å^3, so the range of guests available for study is vast [25].

In a preliminary test of the stability of the C-isobutylpyrogallol[4]arene hexamer, crystals of the compound were slurried in water and sonicated for 30 min. The solid was then collected on a frit and dried under vacuum. The powder pattern of the compound after the water treatment was the same as that before the treatment (in each case, the characteristic low angle peak of the hexamer was observed). The C-isobutylpyrogallol[4]arene hexamer is insoluble in water and the water does not degrade the spherical structure.

A key feature of biological systems is the encapsulation of entities within a structure. A spherical virus is a beautifully complex example of this enclosure of chemical space [28]. Until now, such biological capsules have been beyond the synthetic grasp of the chemist because of the vast size and complexity of the enclosure. A further complication which has not often been addressed in the chemical literature of cell mimics is the very high level of organization found on the interior of enclosures of biological importance.

Indeed, once the enclosure of space has been accomplished, the organization of the guests contained within becomes a key issue. Rebek and coworkers have used steric constraints to organize two guests within a tubular dimer [29], but for those assemblies with large enclosed volumes, both discrete and infinite, the guests are most often disordered [14,17–19,21–24]. The one example of significant order within such an enclosure is the polar core consisting of 30 water molecules and two sodium ions in the second generation 12-p-sulfonatocalix[4]arene assembly [15].

In the first structurally authenticated hexamers (generally represented by **3**), it was not possible to determine any geometric information relating to the included guest molecules. We have subsequently shown that the guest molecules trapped within the host container may adjust their spatial orientation in response to interactions between adjacent nanocapsules. Remarkably, functionalizing the outer shell of the nanocapsule with different alkyl chains leads to highly specific solid-state packing arrangements, which in turn influence the organization of the guest molecules within the capsules. In the course of this work, it was discovered that the nanocapsules may be synthesized reproducibly in quantitative yields under normal laboratory conditions by crystallization from ethyl acetate [21]. TG–IR analysis of the R = heptyl capsules revealed that ethyl acetate molecules are released from the crystalline material at two distinct stages during gradual heating of a dried sample: a weight loss of 7.5% (70–105 °C) corresponds to the loss of only the six ethyl acetate molecules external to the cavity; a further weight loss of 8% immediately prior to decomposition (225–275 °C) corresponds to the loss of both bound ethyl acetate and water molecules from the cavity. In comparison, TG–IR analysis of the monomer rather than the capsule shows no significant weight loss above the boiling point of the solvent, until rapid decomposition occurs at 290 °C. Consistent with the TG–IR studies, the crystal structure of the R = heptyl capsule shows that six ethyl acetate molecules enshroud a disordered water molecule within the host assembly.

With respect to crystal packing, the cases when R = pentyl and heptyl afford simple hexagonal and hexagonal-closest packing arrays, respectively, with capsules separated by their lipophilic tails. However, for R = hexyl the nanocapsules are not forced away from one another by the hexyl tails, but rather congregate to form hydrogen-bonded nanorods (Figure 14.7). Four OH groups on opposite sides of each capsule form hydrogen bonds with two adjacent capsules. Consequently, the walls of the individual nanocapsules are disrupted, which translates through to the packing of the guest species within the capsule.

The distorted intermolecular hydrogen bonding within the host walls enables four of the ethyl acetate guest molecules to undergo hydrogen bonding to the nanocapsule wall through their carbonyl functionalities. The nanocapsules for R = pentyl and heptyl feature ordered arrays of ethyl acetate molecules enshrouding a water molecule [21]. The methyl groups are orientated towards the base of the pyrogallol[4]arene macrocycles and the single water molecule resides at the center of the enclosed space. When the neighboring nanocapsules communicate with each other through the hydrogen bonding interactions observed in the R = hexyl nanocapsule (Figure 14.7), the four ethyl acetate guest molecules hydrogen bonded to the inside of the nanocapsule wall reorient with their ethyl groups in the base of the macrocycle (Figure 14.8). The two remaining ethyl acetate guests position their methyl groups down into the cleft of the macrocycle. It was not possible to locate the exact position of the water molecule inside the nanocapsules owing to extensive disorder. However, the difference in guest interactions is clearly a consequence of communication or lack thereof between the host walls of neighboring nanocapsules. This initial result opens the way to control guest orientation by forces external to the capsules.

In an extension to this work, we have been examining the encapsulation of probe molecules in these large assemblies with a view to reporting on the nature of the "inner phase" of these large assemblies [30–34]. We found that sonication of a hot saturated acetonitrile solution of C-hexylpyrogallol[4]arene and excess pyrenebutyric acid (PBA) [30] afforded dark single crystals (crystals of solvent containing capsules are colorless) [24] that were studied using X-ray crystallography. These studies showed the capsule to encase two pyrene butyric acid molecules that were "bound"

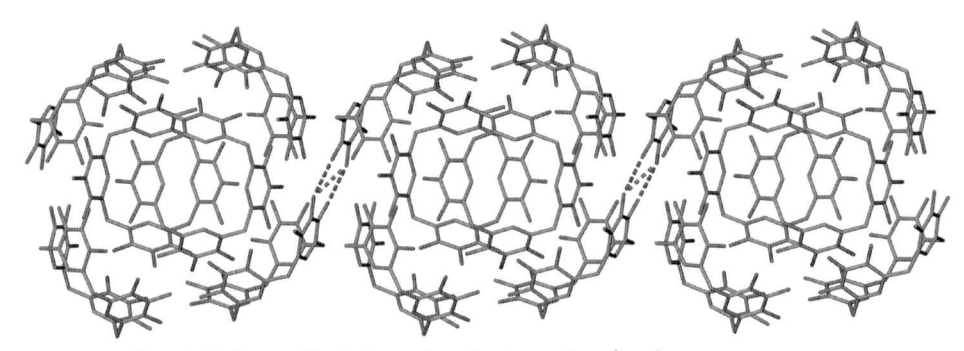

Figure 14.7 Some of the hydrogen bonding interactions found between nanocapsule walls when C-hexylpyrogallol[4]arene is crystallized from ethyl acetate [21].

Figure 14.8 Guest molecules within nanocapsules orient differently depending on the environment external to the overall assembly [21]. In (a) the methyl group of the ethyl acetate is inserted into the pyrogallol[4]arene cavity, whereas in (b) the orientation is reversed, with methyl group insertion observed.

to the capsule wall by π-stacking and CH···π interactions (Figure 14.9). In order to examine whether the assembly was stable in solution, single crystals were dissolved in a non-polar medium (hexane) and the stability of the assembly was followed by spectrofluorimetry in the presence of an appropriate quencher (dimethylaniline). Unfortunately, it was not possible to deconvolute fully the interactions between the

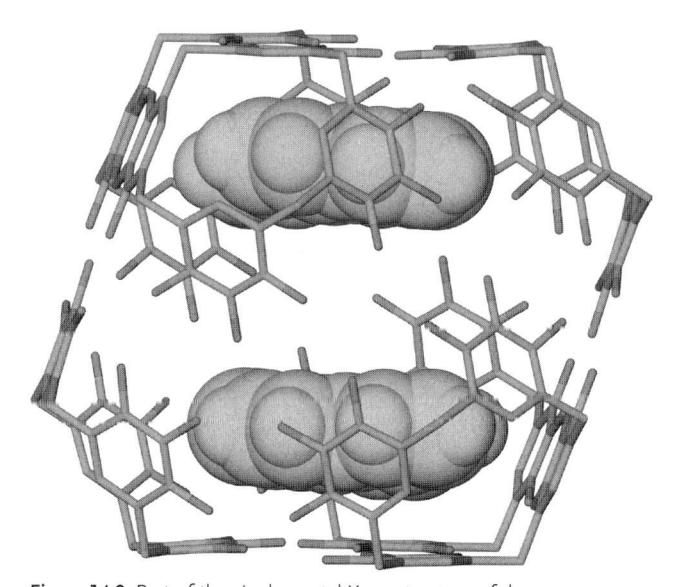

Figure 14.9 Part of the single crystal X-ray structure of the C-hexylpyrogallol[4]arene hexameric nanocapsule shrouding two pyrene butyric acid molecules [30]. The guests are found to stick to the wall through π-stacking and CH···π interactions and are well separated (by around 8 Å) by the butyric acid side-chains and co-encapsulated acetonitrile molecules.

butyric acid functionality of the PBA and the capsule seam hydroxyl groups due to extensive disorder within the large assembly. This was also impeded by the fact that single crystals of these materials are typically small and weakly diffracting. However, both analytical techniques support the retention of the guest-containing assembly and other polyaromatic probes were examined for similar behavior. Perylene, benzo[*a*]pyrene and pentacene have also been successfully encapsulated, although without appropriate side-chain functionality it appears that these probes are not retained even in non-polar media [31]. This feature may also be responsible for lower capsule population that precludes structural characterization due to weak diffraction and high levels of disorder. Through computational studies on a "half-capsule" in the presence of a guest, we believe these guest species to be too large to form the favorable interactions described above for PBA and initial spectrofluorimetry studies appear to show that the probes are rapidly released upon crystal dissolution [32]. In this regard,

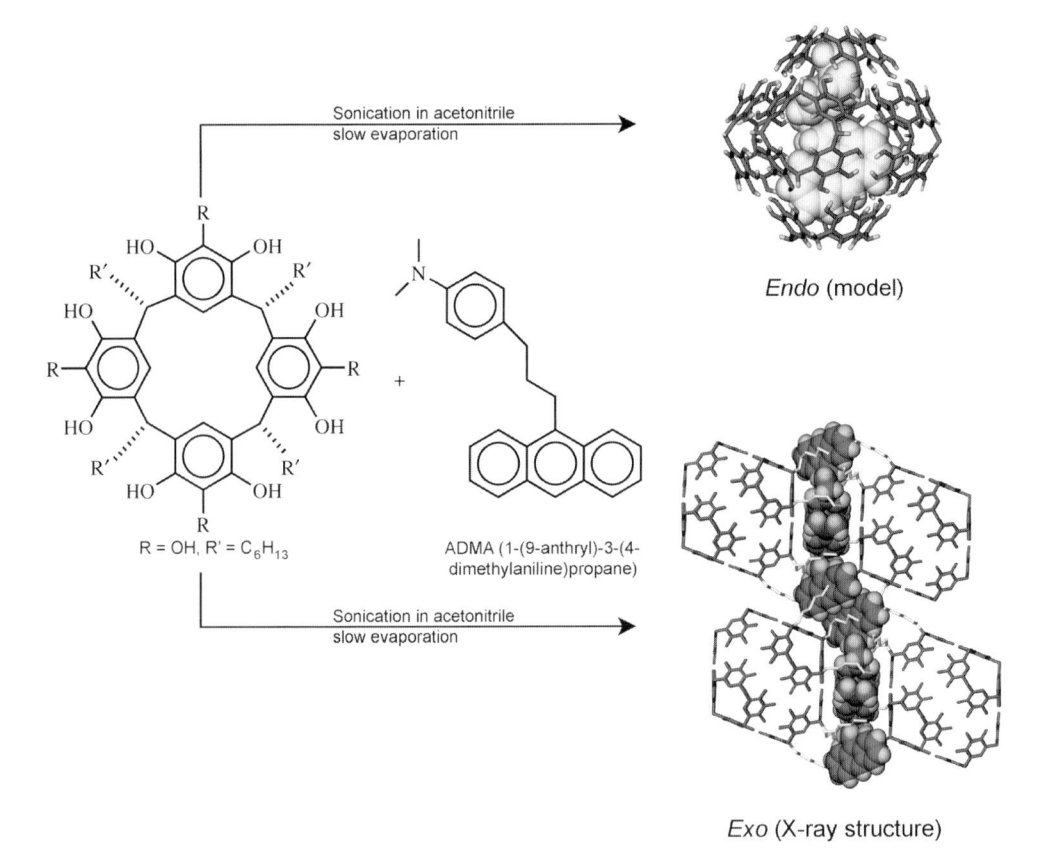

Figure 14.10 Schematic of the formation of C-hexylpyrogallol[4]arene assemblies with the probe ADMA residing either *endo* and *exo* to the nanocapsule motif [33,34]. In the *exo* case, the probe forms channels within the structure, pushing neighboring nanocapsules apart.

we are currently exploring the side-chain functionality as a method of capsule stabilization.

When a probe incorporating a "built-in" quencher, 4-[3-(9-anthryl)propyl]-N,N-dimethylaniline (ADMA), was employed, two very different situations arose with respect to guest interaction with the pyrogallol[4]arene nanocapsules [33,34]. In the majority of cases it was possible to obtain crystals of the probe-containing capsules, but on rare occasions it was possible to obtain crystals of different morphology, all of which were studied using single-crystal X-ray crystallography. The inner phase of the capsule-containing probes was not determinable, which is likely due to the reasons outlined above, but for the different morphology crystals the probe was found to reside *exo* to the capsule and was also found to push the capsules apart through the formation of channels of ADMA in the solid state (Figure 14.10) [33]. Both assemblies were followed by spectrofluorimetry [in tetrahydrofuran (THF)], showing distinctively different spectra. In order to determine whether aggregates of the *exo* probe structure were present in the solution phase, the solution was sonicated to release the guest from the arrangement, producing a characteristic spectrum of the free probe in THF. Various other probe molecules containing quenchers are also being explored for similar behavior.

14.5
Fourth Generation Nanocapsules

The fourth generation of capsules is somewhat related to the previously reported capsule formulated as $[(C\text{-methylresorcin}[4]\text{arene})_6(H_2O)_8]$. Capsule **1** possesses an excess of four hydrogen bond donors, but these donors are positioned such that they project outward from the surface of the enclosure [10]. In this orientation they are incapable of effecting organization of the guests within the capsule. In a similar fashion, all the hydrogen bond donors in $[(\text{pyrogallol}[4]\text{arene})_6]$ are typically used in completing the hydrogen bond pattern that forms the capsule; the guests within are generally not ordered, except in those cases described above for third generation capsules [21]. It was reasoned that a number of mixed systems consisting of different ratios of pyrogallol and resorcinol subunits might form a capsule (or different capsules) with the desired property of excess of hydrogen bond donors (Figure 14.11). Furthermore, such an asembly may offer the possibility to orient guest species through some of the hydrogen donors that would not be involved in self-complementary hydrogen-bonding. The mixed possibilities are shown as **4–7** (Figure 14.11). A modified synthesis of hybrid macrocycle **4** was performed according to the syntheses of resorcin[4]arene and pyrogallol[4]arene and upon recrystallization from diethyl ether (Et_2O), the remarkable structure shown in Figure 14.12 results [35]. The hexamer of **4** takes the shape of a trigonal antiprism with the centers of **4** at the corners of the trigonal antiprism. This assembly, with six hydrogen bond donors positioned toward the interior of the capsule and possesses an internal volume of 860 Å^3. The six Et_2O molecules on the interior of the capsule are ordered by six additional hydrogen bond donors. In addition, there are six more hydrogen bond

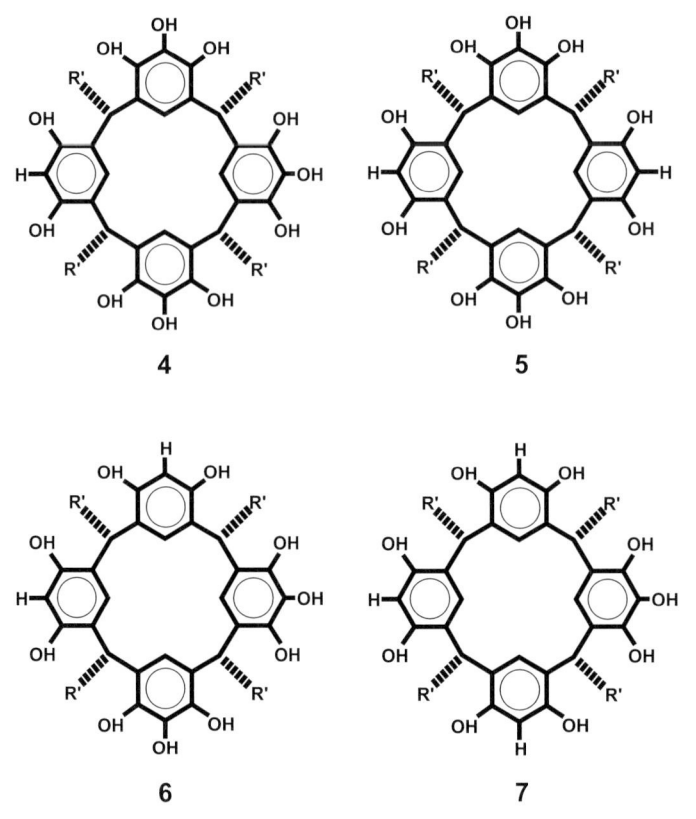

Figure 14.11 Schematic of the mixed resorcinol/pyrogallol[4]arenes formed by performing the literature cyclization in the presence of both subunits [35].

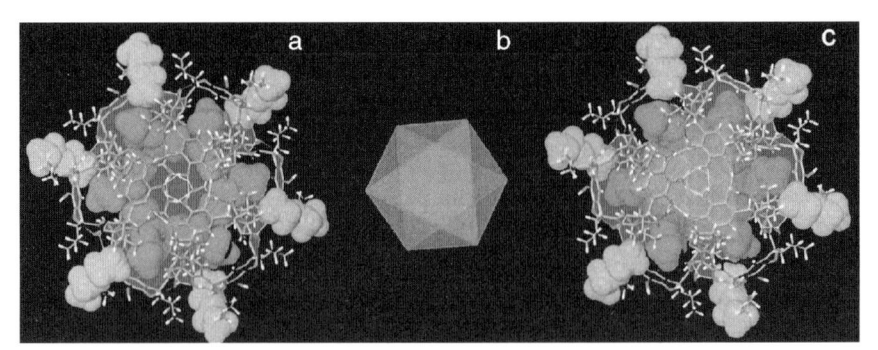

Figure 14.12 The hexameric capsule based on the hybrid macrocycle **4**. (a) The guest diethyl ether molecules and those residing in the shell of the capsule are shown. (b) A trigonal anti-prism. (c) The trigonal anti-prismatic arrangement of guest Et_2O molecules within the capsule and the communication between solvent molecules through the shell [35].

Table 14.1 The mixture of hybrid compounds from the reaction shown in Figure 14.11 [35]. Compounds **3–7** are macrocycles as previously defined. Compounds **8–11** are non-cyclized pyrogallol- and/or resorcinol-containing products.

Compound	%
3	5
4	30
5, 6	33
7	18
Resorcin[4]arene	3
8	5
9	4
10	3
11	1

donors oriented toward the outside and these donors bind six additional Et_2O molecules on the outside of the hexamer, further sealing the capsule. It is worth noting that there is empty volume at the center of the antiprism. This space is about 80 $Å^3$, enough to accommodate an additional guest of appropriate size and hydrophobicity. However, there is also the possibility that this space could be taken by a larger hydrogen bond-accepting guest; for example, it should be possible to bind five Et_2O and one larger guest.

The mixed macrocycles possess remarkable recognition behaviour. In the initial synthesis of **4**, electrospray mass spectrometry revealed that the first precipitate contains at least 10 different compounds with the macrocycles in approximately the percentages shown in Table 14.1.

By simple probability, it would be unlikely that one would be able to isolate only a single compound from the list in Table 14.1 by crystallization alone. Although this is the case, macrocycle **4** continually crystallizes as a hexameric capsule and therefore must possess enough self-complementarity to afford a mono-composite species. The shape of the resulting capsule is more elliptical when compared with the third generation pyrogallol[4]arene capsules and this is a direct consequence of the hybrid nature of the material and the guest species in the capsule. The pursuit of other capsules from hybrid resorcin/pyrogallol[4]arene monomers is under way, as such assemblies *do* offer directional hydrogen bonding to the capsule interior, a feature we consider important for the future use of such assemblies in areas outlined earlier in the chapter.

14.6
Fifth Generation Nanocapsules

The fifth genearation of capsules is based on hydrogen-bonding templates as theoretical binding sites for metal centers, the idea of which was driven by the formation of a copper–cyclodextrin complex [36]. In this regard, the hydrogen-bonded hexamer was considered for complexation (or retro-insertion) with copper and gallium centers and the results of these complexation studies are outlined below, with remarkable results.

Initial results were obtained from the combination of C-propan-3-olpyrogallol[4]arene with 4 equivalents of $Cu(NO_3)_2$ in an acetone–water solution which gave, on standing, single red crystals of $[Cu_{24}(H_2O)_x(C_{40}H_{40}O_{16})_6(acetone)_n]$ [37]. Characterization by single-crystal X-ray diffraction and MALDI-TOF MS revealed that the metal–organic analogue was a combination of 30 components, where six cavitands had all of their available 48 upper rim phenoxy protons replaced with a concomitant square-planar coordination of 24 Cu^{2+} ions to form 96 new Cu–O bonds and the remaining 24 phenoxy protons involved in intramolecular hydrogen bonding (Figure 14.13).

The distances between the oxygens of the hydrogen-bonded capsule and those of the copper capsule are equitable (differing by only 0.002 Å, Figure 14.13), as are the overall size, shape, symmetry and hence volume for these comparative nanocapsules. This clearly showed that the hydrogen-bonded template as a whole was an excellent example of predicting metal–organic analogues. Although this was the case, we were unable to say whether the metal–organic capsule formation process was due to either a templation effect of the metal centers or perhaps a pre-assembled system prior to metal addition.

Our recent studies in this area have given further insight into the mechanism of formation of these nanocapsules. Reaction of C-alkylpyrogallol[4]arenes (C_2–C_{13} alkyl chains) with excess $Cu(NO_3)_2$ in methanol resulted in the instantaneous formation of fine brown precipitates [38]. These precipitates were readily soluble in most organic solvents, however, and it was possible to obtain the single-crystal structure of the C-propylpyrogallol[4]arene metal capsule by crystal growth from an acetone solution (although the crystals were weakly diffracting, it was possible to

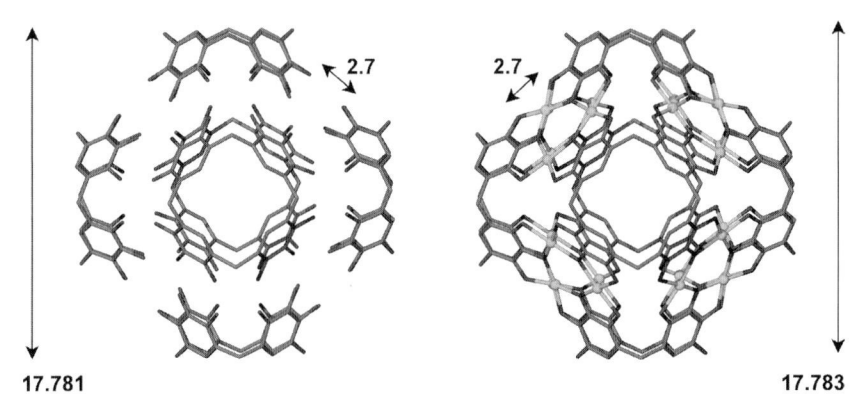

Figure 14.13 Comparison between the hydrogen-bonded (left) and copper bound (right) nanocapsules showing similarities in structure, shape and size of each assembly. Distances shown are in ångstroms [37].

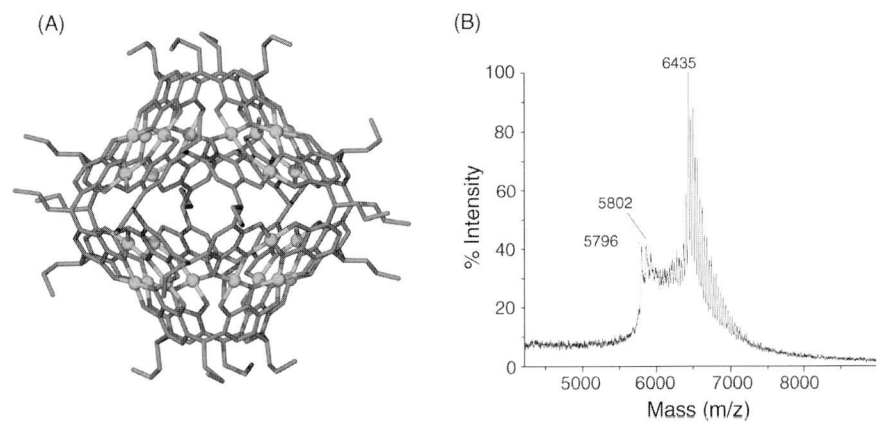

Figure 14.14 (A) Partial crystal structure of the copper C-propylpyrogallol[4]arene. (B) MALDI-TOF mass spectrum of the aforementioned complex showing the bimodal distribution, due to different populations of guest species [38].

obtain a partial structure confirming capsule formation, Figure 14.14A). Analysis using MALDI-TOF MS on the copper nanocapsule precipitates displayed a broad bimodal distribution for each material [38], where the main peaks from either end of the distribution display a consistent mass difference of 633–635 atomic mass units, thereby reflecting the mass difference between a full (empty copper capsule + either 20 MeOH, 35 H_2O or any combination of both) and an empty copper capsule (i.e. one in which there are no apical ligands on the metal centers and no guest species) as shown in Figure 14.14B.

As described for the third generation capsules, ethyl acetate was found to be an excellent solvent for the growth of single crystals of the C-alkylpyrogallol[4]arenes (where alkyl chains range from butyl to undecyl) [21,22]. With longer chains at the lower rim, dodecyl and tridecyl, crystallization from this solvent afforded bilayer structures, indicating that van der Waals forces overcome capsule formation in the solid state at this point. Despite this, a combination of a methanolic solution of these pyrogallol[4]arenes and excess $Cu(NO_3)_2$ in methanol also resulted in the instantaneous formation of a brown precipitate, which was confirmed by MALDI-TOF MS to be the relevant copper nanocapsules. This experiment demonstrated that templation by the hydrogen-bonded hexamers was not a requirement and actually may not play any role whatsoever in the formation of the copper nanocapsule analogues [38].

This was given further credence by reacting a 1:1 mixture of PgC_6 and PgC_{11} as previously described with excess $Cu(NO_3)_2$, again resulting in the instantaneous formation of a brown precipitate [38]. Avram and Cohen had previously reported that there was a degree of self-association among mixtures of C-alkylpyrogallol[4]arenes over an initial 24-h period, after which time a mixed formation of hexameric hydrogen-bonded capsules resulted [39]. The results from our MALDI-TOF MS experiments, however, displayed an instantaneous formation of a statistical mixture of all permutations of metal–organic nanocapsules that could possibly result from

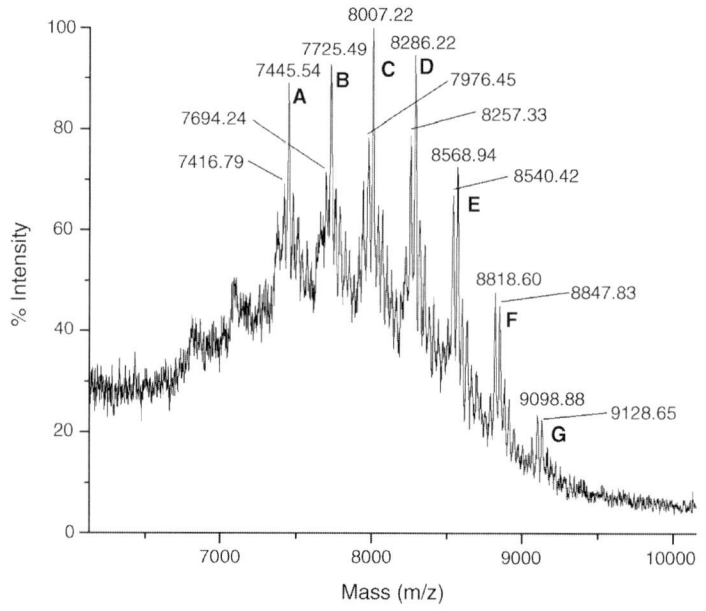

Figure 14.15 MALDI-TOF MS displaying statistical permutations of metal–organic nanocapsules formed from the reaction of copper nitrate and a 1:1 mixture of C-hexyl- and undecylpyrogallol[4]arenes. Peaks A–G reflect PgC_6:PgC_{11} ratios of 6:0, 5:1, 4:2, 3:3, 2:4, 1:5 and 0:6, respectively [38].

the mixture of two pyrogallol[4]arenes, as shown in Figure 14.15 [38]. This suggests a very rapid and indiscriminate reaction of the Cu^{2+} ions with the pyrogallol[4]arenes, thereby showing no inclination to fit within a preordered templation sequence of the hydrogen-bonded hexameric nanocapsules.

In our exploration of other metals that may react to form similar metal–organic nanostructures based on hydrogen-bonded assemblies, gallium nitrate was considered an interesting alternative. The reaction of C-propylpyrogallol[4]arene with 4 equivalents of $Ga(NO_3)_3$ in acetone–water gave single crystals over a matter of hours. Structural analysis revealed the metal–organic capsule $[Ga_{12}(H_{20})_{24}(C_{40}H_{40}O_{12})_6(acetone)_8(H_{20})_6]$ which was assembled from a total of 18 components, i.e. six cavitands and 12 Ga^{3+} ions, as opposed to the expected 24 metal ions in the copper-seamed capsules (Figure 14.16) [40].

With the formation of 48 equatorial Ga–O bonds and the concomitant replacement of 36 protons, there remains a further 36 phenoxy protons which participate in hydrogen bonding, 20 of which are intramolecular and the remaining 16 interact with H_2O in hydrogen bonds that seal up potential surface voids. The vast difference in the hydrogen bonding, induced by the metal centers, results in the formation of a distorted "rugby ball"-like structure, as shown in Figure 14.16. The axial positions of the Ga^{3+} ions are coordinated with 24 H_2O molecules, 12 of which are ligated from the capsules innards, resulting in the ordering of the nanocapsule interior, which eight acetone and 20 water molecules, six of which are non-coordinated waters

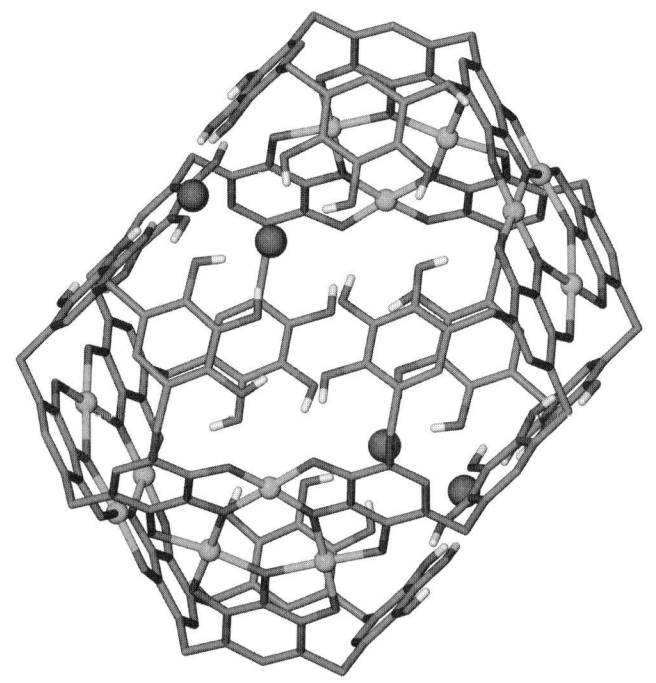

Figure 14.16 The metal–organic nanocapsule formed by reaction of C-propylpyrogallol[4]arene with gallium nitrate. Note the distorted "rugby ball"-like shape of the nanocapsule in addition to the presence of surface water molecules (shown as dark spheres) [40].

involved in two hydrogen-bonded chains $(H_2O)_5$. A variety of other C-alkylpyrogallol[4]arenes have been utilized for capsule formation with Ga^{3+} ions from an acetonitrile–water solvent system [41]. Although in some cases the crystals proved to be unstable to solvent loss once removed from the mother liquor, a suitably stable single crystal for X-ray diffraction formed from C-pentylpyrogallol[4]arene, but in general the interior of these capsules is less well ordered compared with the acetone–water system.

Given the instantaneous nature of the formation of copper pyrogallol[4]arene nanocapsules, we investigated the possibility of exploiting the voids found in the surfaces of the gallium nanocapsules, as shown in Figure 14.16. Gallium C-proylpyrogallol[4]arene capsules were suspended in an acetone–methanol solution and addition of methanolic $Cu(NO_3)_2$ resulted in immediate dissolution of the crystals and a color change from blue–green to red–brown, but notably no precipitate formation. Slow evaporation of the solution gave single red crystals that were analyzed using a synchrotron radiation source, affording the expected "stitched up" nanocapsule (Figure 14.17) [38].

The structure could be solved using either copper or gallium (which is not unsurprising given their proximity in the Periodic Table) and inductively coupled plasma (ICP) analysis was employed to determine the Cu:Ga ratio in the sample.

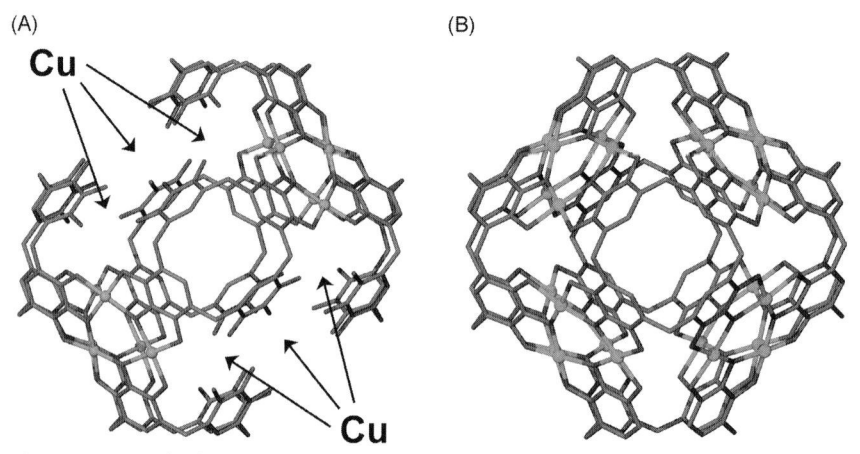

Figure 14.17 (A) The theoretical open binding sites in the gallium nanocapsule shown in Figure 14.16 that are available to copper centers (surface water molecules omitted for clarity). (B) The crystal structure of the mixed metal nanocapsule formed after copper addition [38].

Surprisingly, this was found to be 2:1 (i.e. 16:8 in a nanocapsule), indicating that not only had the Cu centers seamed the surface voids, but also that they expelled and replaced four Ga^{2+} ions, presumably one from each preformed gallium nanocapsule "face". We postulate that the copper exerts a structural correction to the distorted gallium nanocapsule, thus resulting in the expulsion of one gallium center from each array and its replacement by a Cu^{2+} ion [38].

In a similar approach to that used to form the copper nanocapsules, i.e. using hydrogen-bonded arrangements as theoretical blue prints, a search of the Cambridge Structural Database (CSD) revealed a dimeric C-propylpyrogallol[4]arene capsule, reported by Rebek and coworkers [42], which is asymmetric and which possesses a complex polar belt of hydrogen bonds that appeared suitable for metal retro-insertion. Given that a square-planar or octahedral metal center would preclude complex formation with this capsule based on the angles associated with the oxygen atoms of the macrocycles, we investigated zinc (amongst other suitable metals) as a potential complexing agent.

Initial studies with this system found that combination of C-propylpyrogallol[4]arene in pyridine (color change from clear to dark) with 4 equivalents of $Zn(NO_3)_2$ resulted in the growth of colored single crystals of the complex [Zn_8(C-propylpyrogallol[4]arene)$_2$(pyridine)$_8$ ⊂ pyridine] over time (Figure 14.18A) [43]. Characterization by X-ray diffraction and MALDI-TOF MS showed a dimeric metal organic capsule composed of 10 components, two cavitands which have had 16 of their 24 protons replaced by the concomitant coordination of eight Zn^{2+} ions, an encapsulated guest pyridine molecule (disordered due to molecular rotation) and eight axial solvent ligands. Each of the zinc centers is pentacoordinate with four phenoxy groups (arranged in a distorted plane) and an axial pyridine, all of which affords a distorted square-pyramidal configuration (Figure 14.18B).

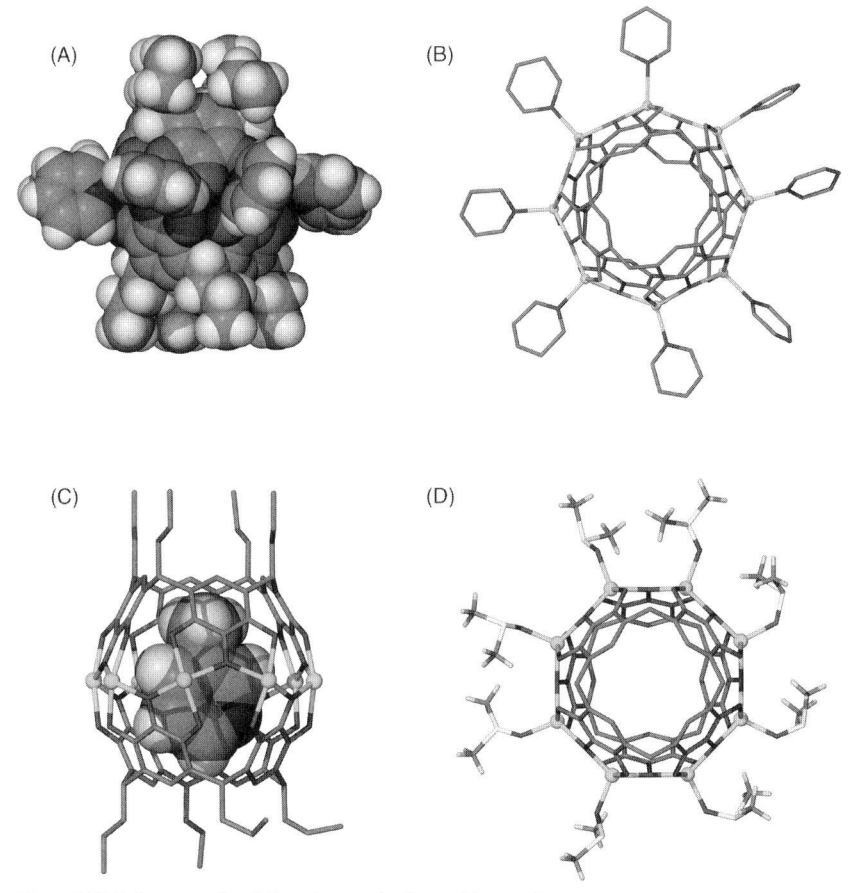

Figure 14.18 Octametallated dimeric capsules formed by reaction of zinc pyridine complexes with C-propylpyrogallol[4]arene. (A) The pyridine-containing capsule in space filling representation. (B) Alternative view of (A) showing pyridine ligation at zinc centers [43]. (C) Space filling model of encapsulated 3-methylpyridine. (D) DMSO ligation at zinc centers demonstrating ligand exchange [44].

As pyridine appeared to play a crucial role in the dimeric metal-seamed capsule formation, the complex [Zn(NO$_3$)$_2$(pyridine)$_3$] was dissolved in MeOH and mixed with a methanolic solution of C-propylpyrogallol[4]arene, giving an instantaneous yellow precipitate which was analyzed and confirmed to be the target metal–organic dimer. This concomitantly allowed for facile variation of guest species and metal ligands including 3-picoline [44], 4-picoline, 4-ethylpyridine and 1-methylimidazole [45]. For purposes of crystallization, the most suitable solvent was found to be DMSO, which also readily exchanged with the heterocycle as a ligand, with complete exchange being achieved via dialysis (Figure 14.18 C and D) [44].

Both ^{1}H and ^{13}C NMR analysis demonstrated the influences of a "tight" or confined space on the guest, a feature that was immediately noticeable by significant upfield shifts for the guest relative to those of the free molecule, a feature that has also been observed for carcerands [46]. For encapsulated 3-picoline, the methyl group experiences a $\Delta\delta$ greater than 4 ppm, reflecting the strong CH···π interaction between host and guest, while the remaining $\Delta\delta$ values for the guests aromatic protons reflect their equatorial positioning within the dimeric capsule, although these are still significant [44]. This correlates with the crystal structure in which the 3-picoline methyl group is shown to be clearly positioned in close proximity to the arene moiety of the capsule (Figure 14.18).

It was observed that upon ligand replacement of DMSO, the ^{1}H NMR upfield shift for the encapsulated guest was sensitive to change. This is most likely a consequence of the electropositive metal centers supplementing for the loss of electron density by drawing from the aryl rings, thus affecting host–guest interactions. A series of titration studies were initiated by the reintroduction of free 3-picoline to a DMSO-d_6 solution of the DMSO-ligated metal–organic dimer, where the guest shows a downfield signal shift by each of its protons (Figure 14.19). This reflects a change in electron density distribution throughout the capsule, which of course would have an influence on the guest's mobility and CH···π interactions.

MALDI-TOF MS analysis on the [Zn_8(C-propylpyrogallol[4]arene)$_2$(pyridine)$_8$ ⊂ pyridine] complex revealed two significant peaks corresponding to the zinc dimeric capsules that are either occupied or vacuous. Notably, this is the third capsular system in which we have recently observed significant void space, $\sim 80\,\text{Å}^3$ in the fourth generation capsules and ~ 1500 and $\sim 140\,\text{Å}^3$ for the copper and zinc capsules described in this section. We are now looking at exploiting these large voids for

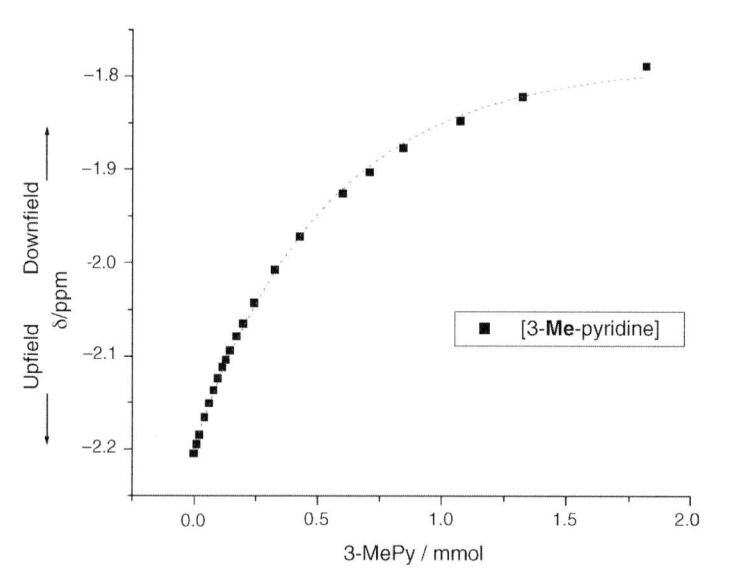

Figure 14.19 Titration plot of ligand concentration versus methyl shift of encapsulated 3-picoline guest, showing an increase in δ owing to the exchange of DMSO for electron donating 3-picoline ligands on zinc centers [44].

molecular entrapment that may be deemed indiscriminate within steric considerations for example.

14.7
Sixth Generation Nanocapsules

The largest closed capsule structurally characterized thus far is the hexamer **3** with an enclosed volume of $\sim$1500 Å^3 [25]. It is anticipated that it may well be possible to enclose (without molecule-sized pores) up to $\sim$2500 Å^3 of chemical space. To enclose completely even more space, one will most likely need larger synthons given that the bulding blocks we have described are highly self-complementary and routinely form those assemblies outlined above. Our first attempts in the area of larger capsules have involved using alkyl chains of varying length and functionality radiating from the hydrogen-bonded and metal–organic nanocapsules. As mentioned above, pyrogallol[4]arenes have been shown to crystallize in either bilayer or hexameric nanocapsule motifs, depending on the crystallization solvent employed [19,21–24,30,33,38]. Although this is the case, little was known of the aggregation and supermolecular assembly of these solution stable entities. To our surprise, a wide variety of packing motifs have been observed for hexameric nanocapsules in the solid state [21,22,24]. Of particular interest is the long-chain motif shown in Figure 14.20 [22]. The alkyl chains effectively pack only in two dimensions, leaving the third dimension free for solvent inclusion. This provided the clue that such simple capsules might indeed form the building blocks for even larger structures. We recently shed light on this behavior through the use of dynamic light scattering (DLS), transmission and scanning electron microscopy (TEM and SEM, respectively) and atomic force microscopy (AFM) techniques to find very large structures/aggregates that are $\sim$80–500 nm in size/diameter (Figure 14.21) [47].

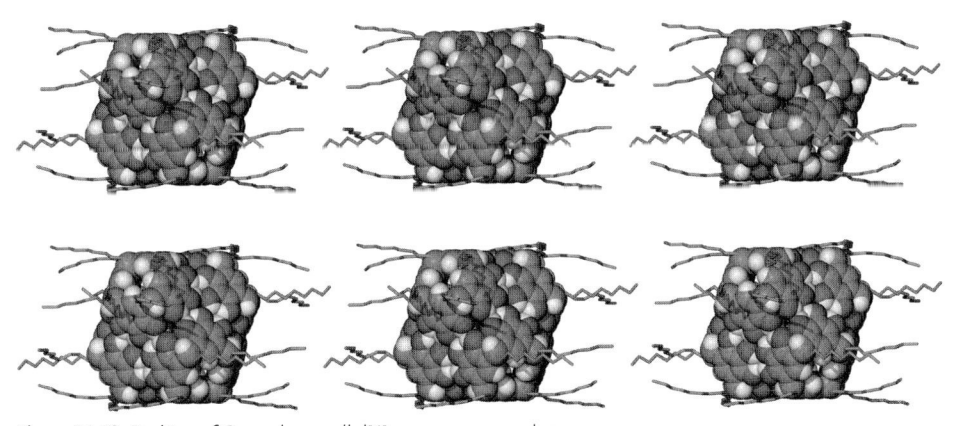

Figure 14.20 Packing of C-nonylpyrogallol[4]arene nanocapsules. Note that the alkyl chains pack in a two-dimensional like arrangement, exposing hydrogen-bonded faces of the nanocapsules in the third dimension [22].

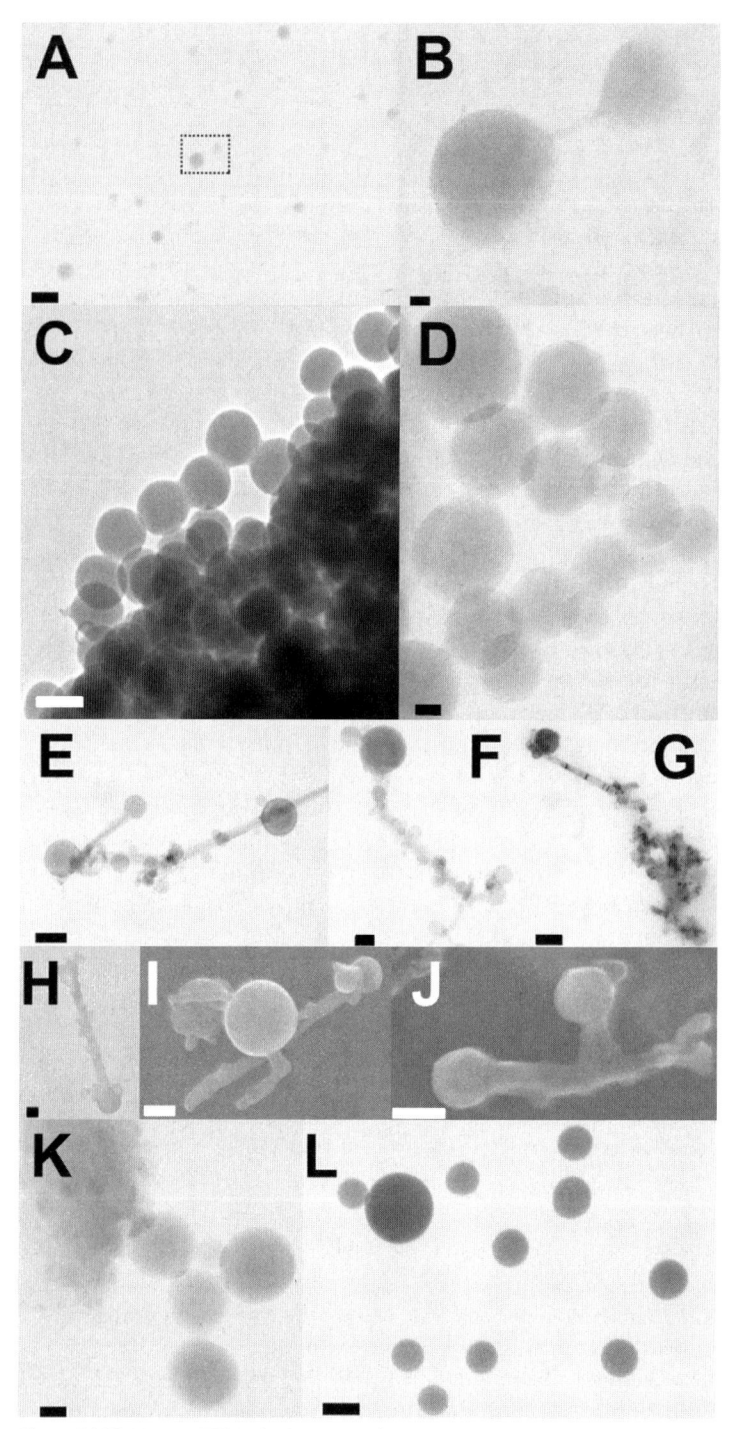

Figure 14.21 Various TEM (A–G, K, L) and SEM (H–J) images showing the large spherical and tubular aggregates formed from hydrogen-bonded and metal–organic nanocapsules [47].

Through the use of DLS, the large supermolecular aggregates were observed for different hydrogen-bonded pyrogallol[4]arene hexamers under dilute conditions (typical concentration $\sim 10^{-3}$ M). Given the stability and recurring nature of the aggregates in solvents other than water, TEM, SEM and AFM were further used to examine the particles formed, the results of which were unexpected [47]. Evaporation of a $\sim 10^{-4}$ M acetone solution of the hydrogen-bonded *C*-isobutylpyrogallol[4]arene hexamer under ambient conditions afforded large, spherical aggregates that were observed by TEM (Figure 14.21A). The aggregates were found to be spherical, of uniform shape and with wide-ranging diameters (92 ± 42 nm) when formed from acetone solutions. Furthermore, the spherical aggregates were (on occasion) found to be linked by interesting tubular architectures as shown in Figure 14.21B. The rate of evaporation was found to play a role in aggregate formation. By employing higher boiling solvents in sample preparation, the number of observed aggregates was reduced and this process also typically led to alternative crystal growth as observed via TEM studies. Similar observations were seen for other solvent systems such as methylene chloride, acetonitrile and chloroform with the visualization of stacked aggregates in the last two systems (Figure 14.21C and D). AFM further corroborated the presence of large spherical aggregates, with similar-sized particles being observed using this additional technique.

Given that lipophilic calix[4]arenes have been shown to form solid lipid nanoparticles that collapse upon solvent removal [48], coupled with the fact that pyrogallol[4]arenes can form such bilayer arrangements [38], it was necessary to determine whether the submicron aggregates were formed from hydrogen-bonded layers or hexamers. In order to probe this, analogous metal-coordinated capsules (fifth generation gallium hexamers that are incapable of disassembly in the solvent systems employed) were studied using TEM. The results of these studies on this system were spherical aggregates with dimensions comparable to those of the hydrogen-bonded analogues. The images produced were significantly darker, presumably due to the high content of gallium in the resultant superstructures (Figure 14.21K and L). This compelling evidence suggests that the spherical aggregates *are* indeed composed of many discrete hexamer building blocks (or hydrogen-bonded analogues) and are *not* a bilayer-related motif.

In general, the tubular connectors were of smaller diameter than the spherical aggregates, which were often found at the tubule ends. These tubules were found to exist in both smooth and rugged morphologies with lengths approaching 1 μm (Figure 14.21F–J). Examination of the smooth and rugged tubes using TEM and SEM, respectively, showed hemispherical swellings that appeared to be spherical aggregates budding from the tubular architectures (Figure 14.21E, F, H–J). These budding aggregates are spread randomly across the length of the tubes and are of comparable diameter. When observing the freshly deposited submicron spheres and tubes using TEM, bands of dark material could be seen to travel the length of the tubular structures (Figure 14.21G). Additional observations showed that spheres connected to tubes had electron dark regions. Upon exposure to the beam, this region diffused from the sphere along the tube as a dark band. Such dark band movement may be attributable to the energy of the electron beam that could cause regional heating, thereby forcing the trapped solvent to undergo translation due to a

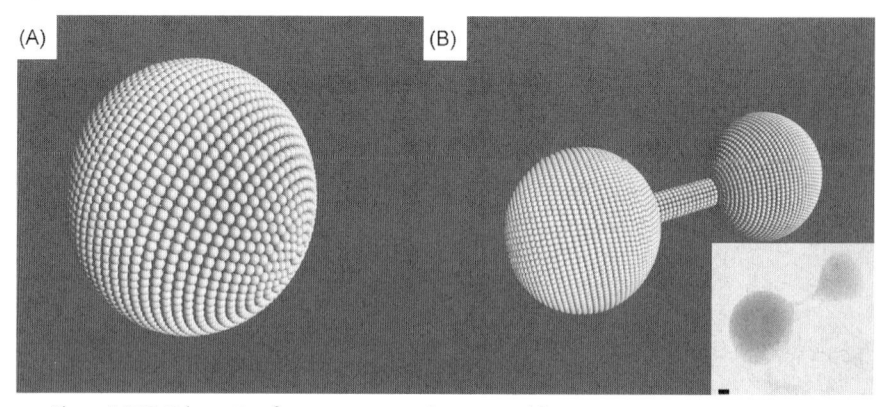

Figure 14.22 Schematic of many nanocapsules arranged into spherical and tubular aggregates (see inset experimental figure) [47].

temperature gradient. This phenomenon was only observed on freshly prepared TEM samples that had residual solvent present, and samples re-examined after a number of days showed no sign of such band movement.

By conducting *in situ* experiments with crystals of pyrogallol[4]arene hexamers with the electron beam (TEM), we were able to discover a direct route to the tubular and spherical aggregates. Using this method, we discovered that as the TEM beam is entrained upon these single crystals, spherical and at times tubular aggregates would also form as the crystalline material decomposes.

Figure 14.22 provides a model of the submicron sphere and tube based on the association of pyrogallol[4]arene nanocapsules. By altering the length of the alkyl chains radiating from the lower rims of the pyrogallol[4]arenes, we aim to introduce further control over the aggregation properties of these nanocapsules to afford larger or desirable nanoparticles for future application.

14.8
From Spheres to Tubes

Spheres and tubes are intimately related, an excellent example of which is fullerene C_{60} and single-walled carbon nanotubes [49,50]. Indeed, these two materials have attracted much interest and it is surprising that much less effort has been invested in developing the synthesis of alternative multicomponent organic nanospheres and tubules. Only a limited number of purely organic nanotubes that have large internal channel volumes have been reported to date, the majority of which are formed through the self-assembly of cyclic oligomers composed of peptides [51], ureas [52] or carbohydrates [53]. However, an important example to be highlighted is the back-to-back solid-state assembly of calix[4]hydroquinone, resulting in a multicomponent organic nanotube [54] which is stabilized by π-stacking interactions between adjacent calixarenes and which was used to form ultrathin silver nanowires.

As mentioned in the section on second generation capsules, it was shown that deviation from the typical anti-parallel bilayer arrangement of *p*-sulfonatocalix[4] arene (Figure 14.2) to form a nanoscale icosahedral arrangement (Figure 14.3) could be achieved through variation of the stoichiometries of the calixarene, pyridine *N*-oxide and metal cation [15]. In addition to forming a near-spheroidal icosahedral arrangement, the *C*-shaped dimer used to form this assembly can also be tailored into nanotubules which contain additional PNO as a spacer molecule. The diameter of these tubules is similar to that of the icosahedral arrangement and the core is composed of hydrated sodium and lanthanum cations.

More recently, and in relation to the first generation of capsules, Rissanen and co-workers showed that *C*-methylresorcin[4]arene can be co-crystallized such that

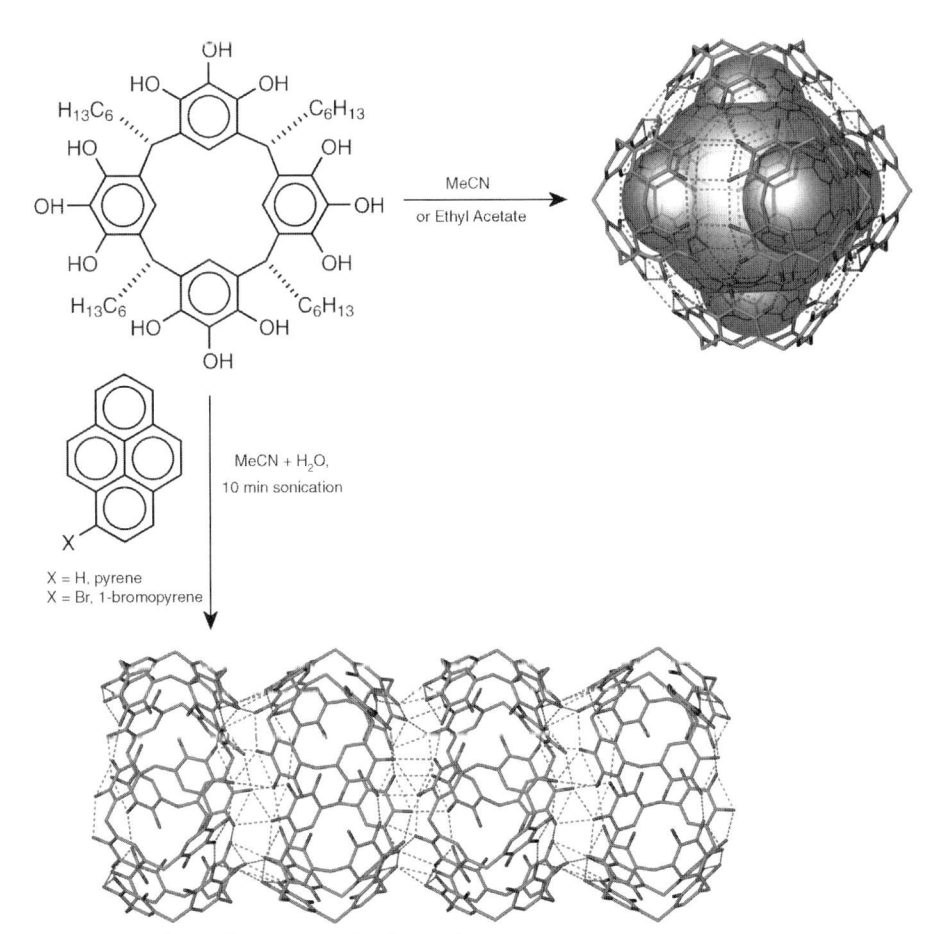

Figure 14.23 The conditions required to form either nanocapsules or nanotubular arrays of *C*-hexylpyrogallol[4]arene. In the latter, pyrene or 1-bromopyrene acts as a type of molecular glue on the exterior of the nanotubular assembly, interacting through many van der Waals interactions [56].

the molecule forms a back-to-back nanotubular arrangement, which contains a small core that is capable of solvent exchange [55]. This is the second example in which a molecule used to form a nanocapsule can be manipulated into a nanotubular array, although in this case the orientation of the host molecule is reversed, i.e. back-to-back rather than face-to-face as in **1**, and differs from the *p*-sulfonatocalix[4]arene examples in which the orientation is the same (both are parallel packing).

We have recently reported the third such example of nanocapsule to nanotube conversion, which is also the second example of packing retention. *C*-Hexylpyrogallol[4]arene, when co-crystallized in the presence of pyrene or 1-bromopyrene from a hot sonicated acetonitrile–water solution and followed by cooling and slow evaporation, forms large single crystals that were structurally analyzed to be pyrogallol[4]arene-based nanotubes. The pyrogallol[4]arene in the nanotube arrays form cyclic or disk-like tetramers through hydrogen bonds. These tetramers are found to link together in a tubular fashion through additional hydrogen-bond interactions, with each tetramer being rotated by approximately 45° relative to an axis running down the nanotube array, all of which shrouds alternating microenvironments of water and acetonitrile that form numerous hydrogen bonds on the interior [56]. In relation to the third generation capsules based on pyrogallol[4]arenes, these tetramers have a diameter nearly identical with that of the corresponding *C*-hexylpyrogallol[4]arene nanocapsule when crystallized from acetonitrile, thereby demonstrating structural retention (Figure 14.23). The pyrene or 1-bromopyrene co-crystallizing species are crucial for nanotube formation, as they interact through numerous van der Waals forces on the exterior of the assemblies behaving like a type of "molecular glue".

It is apparent from these series of results that when a near-spherical calixarene-based nanocapsule is obtained, there should be a strong possibility of isolating the related nanotubular array in either a similar or reversed packing mode. The same should also be true *vice versa*, although this may require a degree of experimentation to achieve.

14.9
Conclusions

We have described a number of generations of organic nanocapsules, and also metal–organic analogues, and have shown that rational design often leads to elegant nanostructures. We have demonstrated the encapsulation of probe molecules in stable hydrogen-bonded hexameric nanocapsules and significant advancements have been made in the area of metal–organic capsules using non-covalent assemblies as theoretical templates. This has been a particularly fruitful approach and the ability to sense ligands on metal centers, while also trapping chosen chemical species indiscriminately (using copper pyrogallol[4]arene nanocapsules, for example) is a very interesting opportunity to explore further this rapidly expanding and burgeoning field of capsule chemistry. It may also be true that a series of metal–organic nanotubular arrays will also be realized, but it is likely that only after significant control is gained over structures such as those described in this chapter will function follow levels of rational design. Some of the above strategies may perhaps be applicable to other systems that contain similar functional properties to some of the macrocycles described herein.

Acknowledgments

We would like to thank all those involved with this work to date, with particular thanks to Professor L. J. Barbour, Dr. Agoston Jerga and Dr. Gareth Cave.

References

1 Collet, A. (1987) *Tetrahedron*, **43**, 5725.
2 Cram, D.J., Choi, H.-J., Bryant, J.A. and Knobler, C.B. (1992) *J. Am. Chem. Soc.*, **114**, 7748.
3 Conn, M.M. and Rebek, J. Jr. (1997) *Chem. Rev.*, **97**, 1647.
4 Rose, K.N., Barbour, L.J., Orr, G.W. and Atwood, J.L. (1998) *Chem. Commun.*, 407.
5 Murayama, K and Aoki, K. (1998) *Chem. Commun.*, 607.
6 Shivanyuk, A., Paulus, E. and Böhmer, V. (1999) *Angew. Chem. Int. Ed.*, **38**, 2906.
7 González, J.J., Ferdani, R., Albertini, E., Blasco, J.M., Arduini, A., Pochini, A., Prados, P. and de Mendoza, J. (2000) *Chem. Eur. J.*, **6**, 73.
8 Shivanyuk, A., Rissanen, K. and Kolehmainen, E. (2000) *Chem. Commun.*, 1107.
9 Atwood, J.L. and Szumna, A. (2002) *J. Am. Chem. Soc.*, **124**, 10646; Atwood, J.L. and Szumna, A. (2003) *Chem. Commun.*, 940.
10 MacGillivray, L.R. and Atwood, J.L. (1997) *Nature*, **389**, 469.
11 Aoyama, Y., Tanaka, Y. and Sugahara, S. (1989) *J. Am. Chem. Soc.*, **111**, 5397.
12 Evan-Salem, T., Baruch, I., Avram, L., Cohen, Y., Palmer, L.C. and Rebek, J. Jr. (2006) *Proc. Natl. Acad. Sci. USA*, **103**, 12296.
13 Avram, L. and Cohen, Y. (2003) *Org. Lett.*, **5**, 1099.
14 MacGillivray, L.R. and Atwood, J.L. (1999) *Angew. Chem. Int. Ed.*, **38**, 1018.
15 Orr, G.W., Barbour, L.J. and Atwood, J.L. (1999) *Science*, **285**, 1049.
16 Atwood, J.L., Hamada, F., Robinson, K.D., Orr, G.W. and Vincent, R.L. (1991) *Nature*, **349**, 683; Coleman, A.W., Bott, S.D., Morley, S.D., Means, C.M., Robinson, K.D., Zhang, H. and Atwood, J.L. (1988) *Angew. Chem. Int. Ed. Engl.*, **100**, 1412; Atwood, J.L., Coleman, A.W., Zhang, H., and Bott, S.D. (1989) *J. Inclusion Phenom. Mol. Recognit. Chem.*, **7**, 203.
17 Atwood, J.L., Barbour, L.J., Dalgarno, S.J., Hardie, M.J., Raston, C.L. and Webb, H.R. (2004) *J. Am. Chem. Soc.*, **126**, 13170.
18 Dalgarno, S.J., Atwood, J.L. and Raston, C.L. (2006) *Chem. Commun.*, 4567.
19 Gerkensmeier, T., Iwanek, W., Agena, C., Frohlich, R., Kotila, S., Nather, C. and Mattay, J. (1999) *Eur. J. Org. Chem.*, 2257.
20 Meissner, R.S., Rebek, J. Jr. and de Mendoza, J. (1995) *Science*, **270**, 1485.
21 Cave, G.W.V., Antesberger, J., McKinlay, R.M. and Atwood, J.L. (2004) *Angew. Chem. Int. Ed.*, **43**, 5263.
22 Cave, G.W.V., Dalgarno, S.J., Antesberger, J., Ferrarelli, M., McKinlay, R.M. and Atwood, J.L. (2007) *Supramol. Chem.*, In press.
23 Dalgarno, S.J., Power, N.P., Antesberger, J., McKinlay, R.M. and Atwood, J.L. (2006) *Chem. Commun.*, 3803.
24 Dalgarno, S.J. Antesberger, J. McKinlay, R.M. Atwood, J.L. (2007) *Chem. Eur. J.*, **13**, 8248.

25 Atwood, J.L., Barbour, L.J. and Jerga, A. (2001) *Chem. Commun.*, 2376.
26 Shivanyuk, A. and Rebek, J. Jr. (2003) *J. Am. Chem. Soc.*, **125**, 3432; Shivanyuk, A. and Rebek, J. Jr. (2001) *Proc. Natl. Acad. Sci. USA*, **98**, 7662; Yamanaka, M., Shivanyuk, A. and Rebek, J. Jr. (2004) *J. Am. Chem. Soc.*, **126**, 2939; Palmer, L.C., Shivanyuk, A., Yamanaka, A. and Rebek, J. Jr. (2005) *Chem. Commun.*, 857.
27 Avram, L. and Cohen, Y. (2002) *J. Am. Chem. Soc.*, **124**, 15148; Avram, L. and Cohen, Y. (2002) *Org. Lett.*, **4**, 4365; Avram, L. and Cohen, Y. (2004) *J. Am. Chem. Soc.*, **126**, 11556.
28 Caspar, D. and Klug, A. (1962) *Cold Spring Harbor Symp. Quant. Biol.*, **27**, 1; Casjens, S. (1995) in *Virus Structure and Assembly*, Jones and Bartlett, Boston.
29 Rudkevich, D.M. and Rebek, J. Jr. (1999) *Eur. J. Org. Chem.*, 1991.
30 Dalgarno, S.J., Tucker, S.A., Bassil, D.B. and Atwood, J.L. (2005) *Science*, **309**, 2073.
31 Dalgarno, S.J., Szabo, T., Siavosh-Haghighi, A., Deakyne, C., Adams, J.E. and Atwood, J.L. (2007) *J. Am. Chem. Soc.*, submitted.
32 Unpublished results, in preparation.
33 Dalgarno, S.J., Bassil, D.B., Tucker, S.A. and Atwood, J.L. (2006) *Angew. Chem. Int. Ed.*, **45**, 7019.
34 Bassil, D.B., Dalgarno, S.J., Cave, G.W.V., Tucker, S.A. and Atwood, J.L. (2007) *J. Phys. Chem. B*, **111**, 9088.
35 Atwood, J.L., Barbour, L.J. and Jerga, A. (2002) *Proc. Natl. Acad. Sci. USA*, **99**, 4837.
36 Fuchs, R., Habermann, N. and Klüfers, P. (1993) *Angew. Chem. Int. Ed. Engl.*, **32**, 852.
37 McKinlay, R.M., Cave, G.W.V. and Atwood, J.L. (2005) *Proc. Natl. Acad. Sci. USA*, **102**, 5944.
38 Dalgarno, S.J., Power, N.P., Warren, J. and Atwood, J.L. (2007) *Angew. Chem. Int. Ed.*, submitted.
39 Avram, L. and Cohen, Y. (2004) *J. Am. Chem. Soc.*, **126**, 11556.
40 McKinlay, R.M., Thallapally, P.K., Cave, G.W.V. and Atwood, J.L. (2005) *Angew. Chem. Int. Ed.*, **44**, 5733.
41 McKinlay, R.M., Thallapally, P.K. and Atwood, J.L. (2006) *Chem. Commun.*, 2956.
42 Shivanyuk, A., Friese, J.C., Doring, S. and Rebek, J. Jr. (2003) *J. Org. Chem.*, **68**, 6489.
43 Power, N.P., Dalgarno, S.J. and Atwood, J.L. (2007) *New J. Chem.*, **31**, 17.
44 Power, N.P., Dalgarno, S.J. and Atwood, J.L. *Angew. Chem. Intl. Ed*, **46**, 8601.
45 Unpublished results, in preparation.
46 Chapman, R.G. and Sherman, J.C. (2000) *J. Org. Chem.*, **65**, 513.
47 Heaven, M.W., McKinlay, R.M., Antesberger, J., Dalgarno, S.J., Thallapally, P.K. and Atwood, J.L. (2006) *Angew. Chem. Int. Ed.*, **45**, 6221.
48 Shahgaldian, P., Da Silva, E., Coleman, A.W., Rather, B. and Zawarotko, M.J. (2003) *Int. J. Pharm.*, **253**, 23, and references therein.
49 Kroto, H.W., Heath, J.R., O'Brien, S.C., Curl, R.F. and Smalley, R.E. (1985) *Nature*, **318**, 162.
50 Ramirez, A.P., Haddon, R.C., Zhou, O., Fleming, R.M., Zhang, J., McClure, S.M. and Smalley, R.E. (1994) *Science*, **265**, 84.
51 Ghadiri, M.R., Granja, J.R., Milligan, R.A., McRee, D.E. and Khazanovich, N. (1993) *Nature*, **366**, 324; Hartgerink, J.D., Granja, J.R., Milligan, R.A. and Ghadiri, M.R. (1996) *J. Am. Chem. Soc.*, **118**, 43.
52 Semetey, V., Didierjean, C., Briand, J.-P., Aubry, A. and Guichard, G. (2002) *Angew. Chem. Int. Ed.*, **41**, 1895.
53 Gattuso, G., Menzer, S., Nepogodiev, S.A., Stoddart, J.F. and Williams, D.J. (1997) *Angew. Chem. Int. Ed. Engl.*, **36**, 1451.
54 Hong, B.H., Bae, S.C., Lee, C.-W., Jeong, S. and Kim, K.S. (2001) *Science*, **294**, 348.
55 Mansikkamäki, H., Nissinen, M. and Rissanen, K. (2004) *Angew. Chem. Int. Ed.*, **43**, 1243.
56 Dalgarno, S.J., Cave, G.W.V. and Atwood, J.L. (2006) *Angew. Chem. Int. Ed.*, **45**, 570.

Index

a

acetone–methanol solution 338
acetonitrile–water solution 346
acid–base reaction 20
aliphatic chain length 164
alkali metal cation 188
alkylammonium site 43
alkyl chains 133
allyl-terminated arms 79
amide hydrogen bond donors 74
amino acidate ligands 182
aminomethyl-substituted ligand 186
ammonium cation 170
ammonium hydrogensulfate 171
amphiphilic dendritic organogelators 135
ancillary ligands 51
anion binding 76
– cavity 74
anion receptors 93
anion-templated interlocked systems 88
– properties 88
anion-templated orthogonal complex 72
anion-templated pseudorotaxane 74
anion-templated rotaxane 88, 93
– synthesis 93
anion templation 64
– limitations 64
– methodology 85
– scope 64
– strategic 63
anionic cyanide linker 258, 259
anthracenyl group 14
anti-gauche conformational transition 105
Archimedean solid 320, 321
aromatic protons 103
aromatic stacking 231
aromatic units 19
– UV fluorescence 19
artificial molecular machines 19

artificial molecular systems 20
– acid–base reactions 20
– isomerization reactions 20
– metal–ligand reactions 20
– redox reactions 20
artificial photochemical devices 1
atomic force microscopy (AFM)
  techniques 341
azobenzene-based pseudorotaxane 21
azobenzene-containing copolymers 216

b

bangle-like structure 296
barbituric acid 168
biochemical networks 293
biomimetic chemistry 206
bis-bidentate bridging ligands 244
bis-bidentate pyrazolylpyridine ligands 242
bowl-shaped molecules
– dimers of 293

c

calixarene-based nanotubes 97, 98, 100
calixarenes 97
calixcrown nanotubes 100
Cambridge Structural Database (CSD) 338
carboxylate-based ligands 260
catenanes 64
– anion-templated synthesis 82
– three-dimensional cavities 64
cation–dipole interactions 103
cationic half-sandwich reagents 179, 189
cationic metallo-prisms 196
cavitand 293, 294
– properties 296
– structure 294
CD signals 123
charge-neutral box 189
charge-transfer band 140, 141

Organic Nanostructures. Edited by Jerry L. Atwood and Jonathan W. Steed
Copyright © 2008 WILEY-VCH Verlag GmbH & Co. KGaA, Weinheim
ISBN: 978-3-527-31836-0

chloride template 81
chloro-bridged dimers 180
cinchomeronic acid 167
cinnamic acids
– crystal-packing effects 307
complex interlocked assemblies 40
– synthesis 40
coplanar hydrogen atoms 81
copper–cyclodextrin complex 334
core–shell structure 207
covalent bond formation 64
cowpea chlorotic mottle virus (CCMV) capside 206
CPK models 300
crew-cut micelles 207
crown ether 14, 39
cryo-electron microscopy techniques 118
crystal engineering 155
crystal isomers 167
crystal polymorphs 167
– phenomenon 167
crystalline ammonium barbiturate salt 168
crystallization flasks 286
crystallization process 113
cyanide ligands 188
cyanide salt 179
cyclic helical structure 238
cyclopentadienyl-pyridyl groups 164
cyclophanediene (CPD) form 12

d

deep-cavity cavitand 294
– synthesis 294
defect boxes 190
dendrimer chemistry 18
dendritic building blocks 139
dendritic gels 133
dendritic hydrogelators 136
dendron rod-coil molecules 145
deoxyribonucleic acid (DNA) 312
diamine 139
– nano-crystalline platelets 139
di(ethylene oxide) linkers 215
differential scanning calorimetry (DSC) 167, 168
diffusion-ordered NMR spectroscopy (DOSY) 214
dihydrogenphosphate anions 170
dihydropyrene photochrome 12
diisocyanides 180
dimethyl sulfoxide 213
dinuclear oxalate complexes 196
dipole–cation interactions 106
dithionate anion 281

ditopic nitrogen donor ligands 260
divergent ligands 156
donor–acceptor interactions 124, 180
donor-to-acceptor energy-transfer steps 18
double clipping catenane synthesis 87
dropping ball method 116
dumbbell-shaped substrate 23
dynamic crystals 170
– motions in the nanoworld 170
dynamic light scattering (DLS) 341

e

electrical extension cables 14
– plug–socket concept 14
electroactive boxes 189
π-electron donor–acceptor 15
– interactions 22, 105
electron microscopy 113, 118
– methods 120
electron spin resonance (ESR) 129
electron-donating macrocycle 25
electron-rich hydroquinone moiety 72
electron-source component 15
electron-transfer processes 4
electron-transfer reaction 6
electrospray mass spectrometry 243, 333
enantioselective hydrogen transfer reactions 182
energy-transfer process 18, 19
energy-transfer rate 10
entropy 188
excited-state lifetimes 3
excited-state reactions 2
exo terminal cyanide 190

f

fatty acid salts 112
FCC boxes 193
ferrocenyl dicarboxylic acid complex 161, 164
fiber–fiber interactions 113
first generation nanocapsules 319
first-order kinetic processes 2, 5
first-row transition metal ions 225
fluorescence spectroscopy 122
fluoroborate complexes 235
Förster-type resonance 9
four-stroke synchronized sequence 25
– electronic process 25
– nuclear process 25

g

gas–solid reactions 168
gelation process 113
gelator molecules 115, 122

– IR spectra 122
gelator systems 124
gel-phase materials 111–114, 120, 130, 137, 139
gel–sol phase boundary 115
gel–sol transition temperature 116
gravity and inertia motions 19
Grubbs' catalyst 79, 85
guanidinium carbonate 161

h

half-sandwich receptors 189
half-sandwich tricyanides 188, 193
heat–cool cycle (sonication) 115
heteroaromatic pyrimidines 130
heterodimerization 70
heteromolecular hydrogen bonded crystals 161
heteropolyanions 279
hexadentate ligand 226
hexafluorophosphate 72, 235
– pyridinium derivatives 84
– pyridinium salts 79
hexamethylbenzene complex possesses 197
high-energy density batteries 170
Hofmann-type clathrates 254, 256, 258
– metal–cyanide sheets 254
Hoskins and Robson mark 258
host–guest interactions 201
hybrid organic–organometallic co-crystals 156, 161
hydrogelators 128
hydrogen-bonded hexameric nanocapsules 336
hydrogen-bonded nanometric adducts 170
hydrogen-bonding interactions 13, 15
hydrogen-bonding templates 333
hydrogen bonds 74, 158
– acceptor 161, 325
– donor 127
– interactions 135
– isomers 169
hydrogen storage 254
hydrogensulfate anion 88, 170
hydrophilic convex surface 294
hydrophobic alkyl chains 287
hydrophobic chain segments 205
hydrophobic concave surface 294
hydroxypyridine ligands 186

i

icosahedral geometry 196
infrared (IR) spectroscopy 122, 168
inorganic–organic hybrid materials 256, 257
inorganic–organometallic co-crystals 156, 161
in situ electro-spray ionization mass spectrometry (ESI-MS) 191
inter-ligand aromatic stacking interactions 244
interlocked molecules 33
– catenanes 33
– rotaxanes 33
intermolecular bonds 155
intermolecular interactions 113
interwoven architectures 64
– anion-directed formation 64
intracalixarene metal tunneling 100
intramolecular deactivation processes 3
ion channels 97
ion-pair binding rotaxanes 66
ionic interactions 156
iron-free ferritin molecules 206
iscrete anionic templates 79
isophthalamide fragment 77
isophthalamide moiety 84
isophthalamide unit 81
– aromatic proton 81
isopolyanions 279
isopolyoxometalate nanoclusters 282

k

Kagomé lattice topology 265
Keplerate clusters 284
– Fe-substituted 286

l

lanthanide(III) ions 225
ligands 46
– pseudorotaxanes 46
– rotaxanes 46
light energy 21
– photochemical reactions 21
light harvesting antennas 17
– chromophoric molecular species 17
– effect 17
– efficiency 17
– redox chemical energy 17
light-induced processes 1
linear difunctional ligands 180
lithium grease 142
lithium salts 142, 184
long-chain alkanes 133
low-intensity ultrasonic radiation 310
luminescent rhenium(I) bipyridyl motif 88

m

macrobicyclic organometallic complexes 187
macrocycles 82, 180

- anion binding properties 82
- isophthalamide cleft 81
- RCM reactions 84
macrocyclization 70
macroscopic single crystals 310
magic angle spinning (MAS) 121
malonic acid 169
mechanochemical methods 156
metal–cyanide materials 253, 256
metal–cyanide structures 254
metal–ligand bonds 46
metal–ligand grid 57
metal–ligand wire 49
metal–organic assemblies (MOAs) 251, 252
metal–organic frameworks (MOF) 33, 49, 156, 260, 265, 279
metal–organic materials 256
metal–organic nanocapsules 335
metal–organic polygons 312, 313
metal–organic structures 260
metal–polyacetylene linkages 49
metal-to-ligand charge-transfer (MLCT) 6, 9
methyl methacrylate (MMA) 214
minimum gelation concentration (MGC) 116
mixed-ligand complexes 244
mixed-metal macrocycles 187
molecular building block (MBB) approach 251, 258
molecular fibrils 128
- helical growth 128
molecular gel 111, 114, 115, 124, 141, 144
- analysis of 115
- applications 141
- building blocks 124
- formation 114
- preparation of 114
molecular glue 346
molecular-level plug–socket devices 14
molecular-level system 20
molecular machines 41
molecular nanotechnology 205
molecular photochemistry 2
molecular rods 258
molecular self-assembly 205
molecular shuttle 42
- single macrocyclic wheel 42
molecular wires 5
- photoinduced electron transfer 5
mono(ethylene oxide) linker 215
MORF 54, 58
- one-dimensional 49
- three-dimensional 51
- two-dimensional 51
multidentate terpyridine group 46
multiple noncovalent interactions 124

n

nanodroplet environment 211
nanoimprinted polymers 146
nano–ion chromatograph 285
nanoporous coordination network crystals 156
nanoscale capsules 277
nanoscale polyoxometalate clusters (POMs) 277
nano-sized reaction vessels 312
nanostructured gel 143
naphthyl esters
- photo-Fries reactivity of 302
nerve regrowth scaffolds 142
nitrogen–donor organic ligands 259, 260
nitromethane 180
NMR analysis 188, 339
NMR methods 120
NMR spectra 121
NMR time-scale 229, 299
NMR titration experiments 72, 195
noncovalent interactions 33
non-crystallographic T symmetry 228, 241
non-hydrogen-bonding solvents 124
nuclear magnetic resonance (NMR) spectroscopy 34, 85, 101, 120, 299
- experiments 84
- measurements 197
- titration 87
nuclear processes 25

o

octane 300
octanuclear cyclic helicate 225
oligo(ethylene oxide) chain 212
oligo(ethylene oxide) linker 215
oligosaccharide 209
one-dimensional fibers 139
organic–inorganic hybrid assemblies 146
organic nanocrystals 310
organic reactions 307
organogels 124, 127
–π-interactions 124
organometallic chemistry 200
oxalato ligands 197

p

penicillin-resistant bacteria 147
penicillin-type antibiotic drugs 147
peptide hydrogelator 143
phenanthroline ligands 70
phenolic oxygen atom 325

phosphorescence processes 3
photochemical molecular devices 26
photoexcited binaphthyl unit 14
photoinduced energy transfer 9, 17, 90
– molecular wires 9
photoinduced solid-state reactions 307
photoisomerization process 21
photoresponsive molecular nanoballs 216
photosynthetic process 17
piperazine-bridged dihydroxypyridine ligand 198
planar-bilayer voltage-clamp techniques 98
planar pyridinium groups 44
plug–socket device 13
– pseudorotaxane 13
plug–socket system 14
polyhedral coordination cages 223
polymer micelles 205, 207
polymorph screening 167
polyoxometalate-based molecular capsules 277
polyoxometalate-containing amphiphiles 288
polyoxometalates
– background 279
– classes 279
porous material 262
porphyrin 209
probe-containing capsules 331
Prussian blue-type materials 254
pseudo-conical cavity 294
pseudorotaxane system 76, 77, 90
pulse gradient spin echo (PGSE) 297
pyridine-based ligands 212, 254
pyridinium cation 70, 76
pyridinium receptor 70, 72
pyridinium-based ligand 70
pyridinium-based macrocyclic precursor 85
pyrrole amide-based catenane 64

r

Raman spectroscopy 167
RCM reaction 79
redox-active rotaxane-SAM 90
robust frameworks 57
rotaxane 33, 49, 64, 88
– anion-templated synthesis 79
– electrochemical properties 88
– molecular shuttle 44
– photo-active anion-sensing
– sensor 88
– three-dimensional cavities 64
Ru(II) complex 48
– fluorescence properties 48
– X-ray structure 48

s

scanning electron microscopy (SEM) 118, 341
second generation nanocapsules 322
second-sphere coordination effects 76
second-sphere coordination interactions 79
self-assembled fibers 146
self-assembled monolayers (SAMs) 90
self-assembled triad 17
self-assembling systems 293
self-assembly nanofabrication methods 146
self-assembly process 113, 131, 134, 170, 277
shell-like molecule 293
single crystal-to-single crystal reactivity 309, 310
single-walled carbon nanotubes (SWNTs) 97, 344
sodium salts 184
sol–gel transition 147
solid-state NMR (SSNMR) spectroscopy 167
solid-state organic–inorganic hybrid compounds 254
solid–state process 170
solid-state reactions 309
solid–state transitions 170
solvent-exchange process 310
solvent–gelator interactions 112
space-filling model 87
spectroscopic methods 168
square-planar palladium(II) ions 208
π–π stacking interaction 34, 76, 87, 113, 135
Stoddart's group 35
Stowell's group 168
straight-chain alkanes 299
structure–affinity investigations 76
suberic acid 164
sulfite-based polyoxotungstate 281
supramolecular isomerism 157
supramolecular metallomacrocycles 164
supramolecular photochemistry 2, 4, 26
– energy transfer processes 4
supramolecular polymerization 113
supramolecular systems 164
surface-encapsulated clusters (SECs) 287
Suzuki coupling reaction 162
synchrotron X-ray studies 217
synthetic nanotubes 97, 98, 108

t

temperature-dependent equilibrium process 285
template-directed synthesis 24
tert-butylbenzyl group 37, 39
tetraethyl orthosilicate (TEOS) 145
tetrafluoroborate anion 228, 234, 254

tetraphenylmethane-type stoppers 79
thermogravimetric analysis 57, 168
three–component systems 6
three-dimensional cavities 64
transmission electron microscopy (TEM) 98, 118, 341
trigonal prismatic coordination geometry 226
triphenylamine electron donor 9
two-component gelation systems 137
two-component gels 139
two-component systems 13
two-dimensional coordination network 158
two-dimensional square net 51
two electron-accepting stations 25

u
Ullmann ether reaction 294, 296
urea–phosphoric acid 170
U-shaped ligands 70
UV spectrophotometry 107

v
van der Waals forces 124
van der Waals interactions 113, 125, 133, 137
variable temperature (VT) experiments 121
vision-related processes 1
VT-NMR spectral data 42

w
water-soluble cavitand 294
– synthesis 294
W-based Keggin ions 279
weaker noncovalent interactions 46
well-characterized geometry 9
Wells–Dawson ($M_{18}O_{54}$) capsules 280
Werner complex formula 254
Williamson-type alkylations 102
wire-type systems 5, 11
– electron-transfer processes 11
– rod-like supramolecular systems 5

x
xerogels 120
X-ray crystallography 208
– analysis 74, 84, 208
X-ray diffraction 159, 167, 168, 319
X-ray methods 120
X-ray structures 41, 46, 77, 84, 87

z
zeolite-like metal–organic frameworks (ZMOFs) 251, 266, 267
zeolite *rho*-like topology 269
zeotype coordination compounds 156
zero–zero excited-state energy 4
zwitterion sandwich complex 164